VOLUME ONE

STRATIGRAPHY & TIMESCALES

VOLUME ONE

STRATIGRAPHY & TIMESCALES

Edited by

MICHAEL MONTENARI

Department of Earth Sciences and Geography
Keele University
Newcastle
United Kingdom

ELSEVIER

AMSTERDAM • BOSTON • HEIDELBERG • LONDON
NEW YORK • OXFORD • PARIS • SAN DIEGO
SAN FRANCISCO • SINGAPORE • SYDNEY • TOKYO

Academic Press is an imprint of Elsevier

Academic Press is an imprint of Elsevier
50 Hampshire Street, 5th Floor, Cambridge, MA 02139, United States
525 B Street, Suite 1800, San Diego, CA 92101-4495, United States
125 London Wall, London EC2Y 5AS, United Kingdom
The Boulevard, Langford Lane, Kidlington, Oxford OX5 1GB, United Kingdom

First edition 2016

ISBN: 978-0-12-811549-7
ISSN: 2468-5178

Publisher: Zoe Kruze
Acquisition Editor: Alex White
Editorial Project Manager: Helene Kabes
Production Project Manager: Radhakrishnan Lakshmanan
Cover Designer: Alan Studholme

Typeset by TNQ Books and Journals

CONTENTS

CONTRIBUTORS

G.L. Albanesi
CICTERRA (CONICET-UNC), Córdoba, Argentina; CONICET – Museo de Paleontología, CIGEA, Universidad Nacional de Córdoba, Córdoba, Argentina

E. Apellaniz
University of the Basque Country (UPV/EHU), Bilbao, Spain

J.I. Baceta
University of the Basque Country (UPV/EHU), Bilbao, Spain

G. Bernaola
University of the Basque Country (UPV/EHU), Bilbao, Spain

F. Caballero
University of the Basque Country (UPV/EHU), Bilbao, Spain

A.S. Collins
University of Adelaide, SA, Australia

J. Dinarès-Turell
Istituto Nazionale di Geofisica e Vulcanologia, Roma, Italy

S. Feist-Burkhardt
SFB Geological Consulting & Services, Ober-Ramstadt, Germany; University of Geneva, Geneva, Switzerland

T. Ferriday
ERCL Ltd – Exploration Reservoir Consultants, Oxon, United Kingdom

A.E. Götz
Keele University, Keele, United Kingdom

S. Monechi
University of Florence, Florence, Italy

M. Montenari
Keele University, Newcastle, United Kingdom

G. Ortega
CONICET – Museo de Paleontología, CIGEA, Universidad Nacional de Córdoba, Córdoba, Argentina

S. Ortiz
PetroStrat Ltd., Conwy, Wales, United Kingdom

X. Orue-Etxebarria
University of the Basque Country (UPV/EHU), Bilbao, Spain

S. Patranabis-Deb
Indian Statistical Institute, Kolkata, India

A. Payros
University of the Basque Country (UPV/EHU), Bilbao, Spain

V. Pujalte
University of the Basque Country (UPV/EHU), Bilbao, Spain

D. Saha
Indian Statistical Institute, Kolkata, India

B. Schmitz
University of Lund, Lund, Sweden

J. Tosquella
University of Huelva, Huelva, Spain

D. Vachard
Université de Lille, UMR CNRS 8198 Evolution, Ecologie et Paléontologie, Villeneuve d'Ascq cedex, France

PREFACE

The academic and industrial need for precise age constraints has made stratigraphy a fundamental and rapidly developing earth scientific discipline. Ranging from evolutionary biology, paleontology via climatology to large scale basin analysis, it is stratigraphy that is providing the time frame for the analysis of any dynamic process. Aside from the historically well-established methods, such as lithostratigraphy and biostratigraphy, relatively young and novel concepts have started and continue to make their impact on our understanding and the constant refinement of stratigraphy. Isotope stratigraphy, magnetostratigraphy, chemostratigraphy, sequence stratigraphy, cyclostratigraphy and astrochronology are only a few to be mentioned. The future in stratigraphy will see the integration of these disciplines into a unified universal stratigraphic framework, which will provide precise and high-resolution stratigraphic information enabling the quantification of complex interacting geological processes as well as regional and global correlations.

The new book series *Stratigraphy and Timescales* aims to form an essential platform for the communication of latest developments and progress within all fields of stratigraphy. It will foster and encourage the research into stratigraphy and present in chapters the latest state-of-the-art reviews and revisions of stratigraphic successions, new research results based upon established methods, and also introduce ground-breaking novel stratigraphic concepts for academia and industry, which still may even be in an experimental phase.

The current volume of *Stratigraphy and Timescales* contains six contributions covering an extensive stratigraphic range as well as a wide array of stratigraphic methods. In the first contribution, **Saha et al.** explore the *"Proterozoic stratigraphy of southern Indian cratons and global context"*. The Proterozoic successions of the entire Indian sub-continent between 1.9 Ga and 0.7 Ga are completely reviewed, revised and integrated into the global geological context. The second chapter by **Albanesi and Ortega** on *"Conodont and Graptolite Biostratigraphy of the Ordovician System of Argentina"* provides a new integrative synthesis of biostratigraphic data and their implications for the Ordovician System of Gondwanan Argentina and beyond. The next chapter by **Ferriday and Montenari** establishes the *"Chemostratigraphy and Chemofacies of Source Rock Analogues: A High-Resolution Analysis of Black Shale Successions from the Lower Silurian Formigoso*

Formation (Cantabrian Mountains, NW Spain)" by analysing chemostratigraphic signatures and patterns derived from a multi-proxy approach. This chapter is followed by the contribution of **Vachard** examining the *"Macroevolution and biostratigraphy of Paleozoic foraminifers"* by detailing the biostratigraphy, evolution and taxonomy of Paleozoic foraminifera with special emphasis on the Devonian to Permian periods. In the fifth chapter, **Feist-Burkhardt and Götz** establish a completely revised and novel *"Ultra-high resolution palynostratigraphy of the Early Bajocian Sauzei and Humphriesianum zones (Middle Jurassic) from outcrop sections in the Upper Rhine area, southwest Germany"*. The final chapter by **Payros et al.** is highlighting *"The Relevance of Iberian Sedimentary Successions for Paleogene Stratigraphy and Timescales"* by reviewing and revising fundamental keysections of the Iberian Paleogene using a wide array of stratigraphic methods and principles.

Michael Montenari
Keele University
United Kingdom

CHAPTER ONE

Proterozoic Stratigraphy of Southern Indian Cratons and Global Context

D. Saha*,[1], S. Patranabis-Deb* and A.S. Collins[§]
*Indian Statistical Institute, Kolkata, India
[§]University of Adelaide, SA, Australia
[1]Corresponding author: E-mails: dsaha@isical.ac.in; sahad.geol@gmail.com

Contents

Stratigraphy & Timescales, Volume 1
ISSN 2468-5178
http://dx.doi.org/10.1016/bs.sats.2016.10.003

Abstract

Precambrian cratons in India host large intracratonic basins with thick Proterozoic successions ranging in age from c.1900 to c.700 Ma, though the tempo of sedimentation and age range vary from basin to basin. The stratigraphy of the major basins in southern India including Cuddapah—Kurnool, Pranhita—Godavari (P—G) valley, Chattisgarh, Kaladgi—Badami, and Bhima basins is reviewed in this chapter in the light of recent sedimentological studies. Earlier attempts of intrabasinal and regional correlations are examined in the light of recently available geochronological data. Unconformity bound sequences in each of these basins show cyclic sedimentation where early rifting stage with coarse alluvial fan, fan-delta, and fluvial deposits, is followed by extensive tidal—intertidal to shallow marine mixed siliciclastic-carbonate sedimentation, with occasional change-over to deep-water carbonate platform or turbidite deposition, during the basin subsidence stage, and apparently influenced by major relative sea-level changes. The oldest sedimentation beginning c.1900 Ma is recorded in the Cuddapah basin in the Eastern Dharwar craton with two cycles of sedimentation in the Paleoproterozoic roughly coinciding with global proliferation in passive margin sedimentation. The older sedimentary sequence in the Kaladgi—Badami basin along the northern margin of the western Dharwar craton probably also started at the same time. The oldest sequence in the P—G valley basin is c.1680 Ma, with at least two more cycles in the Mesoproterozoic, and sedimentation closing in a final cycle dominated by fluvial sedimentation in the Neoproterozoic. Bulk of the sedimentation in the Chattisgarh and other satellite basins in the Bastar craton took place between c.1450 and 1000 Ma, with the development of stromatolite-bearing as well as deep-water carbonate platforms. Precise tectonic models of internal deformation of some of the Paleoproterozoic—Mesoproterozoic sequences, prior to deposition of younger groups, are an unresolved issue, but there are records of subduction-related activity both in the upper Paleoproterozoic and Mesoproterozoic outboard of the Indian cratons, which probably had little influence on the then passive margin sedimentation constituting bulk of the intracratonic basin fills in southern Indian block. Docking of the Eastern Ghats orogenic belt with supposed connections with parts of Antarctica (or Australia) during the amalgamation of Rodinia, along Dharwar—Bastar craton margins led to the cessation of passive margin Neoproterozoic sedimentation as recorded in the Kurnool basin and the P—G valley basin, whereas suturing of the North Indian cratonic block and the southern Indian block led to demise of the sea lane north of Bastar and Dharwar cratons.

1. INTRODUCTION

The end-Neoarchean peak in continental crustal growth as evidenced from global zircon age data distribution is followed by a stasis in the interval 2.7–2.4 Ga (Condie and Aster, 2010; Condie and Kroner, 2013; Hawkesworth et al., 2010). A major spurt in the worldwide development of passive continental margins and shallow seas in the Paleoproterozoic, peaking around 1.9–1.8 Ga, is also a notable feature of the global sedimentary record, which is postulated to be linked to supercontinent break-up (Bradley, 2008, 2011; Nance and Murphy, 2013). The Mesoproterozoic also saw a prolonged sedimentary basin development, possibly controlled by lateral accretion of arcs rather than supercontinent break-up (Bradley, 2008; Pisarevsky et al., 2014). Staged break-up of Rodinia (Rogers and Santosh, 2004) in the Neoproterozoic further paved the way for development of passive margin shallow seas. Against the backdrop of above global record of sedimentary systems and their tectonic drivers, we see in the Proterozoic sedimentary record of India a prolific development of sedimentary basins, traditionally described as Purana basins (Holland, 1909), which generally host thick piles of fluvial to shallow marine sedimentary successions, grading to deep-water successions in some of these basins.

The Proterozoic sedimentary basins in India in general overlie Archean cratonic basements distributed over five major cratons, namely Aravalli–Bundelkhand, Singhbhum, Bastar (Bhandara), Eastern Dharwar, and Western Dharwar, all stabilized roughly 2.5 Ga, and the basins are generally referred to as intracratonic basins, some of which are polycyclic (Saha and Mazumder, 2012; Meert and Pandit, 2015). One of the interesting features of the spatial distribution of these basins is their proximity to trans-Indian middle Proterozoic mobile belts or fold belts (Saha et al., 2016), raising the debate whether basin initiation and/or basin inversion is controlled by tectonic processes at the craton margins. In general, the stratigraphic successions in the Indian Proterozoic basins are flat-lying to very gently dipping except in the proximity of large intrabasinal or basin margin faults. But, older stratigraphic groups in some basins are internally folded and faulted and/or suffered up to low greenschist facies metamorphism prior to deposition of the younger, unconformably overlying successions.

Although the physical stratigraphy of the Indian Proterozoic basins has been studied over more than a century, the early emphasis was more on lithological characterization and arriving at a classification scheme restricted to individual basins (Pascoe, 1973). Toward the end of the 20th century and in the beginning of the 21st century, we see concerted efforts in interpreting the sedimentary environments of the major rock groups as well (Radhakrishna et al., 1987; references therein; Saha and Chaudhuri, 2003; Chakraborti, 2006). Pending robust geochronological constraints, the recognition of unconformity bound successions in all these basins formed the basis of basin-wide as well as regional correlations and suggestions on the tectonic link of the basin initiation and their inversion to craton margin processes, particularly with respect to the southern part of the peninsular India (Chaudhuri et al., 2002; Saha and Chaudhuri, 2003). With growing emphasis on sedimentological studies, and availability of some reliable geochronological data in the last decade or so, we see an integrated approach in regional correlation and postulations on influence of global tectonic processes on the development of the basins, cyclicity within the basinal successions, and connections with sea-level fluctuations (Meert et al., 2010; Patranabis-Deb et al., 2012; Saha and Tripathy, 2012a; Saha and Patranabis, 2014).

The peninsular India can be divided into two major blocks, the North Indian block and the Southern Indian block, separated by the Central Indian tectonic zone (CITZ) and its eastern continuation namely the Satpura mobile belt. In this chapter we focus on the Chattisgarh basin, Pranhita—Godavari (P—G) valley basin, Cuddapah basin, and Kaladgi—Bhima basins in the Southern Indian block (Fig. 1). Of these basins, the Cuddapah basin occurring at the eastern margin of the Eastern Dharwar craton is the largest and also known to host the oldest of the Proterozoic sedimentary successions in India. However, the Cuddapah basin has a prolonged sedimentation history with the stratigraphic development in the younger Kurnool sub-basin straddling into the Neoproterozoic. The Chattisgarh basin occurring in the Bastar craton and the P—G valley basin at the join between the Eastern Dharwar and Bastar cratons have major development during the Mesoproterozoic and straddling with breaks into the Neoproterozoic. Although the Kaladgi basin at the northern edge of the Western Dharwar craton recorded punctuated sedimentation throughout the Proterozoic, the development in the Bhima basin with a relatively thin preserved succession is restricted to the Neoproterozoic. Apart from the above major basins, a number of smaller basins namely the Khariar, the Indravati, the Ampani, and the Sukma (Sabari) basins, occur in the Bastar craton (Fig. 2). The stratigraphic development in each

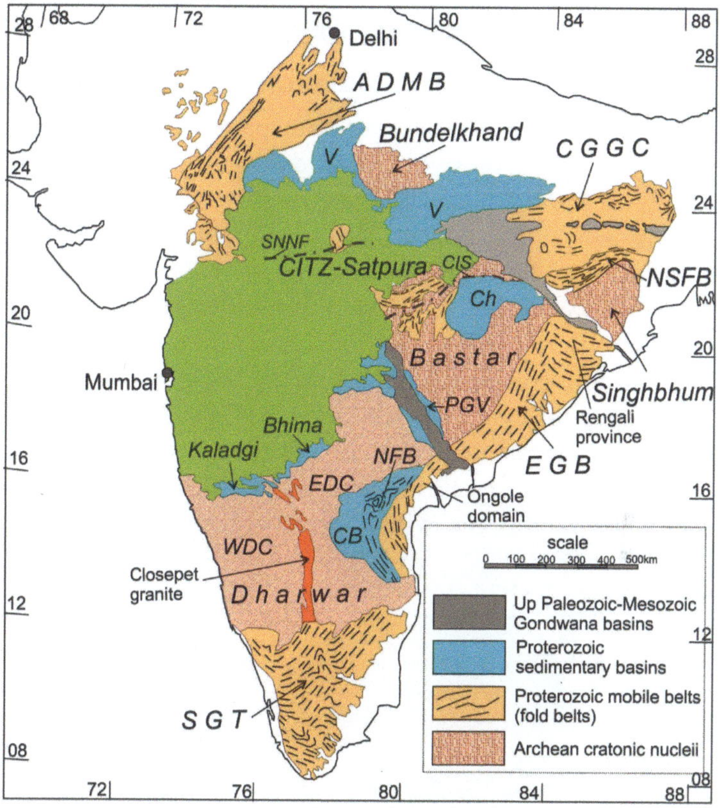

Figure 1 Cratonic blocks in India and major Proterozoic basins. Note the fold belts and mobile belts at the margin of these basins. *ADMB*, Aravalli–Delhi mobile belt; *CB*, Cuddapah basin; *CGGC*, Chotanagpur granite gneiss complex; *Ch*, Chattisgarh basin; *CITZ*, Central Indian tectonic zone; *EDC*, Eastern Dharwar craton; *EGB*, Eastern Ghats belt; *NFB*, Nallamalai fold belt; *NSFB*, North Singhbhum fold belt; *PGV*, Pranhita–Godavari basin; *SGT*, Southern granulite terrain; *V*, Vindhyan basin; *WDC*, Western Dharwar craton. *After Ramakrishna, M., Vaidyanadhan R., 2008. Geology of India: Bangalore, Geological Society of India, vol. I, 556; Saha, D., Bhowmik, S., Bose, S., Sajeev, K., 2016. Proterozoic tectonics and trans-Indian mobile belts: a status report. Proc. Indian Natl. Sci. Acad. 82, 445–460.*

of the above major intracratonic sedimentary basins in Southern India is presented in the next sections and reviewed in the light of the geochronological data that emerged in the recent past. The nature and abundance of available geochronological constraints are not uniform across the basins, and postulated correlations across the southern Indian cratons to be discussed in the latter part of this chapter will at best be speculative on occasions.

2. CUDDAPAH–KURNOOL BASIN

2.1 Basement and Basin Initiation

The Dharwar craton, the largest with preserved Archean nuclei in southern India, consists of two crustal domains, namely the Western Dharwar craton (WDC) and the Eastern Dharwar craton (EDC), sutured roughly along a line that follows the outcrop of the Closepet granite (Fig. 1). The 3.3–3.0 Ga old tonalite–trondhjemite–granodiorite (TTG) of the Peninsular Gneiss with enclaves of an older greenstone succession (Sargur Group), and younger greenstone successions (Bababudan and Chitradurga Groups) constitute the bulk of the WDC (Naha et al., 1991). These older successions are intruded by ~2.6 Ga old granitoids like the Chikmagalur and Chitradurga granites (Chadwik et al., 2007). In the EDC, the oldest preserved schist (greenstone) belts, like Sandur, Ramgiri, Kolar–Kadiri greenstone belts, with majority of the reported ages between

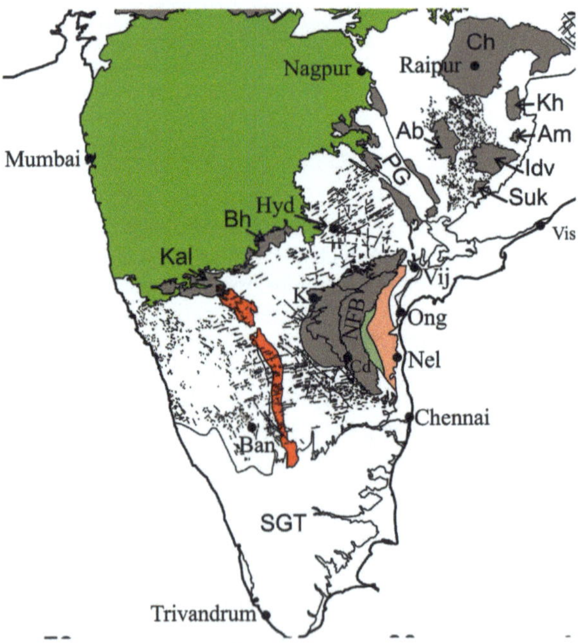

Figure 2 Mafic dyke swarms in Dharwar and Bastar cratons abutting against the Proterozoic Cuddapah, Chattisgarh, and other smaller basins. *Am*, Ampani; *Ab*, Abujhmar; *Bh*, Bhima; *Ch*, Chattisgarh; *Idv*, Indravati; *Kal*, Kaladgi; *Kh*, Khariar; *PG*, Pranhita-Godavari; place names: *Ban*, Bangalore; *Cd*, Cuddpah; *Hyd*, Hyderabad; *Nel*, Nellore; *Ong*, Ongole; *Vij*, Vijaywada; *Vis*, Visakhapatnam.

2.5 and 2.7 Ga are younger compared to those in the WDC (Balakrishnan et al., 1999; Jayananda et al., 2013; Nuttman et al., 1992, 1996; Rogers et al., 2007). The emplacement of the Closepet granite at around 2.5 Ga coincides with the amalgamation of the WDC and the EDC, and stabilization of the Dharwar craton. Granites and granitic gneisses of the EDC, yielding ages in the interval 2.7–2.5 Ga and occurring east of the Closepet granite outcrop constitute the so-called Dharwar batholith (Chadwick et al., 2000). Extensive development of mafic dyke swarms with age clusters between 2.37 and 2.2 Ga in the Dharwar craton as well as in the adjacent Bastar craton (Fig. 2) suggests a new phase of tectonothermal activity in the Indian subcontinental lithospheric mantle culminating in the development of the Paleoproterozoic intracratonic basins. Some authors argue in favor of a c. 1.9 Ga old large igneous province (LIP) connecting some of the mafic dykes swarms in the Dharwar and Bastar cratons and mafic volcanism in the basal part of the Cuddapah succession (French et al., 2008).

2.2 Physical Stratigraphy of the Cuddapah Basin

The crescent-shaped Cuddapah basin developed over the Eastern Dharwar craton has a preserved outcrop extent over 44,000 km^2 (Fig. 1). The sedimentary successions in the Cuddapah basin unconformably overlie granitoid basement including the Archean greenstone belts, traversed by the mafic dyke swarms. Four sub-basins within the Cuddapah basin—the Papaghni sub-basin, the Kurnool sub-basin, the Srisailam sub-basin, and the Palnad sub-basin—have been recognized considering the spatial distribution and thickness variation of the constituent rock groups, and their sedimentation pattern (Nagaraja Rao et al., 1987; Dasgupta and Biswas, 2006). The Nallamalai fold belt (NFB) hosts deformed metasedimentary successions that are thrust above the generally flat-lying sedimentary successions of the Papaghni and Kurnool sub-basins to the west. Relatively older part of the sedimentary succession is restricted to the Papaghni sub-basin and constitutes the bulk of the Cuddapah Supergroup (Table 1). The younger Kurnool Group is best developed in the Kurnool and Palnad sub-basins (stratigraphic classification is shown in Table 1).

The Papaghni sub-basin has an arcuate western boundary where arkosic to feldspathic arenites and conglomerates of the Gulcheru Quartzite unconformably overlie the weathered granitic gneiss of the Dharwar batholith with slivers of greenstone belts. Two unconformity bound successions namely the Papaghni Group and the Chitravati Group represent two successive cycles of sedimentation within the Papaghni sub-basin (Saha and Tripathy, 2012a).

Table 1 Lithostratigraphic subdivisions, Cuddapah basin

Group/Supergroup		Formation	Lithofacies	Sedimentary environment
Kurnool Group *500+ m*		Nandyal Shale	Colour laminated shale, calcareous shale	*Shallow amrine*
		Koilkuntala Limestone	Dolomitic limestone	*Shallow marine*
		Paniam Quartzite	Quartz arenite	*Shallow marine, bars with macrotidal influence*
		Owk Shale	Colour laminated shale with minor felsic volcaniclastics	*Outer shelf*
		Narji Limestone	Thin bedded to flaggy limestone, intercalated glauconitic sandstone near the base	*Shallow to Deep marine below storm wave base*
		Banganapalli Quartzite	Quartz arenite, basal conglomerate	*Gravelly alluvial fan to braid plane, to lower shoreface*
		Unconformity		
Cuddapah Supergroup	*Kimberlite pipes (c.1090 Ma)*	Srisailam Formation	Pebbly grit, quartzite, heterolithic shale - sandstone	*Fluvial to shallow marine*
		Tectonic contact		
	Chelima lamproite/Racherla syenite (c.1350 Ma) **Nallamalai Group** *6000 m*	Cumbum Formation (≈ Pullampet Shale)	Shale, dolomitic limestone, quartzite	*Inner to outer shelf middle part;* *Sub-tidal bar to peritidal in the upper part*
		Bairenkonda Quartzite (≈ Nagari Quartzite)	Pebbly grit, quartzite, heterolithic shale-sandstone	*Fan -delta grading to peritidal to inner shelf, below storm wave base*
		Tectonic contact		
	Chitravati Group *4975 m* *mafic sills (c. 1888 Ma)*	Gandikota Quartzite	Quartzite, pebble beds	*Bar -interbar with tidal influence*
		Tadpatri Formation	Shale, ash fall tuffs, quartzite, stromatolitic dolomite with mafic flows, sills and dykes	*Intertidal to subtidal; local tidal channels*
		Pulivendla Quartzite	Conglomerate and quartzite	*Subtidal bar to intertidal*
		Unconformity		
	Papaghni Group *2110 m*	Vempalle Formation	Stromatolitic dolomite, calc shale; basic flows and intrusive	*Intertidal to subtidal; shoreface to inner shelf*
		Gulcheru Quartzite	Conglomerate, feldspathic sandstone and quartzite	*Alluvial fan in the basal part; tidal flat in the upper part*
		Unconformity		
Peninsular gneiss/Dharwar batholith/greenstone belt(Archean)				

Modified after Saha, D., Tripathy, V., 2012a. Palaeoproterozoic sedimentation in the Cuddapah Basin, South India and regional tectonics — a review. In: Mazumder, R., Saha, D. (Eds.), Paleoproterozoic of India, vol. 365. Geological Society, London, pp. 159–182. Special Publication.; Patranabis-Deb, S.,

These Paleoproterozoic successions, each representing fluvial–shallow-marine psammite–pelite–carbonate deposition, are punctuated at different levels by mafic flow, sill, and dyke and less commonly felsic tuff. Development of extensive carbonate platforms with dolomitic stromatolite and algal laminite marks each of the cycles (Chakarbarti and Shome, 2011; Patranabis-Deb et al., 2012). Thermal anomalies in the mantle led to phased crustal extension and emplacement of mafic flows and dykes in the Papaghni and Chitravati Groups (Bhattacharji and Singh, 1984). Two extensional phases in the Papaghni sub-basin were punctuated by basin inversion, reflected by the basin-wide unconformity between the two groups.

2.2.1 Papaghni Group

The Papaghni Group is divided into two formation rank units—the Gulcheru Quartzite and the Vempalle Formation in ascending order. As the name implies the Gulcheru Quartzite is dominated by psammites with a thin basal interval of conglomerate, lying directly over the weathered granite gneisses of the Dharwar batholith in the Guvvalcheruvu section (type area in Cuddapah district) and other sections like Parnapalle (Anantpur district) or Chinna Tandrapadu (Kurnool district), along the western margin of the basin.

The basal part of the Gulcheru Quartzite consists of matrix to clast supported, thick-bedded polymictic conglomerate with occasional interbeds of gritty feldspathic sandstone (Fig. 3A), together making up about 15 m of the Chinna Tandrapadu section. Pebble size clasts in the basal part include subangular to subrounded pebbles of vein quartz, granite, pegmatite, micaceous sandstone, black chert, jasper, and gray shale, likely derived from the adjoining basement with patches of greenstone. The gritty to coarse sandy matrix consists of subangular to subrounded grains of quartz and pink feldspar, locally with ferruginous patches. The existence of trough cross-strata in the gritty interbeds, channel lags, outsized clasts, and lateral thinning out of the stack of conglomerate beds together with a general fining upward facies has been interpreted to represent alluvial-fan setting for the basal part.

Saha, D., Tripathy, V., 2012. Basin stratigraphy, sea-level fluctuations and their global tectonic connections – evidence from the Proterozoic Cuddapah Basin. Geol. J. 47, 263–283; Saha, D., Patranabis-Deb, S., 2014. Proterozoic evolution of Eastern Dharwar and Bastar cratons, India – an overview of the intracratonic basins, craton margins and mobile belts. J. Asian Earth Sci. 91, 230–251. Salient features of sedimentary environment after Saha, D., Tripathy, V., 2012a. Palaeoproterozoic sedimentation in the Cuddapah Basin, South India and regional tectonics – a review. In: Mazumder, R., Saha, D. (Eds.), Paleoproterozoic of India, vol. 365. Geological Society, London, pp. 159–182. Special Publication. Saha, D., Ghosh, G., Chakraborty, A.K., Chakraborti, S. 2009. Comparable Neoproterozoic sedimentary sequences in Palnad and Kurnool subbasins and their paleogeographic and tectonic implications. Indian J. Geol. 78, 175–192; Patranabis-Deb, S., Saha, D., Tripathy, V., 2012. Basin stratigraphy, sea-level fluctuations and their global tectonic connections – evidence from the Proterozoic Cuddapah Basin. Geol. J. 47, 263–283.

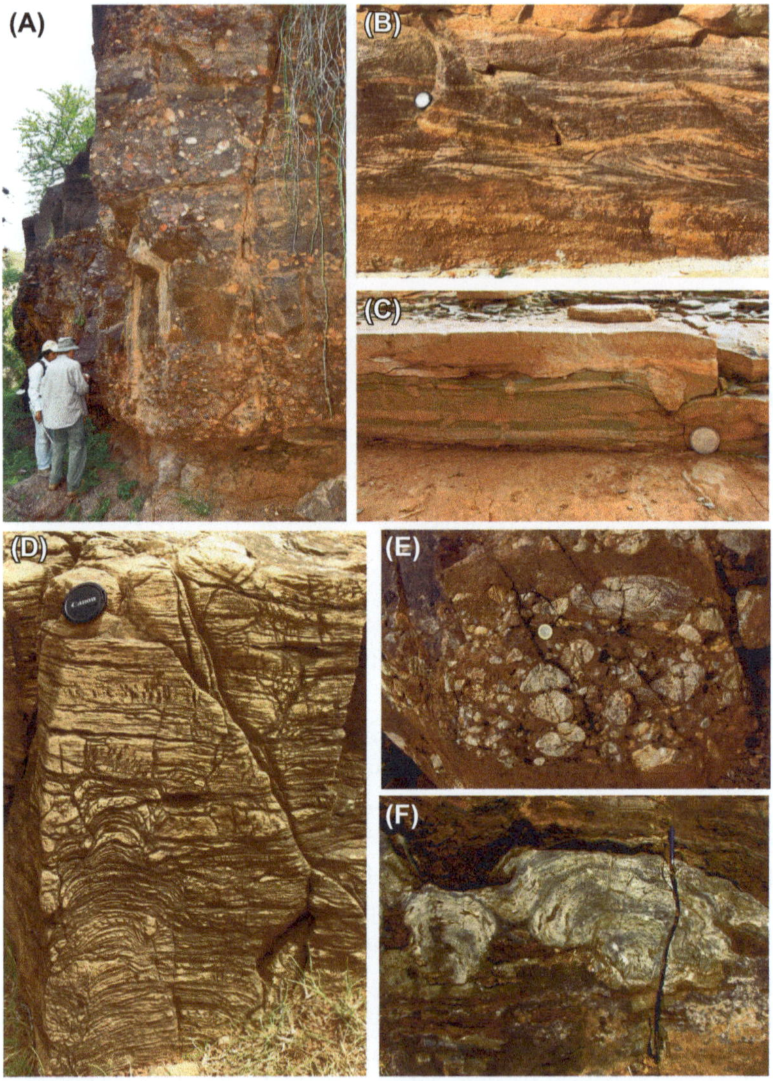

Figure 3 Field photographs showing lithology and sedimentary features in the Cuddapah Supergroup. (A) Basal conglomerate overlying granitic basement, Gulcheru Quartzite, off Parnapalli. (B) Pebbly grit grading to rough cross-stratified coarse sandstone. (C) Heterolithic sandstone-shale, Gulcheru—Vempalle transition. (D) Stromatolitic limestone, Vempalle Formation. (E) Basal conglomerate, Pulivendla Quartzite; note silicified straomatolite remnant in a large pebble. (F) Silicified stromatolite (hemispheroidal forms) intercalated with basaltic flow (dark band), Tadpatri Formation. (G) Tadpatri shale with a thick mafic sill. (H) Plateau forming sheet sandstone, Gandikota Quartzite. (I) Cross-stratified sandstone, Gandikota Quartzite. (J) Interference ripples, Gandikota Quartzite.

Figure 3 ***(continued)*****.**

Basal conglomeratic unit grades to massive to trough cross-stratified gritty—pebbly feldspathic sandstone, which, in turn, is overlain by medium-to-coarse grained, well sorted, rippled to cross-stratified glauconitic sandstone (Fig. 3B). The topmost part consists of heterolithic dark brown micaceous shale-fine sandstone (Fig. 3C) with bipolar trough cross-strata, mud cracks, halite casts, and occasional lag pebbles. Paleocurrent data from measurement of trough cross-stratification in the lower to middle part shows that the dominant paleoflow was toward N through east-north-east to east-south-east, as one follows the arcuate western margin of the basin from south to north (details in Fig. 12 of Saha and Tripathy, 2012a,b). Reports of trace fossils and organic-mat-induced sedimentary structures from the Gulcheru Quartzite also exist (Saha, 2006; Chakrabarti and Shome, 2010). A tectonically controlled alluvial fan system with multiple cycles of basement uplift and erosion during the basin opening stage is suggested by the facies association within the Gulcheru Quartzite. Peneplanation of the source region is indicated by the gradual passage of debris flow to sheet flood (basal unit) and fluvial facies (middle part), ultimately giving way to the development of intertidal flats in the upper part (Patranabis-Deb et al., 2012).

The Gulcheru Quartzite is overlain by the carbonate dominant Vempalle Formation with a gradational contact. The lower part of the Vempalle Formation consists of thin strata of splintery red mudstone alternating with cross-stratified siliciclastic and calc-arenite strata, often with herringbone structure. Other associated structures include tepee structure, desiccation cracks with lime mud or sand filling, molar-tooth structure, and halite casts. Thin- to thick-bedded dolomite with gently convex-up, lens-shaped bed geometry dominate the upper part of the Vempalle Formation. Algal laminites grading through isolated stacked hemispheroid to laterally linked hemispheroidal (LLH) forms (Logan et al., 1964) are common in the stromatolitic dolomite (Fig. 3D). The association of these stromatolitic dolomite and oolite-bearing carbonate in some sections and their cyclic repetition resemble shoaling up bars with fluctuation in the sea level. The Vempalle Formation with bar–interbar dolomite, stromatolite, and mudstone has been interpreted as of subtidal to intertidal origin.

2.2.2 Chitravati Group

Unconformably overlying the Papaghni Group, the Chitravati Group is traditionally divided into three constituent formations—the Pulivendla Quartzite, the Tadpatri Formation, and the Gandikota Quartzite. Sedimentation within the Chitravati Group also saw a strong pulse of mafic igneous activity, indicated by common occurrence of mafic flows and dykes within the Tadpatri Formation. The Pulivendla Quartzite ($\sim$90 m thick), consisting dominantly of medium-to-thick bedded quartz arenite with sparse pebbly sandstone and conglomerate at the basal part, has restricted strike continuity across the Papaghni sub-basin. Pebble size clasts of quartzite, chert, jasper, and occasional silicified stromatolitic dolomite, set in a coarse sandy matrix occur in the basal conglomerate lenses, with 10- to 15-cm-thick massive to normally graded beds (Fig. 3E). In the Yagantipalle section, the conglomerate grades upward to well-sorted quartz arenite mostly with trough cross-strata, planar cross-strata, and plane parallel stratification, deposited as shoaling-up bars with slight pinch and swell beds. Poorly sorted fine-grained sandstone and siltstone constitute the interbar areas between shoal bars. Locally, deformed cross-strata are also present. Presence of symmetric to slightly asymmetric sinuous to straight crested ripples with tuning fork bifurcation, and bimodal, bipolar cross-beds, together with desiccation cracks on the upper surfaces of well-sorted quartz arenite beds indicate wave–tide dominated setting of deposition with intermittent exposure. The high maturity of the sandstones indicates that the sediment accumulation was in critical balance with extensive sedimentary recycling, and deposition was at or very close to mean sea level (Patranabis-Deb et al., 2012; Saha and Tripathy, 2012a).

Heterolithic shale-calcareous sandstone with dolomitic limestone interbeds constitutes the lower part of the Tadpatri Formation conformably overlying the Pulivendla Quartzite. Although symmetric to asymmetric ripples are common in the sandy units, algal laminite and stromatolitic mound characterize the dolomitic interbeds (Fig. 3F). Sections near Yagantipalle show intercalations of carbonaceous shale within the dolomitic limestone which grades upward into a thick (40–50 m) succession of plane laminated, splintery gray-green shale. Throughout the Papaghni sub-basin, the dolomitic limestone is commonly intruded by up to 4-m-thick dolerite sills with chilled margin and local contact metamorphic effect in the host dolomite (Fig. 3G). Rhyolitic ignimbrites are locally intercalated with silicified dolomite in the uppermost part of the Tadpatri Formation. Stromatolites in the Tadpatri Formation have low amplitude hemispheroidal forms with diameter up to 1 m, or occur as laterally linked hemispheroids. The lithofacies association of algal laminite, stromatolite, and rippled fine calcareous sandstone has been interpreted as representing tidal–subtidal regime of sedimentation. The development of plane laminated thick shale in the middle–upper part suggest sea-level rise suppressing the carbonate factory (Saha and Tripathy, 2012a).

The Tadpatri Formation is overlain by the Gandikota Quartzite with a gradational contact in the east-central part of the Papaghni sub-basin, though new geochronological data discussed later raises doubts on its inclusion within the Paleoproterozoic Chitravati Group. The sand-shale intercalation in the transitional zone is overlain by amalgamated quartz arenite beds with sheet geometry (Fig. 3H). Medium- to coarse-grained quartz arenite to feldspathic arenite constitute the bulk of the formation. Large planar tabular to large trough cross-stratified units are intercalated with plane-parallel units and rippled units having straight or bifurcated crests, or interference ripples (Fig. 3I and J). Deformed cross-strata and ball-and-pillow structures are common in the upper part of the formation. Cross-stratified beds with opposite paleocurrent directions, hummocky cross-stratification and massive beds with local abundance of mud flakes occur throughout the Gandikota succession. Sediments were deposited primarily as high energy shallow wide bars and low energy interbars, which experienced frequent storms in open marine condition. On the whole, the Gandikota Quartzite represents subtidal to intertidal environment with well-preserved tidal flat showing frequent emergence of the depositional interface (Patranabis-Deb et al., 2012).

2.2.3 Kurnool Group

A major unconformity separates the components of the Cuddapah Supergroup from the younger Kurnool Group developed in two separate sub-basins, the Kurnool sub-basin in the west and the Palnad sub-basin in the northeastern part of the crescent shaped intracratonic Cuddapah basin (Saha et al., 2009). The plateau forming Srisailam Quartzite separates the Kurnool and Palnad sub-basins. Although the tilted beds of the Tadpatri Formation (Chitravati Group) occur beneath the basal Kurnool succession in the Erraguntla area (Kurnool sub-basin), further west the Kurnool Group rocks directly overlie the Peninsular gneiss. In the Palnad sub-basin, the basal units of the Kurnool Group unconformably overlie the granite gneisses (Eastern Dharwar craton), but folded and faulted Nallamalai Group of rocks constituting the Palnad klippe, are thrust over the Kurnool Group along the northeastern margin of the sub-basin (Natarajan and Nair, 1977; Saha and Chakraborty, 2003; Saha et al., 2009). The Kurnool Group is much thinner (cumulative thickness of about 500—600 m), compared to the older Chitravati and Papaghni Groups, but the outcrop extent of the flat-lying Kurnool rocks is quite extensive.

2.2.3.1 Banganapalle Quartzite

The Kurnool Group is subdivided into five constituent formation rank units (Table 1). Of these, the basal unit, the Banganapalle Quartzite (40—50 m), includes a massive, polymictic, matrix- to clast-supported basal conglomerate intercalated with trough cross-stratified, pebbly to gritty feldspathic sandstone or subarkose. Subangular to subrounded pebble to boulder size clasts of vein quartz, feldspar, chert, quartzite, occasional granite (granite-gneiss), and slate/phyllite, derived from the underlying basement, are set in variable proportions of sandy to gritty matrix in the basal conglomerate with local crude layering defined by clast alignment. The sheet-like massive conglomerate to pebbly sandstone with wavy upper bounding surface, grade upward to a medium to fine quartz arenite with thin muddy intercalation. Sedimentary structures include profuse wave ripples, interference ripples, flaser bedding, and syneresis/desiccation cracks (Fig. 4A and B; Saha et al., 2009).

The upper 30—40 m of the Banganapalle succession around Jaggayapeta in the Paland sub-basin consists of medium to fine grained quartz arenite beds with tabular geometry commonly showing wavy planar or hummocky—swaley stratification, locally with intervals of trough cross-stratified beds. The gentle wavy laminae within the wavy planar beds conform to the upper bounding surface and successive beds are separated by several

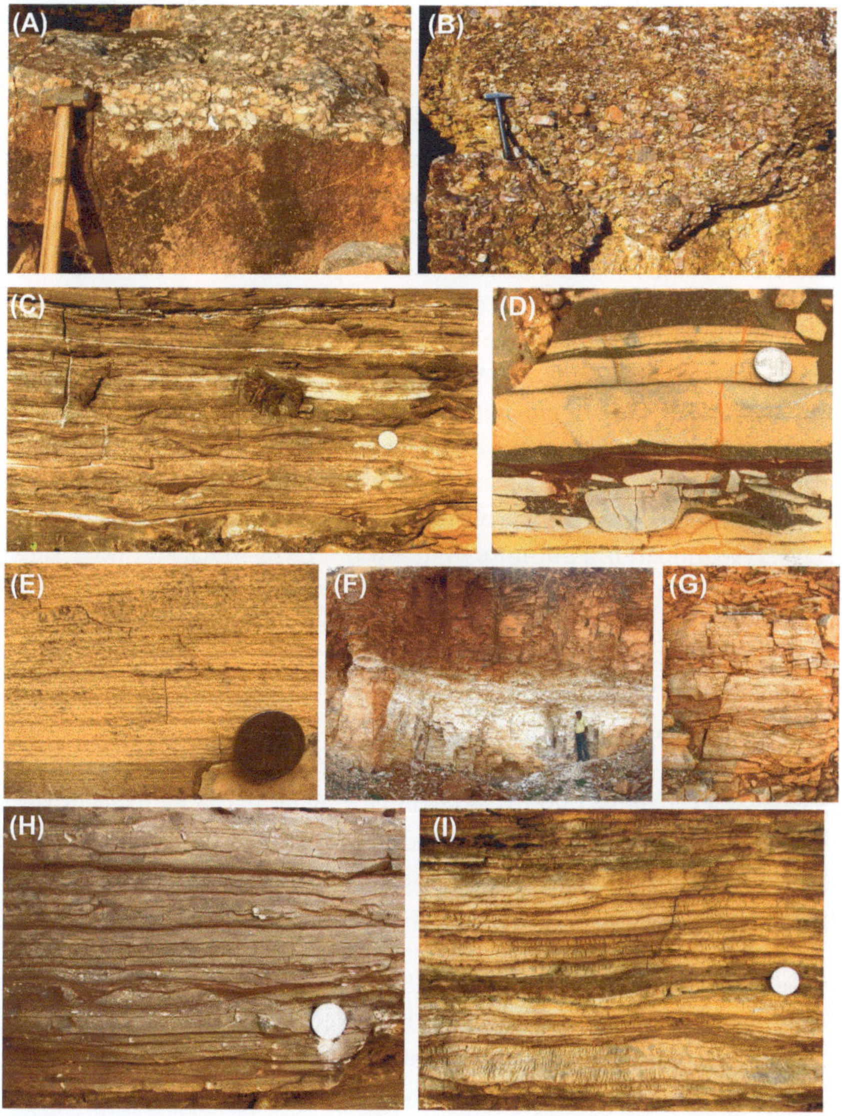

Figure 4 Field photographs showing lithology and sedimentary features in the Kurnool Group. (A) Basal conglomerate overlying granitic basement, Banganapalle Quartzite. (B) Thick matrix supported polymict conglomerate with apparent size grading. (C) Thin wavy bedded impure limestone, lower part of Narji Limestone. (D) Glauconitic sandstone interbeds in limestone; note pockets of lime pebble conglomerate, Narji Limestone. (E) Thinly laminated ash beds, Owk Shale. (F) Transition from Owk Shale (lower half of photo) to Paniam Quartzite. (G) Truncated wavy lamination, Paniam Quartzite. (H) Thin bedded argillaceous limestone, Koilkuntla Limestone. (I) Impure limestone-shale intercalation, Koilkuntla Limestone–Nandyal Shale transition.

millimeter thick mud partings showing ripple lamination. The wavy bedded to hummocky cross-stratified units grade upward to a plane laminated, heterolithic siltstone-shale, the latter being calcareous in the uppermost part.

The basal part of the Banganapalle Quartzite with massive to crudely stratified debris-flow conglomerate and interbedded subarkosic coarse to gritty sandstone has been interpreted as gravelly alluvial fan to braid plane deposits at tectonically active semiarid basin margin (Collinson, 1996; Saha et al., 2009). The sheet-like geometry of the conglomeratic units and pebbly sandstone beds with wavy upper bounding surface imply wave dominated foreshore to shoreface depositional setting, with wave reworking of relict fluvial sediments in a transgressive sequence (Bourgeois and Leithold, 1984). The ripple laminated quartz arenite with thin muddy partings showing profuse wave interference ripple structures, desiccation cracks and flaser bedding indicate deposition in a foreshore environment with intermittent subaerial exposure. The association of low-angle cross-strata, wavy laminations, hummocky-to-swaley cross-strata, and wave ripple cross-lamination in the quartz arenite has been interpreted as intermittent storm deposits in the lower shoreface region under influence of combined flow (Myrrow and Southard, 1991).

2.2.3.2 Narji Limestone

The Narji Limestone, about 500 m thick, conformably overlies the Banganapalle Quartzite along a gradational contact in the Palnad sub-basin as well as the Kurnool sub-basin, and is the main repository of cement grade limestone in the Cuddapah basin. The plane laminated heterolithic siltstone-calcareous shale in the upper most Banganapalle Quartzite gives way to laterally persistent beds of gray and red lime mudstone alternating with thin glauconitic sandstone with occasional pockets of lime-clast conglomerate in the basal part of the Narji Limestone. Cherty and argillaceous lenses and bands are common in the lower part, so are thin sandstone dykes. Plane parallel and wave ripple laminations are common in the gray limestone (lime mud) beds (Fig. 4C and D). The middle to upper part of the Narji Limestone consists of dark gray to black limestone with sparse calcareous shale, showing thin persistent laminations laterally traceable for tens of meters. Pyrite cubes are common in the black limestone and at places coherent to incoherent slumps on meter to decameter scale are reported (Saha et al., 2009). The shallow water wave features and intercalation with glauconitic sandstone are restricted to the lower part of the Narji Limestone. Horizontal continuity of thin limestone beds for long distances, sharp planar boundaries, and rarity of sand and silt size material are indicative deep basin deposition for bulk of the Narji Limestone. Presence of pyrite indicates euxinic condition and deposition in deeper isolated basin free from terrigenous influx.

2.2.3.3 Owk Shale

The development of the Kurnool Group is apparently truncated in the Palnad sub-basin with preserved outcrops of the Banganapalle Quartzite and the Narji Limestone formations only, compared to that in the Kurnool sub-basin where all the five constituent formations are outcropped. The Owk Shale, though having only 10–12 m of thickness is laterally extensive across the Kurnool district and often marked by clayey horizons near its lower and upper contacts. Conformably overlying the Narji Limestone with a sharp transition, the Owk Shale consists of red and brown weathering thinly laminated shale (Fig. 4E) with a 2-m-thick zone in the lower part which includes intercalation of ochreous shale and more compact cm thick welded coarse to fine rhyolitic to rhyodacitic tuff and volcaniclastic sandstone layers (Saha and Tripathy, 2012b). Some of the tuffaceous horizons also show ferruginous enrichment. The shale beds have plane parallel laminations. The Owk Shale grades upward into of the overlying Paniam Quartzite.

2.2.3.4 Paniam Quartzite

The Paniam Quartzite is also referred to as the "Plateau Quartzite" or "Pinnacled Quartzite" (King, 1872), because of the special geomorphic features in the Kurnool district, with which the formation is associated. The transitional zone with the underlying Owk Shale is marked by thin, wavy bedded to rippled, medium grained quartzite (Fig. 4F). Very well sorted medium-to-fine quartz arenite constitute the bulk of the Paniam Quartzite. Amalgamated wavy planar to lenticular beds give rise to meter-thick sheet like geometry. Wavy planar laminations, trough cross-strata, and large planar tabular cross-strata are common in the quartz arenite (Fig. 4G). Very well-sorted quartz arenite with amalgamated cross-stratified beds have been interpreted as products of wave reworked bars in a near shore open environment (Patranabis-Deb et al., 2012).

2.2.3.5 Koilkuntla Limestone

The Koilkuntla Limestone overlies the Paniam Quartzite with a sharp transition. Thin-bedded gray micritic limestone with marly intercalations constitute the Koilkuntla Limestone with common development of cherty nodules (Fig. 4H). Local development of algal laminate is also seen.

2.2.3.6 Nandyal Shale

The Koilkuntla Limestone grades upward into a brown-gray, color-laminated shale-calcareous shale (Fig. 4I). Internally 1–15 cm beds are with plane parallel laminations with streaks of fine sands and fading ripples. Thicker beds exhibit normal grading, and local development of mud clast conglomerate with sandy matrix. The sand deficient, mud dominated Nandyal Shale (50–100 m thick) possibly represents deposition below storm wave base in a wide shelf, under tectonic quiescence.

2.2.4 Nallamalai Group

Outcropped in the NFB in eastern part of the Cuddapah basin (Fig. 2) and traditionally included within the Cuddapah Supergroup (Nagaraja Rao et al., 1987), the stratigraphic position of the Nallamalai Group has been revised in recent years. The NFB is considered as an allochthonus unit separated from the flat lying to gently inclined strata of the Papaghni and Chitravati Groups, or the unconformably overlying Kurnool Group by a major thrust (Chakraborti and Saha, 2006; Saha et al., 2010; Saha and Tripathy, 2012a). The folded, faulted and cleaved low grade metasedimentary succession within the Nallamalai Group is divided into two formation rank units—the sandstone dominated Bairenkonda Quartzite and the dominantly argillaceous Cumbum Formation in ascending order. In the western part of the NFB (Sanipai–Balrajupalle section), the Nagari Quartzite considered as equivalent of the Bairenkonda Quartzite, consists of coarse to pebbly trough cross-stratified and rippled sandstone overlain by hummocky cross-stratified sandstone–siltstone. In the lower part straight crested ripples are common in the coarse sandstone with common desiccation features. In the upper part of the formation plane laminated shale with glauconitic medium-fine sand interbeds (Fig. 5A) give way to shaley rocks with thin dolomite and ferruginous quartzite, and thin oolitic ironstone, marking the transition to the overlying Pullampet Formation, the local equivalent of the Cumbum Formation. Plane laminated shale with thin dolomite constitutes the bulk of the Pullampet Formation (Fig. 5B). In the Rajampet area, the shale with local slump horizons is intercalated with massive to graded, gritty to coarse arenite with common carbonate clasts and rippled calcareous sandstone (Saha, 2004). The Pullamapet Formation is topped by a thin cross-stratified to rippled quartzite.

In the east central part of the NFB, the Nallamalai Group consists of quartzite, quartz phyllite, slate, and minor dolomitic limestone, with cumulative thickness of 1200 m (Tripathy and Saha, 2010). The quartzite dominant part represents the lower Bairenkonda Quartzite and the slates/phyllites with thin quartzite intercalations constitute the Cumbum

Formation (assigned a subgroup status by Meijerink et al., 1984). The folded succession around Porumamilla shows axial planar cleavage, partly obscuring the sedimentary structures of the protolith (sandstone-shale). However, locally preserved sedimentary structures include thin wavy bedding, ripples, trough cross-strata, low-angle stratification in the medium grained sandstone in the lower part. The Cumbum Formation consists of gray-green slate with thin, fine grained sandstone and local dolomitic interbeds. The slate (shale siltstone) is topped by medium grained quartzite with plane parallel laminations and large planar tabular cross-strata with tidal bundles. South and west of the Iswarakuppam dome, thick shale siltstone (slates of the Cumbum Formation) show repeated cycles of massive to normally graded siltstone, followed upwards by plane-parallel units and, finally, plane-laminated shale (Tripathy and Saha, 2010).

The lower part of the Bairenkonda (Nagari) Quartzite shows features of a fan-delta grading upwards to a peritidal depositional setting. The oolitic ironstone together with mixed siliciclastic-carbonate in the transition zone to Pullampet Formation represents inner shelf deposition. The shale-siltstone

Figure 5 Field photographs showing lithology and deformation features of the Nallamalai Group. (A) Thin bedded, intercalated sandstone-shale alternating with thicker sandstone, upper part of Bairenkonda Quartzite. (B) Plane laminated shale, Cumbum Formation. (C) West vergent folds and faults truncating fold limbs, west margin of Nallamalai fold belt (NFB). (D) Tight folds and axial plane cleavage in intercalated phyllite–quartzite, eastern margin of NFB.

dominant middle part has been interpreted as a turbidite succession, although the quartzite-dominant upper Cumbum Formation represents subtidal bars–peritidal deposits (Dasgupta and Biswas, 2006; Saha and Tripathy, 2012a).

3. P–G VALLEY BASIN

Extending for over 450 km along strike from CITZ in the north to Eastern Ghats belt (EGB) in the south, the Proterozoic P–G valley basin developed as a major rift system along the Northwest–Southeast (NW–SE) trending Karimnagar granulite belt which delineates the Neoarchean suture between the Eastern Dharwar and Bastar cratonic nuclei (Fig. 1). Multiple opening and closure along this join are recorded in the different Proterozoic successions representing three different cycles in the Proterozoic as well as a much younger Upper Paleozoic–Mesozoic Gondwana succession (Basumallick, 1967; Chaudhuri, 1985; Chaudhuri et al., 2012; Saha and Patranabis-Deb, 2014; Srinivasa Rao et al., 1979; Subba Raju et al., 1978).

Two NW–SE trending outcrop belts of the Proterozoic successions of the P–G valley basin are now separated by the axial outcrop of the Upper Paleozoic–Mesozoic Gondwana succession. Several classification schemes for the Proterozoic successions have been proposed, with conflicts emanating from recognition of unconformities, unconformity bound successions and nomenclature (Basumallick, 1967; Chaudhuri, 2003; Chaudhuri et al., 2012; King, 1881; Saha and Ghosh, 1998; Srinivasa Rao, 1987; Subba Raju et al., 1978). More recently, a reconstructed stratigraphic division has been proposed which provides a summary for the entire P–G valley basin, based on up-to-date field observations, age-data and concept of sequence (Chaudhuri et al., 2012; Conrad et al., 2011; Saha and Patranabis-Deb, 2014). Three major unconformity bound successions namely the Pakhal (Super)group with Mallamapalli and Mulug Groups, Penganga and Sullavai groups in the western belt are clubbed together under the Godavari Supergroup (Chaudhuri, 2003), with correlatives in the eastern belt consisting of Devalmari, Somanpalli, Albaka, and Usur Groups in ascending order (Table 2).

3.1 Pakhal Supergroup

Unconformably overlying the basement granites and gneisses including the Karimnagar granulite belt, the Pakhal Supergroup consists mainly of dolomitic limestone and minor conglomerate in the lower part and calcareous shales in the upper part constituting the Mallampalle Group (1400 m), and conglomerate, feldspathic sandstone, dolomitic limestone, calcareous shale, and minor chert constituting the Mulug Group (700 m). The sandstones of the Bolapalle Formation formed a bar–interbar association (Fig. 6A) with very well-developed tidal flat ripples, intermittent subaerial exposure marked by halite casts and desiccation cracks. The Bolapalle Formation gives way

Table 2 Stratigraphic subdivisions of the Proterozoic succession of the P–G valley basin

		Western belt	*Depositional environment*		*Eastern belt*	*Depositional environment*
		Gondwana Supergroup			**Gondwana Supergroup**	
		~ ~ ~ Unconformity ~ ~ ~			~ ~ ~ Unconformity ~ ~ ~	
	Sullavai Group			Usur Group	Usur Sandstone	
		Venkatpur Sandstone	Erg deposit		Doli Sandstone	Fluvial to shallow coastal
		Encharani Formation	Alluvial fan to braid plain		Nambi Breccia	Alluvial fan
		~ ~ ~ Unconformity ~ ~ ~			~ ~ ~ Unconformity ~ ~ ~	
	Pengnaga Group	Sat Nala Shale	Outer shelf to basinal	Albaka Group		
		Chanda Limestone	Deep water carbonate platform		Chalamala Sandstone	Storm and tide dominated shallow shelf
		Pranhita Sandstone	Storm and tide dominated shoreface to inner shelf		Tippapuram Shale	Inner shelf
		Nalla Gutta Sandstone	Alluvial fan, braid plain		Somandevra Quartzite	Storm dominated shoreface
		~ ~ ~ Unconformity ~ ~ ~			~ ~ ~ Unconformity ~ ~ ~	
Pakhal Supergroup	Mulug Group			Somanpalli Group	Po Gutta Sandstone	Lower to upper shoreface
		Rajaram Formation	Back-lagoon, tidal flat		Kopela Shale	Deep shelf to intertidal to shallow subtidal
		Enchencheruvu Chert	Shelf		Tarur Nala Formation/pedda Gutta Chert Formation	Unstable shelf, slope to basinal deposit
		Ramgundam Sandstone	Tidal flat to shoreface		Somnur Formation	Intertidal flat, shoreface to shallow shelf
		Jakaram/Damala Gutta Conglomerate	Coastal alluvial fan; tidal flat and shoreface		Bodela Vagu Formation	Subtidal to intertidal flat
		~ ~ ~ Unconformity ~ ~ ~			~ ~ ~ Unconformity ~ ~ ~	
	Mallamapalle Group	Pandikunta/Karlai Shale	Shelf	Devalmari Group	Kotapalle Limestone	
		Pandikunta Limestone	Tidal flat to stable shallow shelf		Madhumtorna Shale	Tidal shelf
		Bollapalli Formation	Small alluvial fan to tidal flats and sabkha		Amba Gutta Formation	
		~ ~ ~ Unconformity ~ ~ ~			~ ~ ~ Unconformity ~ ~ ~	
		Karimnagar granulite & granite gneiss			Bhopalpatnam granulite & granite gneiss	

Compiled mainly from Chaudhuri, A.K., 2003. Stratigraphy and palaeogeography of the Godavari Supergroup in the South-central Pranhita–Godavari Valley, South India. J. Asian Earth Sci. 21, 595–611; Chaudhuri, A.K., Deb, G.K., Patranabis-Deb, S., 2015. Conflicts in stratigraphic classification of the Puranas of the Pranhita–Godavari Valley: review, recommendations and status of the 'Penganga' sequence. In: Mazumder, R., Eriksson, P.G. (Eds.), Precambrian Basins of India: Stratigraphic and Tectonic Context. Geological Society, London, Memoirs, vol. 43, 165–183; Saha, D., Ghosh, G., 1998. Lithostratigraphy of deformed proterozoic rocks from around the confluence of Godavari and Indravati rivers, South India. Indian J. Geol. 70, 217–230; Sreenivasa Rao, T., 1987. The Pakhal basin – a perspective. In: Radhakrishna, B.P. (Ed.), Purana Basins of Peninsular India, vol. 6. Geological Society of India, Bangalore, pp. 161–187 (Memoir).

Figure 6 Field photographs illustrating lithology and sedimentary structures, P-G valley basin. (A) Sandstones, Bollapalle Formation. (B) Stromatolites, Pandikunta Limestone. (C) Intercalated conglomerate and sandstone, lower part of Mulug Group. (D) Limestones, Rajaram Formation. (E) Large aeolian cross-strata, Sullavai sandstone. (F) Cross-stratified Albaka sandstone. (G) Rippled unit, Somandevara Quartzite. (H) Laminated lime mudstone, Bodela Vagu Formation. (I) Heterolithic sandstone—shale with profuse desiccation crack in-fills, Somnur Formation. (J) Ash beds of Kotturu Somanpalli, Tarur Nala Formation. (K) Graywacke-shale turbidite with cherty intervals, Tarur Nala Formation.

Figure 6 (***continued***).

abruptly to the carbonate dominant Pandikunta Limestone (thickness about 1270 m) consisting of limestone, dolomitic limestone, dolomite with common nonbranching stromatolites (Fig. 6B), and thin lenses of glauconitic sandstones (Chaudhuri, 2003). In the upper part of the Malampalle Group, a succession of shales constitutes the Karlai (Pandikunta) Shale. The depositional environment in the Pakhal Supergroup shows a transition from an alluvial fan through tidal flat to a stable shelf with carbonate platform, the last inundated by sea-level rise culminating in the thick shales of the Karlai Formation (Chaudhuri et al., 2012).

3.2 Albaka Group

Traditionally, a thick succession of shelf sandstone, named as the Albaka Sandstone, was thought to be restricted to the eastern Proterozoic belt in the P–G valley (King, 1881; Sreenivasa Rao, 1987). The Albaka Sandstone unconformably overlies a deformed succession recognized as the Somanpalle Group in the eastern belt (Saha and Ghosh, 1998). The sub-Albaka unconformity was also later traced in the western belt, and based on distinct lithostratigraphy and depositional motif the sandstone-shale succession overlying the Mulug Subgroup comprising Mulug Orthoquartzite–Mulug Shale succession in the Mallampalli–Mulug area

(Basumallick, 1967) has been assigned to the Albaka Group (Chaudhuri, 2003). In the eastern belt, the Albaka Group is unconformably overlain by the Usur Group correlated with the Sullavai Group in the western belt (Chaudhuri et al., 2012). In the southern part of the eastern belt, a thin (150 m) succession consisting of deformed limestone and sandstone-shale, named as the Cherla Formation (Saha, 1988; Srinivasa Rao et al., 1979), has been correlated with the Mallampalli Group of the western belt (Chaudhuri et al., 2012).

In the eastern belt, the Albaka Group is divided into three formation rank units—the Somandevara Quartzite, the Tippapuram Shale and the Chalamala Sandstone in ascending order, with cumulative thickness of ~3200 m. Subarkosic sandstone to quartz arenite with small lenses of pebbly arenite and conglomerate constitute the Somandevra Quartzite, where profuse cross-strata and ripple marks are preserved (Fig. 6F and G). Overlying the Somandevara Quartzite with gradational contact, the Tippapuram Shale consists of thinly laminated gray-green shale with intercalations of sand lenses with megaripples. The Chalamala Sandstone consists of medium to fine grained subarkosic sandstone with common ripple laminations, cross-stratification, wrinkle marks and current crescents. The depositional environment varies from storm dominated shoreface in the lower part through muddy inner shelf to storm and tide dominated shallow self (Chaudhuri, 2003).

3.3 Penganga Group

In the northern part of the western belt, the thick succession of sandstone–shale–limestone unconformably overlying the gneissic basement around Adilabad has been classified as the Penganga Group as distinct from the Albaka Group which lacks the carbonate facies; moreover, in contrast to dolomitic limestone or dolomites in the Mulug Group, the Penganga Group includes a true platformal limestone (Chaudhuri et al., 1989). The Penganga succession can be traced right up to Ramgundam in the south and unconformable relationship with the Mulug Group or the Mallampalli Group has been recognized (Deb, 2003; Chaudhuri et al., 2012; Sreenivasa Rao, 1985).

The Penganga Group with a cumulative thickness in excess of 1200 m around Adilabad is subdivided into the Pranhita Sandstone, the Chanda Limestone, and the Sat Nala Shale in ascending order. Thick wedges of red to purplish conglomerate, pebbly sandstone, and arkose, organized into successive coarsening up sequence, occurring in the basal part of the Penganga rocks in the Nalla Gutta section has been recognized as a formation rank unit, the Nalla Gutta Sandstone (Table 2). The Pranhita Sandstone consists of gray to buff, cross-stratified, subarkosic to quartzose sandstone grading upward to silty shale. The overlying Chanda Limestone consists of bedded lithographic limestone with intervals of siliceous limestone and chert with local Mn-deposits, and a very distinctive marker horizon of black limestone. Debris flow deposits of lime pebble conglomerate are also common (Chaudhuri, 2003; Mukhopadhyay and Chaudhuri, 2003). Red to purple shale with local intercalations of siliceous limestone constitutes the Sat Nala Shale. The Nalla Gutta Sandstone has been interpreted as representing alluvial fan to braid plane deposits in a fault controlled setting, and the Pranhita Sandstone as storm to tide dominated inner shelf deposit; the Chanda Limestone was deposited in a deep-water carbonate platform with oxic to anoxic condition (Chaudhuri, 2003).

3.4 Somanpalli Group

A deformed mixed carbonate-siliciclastic succession occurring around the confluence of Indravati and Godavari rivers in the eastern Proterozoic belt and lying unconformably below the Albaka Group has been designated as the Somanpalli Group (Table 2; Saha and Ghosh, 1998). The Somanpalli Group has been divided into six constituent formation rank units—Bodela Vagu Formation, Somnur Formation, Tarur Nala Formation, Pedda Gutta Chert Formation, Kopela Shale Formation and Po Gutta Formation.

An association of argillaceous limestone, dolomitic limestone, dolomite with common siliciclastic intercalation in the lower half, constitutes the Bodela Vagu Formation (~200 m). Dolomitic algal laminate and stromatolitic limestone predominate in the upper part of the formation. Siliciclastic intercalations include thick, massive beds of breccia-conglomerate, gritty quartzite to fine grained quartz arenite showing trough cross-strata, ripple laminations and salt pseudomorphs. Partings of calcareous mudstone or siliceous limestone often show asymmetric ripples, polygonal shrinkage cracks and water escape features (Fig. 6H). The algal laminite–stromatolite-bearing dolomitic limestone is marked by molar-tooth structure, stylolite, syneresis crack filled with sparry calcite, and LLH stromatolite heads. Carbonate breccia complexes with circular to elliptical outline and varying in size from a few tens of meters to a few kilometers, locally cross-cut the bedded units of the Bodela Vagu Formation.

The Somnur Formation consists of plane laminated shale, calcareous shale with sparse, thin sandstone intercalations in the lower part grading upwards into coarse- to fine-grained, well-sorted quartz arenite with thin shale partings, and common occurrence of trough and herringbone cross-strata, asymmetric to interference ripples, and desiccation cracks. Heterolithic sandstone-shale with wavy and flaser bedding and synresis cracks (Fig. 6I), separate the two quartzite members within the formation. The upper quartzite grades into argillaceous limestone with thin stromatolite biostromes with algal lamination and locally abundant domal stromatolite, constituting the upper part of the Somnur Formation.

The Tarur Nala Formation consists of a thick succession of ash beds and graywacke-shale grading to intercalated shale and limestone, with a basal succession of volcaniclastics, litharenite, and slumped shales (Fig. 6J and K). The ash beds laterally grade into bedded chert interstratified with siliceous black shale and thin beds of limestone. The graywacke beds are massive to normally graded, while the shales are plane laminated. The Pedda Gutta Chert formation consists of calcareous shale grading to bedded chert.

The Kopela Shale Formation consists of gray-green to black, plane-laminated shale with sparse flat lenses of cross-bedded medium to fine grained sandstone with a capping of heterolithic sandstone-shale having wavy and flaser bedding, asymmetric ripples and desiccation cracks.

Representing the topmost unit of the Somanpalli Group, the Po Gutta Formation consists of a thin basal conglomerate with pebbles of shale, sandstone, vein quartz and feldspar set in a micaceous sandy matrix, giving way to a massive to trough and planar cross-stratified subarkosic to quartzose sandstone through sandstone—shale intercalation with common wavy bedding and asymmetric ripples.

Although the mixed carbonate-siliciclastic successions of the Bodela Vagu Formation and the Somnur Formation represent tidal shelf deposition, the Tarur Nala Formation and the Pedda Gutta Chert Formation represent deep shelf setting with common turbidite deposits (Chaudhuri et al., 2015; Saha and Ghosh, 1987). The thick succession of black shales of the Kopela Shale Formation has been interpreted as representing pelagic mud deposition indicating maximum sea-level rise; a reversal to tidal shelf environment of deposition is attested by the Po Gutta Formation (Chaudhuri et al., 2012).

3.4.1 Sullavai Group

The Sullavai Group in the western Proterozoic belt unconformably overlies the Mulug, Mallamapalli, and Penganga Groups and represents an extensive red sandstone deposit with subordinate conglomerate, practically devoid of any shale or carbonates unlike the older groups. Divided into two formations—Encharani Sandstone and Venkatpur Sandstone—the Sullavai

Group consists of medium-to-coarse grained arkosic to quartzose sandstone in the lower part and very well-sorted medium-to-fine quartzose sandstone with large, aeolian cross-strata in the upper part (Fig. 6E). Although the lower formation is deposited under braid fluvial systems, the upper formation represents vast sand seas (Chakraborty, 1991, 1999). The sandstone dominant successions Mancheral Quartzite and Kapra Sandstone are considered to be lateral equivalents of the Encharani Quartzite (Chakraborty and Chaudhuri, 1993; Chaudhuri et al., 2012).

3.4.2 Usur Group

Considered to be equivalent of the Sullavai Group, the Usur Group in the eastern Proterozoic belt is divided into three members—Nambi breccia (conglomerate), Doli sandstone, and Delam sandstone (Srinivasa Rao, 1987; Srinivasa Rao et al., 1979). Development of 200-m-thick bouldery, breccia-conglomerate with gritty sandstone in the lower part of the Usur Group represent a new cycle of deposition over the Albaka Group. The Doli Sandstone consists of profusely cross-bedded sandstone and pebble conglomerate, while reddish brown feldspathic sandstone representing fluvial channel bar to channel lag, predominate in the uppermost member. The thick breccia conglomerate represents graben-fill deposit (alluvial fan), which rapidly evolved into fluvial to shallow coastal depositional system in the upper part (Chaudhuri et al., 2012).

The Sullavai and Usur Groups represents the youngest Proterozoic successions in the P—G valley basin and following long hiatus in sedimentation, upper Paleozoic to Mesozoic Gondwana rocks were deposited unconformably over these.

4. CHATTISGARH AND OTHER SMALL INTRACRATONIC BASINS

4.1 Basement and Basin Initiation

The Bastar craton is delimited by Mahanadi Rift and Pranhita Godavari Valley Rift in the northeast and southwest respectively, CITZ in the northwest and Eastern Ghat mobile belt in the southwest (Fig. 1). The craton was subjected to major events of extensional tectonics during the Mesoproterozoic with the opening of several cratonic rift basins in different parts of the craton (Chaudhuri et al., 2002). These basins are repositories of widespread cratonic successions of virtually unmetamorphosed and mildly deformed arkose—quartzarenite, limestone—dolomite, and shale with subordinate conglomerate, evaporite and felsic ignimbrite. The largest of these basins of the Bastar craton, the Chattisgarh basin occurs in the northern part of the craton (Fig. 1). Khariar, Indravati, Ampani, and Sukma are the other relatively smaller basins which occur mostly in the southeastern sector of the craton.

4.2 Physical Stratigraphy of the Chattisgarh Basin

The Chattisgarh basin developed as a north–south trending rift basin, which consists of two major unconformity bound sequences (Das et al., 1992; Datta, 1998; Dhang and Patranabis-Deb, 2011; Moitra, 1995; Murti, 1987; Patranabis-Deb and Chaudhuri, 2008), the lower Chandarpur–Raipur (CR) sequence and the upper, Kharsiya sequence. The CR sequence constitute major part of the preserved succession, and SHRIMP zircon ages from the base and top of the CR sequence attest that the CR basin, and by default the Chattisgarh basin opened at c.1400 Ma (Bickford et al., 2011b, Das et al., 2009) and inverted at c.1000 Ma (Patranabis-Deb et al., 2007). The CR sequence has long been classified into two groups, the lower, Chandarpur Group comprising an assemblage of coarse siliclastics, and the upper, Raipur Group including an assemblage of mixed carbonate and shale (Dutt, 1964; Moitra, 1995; Murti, 1987, 1996; Patranabis-Deb, 2004; Patranabis Deb and Chaudhuri, 2008). The upper sequence, the Kharsiya Group consists of two formations namely Sarnadih Formation and Maniari/Nandeli Formation, composed of extensive red sandstone with minor conglomerate and shale, respectively (Table 3).

4.2.1 Chandarpur Group

The siliciclastics of the Chandarpur Group exhibit remarkable variation in facies and thickness in different parts of the basin from east to west. The basin initiates with deposition of a wedge of immature clastics (Fig. 7A), followed by deposition of very thick heterolithic succession (Fig. 7B), the Lohardih–Gomarda Formations, which grades up to a sheet of texturally and compositionally mature arenite, the Kansapathar Sandstone (Fig. 7C), with minor arkose/subarkose. The sequence shows strong lateral facies variations between different lithologies, and is inferred as a tectonically controlled fan-delta–pro-delta succession which developed at the initial stage of basin opening (Patranabis-Deb and Chaudhuri, 2007, 2008). The Kansapathar Sandstone was deposited in a wide tidal shelf (Patranabis-Deb, 2005), and represents the transition from the unstable rifting phase to the stable subsidence phase of the basin evolution (Patranabis-Deb and Chaudhuri, 2008).

Table 3 Lithostratigraphic subdivisions, Chattisgarh basin

Supergroup	Group	Formation	Lithology	Depositional environment
		Gondwana Supergroup		
		Unconformity		
Chattisgarh Supergroup	Kharsiya Group	Nandeli Formation	Shale, siltstone	Muddy shelf
		Sarnadih Formation	Conglomerate, red-brown sandstone, shale	Fluvial to tidal
		~ ~ ~Unconformity~ ~ ~		
	Raipur Group	Churtela/Tarenga Shale	Slate, local dolerite dyke	Pelagic shale
		Saradih/Chandi Lst.	Slate, chert, dolomite	Tidal flat to minor fluvial
		Saradih/Chandi Lst.	Conglomerate, quartz arenite	Alluvial fan to tidal shelf
		Gunderdehi Shale	Shale/slate	Offshore muddy shelf
		Sarangarh/Charmuria Lst.	Flaggy limestone with mixed siliciclastic—carbonate	Non rimmed carbonate platform (ramp)
		Bijepur Shale	Shale	Muddy tidal flat
	Chandarpur Group	Kansapathar Formation	Quartz arenite and feldspathic sandstone	Shelf sandstone bar—interbar, subtidal to intertidal.
		Gomarda/Chaporadih Formation	Sand—mud heterolithic and shale	Nearshore, tidal, subtidal shelf
		Lohadih Formation	Basal conglomerate, feldspathic sandstone	Fan-delta to shallow shelf
		~ ~ ~Unconformity~ ~ ~		
		Basement granite gneiss and greenstone belts (Archean)		

Modified after Saha, D., Patranabis-Deb, S., 2014. Proterozoic evolution of Eastern Dharwar and Bastar cratons, India — an overview of the intracratonic basins, craton margins and mobile belts. J. Asian Earth Sci. 91, 230—251. Salient features of sedimentary environment after Patranabis-Deb, S., 2005. Tidal shelf sedimentation in the Neoproterozoic Chattisgarh succession of Central India. J. Earth Syst. Sci. 11, 211—226. Patranabis-Deb, S., Chaudhuri, A.K., 2008. Sedimentological products and processes in the Mesoproterozoic Chattisgarh basin and contemporary tectonics in Central India: Indian J. Geol. 80, 139—155 (published 2010); Patranabis-Deb, S., Chaudhuri, A.K., 2008. Sedimentological products and processes in the Mesoproterozoic Chattisgarh basin and contemporary tectonics in Central India. Indian J. Geol. 80, 139—155 (published 2010).

4.2.2 Raipur Group

The Raipur Group consists of two major clastic-carbonate cycles, the lower one consisting of Bijepur Shale and Charmuria/Sarangarh Limestone, and the upper one consisting of Gunderdehi Shale, and the overlying stromatolite-bearing carbonates of the Chandi/Saradih Limestone Formation. The latter

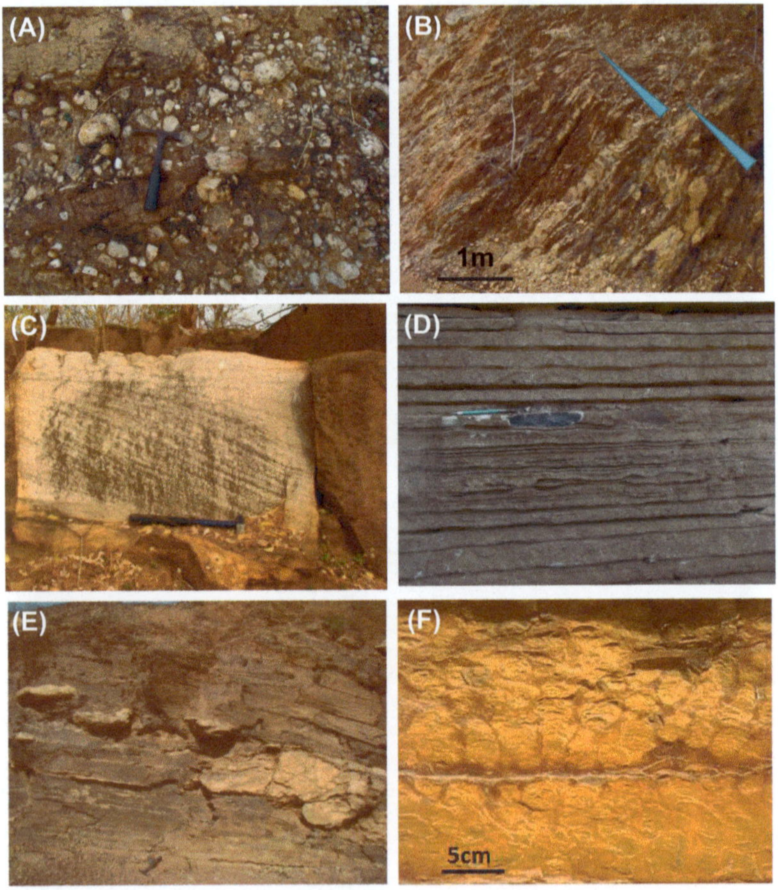

Figure 7 Field photographs illustrating lithology and sedimentary structures, Chattisgarh basin. (A) Poorly sorted, matrix supported ungraded, conglomerate, Lohardih Formation. Note the occurrence of large fairly rounded clast with relatively smaller angular clasts. (B) Sand-mud heterolithic rocks, Gomarda Formation. Note the successive fining upward cycles marking cyclic change in depositional environment. (C) Cross-stratified sandstone, Kansapathar Formation. (D) Limestone-marl rhythmite, Charmuria/Sarangarh Formation. Bed thickness: 2–10 cm and separated by mm thin marl layer. (E) Upper part of the Gunderdehi Shale with stromatolite bioherms appearing as isolated pockets. (F) Stromatolitic limestone of the Chandi Formation. Note that stromatolite columns are nonbranching and branching with two to six branches. Column height 5–15, width 1–3 cm.

passes up to the Churtela Shale, a shallow shelf to offshore mud deposit, with intermittent volcanism (Patranabis-Deb, 2007; Patranabis-Deb and Chaudhuri, 2008), which in all likelihood represents a truncated cycle where carbonates were either not deposited or were removed by erosion during the sub-Kharsiya unconformity. The carbonates are mostly micrite with minor dolomitization at places. Early lithified intrabasinal lime clasts occur as a subordinate constituent, and calcareous peloids occur in a very minor amount.

The Sarangarh/Charmuria Limestone (Fig. 7D) occurs as an extensive platform overlying the Kansapathar Sandstone through a transitional zone of brown shale, the Bijepur Shale. The platform is totally devoid of algal stromatolite or any related type of microbial laminites. However, small isolated bioherms occur at the upper part of the Gunderdehi Shale (Fig. 7E), and are profusely developed in the Chandi Limestone (Fig. 7F) (or the Saradih Limestone in the eastern part of the Chattisgarh basin). The carbonate factory was interrupted by deposition of the Tarenga/ Churtela Shale which is mainly composed of pyroclastics, both coarse and fine ash tuff (Fig. 8A). The Sarangarh Limestone is characterized by a remarkable color defined intervals, brown to gray to black to mauve in an ascending order, with the same order of superposition in different sections. The brown and gray intervals show significant admixture of siliciclastic sands, either as dispersed grains within micrites or as thin laminae intercalated with sandy carbonates or as small isolated sandstone lenses enclosed within limestones. The black and the mauve intervals, by contrast, are marked by virtual absence of sand-silt grade siliciclastics. The sand-free, below storm wave base deposits of black limestone points to maximum flooding level, which is being used as line of interbasinal and intrabasinal correlation.

Figure 8 Volcaniclastics from the Raipur and Kharsiya Groups, Chattisgarh basin. (A) Planar tabular-bedded, cross-stratified, coarse ash tuff bed within Churtela/Tarenga Shale. (B) Pebbly sandstone from the Sarnadih Formation with clast derived from underlying tuff units and from the basement.

4.2.3 Kharsiya Group

The Sarnadih Sandstone, is the basal formation of the Kharsiya Group, which overlies different formations of the Raipur Group with an erosional unconformity, and comprises an extensive interval of red sandstone, with a thin conglomerate horizon mantling the unconformity surface. The conglomerate grades up to pebbly sandstone (Fig. 8B) and very coarse-grained arkosic sandstone, deposited in fluvial to shallow marine environment. The Nandeli Shale, gradationally overlying the Sarnadih Sandstone, is composed mainly of sandstone mudstone heterolith, and is characterized by intense soft sediment deformation in several stratigraphic levels. The deformation is manifested by ball and pillow structures, detached sandstone balls enclosed within red shale, or small sand volcano type structures protruding through shale. Many undeformed thin sheets of sandstone exhibits straight crested ripples, interference ripples and rain prints covering large areas of sandstone-pavements. The upper part of the sequence is shale dominated, inferred to be deposited in tidal flat to muddy shelf environment.

4.3 Physical Stratigraphy of the Khariar Basin

The N—S trending Khariar basin, situated in the SE corner of the Chattisgarh basin, unconformably overlies granites and basic rocks of the Bastar craton. Khariar succession was subdivided into lower sandstone, middle shale and upper sandstone units, deposited in a shallow shelf environment, and roughly corresponding with the three formations of the Chandarpur Group. Das et al. (2001, 2003) subdivided the succession into six units of formation status and clubbed them into a single group, the Pairi Group, after the Pairi river running across the basin (Table 4). Das et al. (2009) obtained a 1455 $\pm$ 47 Ma age through U—Th—Pb electron probe micro-analyzer geochronology of monazite and zircon grains from porcellanitic tuffaceous units from the lowermost part of the succession and equated it with the Singhora Group of rocks of the Chattisgarh Supergroup.

4.4 Physical Stratigraphy of the Ampani Basin

Pascoe (1973) first mentioned about the Ampani succession and correlated the succession with those in the Vindhyan and Cuddapah basins. The siliciclastic-carbonate succession, with only one cycle of sedimentation, unconformably overlies the Bastar granite gneiss complex (Table 5) and shares a tectonic contact with the EGB. The eastern contact of the basin shows broad warping and shearing as a result of docking of the Eastern

Table 4 Lithostratigraphic subdivisions of the Khariar basin

Group	Formation	Lithology	Depositional environment
Pairi Group	Ling Dongri Formation	Quartz arenite and feldspathic sandstone	Shelf sandstone bar intertidal.
	Tarjhar Formation	Sand–mud heterolithic and shale	Nearshore, tidal, subtidal shelf
	Galighat Formation	Conglomerate (at base only), quartz arenite	Alluvial fan to tidal shelf
	Neor Formation	Sandstone–mudstone heterolithic	Offshore muddy shelf
	Kulharighat Formation	Shale with limestone pockets	Muddy tidal flat
	Devdahra Formation	Conglomerate and sandstone	Braid delta, shallow marine shelf
~ ~ ~Unconformity~ ~ ~			
Basement granite gneiss and greenstone belts			

Modified after Das, D.P., Dutta, N.K., Dutta, D.R., Thanavelu, C. and Baburao, K., 2003. Singhora Group- the oldest Proterozoic lithopackage of Eastern Bastar Craton and its significance. Indian Min. 57, 127–138. Salient features of depositional environments after Datta, B., 1998. Stratigraphic and sedimentologic evolution of the Proterozoic siliciclastics in the southern part of Chhattisgarh and Khariar, Central India. J. Geol. Soc. India 51, 345–360.

Ghats against the eastern margin of the Ampani basin (Balakrishnan and Mahesh Babu, 1987). However, based on lithostratigraphic correlation, subsequent workers equated the Ampani with the Chattisgarh–Indravati master basin.

Table 5 Lithostratigraphic subdivisions of the Ampani basin

Lithology	Depositional environment
Flaggy limestone and calcareous shale	Carbonate platform
Purple to gray shale and siltstone	Offshore muddy shelf
Siltstone	Muddy tidal flat
Conglomerate and subarkose	Coastal fluvial and shallow marine shelf
~ ~ ~Unconformity~ ~ ~	
Basement granite gneiss and greenstone belts	

Modified after Balakrishnan, P., Mahesh Babu, M., 1987. The geology of the Ampani outlier, Kalahandi, Koraput district, Orissa. In: B.P. Radhakrishna (Ed.) Purana Basins of Peninsular India (Middle to Late Proterozoic). Geological Society of India (Memoir), vol. 6, pp. 281–286.

4.5 Physical Stratigraphy of the Indravati Basin

Further south in the line of Chattisgarh and Khariar basins lie the Indravati basin, with a tectonic contact with the EGB along the eastern margin. The eastern part of the succession thus developed tight folds and axial plane cleavage (Das et al., 2001) while the rest of the basin remains undeformed. The basin filling succession preserves thick succession of shallow coastal sediments, mainly represented by flat lying laterally persistent well sorted glauconitic sandstone beds (Fig. 9A) which passes up to thick shale–limestone unit with a transitional contact.

Ball (1877), King (1881), Walker, 1902, and Crookshank (1963) were among the early workers to propose the informal stratigraphic division of the 500 m thick succession into Lower Beds (mainly arenaceous) and Upper Beds (mainly argillaceous) with calcareous intercalations and proposed a correlation with the Kurnool Group. Schnitzer (1971) and Murti (1987) correlated Indravati succession with the Chattisgarh succession and proposed a connection through the outliers of Ampani, Keshkal, and Khariar basins (Fig. 2). Dutt (1963) proposed the name Indravati series and subdivided the succession into three mappable units of stage status, namely, Tiratgarh, Kanger, and Jagdalpur, which were correlatable with the Chattisgarh succession. Schnitzer (1969) established four carbonate–shale cycles in the Kanger and Jagdalpur stages and separated them from the Tiratgarh by a disconformity. Ramakrishnan (1987) formally divided the succession into four mappable units namely Tiratgarh, Cherakur, Kanger, and Jagdalpur Formations making up the Indravati Group (Table 6).

Figure 9 Field photographs illustrating lithology and sedimentary structures in different formations of the Indravati basin. (A) Laterally persistent wavy bedded mature quartz arenite, Chitrakot Sandstone Member, Tiratgarh Formation. (B) Birsaguda tuff, Jagdalpur Formation.

Table 6 Lithostratigraphic subdivisions of the Indravati basin

Group	Formation	Lithology	Depositional environment
Indravati Group	Jagadalpur Formation	Calcareous shale	Offshore muddy shelf
	Kanger Formation	Limestone	Carbonate platform
	Cherakur Formation	Silt shale	Shelf sand to mud transition.
	Tirathgarh Formation	Conglomerate and subarkose, quartz arenite	Alluvial fan to shallow shelf mainly beach and tidal flat
~ ~ ~ Unconformity ~ ~ ~			
Basement granite gneiss and greenstone belt			

Modified after Ramakrishnan, M., 1987. Stratigraphy, sedimentary environment and evolution of the Late Proterozoic Indravati basin, central India. In: Radhakrishna, B.P. (Ed.), Purana Basins of Peninsula India. Geological Society of India (Memoir), vol. 6, pp. 139–160.

The Tiratgarh Formation, unconformably overlying the basement, consists of conglomerate, subarkose and quartz arenite deposited in shallow coastal environment ranging from fluvial to tidal bars. The shelf sandstone passes up to thick shale–limestone succession of the Kanger and Jagdalpur Formations with a transition of silt–shale unit, the Cherakur Formation. On the basis of the stratigraphic position of the Tokapal and Bhejripadar kimberlite pyroclastics (Laser ablation inductively coupled mass spectrometry, LA-ICPMS U–Pb age 620 ± 30 Ma) within Kanger Formation, Mainkar et al. (2004) proposed an event-break between the Kanger and the overlying Jagdalpur Formations. U–Pb isotopic analyses (LA-ICPMS) of the zircons from the Birsaguda tuff (Fig. 9B), within the Jagdalpur Formation point to closure of the basin at 1001 ± 7 Ma (Mukherjee et al., 2012).

5. KALADGI AND BHIMA BASINS

5.1 Physical Stratigraphy of the Kaladgi Basin

The Kaladgi basin, nearly 200 km long and about 100 km wide, occupies an area of about 8000 km^2, in the northern margin of the western Dharwar craton (Fig. 1). The maximum aggregate thickness of the sedimentary succession is ~3900 m (Jayaprakash et al., 1987; Radhakrishna and Vaidyanadhan, 1994). Parts of the northerly and westerly extension of the basin is concealed under the Deccan Traps but locally are exposed as inliers within the trap (Raha and Sastry, 1982). Metasediments of Dharwar Supergroup, Hungund schists and granite gneisses of the western Dharwar craton form the basement for the Kaladgi basin. The succession is divided into two groups, the Kaladgi Group and the Badami Group, separated by an angular unconformity (Table 7). The Kaladgi Group is further subdivided into two sub groups, the Lokapur Subgroup (2750 m) and Simikeri Subgroup (1150 m), each subdivided into a number of units of formation status. No definite ages for the basins were available, but tentative age from the stromatolites is suggested to be Meso-to-Neoproterozoic for Kaladgi (Sharma and Pandey, 2012). The lithology mainly includes conglomerate, quartz arenite, shale, micritic limestone, and dolomite (Table 7).

The lowermost unit, the Ramdurg Formation of the Kaladgi Group unconformably overlies the TTG gneisses and Archean greenstone belts (Dharwar schist, Hungund schist belt). The Ramdurg Formation is represented by conglomerate, immature sandstone and quartz arenite (Fig. 10A–C), deposited in fan-delta, pro-delta and shallow shelf setting. The rifting stage of the basin evolution is attested by the deposition of texturally and mineralogically immature sandstone deposited in a major fault-controlled basin, and was followed by the early subsidence stage. Extensive tidal shelf sandstone bars were deposited when the basin went through quiescent phase of basin development. Major transgression inundated the sea, and shale, chert and limestones were deposited. Development of shallow water carbonate platform with microbial mats on top of fairly deep-water shale points to upbuilding of the carbonate platform, and shallowing up phase of basin development (Fig. 11A and B). Basement uplift, folding and a hiatus in deposition follow the first cycle of sedimentation. The flat lying Badami Group sediments preserve signature of fluvial to shallow marine tidal environment (Fig. 12A and B), which overlies the Kaladgi succession with an angular unconformity (Table 7). Texturally and mineralogically mature sandstone of the Kerur Formation, popularly known as Temple Arenite, points to an environment where rate of sediment input was in balance with the rate of creation of accommodation. Paleocurrent analysis of the Ramdurg Formation of the Kaladgi basin indicates northwest to westerly flow. However, during deposition of the Kerur Formation (Temple arenite) of the Badami Group, the flow pattern changed a little with strong bimodality toward northeast and southwest indicating tidal activity, which played a major role in sorting and sculpturing the Temple arenite. The Rabanpalli Formation of the Badami Group shows flow towards northeast and southeast. Overall changes in the palaeocurrent direction (unpublished data) through time indicate a major shift of the provenance from southeast to northwest.

Table 7 Lithostratigraphic subdivisions, Kaladgi basin

Group	Subgroup	Formation	Lithology	Depositional environment
		Deccan basalt (end Cretaceous)		
		~ ~ ~ Unconformity ~ ~ ~		
Badami Group		Katageri Formation	Quartz arenite, shale, limestone	Tidal shelf
		Kerur Formation	Conglomerate, red-brown sandstone, shale	Fluvial to tidal
		~ ~ ~ Unconformity ~ ~ ~		
Kaladgi Group	Simikere Subgroup	Hoskatti Formation	Slate, local dolerite dyke	Pelagic shale
		Arlikatti Formation	Slate, chert, dolomite	Tidal flat to minor fluvial
		Kundargi Formation	Conglomerate, quartz arenite	Alluvial fan to tidal shelf
		—Disconformity—		
	Lokapur Subgroup	Yadhalli Formation	Shale/slate	Offshore muddy shelf
		Muddapur Formation	Shale/slate, limestone, dolomite	Muddy tidal flat
		Yendigere Formation	Shale/slate, dolomite, and dolomitic limestone	Muddy tidal flat
		Yargatti Formation	Shale/slate, dolomite	Muddy tidal flat
		Malaprabha Formation	Shale, chert, chert breccia	Offshore subtidal deposit
		Ramdurg Formation	Basal conglomerate, feldspathic-to-quartz arenite	Fan-delta to shallow shelf
		~ ~ ~ Angular unconformity ~ ~ ~		
		Dharwar schist, Hungund schist belt, Closepet granite (Archean)		

After Jayaprakash, A.V., Sundaram, V., Hans, S.K., Mishra, R.N., 1987. Geology of Kaladgi- Badami basin, Karnataka. Mem. Geol. Soc. India 6, 201–206. Salient features of depositional environment are given for comparison.

5.2 Physical Stratigraphy of the Bhima Basin

The NE–SW trending Bhima basin is the smallest among all the Proterozoic basins of India, whose outcropping extent is only about 5200 km^2. The

Figure 10 Field photographs illustrating lithology and sedimentary structures in the Ramdurg Formation, Kaladgi Group. (A) Basal conglomerate, massive, ungraded with subrounded to well-rounded clasts floating in very coarse-grained sandstone matrix, (B) coarse-grained feldspathic sandstone from the lower part of the succession, (C) planar parallel beds of quartz arenite alternating with mud rich intervals, representing successive coarsening up cycles.

Figure 11 Field photographs illustrating lithology and sedimentary structures in the upper part of the Lokapur Subgroup. (A) Microbial mat structure in Yargatti Formation, dolomite, and limestone. (B) Arlikatti Formation, shale, siltstone, and chert exposed in railway cutting. Width of the photograph is ~5 m.

Figure 12 Field photographs illustrating lithology and sedimentary structures in the Badami Group. (A) Temple arenite of the Badami Group, Kerur Formation. (B) Close-up view of the sandstone showing widely varying palaeocurrent direction. Lens cap as scale in the center of photo.

basin preserves 273-m-thick sediments (Mishra et al., 1987) and unconformably overlies the Archean basement of the Dharwar craton (Fig. 1). The top of the succession is mostly covered by Deccan Trap flows and intertrappean beds. The Bhima Group comprises two subgroups, Sedam and Andola, which may be subdivided into five units of formation status representing alternate clastic and carbonate cycles (Table 8). However, in view of

Table 8 Lithostratigraphic subdivisions, Bhima basin

Group	Formation	Lithology	Depositional environment
	Deccan basalt (end Cretaceous)		
	~ ~ ~ Unconformity ~ ~ ~		
Bhima Group	Harwal Formation	Shale	Offshore muddy shelf
	Katamdevarahalli Limestone	Flaggy limestone	Nonstromatolitic carbonate platform
	Halkal Formation	Shale	Turbidite and pelagic shale
	—Disconformity—		
	Shahabad Limestone	Flaggy limestone	Nonstromatolitic carbonate platform
	Rabanpalli Formation	Conglomerate, pebbly sandstone, quartz arenite	Fan-delta to shallow shelf
	~ ~ ~ Angular unconformity ~ ~ ~		
	Greenstone belt, Closepet granite (Archean)		

After Jayaprakash, A.V., Sundaram, V., Hans, S.K., Mishra, R.N., 1987. Geology of Kaladgi- Badami basin, Karnataka. Mem. Geol. Soc. India 6, 201–206. Salient features of depositional environment are presented here for comparison.

difference in altitudes of outcrops of the flat lying beds of Bhima basin, Kale (1990) interpreted Andola Subgroup (upper Bhima) to be lateral equivalent of the Sedam Subgroup (lower Bhima). The sedimentation started with deposition of coarse clastic sediments, mainly granule to pebble sandstone and minor conglomerates of the Rabanapalli Formation (Fig. 13A and B), passing upward to sand-free flaggy limestone of the Shahabad Formation (Fig. 13C). Detailed facies classification and cyclicity analysis of the Shahabad Limestone of the Sedam Subgroup shows that it represents a nonrimmed carbonate platform. On the basis of field evidences the succession may be classified into two major facies associations, representing a shallow water mixed carbonate-siliciclastic stage, and a major transgressive stage depositing sand-free black limestone followed by deposition of second cycle shale—limestone—shale succession respectively of the Halkal Formation, Katamdevarahalli Limestone (Fig. 13D), and Harwal Formation.

Figure 13 Field photographs illustrating lithology and sedimentary structures in the Bhima basin. (A) Granule conglomerate and coarse grained sandstone alternation, Rabanpalli Formation. (B) Thick-bedded, planar cross-stratified beds alternating with planar parallel beds, Rabanpalli Formation. (C) Limestone-marl rhythmite Shahabad Limestone. (D) Gray flaggy limestone (Katamdevarahalli Limestone), over brown shale of Halkal Formation. Note sharp contact (*black arrow*) between the two.

6. NEW GEOCHRONOLOGICAL DATA, PROVENANCE AND REGIONAL CORRELATION

Various formations within the Proterozoic basins of southern India have been the subject of considerable geochronological and isotopic provenance studies over the last decade, which is yielding a much better understanding of the timing of basin formation, of correlations between basins, and of the tectonic geography of southern India throughout the Proterozoic. In a number of cases, stratigraphic assumptions are being challenged, or completely overturned, by the new age constraints. Age and isotopic provenance constraints provide new tests to previously suggested tectonic hypotheses for the evolution of southern India; in many cases these older models fail and new models are required. Work of this sort is far from complete, though, as many formations, and even whole basins, have barely been studied, and even in the better studied basins, it has proved hard to constrain the depositional age of many of the formations. Here we discuss some of the more recent geochronological data, along with new isotopic provenance studies, and discuss the implications of these for regional correlation and tectonic geography of Proterozoic southern India.

6.1 Detrital Zircon Geochronology and Provenance—Cuddapah Basin

The Cuddapah basin is one of the largest Proterozoic basins in southern India, yet until recently, no systematic geochronological study of the basin had been undertaken. Excellent early work using Rb—Sr age of the Papaghni and Chitravati Groups, which forms the basal units of the Cuddapah basin, are reasonably well constrained. The Papaghni Group nonconformably overlie crystalline rocks of the Eastern Dharwar Province of Peucat et al. (2013), which include granulite-facies metamorphism and migmatites that crystallized at 2.53—2.51 Ga (Chardon et al., 2002; Glorie et al., 2014). The overlying Chitravati Group includes 1891—1893 Ma volcanic rocks and dykes within the Tadpatri Formation (Anand et al., 2003; Bhaskar Rao et al., 1995; French et al., 2008) that provide an excellent minimum age of deposition. U—Pb LA—ICPMS ages of near-concordant detrital zircons within the Gulcheru and Vempalle Formations, of the Papaghni Group, better constrain the maximum depositional age of these formations to younger than 2524 ± 9 and 2422 ± 17 Ma, respectively (Collins et al., 2015; Lancaster et al., 2015). But still the time window for deposition of these formations is c.500 million years, which illustrates that during Papaghni Group deposition, there was little active magmatism in the surrounding sediment-source regions.

The lowermost formation of the Chitravati Group, the Pulivendla Formation, yielded some younger detrital zircons that more closely constrain this formation's age of deposition to being younger than 1923 ± 22 Ma (Collins et al., 2015). These herald the large igneous province magmatism that is represented by the volcanic deposits and intrusions in the overlying Tadpatri Formation as well as in the extensive dyke swarms that intrude the Eastern Dharwar and Bastar cratons (French et al., 2008). Traditionally, the Gandikota Formation is included as the uppermost unit of the Chitravati Group. However, recent U–Pb detrital zircons from the formation have yielded a maximum depositional ages of 1129 ± 29 Ma (Collins et al., 2015) that question the traditional assignment of this formation.

The Srisailam Formation crops out in the north of the Cuddapah basin and nonconformably overlies the Eastern Dharwar Province to its northwest. It is overthrust by the Nallamallai Group to the southeast and is unconformably overlain by the Kurnool Group to its south and north. The relationship between it and the Papaghni and Chitravati Groups is unknown. Detrital zircons within the extensive arenites of the Srisailam Formation yield a maximum depositional age of 1787 ± 22 Ma (Collins et al., 2015) that is confirmed and slightly better constrained by five detrital muscovite ^{40}Ar–^{39}Ar total fusion ages that yielded a mean of 1773 ± 18 Ma (Collins et al., 2015).

The Kurnool Group unconformably overlies the Papaghni and Chitravati Groups (but no relationship is seen with the Gandikota Formation) and the Srisailam Formation. Sandstones within the group are relatively poor in detrital zircon, with those obtained from the basal Banaganapalle Quartzite exclusively Archaean in age. The Paniam Quartzite lies higher in the stratigraphy, with detrital zircon ages still dominated with Archaean ages, yet also with a few Paleoproterozoic near-concordant results and one Tonian zircon dated at 913 ± 11 Ma (Collins et al., 2015). Bickford et al. (2013) reported similar Archaean zircon U–Pb ages and a single Paleoproterozoic age from zircons extracted from a tuffaceous arenite collected from within the Owk Shale. We interpret these data to show that the Kurnool Group is definitely younger than 1773 ± 18 Ma (the maximum depositional age of the Srisailam Formation) and probably Neoproterozoic, with the caveat that only one zircon has so far yielded a Neoproterozoic age.

The tectonically bound Nallamalai Group has 416 near-concordant detrital zircon ages published from sandstones throughout the NFB

(Collins et al., 2015). As such, it is one of the best characterized groups in the Proterozoic basins of India. The detrital zircon ages from this group form a near-continuous spread from c.2750 to 1700 Ma with major concentrations at c.2510 and c.1850 Ma (Collins et al., 2015). The maximum depositional age for the group is interpreted to be 1659 ± 22 Ma based on the youngest near-concordant detrital zircon age. The Nallamalai Group is cut by the Chelima lamproite and the Racherla alkali syenite, whose ages are constrained to c.1350 Ma (Chalapathi Rao et al., 1999, 2012; Kumar et al., 2001). A series of alkaline granites intrude close to the eastern boundary of the Nallamalai Group, where it is in shear contact with the Nellore Schist Belt. These granites are interpreted to postdate deformation in both the Nellore Schist Belt and the NFB (Dobmeier et al., 2006; Saha and Chakraborty, 2003). The Vinukonda granite is dated at c.1590 Ma (Dobmeier et al., 2006) and lies only 4 km to the east of the NFB, whereas the Vellaturu granite intrudes the Nallamalai Group and has a Rb—Sr whole rock age of c.1575 Ma (Crawford and Compston, 1973). We therefore interpret the age of the Nallamalai Group to between 1659 ± 22 and c.1590 Ma.

6.2 Tuff Ages and Detrital Zircon Ages From the Chattisgarh Basin

Some of the most successful dating of the time of deposition has been achieved recently from the Chattisgarh basin due to the identification of volcanic tuffs and their subsequent U—Pb zircon dating. Patranabis-Deb et al. (2007) described and dated tuffs from Sukhda and Sapos in the upper Chhattisgarh basin stratigraphy (top of Raipur Group). They showed that these were deposited at 1020 ± 5, 1011 ± 19, and 990 ± 23 Ma, forcing a significant revision of the age of the basin stratigraphy. Similar-aged tuffs were later described and dated from the Dhamda area in the central part of the basin (Bickford et al., 2011a). In addition, Bickford et al. (2011b) dated zircons from the Singhora tuff, within the Chandarpur Group, the lowest group in the Chhattisgarh basin stratigraphy. They demonstrated that these basal units were deposited after 1405 ± 9 Ma. Detrital zircons from the Kharsiya Group also yielded considerable Stenian—Tonian zircon with the youngest recorded at 928 ± 8 Ma (Bickford et al., 2011b).

The smaller Indravati basin, to the southeast of the Chhattisgarh basin, hosts similar tuffs near the top of its stratigraphy. Here zircons from the Jagdalpur Formation tuffs yielded a precise LA-ICPMS-MC $^{207}Pb-^{206}Pb$ age of 1001 ± 7 Ma (Mukherjee et al., 2012), illustrating the similarity in the basin stratigraphy of the Chhattisgarh and Indravati basins and the widespread extent of c.1000 Ma volcanic deposits in these basins.

6.3 ^{40}Ar-^{39}Ar Glauconite and Detrital Zircon Data From the P–G Valley

The stratigraphy of this basin has long caused problems due to the gap in outcrop between the eastern and western side of the basin. Geochronological studies have considerable potential to sort out some of the stratigraphic uncertainties, but to date there are only a few studies, which beginning to assist. Recent U–Pb dating of detrital zircon from sandstones within the P–G basin has focused on the Somanpalli Group of the eastern side of the basin and the Sulluvai Group exposed on the west (Amarasinghe et al., 2015). The Somanpalli Group forms part of the lowest cycle of deposition in the eastern belt and has yielded a maximum depositional age of 1613 ± 19 Ma (Amarasinghe et al., 2015). Authigenic glauconite from the same group was dated by step heating individual crystals and yielded an age of 1620 ± 6 Ma (Conrad et al., 2011). These overlapping maximum and minimum depositional ages constrain the Somanpalli Group to being deposited at c.1620 Ma. The Mallampalli Group, at the base of the Pakhal Supergroup in the western belt, has also yielded authigenic glauconite ^{40}Ar-^{39}Ar age of 1686 ± 6 Ma (Conrad et al., 2011), whilst the Mulug Group has yielded an age of 1565 ± 6 Ma. The Mulug Group is usually correlated with the Somanpalli Group in the eastern belt (Chaudhuri et al., 2012). The 80-million-year difference in ages from the supposed correlative groups either side of the valley does question the link. However, although glauconite is usually a very early diagenetic phase (Odin and Fullagar, 1988), the ages are strictly minimum ages of deposition and therefore the correlation is still possible.

Glauconite dating was attempted on the basal part of the carbonate sequence of the third depositional cycle (Penganga Group), but no plateau was found (Conrad et al., 2011). A provisional minimum age of c.1200 Ma was, however, interpreted from the results. The Sullavai Group, on the west side of the P–G valley, has yielded earliest Neoproterozoic detrital zircon ages that demonstrate that this group was deposited after 972 ± 20 Ma (Amarasinghe et al., 2015). Joy et al. (2015) also reported several age peaks of detrital zircon from the Vekatpur Sandstone (Sullavai Group), the oldest of which is around 2500 Ma, suggesting derivation from granites of the Eastern Dharwar craton, and the youngest being 704 Ma. Detrital zircon data also exist from the Albaka Group (Collins et al. unpublished). The youngest detrital zircon that yielded ages within 10% of concordance preserved a $^{207}Pb/^{206}Pb$ age of 1474 ± 46 Ma (2 standard deviation error), which is consistent with the eastern belt Albaka Group being older than the western belt Sullavai Group.

6.4 Regional Correlation: Unconformities, Geochronologic Constraints

Prior to the availability of the new geochronological data, recognition of unconformities and several unconformity bound sequences with distinctive lithostratigraphic characters provided the most crucial criteria for correlation within individual basins (Chaudhuri et al., 2012). However, correlation across several Proterozoic basins with stratigraphic successions of various thicknesses, and basins straddling over three different cratonic blocks remained at best tentative (Saha and Chaudhuri, 2003). The new geochronological data including detrital zircon, zircon tuff ages, Ar—Ar ages from several of the basins as discussed above pave the foundation for a better stratigraphic correlation. The tentative correlation of the major stratigraphic groups from the eastern and western Dharwar, and Bastar cratons is shown in Fig. 14. We emphasize that the density of available data is not uniform across the basins, and the present correlation chart is only a modification of similar attempts in recent times (Amarasinghe et al., 2015; Collins et al., 2015; Conrad et al., 2011; Saha and Patranabis-Deb, 2014).

7. DISCUSSION AND CONCLUSION

Sedimentation cycles, relative sea-level changes and supercontinent connections.

7.1 Paleoproterozoic Scenario

Three major cycles of sedimentation in the Cuddapah basin is represented, respectively, by the unconformity bound Papaghni Group, Chitravati Group, and the Kurnool Group. Of these, the Kurnool Group sedimentation in the Neoproterozoic started after a prolonged hiatus, long after the cessation of Paleoproterozoic sedimentation in the Papaghni sub-basin and its inversion, as evidenced by the erosional unconformity at the base of the Kurnool Group resting variously on the Archean basement, and faulted and tilted strata of the Papaghni Group or the Chitravati Group. Based on the nature of the stratigraphic succession and sedimentological details, Saha and Tripathy (2012a,b) interpreted the succession of the Papaghni sub-basin in terms of

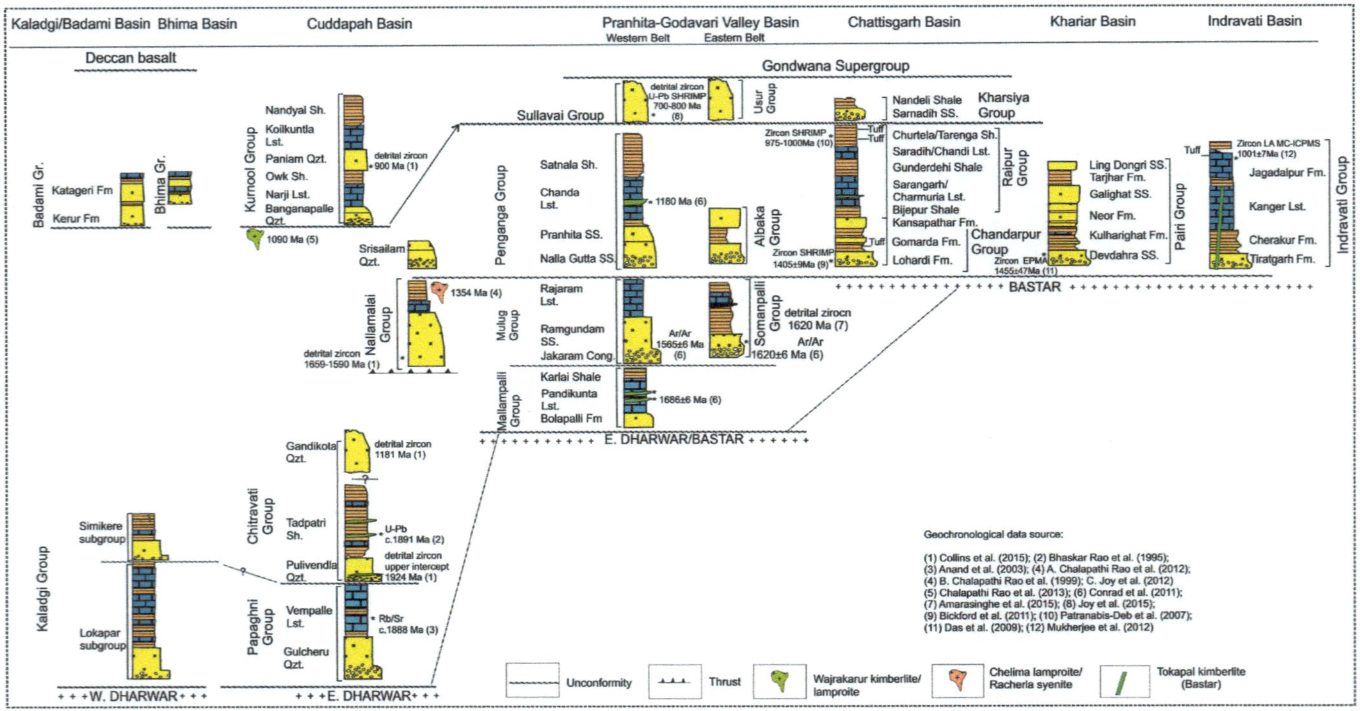

Figure 14 Tentative correlation of stratigraphic successions in the major Proterozoic basins in southern India. Note unconformity bound successions in each basin, representing cycles of sedimentation. Source of recent geochronological data used in correlation are indicated by numerals in parenthesis.

two stratigraphic sequences reflecting first order changes in the relative sea level.

Paleoproterozoic sequence-I in the Cuddapah basin begins with the alluvial fan of the Gulcheru Quartzite with basal polymict, matrix to clast supported conglomerate lying directly over the Peninsular gneiss (Fig. 3). The north-northwest trending Archean greenstone belts are truncated at the western outcrop margin of the Gulcheru Quartzite indicating an angular unconformity (nonconformity) at the base. The fluvial system gradually gave way to a supratidal–intertidal system possibly within a shallow embayment leading to the establishment of a carbonate platform of the Vempalle Formation. Gray-brown shales with occasional thin carbonate intercalations in the upper part of the Vempalle Formation has been interpreted as indicating maximum flooding under a high stand systems tract (Patranabis-Deb et al., 2012; Saha and Tripathy, 2012a). The contact between the siliciclast-dominant Pulivendla Quartzite with a thin basal conglomerate having materials derived from the underlying formations, and the underlying Vempalle Formation constitute the other sequence boundary for sequence-I. The presence of mafic dykes and sills both in the upper part of the Vempalle Formation (Fig. 3) was possibly linked to regional tectonothermal disturbance affecting the Eastern Dharwar craton, partly controlling the basin subsidence during Papaghni sedimentation.

The wider and extensive outcrop belt of the stromatolitic limestone and dolostone of the Tadpatri Formation has been interpreted as representing a stable carbonate platform with very little input of terrigenous coarser clastics, but showing laminated shales at the top. Saha and Tripathy (2012a) suggested a second maximum flooding event marking the deposition of extensive shales in the upper part of the Chitravati Group (Paleoproterozoic sequence-II).

Taking into consideration dyke swarms of the Eastern Dharwar and Bastar cratons (Fig. 2), together with c.1900 Ma mafic dykes and sills in the Cuddapah basin, some authors argue in favor of a Paleoproterozoic LIP (French et al., 2008). Although plume control of the proposed LIP is still hypothetical, the break-up of a hitherto unknown supercontinent and its fall out in terms of generation of passive margins (Bradley, 2008, 2011) and potential shallow marine sedimentation strike a chord on early fluvial sedimentation giving way to extensive shallow marine sedimentation within the Cuddapah basin c.1900 Ma.

Although a little younger than the earliest cycle of sedimentation in the Cuddapah basin, the sedimentation of the Mallampalli–Devalmari groups in the P—G valley basin constitute the earliest (c.1700 Ma) extensive passive margin quartz arenite-carbonate deposition on a slowly subsiding stable shelf (sequence I of Chaudhuri et al., 2012). In the eastern belt of P—G valley, the tidal intertidal deposits of the Somnur Formation quickly grade into the thick succession of graywacke—carbonate—fine sandstone—mudstone—shale turbidite of the Tarur Nala Formation (Table 2), which has been interpreted as rapid drowning of the shelf, at a rate much higher than the sediment supply, and expansion of the shelf.

The Bagalkot Group in the Kaladgi basin is another candidate that shows two cycles of Paleoproterozoic passive margin sedimentation in the Lokapur and Simikere Supbgroups respectively. Each cycle begins with local conglomerate, extensive arenite giving way to repeated intercalations of shale with limestone—dolomite—chert. Although a complete sequence stratigraphic interpretation is pending (Dey, 2015), the basal Salgundi conglomerate and pebbly beds of the Saundatti Quartzite has been interpreted as a fan-delta under low stand systems tract, later replaced by a transgressive system tract represented by the upper part of the Ramdurg and Yargatti Formations (Bose et al., 2008).

On the whole, extensive passive margin sedimentation is recorded in all the southern Indian cratons beginning around 1900 Ma. The tempo of sedimentation apparently reached acme in the Eastern and Western Dharwar cratons earlier than the P—G valley basin. One can recognize more than one cycle of sedimentation in the Paleoproterozoic, and erosional unconformity separating successive sequences, suggesting periodic, first order relative sea-level changes. The Proterozoic intracratonic basins of southern India apparently had connections with the open sea as demonstrated by the lithostratigraphic successions and their interpretation. As reviewed by Bradley (2008), acme of passive margin sedimentation in the time interval from 2050 to 1850 Ma is evidenced by Paleoproterozoic successions in the Nain, Slave, Superior, Wyoming cratons in Northern America, Kola craton in Russia, Aravalli—Delhi orogen in northern India (Bundelkhand craton), and Halls creek orogen and Gawler craton, Australia. The new geochronological evidence discussed here suggests that early sedimentation in the Cuddapah basin would probably add to the above list, and that the passive margin sedimentation occurred prior to assembly of the supercontinent Columbia/Nuna around c.1800 Ma, although there are lingering debates on the supercontinent configuration (Hoffman, 1997; Rogers and Santosh, 2003; Zhao et al., 2004). Although subduction-related activity has been suggested from the occurrence of supra-subduction zone c.1900 Ma Kandra ophiolite in the Nellore schist belt, this tectonic activity had little influence on the sedimentation in the EDC passive margin, and probably occurred further offshore.

The allochthonus Nallamalai Group is strongly folded (Fig. 5C and D) and sedimentological interpretations suggest early fan-delta deposition grading into peritidal sedimentation in the Bairenkonda (Nagari) Quartzite (Saha and Tripathy, 2012a). Transition to the Cumbum (Pullampet) Formation has been interpreted as a major transgressive event and occurrence of turbidites in the middle-upper part of the formation suggest deep-water deposition. However, presence of volcaniclastics, slumped horizons, and sysnsedimentary faults indicates slope instability and proximity to eruptive centers. Considering the detrital zircon ages and similarity of Hf isotopic data from NFB rocks and the Ongole domain granulite facies metasediemnts, the NFB has been interpreted as constituting the foreland of late Paleoproterozoic Krishna orogen, now represented by the Ongole domain in the southern EGB (Collins et al., 2015; Henderson et al., 2014). However, the Krishna orogen was separated from the EDC craton margin possibly till the Tonian (early Neoproterozoic). Henderson et al. (2014) suggested that the ultimate source of the zircons from the Ongole domain metasedimentaries could be the Napier complex of Antarctica or even the Proterozoic Australia. The juxtaposition of the Ongole domain with the craton margin occurred probably in the Neoproterozoic, as the Kurnool basin on the EDC margin is shown to have connection with open sea till that time (Saha et al., 2009).

7.2 Mesoproterozoic Scenario

The Albaka Group in the P—G valley basin (sequence V of Chaudhuri et al., 2012) overlies the folded Somanpalli Group along a major erosional unconformity. This tectonically modulated break in sedimentation is probably linked to Mesoproterozoic contractional deformation in the P—G valley (Ghosh and Saha, 2003, 2005). Contrary to the siliciclast-dominant Albaka Group, the stratigraphically similarly placed Penganga Group in the western belt has significant development of limestone and dolomite. The development of extensive deep-water carbonate platform and thick shale in the upper part of the Penganga Group involves changeover from shallow tidal/sub-tidal sedimentation in the siliciclastic lower part to deep-water deposition (Bandyopadhyay, 1996; Mukhopadhyay and Chaudhuri, 2003).

The interpreted paleoslope in the Chattisgarh basin is toward north and northeast, and the detrital zircon provenance favors derivation of the clastic sediments from the Bastar craton (Bickford et al., 2011a). Early sedimentation in fan-delta, pro-delta to prograding shelf environment represented by the coarse siliciclasts of the lower Chandarpur Group grades into the limestone shale-dominant Raipur Group developed on a subsiding passive margin bordering the northern/northeastern margin of the Bastar craton.

Sedimentation in the Chattisgarh basin and in the P–G valley basin apparently spanned over the entire Mesoproterozoic with several regionally extensive erosional unconformites particularly in the P–G valley (Chaudhuri et al., 2012) and deformation of relatively older successions like the Somanpalli Group suggesting gradual accretion. Although Ghosh and Saha (2005) suggested Mesoproterozoic crustal convergence for the evolution of the Somanpalli fold-and-thrust structures, it is only in recent times that better age constraints on subduction-related crustal convergence outboard of the southeastern (present day co-ordinates) margin of Proterozoic India is demonstrated from the c.1334 Ma Kanigiri ophiolitic mélange from the Nellore schist belt (Dharma Rao et al., 2011; see also discussion in Saha et al., 2016). It remains to be demonstrated, however, that the prolonged Mesoproterozoic sedimentation in southern Indian intracratonic basins emulates lateral accretion of arcs as proposed by Pisarevsky et al. (2014).

7.3 Neoproterozoic Scenario

There is extensive development of carbonate platform in the Kurnool sub-basin or in the Badami and Bhima successions in the Eastern and Western Dharwar cratons. Detrital zircon provenance studies (Collins et al., 2015) suggest that bulk of the siliciclastic sediments in the Kurnool Group were sourced from the Dharwar craton, rather than from an emerging orogen to the east, though a suggestion for a "long wavelength" foreland basin setting in relation to the Tonian Eastern Ghats orogeny needs to be tested further. The preliminary paleocurrent data (unpublished) from the Badami sandstones show derivation of detritus from the south, i.e., northern margin of the Dharwar craton, but pending reliable age constraints it is only tentative that a Neoproterozoic epieric sea extended to the northern margin of the Dharwar craton.

The fluvial to aeolian successions of the Sullavai Group and the stratigraphically equivalent Usur Group suggest that at least in the P–G valley basin connection with the open ocean was cut off by the end of Neoproterozoic. By the time the Kharsiya Group in the Chattisgarh basin (Table 3) was deposited, the sea to the north/northeast of the Bastar craton was also closing as interpreted from geochronological studies in the Central Indian tectonic zone, which yielded metamorphic and magmatic ages between 1030 and 960 Ma, interpreted as the age of final suturing of the northern and southern Indian blocks coinciding with the amalgamation of the Rodinia supercontinent (Bhowmik et al., 2012; Chattopadhyay et al., 2015).

Although there are remaining debates, the EGB clearly shows multiple orogenic imprints (Saha et al., 2016). North of the Godavari graben (P–G valley), c.950–900 Ma metamorphic events in the so-called Eastern Ghats Province match with the Rayner Province of east Antarctica and these two belts probably evolved simultaneously as a part of Rodinia (Dasgupta et al., 2013). The younger ages (c.780 and 520 Ma) reported from the northern EGB including the Rengali province in the extreme north may apparently link up with Rodinia break-up and subsequent Pan-African assembly of the East Gondwana (Bose et al., 2016). Cessation of sedimentation in the Kurnool basin or the P–G basin or other satellite basins in the Bastar craton and timing(s) of final closure and inversion of these basins are apparently linked to the final docking of the EGB segments on to the Bastar and Dharwar craton margins, but a precise tectonic model remains an unresolved issue.

ACKNOWLEDGMENTS

We thank Mary Ann Zimmerman (Elsevier) for inviting us to write this chapter. The late Asru Chaudhuri who spent his life time working on Proterozoic basins of India, has been an inspiration for SPD and DS, and some of the field photos from the P–G valley were taken by him, while on joint campaign. Photograph for aeolian sandstone from the Sullavai Group is kindly provided by Tapan Chakraborty. Some of the geochronological data used here were generated during execution of DST-AISRF project (ST030046). Infrastructure at Indian Statistical Institute helped us to complete the manuscript.

REFERENCES

Amarasinghe, U., Chaudhuri, A., Collins, A.S., Deb, G., Patranabis-Deb, S., 2015. Evolving provenance in the proterozoic Pranhita–Godavari basin, India. Geosci. Front 6, 453–463.

Anand, M., Gibson, S.A., Subbarao, K.V., Kelley, S.P., Dickin, A.P., 2003. Early Proterozoic melt generation processes beneath the intracratonic Cuddapah Basin, Southern India. J. Petrol. 44, 2139–2171.

Balakrishnan, S., Hanson, G.N., Rajamani, V., 1999. U–Pb ages for zircon and titanite from the Ramagiri area, Southern India: evidence for accretionary origin of the Eastern Dharwar craton during the late Archean. J. Geol. 107, 69–86.

Ball, V., 1877. On the geology of Mahanadi Basin and its vicinity. Geol. Surv. India Rec. 10, 167–186.

Balakrishnan, P., Mahesh Babu, M., 1987. The geology of the Ampani outlier, kalahandi, Koraput district, Orissa. In: Radhakrishna, B.P. (Ed.), Purana Basins of Peninsular India (Middle to Late Proterozoic), vol. 6. Geological Society of India, pp. 281–286 (Memoir).

Bandyopadhyay, P.C., 1996. Facies association and depositional environment of the Proterozoic carbonate-hosted microbanded manganese oxide ore deposit, Penganga Group, Godavari Rift basin, India. J. Sed. Res. 66 http://dx.doi.org/10.1306/D42682F7-2B26-11D7-8648000102C1865D.

Basumallick, S., 1967. Problems of the Purana stratigraphy of the Godavari Valley with special reference to the type area in Warangal district, Andhra Pradesh, India. Q. J. Geol. Min. Metal. Soc. India 39, 115–127.

Bhaskar Rao, Y.J., Pantulu, G.V.C., Damodar Reddy, V., Gopalan, K., 1995. Time of Early sedimentation and volcanism in the Proterozoic Cuddapah basin, South India: evidence from Rb–Sr age of Pulivendla mafic sill, vol. 33. Geological Society of India, Bangalore, pp. 329–338 (Memoir).

Bhattacharji, S., Singh, R.N., 1984. Thermomechanical structure of the southern part of the Indian Shield and its relevance to Precambrian basin evolution. Tectonophysics 105, 103–120.

Bhowmik, S.K., Wilde, S.A., Bhandari, A., Pal, T., Pant, N.C., 2012. Growth of the greater Indian landmass and its assembly in Rodinia: geochronological evidence from the central Indian tectonic zone. Gondwana Res. 22, 54–72.

Bickford, M.E., Basu, A., Mukherjee, A., Hietpas, J., Schieber, J., Patranabis-Deb, S., Ray, R.K., Guhey, R., Bhattacharya, P., Dhang, P.C., 2011a. New U-Pb SHRIMP zircon ages of the Dhamda tuff in the Mesoproterozoic Chhattisgarh basin, peninsular India: stratigraphic implications and significance of a 1-Ga thermal-magmatic event. J. Geol. 119, 535–548.

Bickford, M.E., Basu, A., Patranabis-Deb, S., Dhang, P.C., Schieber, J., 2011b. Depositional history of the Chhattisgarh basin, Central India: constraints from new SHRIMP zircon ages. J. Geol. 119, 33–50.

Bickford, M.E., Saha, D., Schieber, J., Kamenov, G., Russell, A., Basu, A., 2013. New U-Pb ages of zircons in the Owk Shale (Kurnool Group) with reflections on proterozoic porcellanites in India. J. Geol. Soc. India 82, 207–216.

Bose, P.K., Sarkar, S., Mukhopadhyay, S., Saha, B., Eriksson, P., 2008. Precambrian basin margin fan deposits: Mesoproterozoic Bagalkot Group, India. Precambrian Res. 162, 264–283.

Bose, S., Das, K., Torimoto, J., Arima, M., Dunkley, D.J., 2016. Evolution of the Chilka Lake Granulite Complex, Northern Eastern Ghats Belt, India: First Evidence of ~780 Ma Decompression of the Deep Crust and its Implication on the India–Antarctica Correlation. Lithos. http://dx.doi.org/10.1016/j.lithos.2016.01.017.

Bourgeois, J., Leithold, E., 1984. Wave-worked conglomerates-depositional process and criteria for recognition. In: Koster, H.E., Steel, R.J. (Eds.), Sedimentology of Gravels and Conglomerates, Canadian Society of Petroleum Geologist Memoir 10, pp. 331–343.

Bradley, D.C., 2008. Passive margins through earth history. Earth Sci. Rev. 91, 1–26.

Bradley, D.C., 2011. Secular trends in the geologic record and the supercontinent cycle. Earth Sci. Rev. 108, 16–33.

Chadwick, B., Vasudev, V.N., Hegde, G.V., 2000. The Dharwar craton, Southern India, interpreted as the result of late Archaean oblique convergence. Precambrian Res. 99, 91–101.

Chadwick, B., Vasudev, V.N., Hegde, G.V., Nutman, A.P., 2007. Structure and SHRIMP, U/Pb zircon ages of granites adjacent to the Chitradurga schist belt: implications for Neoarchaean convergence in the Dharwar craton, Southern India. J. Geol. Soc. India 69, 5–24.

Chakraborti, C.C., 2006. Proterozoic intra-continental basin: the Vindhyan example. J. Earth Syst. Sci. 115, 3–22.
Chakraborty, T., 1991. Sedimentology of a proterozoic erg: the Venkatapur sandstone, Pranhita-Godavari Valley, South India. Sedimentology 38, 301–322.
Chakraborty, T., 1999. Reconstruction of fluvial bars from the Proterozoic Mancherial Quartzite, Pranhita-Godavari Valley, India. In: Smith, W.D., Rogers, J. (Eds.), Fluvial Sedimentology VI: International Association of Sedimentologists, vol. 28. Special Publication, pp. 449–464.
Chakraborty, T., Chaudhuri, A.K., 1993. Fluvial-aeolian interactions in a proterozoic alluvial plain: example from Mancheral quartzite, Sullavai group, Pranhita-Godavari Valley, India. In: Pye, K. (Ed.), Dynamics and Environmental Context of Aeolian Sedimentary Systems, vol. 72. Geological Society, London, pp. 127–141. Special Publications.
Chakrabarti, G., Shome, D., 2010. Interaction of microbial communities with clastic sedimentation during Palaeoproterozoic time – an example from basal Gulcheru Quartzite. Sed. Geol. 226, 22–28.
Chakraborti, S., Saha, D., 2006. (Pub.2009). Tectonic stresses and thin-skinned tectonics in a Proterozoic fold-and- thrust belt read from calcite mylonites in the Cuddapah basin, South India. Indian J. Geol. 78, 37–54.
Chakrabarti, G., Shome, D., 2011. Carbonate facies and depositional environment of the Palaeoproterozoic Vempalle Formation, Cuddapah Basin, India. Indian J. Geol. 81, 27–37.
Chalapathi Rao, N.V., Miller, J.A., Gibson, S.A., Pyle, D.M., Madhavan, V., 1999. Precise $^{40}Ar/^{39}Ar$ dating of Kotakonda kimberlite and Chelima lamproite, India: implication to the timing of mafic dyke swarm activity in the Eastern Dharwar Craton. J. Geol. Soc. India 53, 425–433.
Chalapathi Rao, N.V., Wu, F.Y., Srinivas, M., 2012. Mesoproterozoic emplacement and enriched mantle derivation of the Racherla alkali syenite, Palaeo- Mesoproterozoic Cuddapah basin, Southern India: insights from in situ Sr-Nd isotope analysis on apatite. In: Majumdar, R., Saha, D. (Eds.), Palaeoproterozoic of India, vol. 365. Geological Society, London, pp. 185–195. Special Publications.
Chardon, D., Peucat, J.J., Jayananda, M., Choukroune, P., Fanning, C.M., 2002. Archean granite- greenstone tectonics at Kolar (South India): interplay of diapirism and bulk inhomogeneous contraction during juvenile magmatic accretion. Tectonics 21, 1–17.
Chaudhuri, A.K., Saha, D., Deb, G.K., Patranabis-Deb, S., Mukherjee, M.K., Ghosh, G., 2002. The Purana basins of Southern Cratonic province of India—A case for Mesoproterozoic fossil rifts. Gondwana Res. 5, 23–33.
Crookshank, H., 1963. Geology of Southern Bastar and Jeypore from Bailadila to Eastern Ghats, vol. 87. Geological Survey of India, Kolkata (Memoir).
Chaudhuri, A.K., 1985. Stratigraphy of the Purana Supergroup, Andhra Pradesh. J. Geol. Soc. India 26, 301–314.
Chaudhuri, A.K., 2003. Stratigraphy and palaeogeography of the Godavari Supergroup in the South-central Pranhita–Godavari Valley, South India. J. Asian Earth Sci. 21, 595–611.
Chaudhuri, A.K., Dasgupta, S., Bandopadhyay, G., Sarkar, S., Bandopadhyay, P.C., Gopalan, K., 1989. Stratigraphy of Penganga group around Adilabad, Andhra Pradesh. J. Geol. Soc. India 34, 291–302.
Chaudhuri, A.K., Deb, G.K., Patranabis-Deb, S., Sarkar, S., 2012. Paleographic and tectonic evolution of the Pranhita-Godavari valley, Central India: a stratigraphic perspective. Am. J. Sci. 312, 766–815.
Chaudhuri, A.K., Deb, G.K., Patranabis-Deb, S., 2015. Conflicts in stratigraphic classification of the Puranas of the Pranhita–Godavari Valley: review, recommendations and

status of the 'Penganga' sequence. In: Mazumder, R., Eriksson, P.G. (Eds.), Precambrian Basins of India: Stratigraphic and Tectonic Context, vol. 43. Geological Society, London, Memoirs, pp. 165–183.

Chattopadhyay, A., Das, K., Hayasaka, Y., Sarkar, A., 2015. Syn- and post-tectonic granite plutonism in the Sausar Fold Belt, central India: age constraints and tectonic implications. J. Asian Earth Sci. 107, 110–121.

Collins, A.S., Patranabis-Deb, S., Alexander, E., Bertram, C.N., Falster, G.M., Gore, R.J., Mackintosh, J., Dhang, P.C., Saha, D., Payne, J., Jourdan, F., Backe, G., Halverson, G.P., Wade, B.P., 2015. Detritalmineral age, radiogenic isotopic stratigraphy and tectonic significance of the Cuddapah Basin, India. Gond. Res. 28, 1294–1309.

Collinson, J.D., 1996. The coast. In: Reading, H.G. (Ed.), Sedimentary Environments: Processes, Facies and Stratigraphy, third ed. Blackwell Science, London, pp. 37–81.

Condie, K.C., Aster, R.C., 2010. Episodic zircon age spectra of orogenic granitoids: the supercontinent connection and continental growth. Precambrian Res. 180, 227–236.

Condie, K.C., Kroner, A., 2013. The building blocks of continental crust: evidence for a major change in the tectonic setting of continental growth at the end of the Archean. Gondwana Res. 23, 394–402.

Conrad, J.E., Hein, J.R., Chaudhuri, A.K., Patranabis-Deb, S., Mukhopadhyay, J., Deb, G.K., Beukes, N.J., 2011. Constraints on the development of Proterozoic basins in central India from $^{40}Ar/^{39}Ar$ analysis of authigenic glauconitic minerals. Geol. Soc. Am. Bull. 123, 158–167.

Crawford, A.R., Compston, W., 1973. The age of the Cuddapah and Kurnool systems, Southern India. J. Geol. Soc. Aust. 19, 453–464.

Das, N., Dutta, D.R., Das, D.P., 2001. Proterozoic Cover Sediments of Southeastern Chhattisgarh State and Adjoining Parts of Orissa. In: Geological Survey of India, vol. 55. Special Publication, Kolkata, pp. 237–262.

Das, D.P., Dutta, N.K., Dutta, D.R., Thanavelu, C., Baburao, K., 2003. Singhora Group-the oldest roterozoic lithopackage of Eastern Bastar Craton and its significance. Indian Min. 57, 127–138.

Das, K., Yokoyama, K., Chakraborty, P.P., Sarkar, A., 2009. Basal tuffs and contemporaneity of the Chattisgarh and Khariar basins based on new dates and geochemistry. J. Geol. 117, 88–102.

Das, D.P., Kundu, A., Das, N., Dutta, D.R., Kumaran, K., Ramamurthy, S., Thanavelu, C., Rajaiya, V., 1992. Lithostratigraphy and sedimentation of Chhattisgarh basin. Indian Min. 46, 271–288.

Datta, B., 1998. Stratigraphic and sedimentologic evolution of the Proterozoic siliciclastics in the southern part of Chhattisgarh and Khariar, Central India. J. Geol. Soc. India 51, 345–360.

Dhang, P.C., Patranabis-Deb, S., 2011. Lithostratigraphy of the basal part of the Chattisgarh Supergroup around Singhora–Saraipali area and its tectonic implication. In: Tiwari, R.P. (Ed.), Cenozoic Tectonics, Seismology, and Palaeobiology of the Eastern Himalayas and Indo-Myanmar Range, vol. 77. Geological Society of India, pp. 493–515 (Memoir).

Dasgupta, P.K., Biswas, A., 2006. Rhythms in Proterozoic Sedimentation: An Example from Peninsular India. Satish Serial Publishing, Delhi.

Dasgupta, S., Bose, S., Das, K., 2013. Tectonic evolution of the Eastern Ghats belt. Precambrian Res. 227, 247–258.

Deb, G., 2003. Deformation pattern and evolution of the structures in the Penganga group, the Pranhita–Godavari Valley, India: probable effects of Grenvillian movement on a Mesoproterozoic basin. J. Asian Earth Sci. 21, 567–577.

Dey, S., 2015. Geological history of Kaladgi-Badami and Bhima basins, South India: sedimentation in a Proterozoic intracratonic setup. In: Mazumder, R., Saha, D. (Eds.), Paleoproterozoic of India, vol. 43. Geological Society, London, pp. 283–296 (Memoirs).

Dharma Rao, C.V., Santosh, M., Yuan-Bao, W., 2011. Mesoproterozoic ophiolitic melange from the SE periphery of the Indian plate: U–Pb zircon ages and tectonic implications. Gond. Res. 19, 384–401.

Dobmeier, C., Lütke, S., Hammerschmidt, K., Mezger, K., 2006. Emplacement and deformation of the Vinukonda meta-granite (Eastern Ghats, India)—Implications for the geological evolution of peninsular India and for Rodinia reconstructions. Precambrian Res. 146, 165–178.

Dutt, N.B.V.S., 1963. Stratigraphy and correlation of Indravati series (Purana group), Bastar district, M.P. J. Geol. Soc. India 4, 35–49.

Dutt, N.V.B.S., 1964. A suggested succession of the Purana Formations of the Southern part of Chhattisgarh, M.P. Rec. Geol. Surv. India 93 (II), 143–148.

French, J.E., Heaman, L.M., Chacko, T., Srivastava, R.K., 2008. 1891–1883 Ma Southern Bastar–Cuddapah maficigneous events, India: a newly recognized large igneous province. Precambrian Res. 160, 308–322.

Ghosh, G., Saha, D., 2003. Deformation of the proterozoic Somanpalli group, Pranhita-Godavari Valley, South India—implication for a Mesoproterozoic basin inversion. J. Asian Earth Sci. 21, 579–594.

Ghosh, G., Saha, D., 2005. Kinematics of large scale asymmetric folds and associated smaller scale brittle-ductile structures in the Proterozoic Somnur Formation, Pranhita-Godavari valley, South India. J. Earth Syst. Sci. 114, 125–142.

Glorie, S., De Grave, J., Singh, T., Payne, J.L., Collins, A.S., 2014. Crustal root of the Eastern Dharwar craton: zircon U-Pb age and Lu-Hf isotopic evolution of the East Salem block, Southeast India. Precambrian Res. 249, 229–246.

Hawkesworth, C., Dhuime, B., Pietranik, A., Cawood, P., Kemp, T., Storey, C., 2010. The generation and evolution of the continental crust. J. Geol. Soc. Lond. 167, 229–248.

Henderson, B., Collins, A.S., Payne, J., Forbes, C., Saha, D., 2014. Geologically constraining India in Columbia: the age, isotopic provenance and geochemistry of the protoliths of the Ongole domain, Southern Eastern Ghats, India. Gondwana Res. 26, 888–906.

Hoffman, P.F., 1997. Tectonic genealogy of North America. In earth structure. In: van der Pluijm, B.A., Marshak, S. (Eds.), An Introduction to Structural Geology and Tectonics. McGraw-Hill, New York, pp. 459–464.

Holland, T.H., 1909. Indian empire. Geology. Imperial Gazetteer India 1, 50–103.

Jayananda, M., Peucat, J.-J., Chardon, D., Krishna Rao, B., Fanning, C.M., Corfu, F., 2013. Neoarchean greenstone volcanism and continental growth, Dharwar craton, Southern India: constraints from SIMS U–Pb zircon geochronology and Nd isotopes. Precambrian Res. 227, 55–76.

Jayaprakash, A.V., Sundaram, V., Hans, S.K., Mishra, R.N., 1987. Geology of Kaladgi-Badami basin, Karnataka. Mem. Geol. Soc. India 6, 201–206.

Joy, S., Hielke, J., Tappe, S., Armstrong, R., 2015. Shrimp U–Pb zircon provenance of the Sullavai group of Pranhita– Godavari basin and Bairenkonda quartzite of Cuddapah basin, with implications for the Southern Indian Proterozoic tectonic architecture. J. Asian Earth Sci. 111, 827–839.

Kale, V., 1990. Problem of Purana basins, vol. 18. Geological Society of India, Bangalore, pp. 77–93 (Memoir).

King, W., 1872. Kudapah and Karnul Formations in the Madras Presidency. Geological Survey of India, vol. 8. Calcutta, 346 pp. (Memoir).

King, W., 1881. The geology of the Pranhita–Godavari Valley, vol. 18. Geological Society of India, Bangalore, pp. 151–311 (Memoir).

Kumar, A., Gopalan, K., Rao, K.R.P., Nayak, S.S., 2001. Rb-Sr ages of kimberlites and lamproites from Eastern Dharwar craton, South India. J. Geol. Soc. India 58, 135–142.

Lancaster, P.J., Dey, S., Storey, C.D., Mitra, A., Bhunia, R.K., 2015. Contrasting crustal evolution processes in the Dharwar craton: insights from detrital zircon U-Pb and Hf isotopes. Gondwana Res. 28, 1361–1372.

Logan, B.W., Rezak, R., Ginsburg, R.N., 1964. Classification and environmental significance of algal stromatolites. J. Geol. 72, 68–83.

Mainkar, D., Lehmann, B., Haggerty, S.E., 2004. The crater-facies kimberlite system of Tokapal, Bastar district, Chhatisgarh, India. Lithos 76, 201–217.

Meert, J.G., Pandit, M.K., 2015. The Archaean and Proterozoic history of Peninsular India: tectonic framework for Precambrian sedimentary basins in India. In: Mazumder, R., Eriksson, P.G. (Eds.), Precambrian Basins of India: Stratigraphic and Tectonic Context, vol. 43. Geological Society Memoir, London, pp. 29–54.

Meert, J.G., Pandit, M.K., Pradhan, V.R., Banks, J., Stroud, M., Newstead, B., Gifford, J., 2010. Precambrian crustal evolution of Peninsular India: a 3.0 billion year odyssey. J. Asian Earth Sci. 39, 483–515.

Meijerink, A.M.J., Rao, D.P., Rupke, J., 1984. Stratigraphic and structural development of the Precambrian Cuddapah Basin, SE India. Precambrian Res. 26, 57–104.

Mishra, R.N., Jayaprakash, A.V., Hans, S.K., Sundaram, V., 1987. Bhima group of upper proterozoic – a stratigraphic Puzzle. In: Purana basins of peninsular India (middle to late proterozoic). Mem. Geol. Soc. India 6, 227–237.

Moitra, A.K., 1995. Depositional environmental history of the Chattisgarh Basin, M.P., based on stromatolites and microbiota. J. Geol. Soc. India 46, 359–368.

Mukherjee, A., Bickford, M.E., Hietpas, J., Schieber, J., Basu, A., 2012. Implications of a newly dated ca. 1000-Marhyolitic tuff in the Indravati basin, Bastar craton, India. J. Geol. 120, 477–485.

Mukhopadhyay, J., Chaudhuri, A.K., 2003. Stratigraphy of the Chanda limestone from the proterozoic Penganga group, Adilabad, Andhra Pradesh, India: implications for understanding depositional setting and palaeogeography. J. Geol. Soc. India 62, 356–368.

Murti, K.S., 1996. In: Geology, Sedimentation and Economic Mineral Potential of the South-Central Part of Chattisgarh Basin, vol. 125. Geological Survey of India, 139 pp. (Memoir).

Murti, K.S., 1987. Stratigraphy and sedimentation in Chhattisgarh basin. In: Radhakrishna, B.P. (Ed.), Purana Basins of Peninsular India, vol. 6. Geological Society of India, pp. 239–260 (Memoir).

Myrrow, P.M., Southard, J.B., 1991. Combined flow model for vertical stratification sequences in shallow marine storm-deposited beds. J. Sed. Petrol 61, 202–210.

Naha, K., Srinivasan, R., Jayaram, S., 1991. Sedimentological, structural and migmatitic history of the Archaean Dharwar tectonic province, Southern India. Proc. Indian Acad. Sci. (Earth Planet. Sci. 100, 413–433.

Nance, R.D., Murphy, J.B., 2013. Origins of the supercontinent cycle. Geosci. Front. 4, 439–448.

Nutman, A.P., Chadwick, B., Krishna Rao, B., Vasudev, V.N., 1996. SHRIMP U/Pb zircon ages of acid volcanic rocks in the Chitradurga and Sandur groups, and granites adjacent to the Sandur schist belt, Karnataka. J. Geol. Soc. India 47, 153–164.

Nutman, A.P., Chadwick, B., Ramakrishnan, M., Viswanatha, M.N., 1992. SHRIMP U–Pb ages of detrital zircons in Sargur supracrustal rocks in Western Karnataka, Southern India. J. Geol. Soc. India 39, 367–374.

Natarajan, V., Nair, S.R., 1977. Post-Kurnool thrust and structural features in the northeast part of the Palnad basin, Krishna district, A.P. J. Geol. Soc. India 18, 111–116.

Odin, G.S., Fullagar, P.D., 1988. Geological Significance of the glaucony facies. In: Odin, G.S. (Ed.), Green Marine Clays, Developments in Sedimentology Series, vol. 45. Elsevier, Amsterdam, pp. 295–332.

Pascoe, E.H., 1973. A Manual of the Geology India and Burma, third ed., vol. 1. Geological Survey of India, Calcutta.

Patranabis-Deb, S., 2004. Lithostratigraphy of the Neoproterozoic Chattisgarh sequence, its bearing on the tectonics and palaeogeography. Gondwana Res. 7, 323–337.

Patranabis-Deb, S., 2005. Tidal shelf sedimentation in the Neoproterozoic Chattisgarh succession of Central India. J. Earth Syst. Sci. 11, 211–226.

Patranabis-Deb, S., Chaudhuri, A.K., 2007. A retreating fan-delta system in the Neoproterozoic Chattisgarh rift basin, central India: major controls on its evolution. Am. Assoc. Petrol. Geol. Bull. 91, 785–808.

Patranabis-Deb, S., Chaudhuri, A.K., 2008. Sedimentological products and processes in the Mesoproterozoic Chattisgarh basin and contemporary tectonics in Central India. Indian J. Geol. 80, 139–155 (published 2010).

Patranabis-Deb, S., Bickford, M.E., Hill, B., Chaudhuri, A.K., Basu, A., 2007. SHRIMP ages of zircon in the uppermost tuff in Chattisgarh Basin in central India require up to ~500 Ma adjustment in Indian Proterozoic stratigraphy. J. Geol. 15, 407–415.

Patranabis-Deb, S., Saha, D., Tripathy, V., 2012. Basin stratigraphy, sea-level fluctuations and their global tectonic connections – evidence from the Proterozoic Cuddapah Basin. Geol. J 47, 263–283.

Peucat, J.-J., Jayananda, M., Chardon, D., Capdevila, R., Fanning, C.M., Paquette, J.-L., 2013. The lower crust of the Dharwar Craton, Southern India: patchwork of Archean granulitic domains. Precambrian Res. 227, 4–28.

Pisarevsky, S.A., Elming, S.-A., Pesonen, L.J., Li, Z.X., 2014. Mesoproterozoic paleogeography: supercontinent and beyond. Precambrian Res. 244, 207–225.

Nagaraja Rao, B.K., Rajurkar, S.T., Ramalingaswamy, G., Ravindra Babu, B., 1987. Stratigraphy, structure and evolution of the Cuddapah Basin. In: Radhakrishna, B.P. (Ed.), Purana Basins of Peninsular India (Middle to Late Proterozoic), vol. 6. Geological Society of India, pp. 33–86 (Memoir).

Radhakrishna, B.P. (Ed.), 1987. Purana basins of Peninsular India: Geological Society of India, vol. 6 (Bangalore, India, Memoir).

Radhakrishna, B.P., Vaidyanadhan, R., 1994. Geology of Karnataka. Geological Society of India, Bangalore.

Raha, P.K., Sastry, M.V.A., 1982. Stromatolites and Precambrian stratigraphy in India. Precambrian Res. 18, 293–318.

Ramakrishnan, M., 1987. Stratigraphy, sedimentary environment and evolution of the Late Proterozoic Indravati basin, central India. In: Radhakrishna, B.P. (Ed.), Purana Basins of Peninsula India, vol. 6. Geological Society of India, pp. 139–160 (Memoir).

Ramakrishna, M., Vaidyanadhan, R., 2008. Geology of India, vol. I. Geological Society of India, Bangalore, p. 556.

Rogers, A.J., Kolb, J., Meyer, F.M., Armstrong, R.A., 2007. Tectono-magmatic evolution of the Hutti–Maski Greenstone Belt, India: constrained using geochemical and geochronological data. J. Asian Earth Sci. 31, 55–70.

Rogers, J.J.W., Santosh, M., 2003. Supercontinents in earth history. Gondwana Res. 6, 357–368.

Rogers, J.J.W., Santosh, M., 2004. Continents and Supercontinents. Oxford University Press, Oxford.

Saha, D., 1988. Deformational controls in and lithostratigraphy of Cherla Formation, Albaka belt, South India. Indian J. Geol. 60, 92–103.

Saha, D., 2004. Structural asymmetry and plate tectonic set-up for a proterozoic fold and thrust belt: Nallamalai Fold belt and adjoining Terrane, South India. In: Roy, A., Jain, S.C., Chakrabarti, U. (Eds.), Uniformitarianism Revisited: Comparison between Ancient and Modern Orogens of India: Proceedings of the Workshop on IGCP 453, vol. 84. Geological Survey of India, pp. 101–119. Special Publications.

Saha, D., 2006. Rhysonetron and probable trace fossils in the Palaeoproterozoic to Mesoproterozoic quartzites, Cuddapah basin, South India – debate on early metazoan antiquity.

In: Proceedings of the National Seminar on Precambrain Life: Indian Scenario. Durgapur Government College, Durgapur, pp. 25–26.
Saha, D., Chaudhuri, A.K., 2003. Deformation of Proterozoic successions in the Pranhita-Godavari basin – regional perspective. J. Asian Earth Sci. 21, 557–564.
Saha, D., Chakraborty, S., 2003. Deformation pattern in the Kurnool and Nallamalai groups in the northeastern part (Palnad area) of the Cuddapah basin, South India and its implication on Rodinia. Gondwana Res. 6, 73–83.
Saha, D., Ghosh, G., 1987. Tectonic setting of proterozoic sediments around Somanpalli, Godavari Valley. J. Indian Assoc. Sedimentologists 7, 29–46.
Saha, D., Ghosh, G., 1998. Lithostratigraphy of deformed proterozoic rocks from around the confluence of Godavari and Indravati rivers, South India. Indian J. Geol. 70, 217–230.
Saha, D., Chakraborti, S., Tripathy, V., 2010. Intracontinental thrusts and inclined transpression along eastern margin of the East Dharwar Craton, India. J. Geol. Soc. India 75, 323–337.
Saha, D., Ghosh, G., Chakraborty, A.K., Chakraborti, S., 2009. Comparable Neoproterozoic sedimentary sequences in Palnad and Kurnool subbasins and their paleogeographic and tectonic implications. Indian J. Geol. 78, 175–192.
Saha, D., Mazumder, R., 2012. An overview of the Paleoproterozoic geology of peninsular India, and key stratigraphic and tectonic issues. In: Mazumder, R., Saha, D. (Eds.), Paleoproterozoic of India, vol. 365. Geological Society, London, pp. 5–29. Special Publication.
Saha, D., Patranabis-Deb, S., 2014. Proterozoic evolution of Eastern Dharwar and Bastar cratons, India – an overview of the intracratonic basins, craton margins and mobile belts. J. Asian Earth Sci. 91, 230–251.
Saha, D., Bhowmik, S., Bose, S., Sajeev, K., 2016. Proterozoic tectonics and trans-Indian mobile belts: a status report. Proc. Indian Natl. Sci. Acad. 82, 445–460.
Saha, D., Tripathy, V., 2012a. Palaeoproterozoic sedimentation in the Cuddapah Basin, South India and regional tectonics – a review. In: Mazumder, R., Saha, D. (Eds.), Paleoproterozoic of India, vol. 365. Geological Society, London, pp. 159–182. Special Publication.
Saha, D., Tripathy, V., 2012b. Tuff beds in Kurnool subbasin, Southern India, and implications for felsic volcanism in intracratonic basins. Geosci. Front. 3, 429–444.
Schnitzer, W.A., 1969. Zur Stratigraphie and Lithologie des nrdlichen Chhattisgarh-Beckens (Zentral-Indien) Unter besonderer Berckschtigung von Algenriff Komplexen. Dtsch. Geol. Geellschaft 118, 290–295.
Schnitzer, W.A., 1971. Das Jungpra kambrium Indiens ("Purana system"). Erlanger Geol. Abh. 85, 44.
Sharma, M., Pandey, S.K., 2012. Stromatolites of the Kaladgi Basin, Karnataka, India: their systematics, biostratigraphy and age implications. Palaeobotanist 61, 103–121.
Srinivasa Rao, K., 1987. Depositional sedimentary environment of the Albaka belt, Godavari Valley, Andhra Pradesh and Madhya Pradesh. In: Radhakrishna, B.P. (Ed.), Purana Basins of Peninsular India, vol. 6. Geological Society of India, Bangalore, pp. 261–280 (Memoir).
Srinivasa Rao, K., Sreenivasa Rao, T., Nair, R., 1979. Stratigraphy of the upper Precambrian Albaka belt, east of the Godavari river in Andhra Pradesh and Madhya Pradesh. J. Geol. Soc. India 20, 205–213.
Sreenivasa Rao, T., 1987. The Pakhal basin – a perspective. In: Radhakrishna, B.P. (Ed.), Purana Basins of Peninsular India, vol. 6. Geological Society of India, Bangalore, pp. 161–187 (Memoir).
Sreenivasa Rao, T., 1985. A note on the stratigraphy of the upper Precambrian sedimentaries around Ramagundam, Andhra Pradesh. Indian Miner. 39, 9–11.

Subba Raju, M., Sreenivasa Rao, T., Setti, D.N., Reddi, B.S.R., 1978. Recent advances in our knowledge of the Pakhal Supergroup with special reference to the Central part of the Godavari Valley. Geol. Surv. India, Rec. 110, 39–59.

Tripathy, V., Saha, D., 2010. Structure and low grade metamorphism of the east central part of the Proterozoic Nallamalai fold belt, South India – thrust stacking and discontinuous metamorphic gradients along eastern margin of east Dharwar craton. Indian J. Geol. 80, 173–188.

Walker, T.L., 1902. The geology of kalahandi state, central provinces. Mem. Geol. Surv. India 33 (Part 3).

Zhao, G., Sun, M., Wilde, S.A., Li, S., 2004. A Palaeo–Mesoproterozoic supercontinent: assembly, growth and breakup. Earth Sci. Rev. 67, 91–123.

CHAPTER TWO

Conodont and Graptolite Biostratigraphy of the Ordovician System of Argentina

G.L. Albanesi*,§,1 and G. Ortega§
*CICTERRA (CONICET-UNC), Córdoba, Argentina
§CONICET – Museo de Paleontología, CIGEA, Universidad Nacional de Córdoba, Córdoba, Argentina
[1]Corresponding author: E-mail: guillermo.albanesi@unc.edu.ar

Contents

Stratigraphy & Timescales, Volume 1
ISSN 2468-5178
http://dx.doi.org/10.1016/bs.sats.2016.10.002

Abstract

The state of the art on conodont and graptolite biostratigraphy of the Ordovician System of Argentina is introduced. A composite biostratigraphic chart is assembled by precise ties of conodont and graptolite index species throughout the system. The conodont faunas of the Precordillera of western Argentina are dominated by components of the Tropical Domain during the Early Ordovician, but an increasing interplay of temperate to cold-water taxa is developed along the period. The conodont associations from northwestern Argentine basins (with major faunas in Cordillera Oriental and Sistema de Famatina) consist of mixed assemblages from both Domains characterizing transitional environments, always from the marine Shallow-Water Realm. Two series of biostratigraphic units, 29 conodont zones and 35 graptolite zones, are determined through the Ordovician System of Argentina, with several bizones divided into subzones. In the Argentine Precordillera, 17 conodont zones and 19 graptolite zones are determined. A number of these biozones are recognized in neighboring areas, such as San Jorge, Bloque de San Rafael, and Cordillera Frontal. In northwestern Argentine basins, the biozones are distributed in the Sistema de Famatina, Puna, Cordillera Oriental, and Sierras Subandinas, totalizing 12 conodont zones and 16 graptolite zones. Detailed graptolite zonations are achieved for the Lower Ordovician of northwestern Argentina and for the Middle and Upper Ordovician of the Precordillera. The conodont faunas from the Upper Ordovician are not well documented in Argentine basins instead. The graptolite assemblages of the Cordillera Oriental are characterized by faunas of high and medium paleolatitudes. Conversely, Middle and Upper Ordovician graptolite faunas of the Precordillera are referred to as low paleolatitudes.

1. INTRODUCTION

The present contribution approaches the state of the art on conodont and graptolite biostratigraphy from the Ordovician System of the Precordillera of western Argentina, through the Mendoza, San Juan, and La Rioja provinces, and northwestern Argentine basins, in the Catamarca, Salta, and Jujuy provinces, including the Sistema de Famatina and Cordillera Oriental, as well as relevant data from the Puna region and Sierras Subandinas (Fig. 1). The first information on graptolites from the Argentine territory corresponds to Brackebush (1883), related to didymograptids from the Cordillera Oriental, later described by Kayser (1897). Argentine conodont studies began with a preliminary report by Iglesias (1949) on Middle Ordovician conodonts from the Sierras Subandinas, later published by Youngquist and Iglesias (1951). Ordovician conodonts from the Precordillera were discovered by Hünicken and Gallino (1970); since this contribution, and particularly during the last three decades, the knowledge on conodont and graptolite faunas from the Ordovician System of Argentine received increased attention. There was sustained interest by researchers in the area where a series of biozonal schemes was conducted to propose as reference standards for the Ordovician of South America. This was introduced by Albanesi et al. (1995a, 2008), Albanesi and Ortega (2002), and Brussa et al. (2003, 2008) and all provided descriptions and intercontinental correlations.

Lower Paleozoic successions of Argentina were traditionally described with reference to the British chronostratigraphic schemes (Fortey et al., 2000). Although Aceñolaza (1992) attempted a local scheme with original chronostratigraphic units, this first endeavor was unsuccessful to widespread use. Recent advances by the International Subcommission on Ordovician Stratigraphy (ICS, IUGS) established complete global schemes for the Ordovician System, based on global stratigraphic sections and points formally approved (Webby et al., 2004; Bergström et al., 2009).

Widespread exposures of lower Paleozoic marine strata in the Argentine territory are affected locally by diastrofism, which causes problems for a precise regional correlation. Difficulties for good correlations are also caused by stratigraphic discontinuities in the sedimentary successions and a variety of lithofacies patterns restricting the fossil record. These physical conditionings are in turn dependant on the paleogeographical location and geological settings of the basins, as well as global sea level changes through time. The usage of important index fossils for Ordovician biostratigraphy, i.e., conodonts and graptolites, allows for resolving the complex stratigraphic architecture of Ordovician terrains in Argentina.

Our contribution resolves this problem by obtaining the most precise ties between key species of orthochronologic fossil groups across successive time planes. This procedure permits the most refined composite biozonation up to date, according to the time slices (TS) defined by Bergström et al. (2009) for each stage of the Ordovician System. All conodont—graptolite linkages referred in each biozone either from the Precordillera or northwestern Argentine basins (mainly Sistema de Famatina and Cordillera Oriental) are thoroughly discussed by the authors,; in view of that, no further comments are incorporated herein. The biostratigraphic chart (Fig. 2) was organized taking into account biozones defined formally following the Argentine Stratigraphic Code (1992), mainly interval zones defined by the first appearance datum (FAD) of index species, and assemblage zones by a characteristic group of associated species, such as those described by Albanesi et al. (1998, 1999a, 2006, 2015b), Ortega and Albanesi (1999, 2000, 2005), Ortega et al. (2007a, 2008), Zeballo and Albanesi (2013b), Serra et al. (2015),

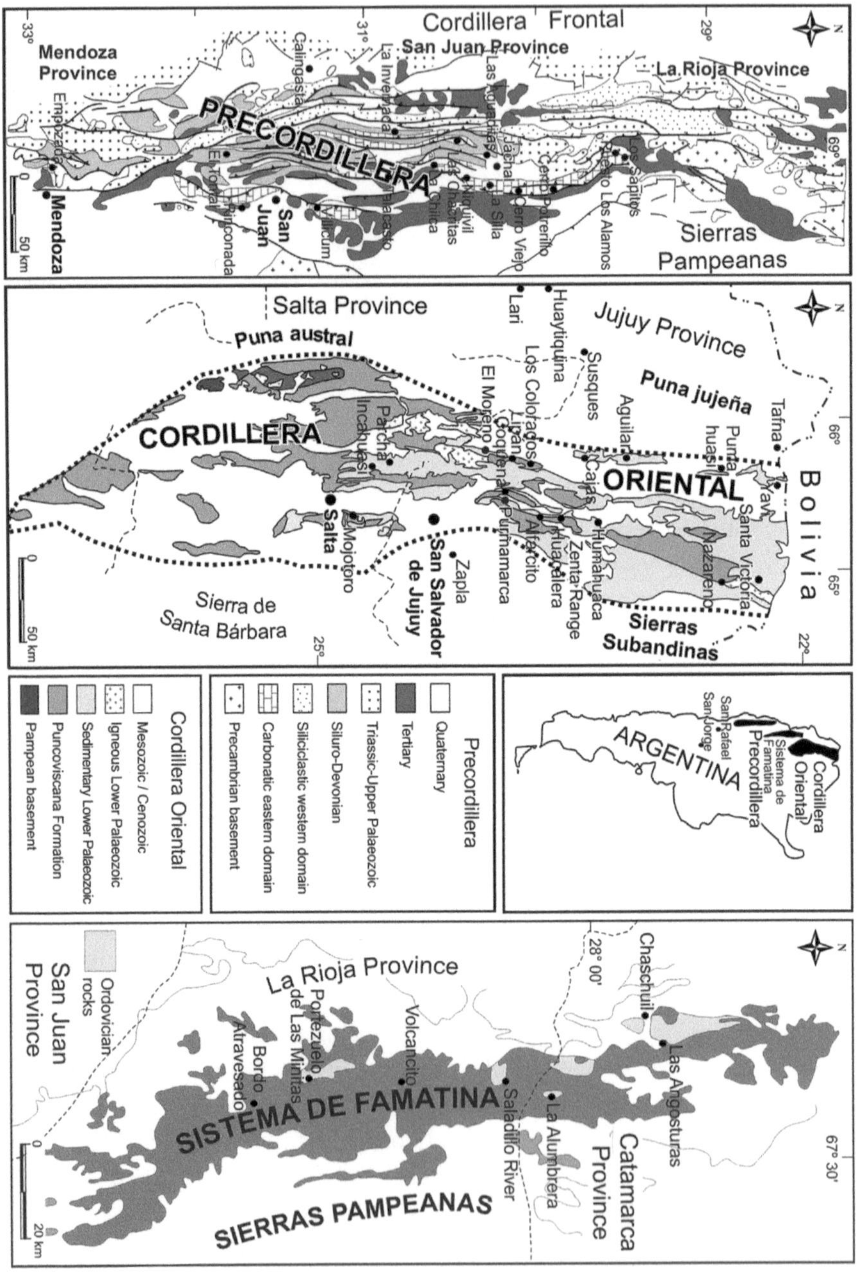
Cordillera Frontal
San Juan Province
Mendoza Province
La Rioja Province
PRECORDILLERA
Sierras Pampeanas
Mendoza
San Juan
50 km
Salta Province
Jujuy Province
Puna austral
Puna jujeña
CORDILLERA
ORIENTAL
Bolivia
Salta
San Salvador de Jujuy
Sierra de Santa Bárbara
Sierras Subandinas
Precordillera
Quaternary
Tertiary
Triassic-Upper Palaeozoic
Siluro-Devonian
Siliciclastic western domain
Carbonatic eastern domain
Precambrian basement
Cordillera Oriental
Mesozoic / Cenozoic
Igneous Lower Palaeozoic
Sedimentary Lower Palaeozoic
Puncoviscana Formation
Pampean basement
ARGENTINA
Cordillera Oriental
Sistema de Famatina
Precordillera
San Rafael
San Jorge
La Rioja Province
Catamarca Province
San Juan Province
Ordovician rocks
SISTEMA DE FAMATINA
SIERRAS PAMPEANAS
20 km

Figure 1 Sketch maps of the geological provinces of west and northwest Argentina with main Ordovician outcrops as discussed in the text; i.e., Precordillera, Cordillera Oriental, and Sistema de Famatina, showing the location of a number of localities with conodont and graptolite records. *Maps modified after Albanesi, G.L., Hünicken, M.A., Barnes, C.R., 1998. Bioestratigrafía de conodontes de las secuencias ordovícicas del cerro Potrerillo, Precordillera Central de San Juan, R. Argentina. Actas XII Acad. Nac. Cienc. Córdoba, 7—72; Albanesi G.L., Esteban, S.B., Ortega, G., Hünicken, M.A., Barnes, C.R., 2005. Bioestratigrafía y ambientes sedimentarios de las formaciones Volcancito y Bordo Atravesado (Cámbrico Superior — Ordovícico Inferior), Sistema de Famatina, provincia de La Rioja, Argentina. In: Dahlquist, J.A., Baldo, E.G., Alasino, P.H. (Eds.), Geología de la provincia de La Rioja: Precámbrico — Paleozoico inferior. Asoc. Geol. Argent., Serie D: Publ. Especial, vol. 8, pp.41—64; Albanesi, G.L., Ortega, G., Zeballo, F.J., 2008. Faunas de conodontes y graptolitos del Paleozoico inferior de la Cordillera Oriental Argentina. In: Coira, B., Zappettini, E.O. (Eds.), Geología y Recursos Naturales de la Provincia de Jujuy, Relatorio XVII Congreso Geológico Argentino, Jujuy, pp. 98—118; Voldman, G.G., Genge, M.J., Albanesi, G.L., Barnes, C.R., Ortega, G., 2012-13. Cosmic spherules from the Ordovician of Argentina. Geological Journal 48, 222-235. DOI: 10.1002/gj.2418.*

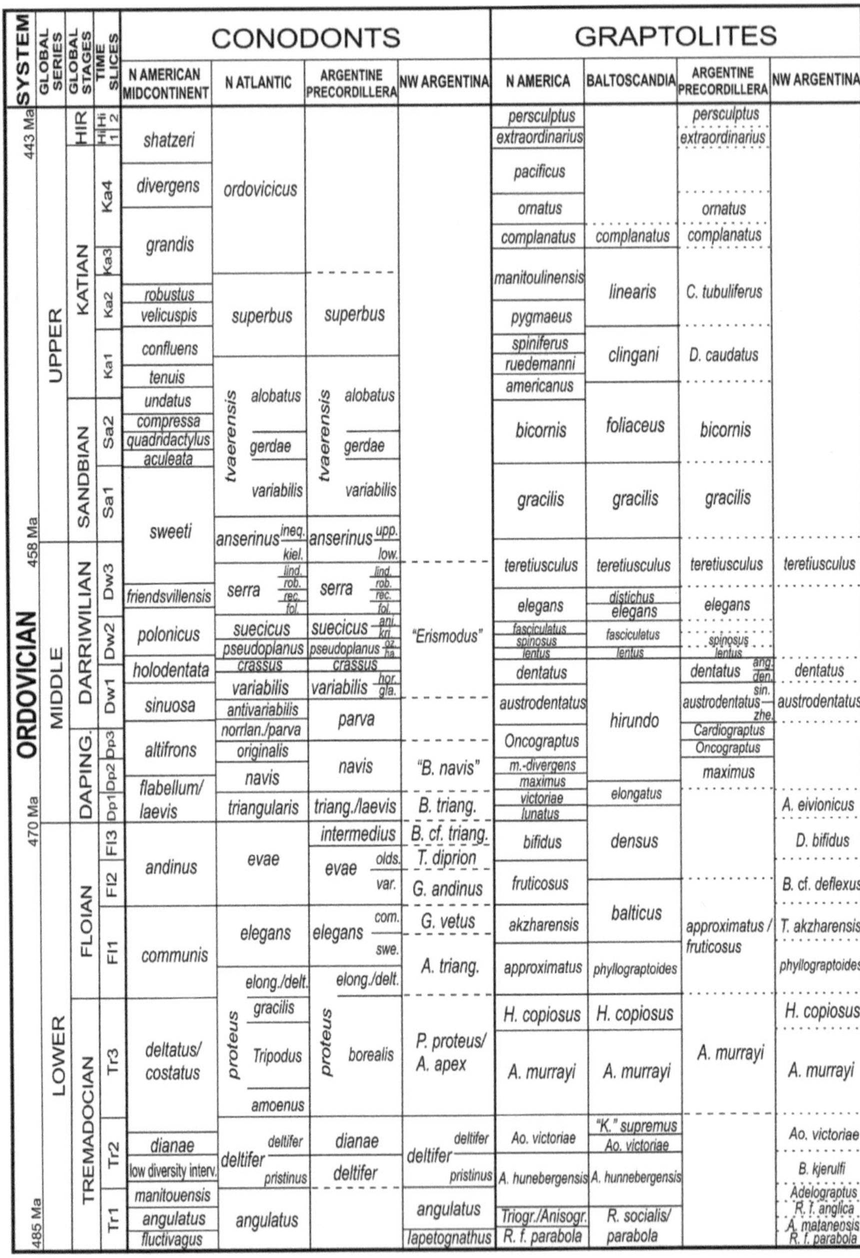

Figure 2 Conodont and graptolite biostratigraphic chart of the Ordovician System of Argentina (North American Midcontinent conodont zones: Ethington and Clark, 1981; Sweet, 1984; Ross et al., 1997. North Atlantic conodont zones: Lindström, 1971; Bergström, 1971; Bagnoli and Stouge, 1997; Löfgren and Zhang, 2003. North America graptolite zones: Cooper, 1999; Goldman et al., 2007. Baltoscandia graptolite zones: Maletz and Egenhoff, 2001; Maletz, 2009. Global high-resolution time scale: Webby et al., 2004; Bergström et al., 2009; Sadler et al., 2009; Gradstein et al., 2012). *Updated after Albanesi, G.L., Ortega, G., 2002. Advances on conodont-graptolite biostratigraphy of the Ordovician System of Argentina. In: Aceñolaza, F.G. (Ed.), Aspects of the Ordovician System of Argentina. Ser. Correl. Geol. 16, 143–165.*

Toro et al. (2015), Feltes et al. (2016), and Voldman et al. (2016a). The nature of the fossil record in the Argentine Ordovician basins is restrictive in particular time spans of the chart. These intervals with scarce information are analyzed after a number of studies by various authors, who recognized biostratigraphic units by the presence of index taxa from standard biozonations of well-known regions like Baltoscandia and North America.

2. PALEOENVIRONMENTS AND PALEOBIOGEOGRAPHY OF CONODONTS AND GRAPTOLITES

Zhen and Percival (2003) presented a new scheme considering paleoecological conditions as well as paleogeographic and paleoceanographic factors involved in the definition of paleobiogegraphical units on the basis of the conodont paleobiogeographical model proposed by Pohler and Barnes (1990). Accordingly, the Shallow-Sea Realm (depth of <200 m) includes the tropical (similar to the Midcontinent Realm of previous authors, cf. Sweet and Bergström, 1974), temperate, and cold domains with different provinces, and the Open-Sea Realm with deep-cold waters in the same three domains (equivalents to the earlier Atlantic Realm). The presence of certain endemic nectobenthic conodonts (Barnes and Fåhraeus, 1975) indicates a higher or lower relative proximity between basins of different provinces.

Following this approach, the conodont faunas of northwestern Argentina belong to the Shallow-Sea Realm and Cold Domain, representing the Southwestern Gondwana Province, which is characterized by a predominance of endemic taxa and lower proportions of Baltoscandian and North American conodont genera (Albanesi et al., 2007a; Zeballo and Albanesi, 2013a). The conodont zones recognized in the Cordillera Oriental could also be referred conceptually to the Atlantic Realm (Pyle and Barnes, 2002), to which pertains this Gondwana Province, characterized by its South American endemic conodonts. On the basis of the faunal interplays, the paleoenvironment of the Cordillera Oriental can be interpreted as an outer continental shelf to open sea of the Transitional Faunal Realm (Dubinina, 2000).

The conodont faunas of the Argentine Precordillera reveals compositional changes through time in the Ordovician Period, with the incorporation of an increasing percentage of taxa from the Cold Domain after the middle Darriwilian, which accounts for a transition of the Precordilleran Province from the Shallow-Sea Realm and tropical domain in the Early Ordovician to the temperate and cold domain of the same realm in the

Middle and Late Ordovician, respectively (cf. Albanesi and Bersgtröm, 2004, 2010). This evolutionary faunal trend of the Precordillera is in agreement with a progressive sea-water cooling through the Ordovician Period as demonstrated by oxygen isotope analyses from conodonts of the Precordillera and other basins as external paleoceanographic controls (Albanesi et al., 2015a). A combination of physical and biotic conditions caused the two highest conodont diversity pulses during the middle Floian and middle Darriwilian in the Argentine Precordillera as verified worldwide (Albanesi and Bergström, 2004).

The paleogeographical distribution of planktic graptolites and their paleoecological preferences were carried out by many authors since the last century (see Cooper et al., 2012; Goldman et al., 2013, and references therein). A combination of the surface water temperature model and the depth stratification model was considered by Goldman et al. (2013), who referred the graptolite faunas to the high, medium, and low paleolatitudes, which represent the earlier Atlantic and Pacific graptolite provinces of Skevington (1974).

During the Tremadocian most part of graptolite faunas from northwestern Argentina are of low diversity and worldwide distribution. However, the *Bryograptus kjerulfi* fauna is thought to be restricted to shallow-water environments and high paleolatitudes (Maletz et al., 2010).

An outer platform is considered for the lower Acoite Formation, Cordillera Oriental, with a deep-water biofacies of early Floian (*Tetragraptus phyllograptoides* and *Tetragraptus akzharensis* zones), shallowing to the upper part of the formation, in transition to inner platform (*Baltograptus* cf. *B. deflexus* and *Didymograptellus bifidus* zones) (Toro, 1994, 1999; Vento et al., 2014). The statistical study of graptolite faunas by Vento et al. (2012, 2014, and references therein) from the Cordillera Oriental shows affinities with Baltoscandia during the early Floian, which suggests that this basin was located in middle to high latitudes, representing the Atlantic Faunal Realm. On the other hand, these authors point out that the record of species of the *Baltograptus turgidus* group in the two upper biozones of the Acoite Formation, which are restricted to the inner shelf environment in southwest China, indicates affinity with this region in the late Floian. Post-Floian graptolite faunas (*Azygograptus eivionicus*, *Levisograptus austrodentatus*, *Levisograptus dentatus*, and *Hustedograptus teretiusculus*) from the Puna region, scarcely investigated up to the present, correspond to offshore and slope environments. Pendent didymograptids, such as *Aulograptus cucullus* and *Didymograptus* cf. *D. artus*, and the biserial Axonophora *Pseudamplexograptus* cf.

P. distichus from Puna indicate that the basin occupied high paleolatitudes during the Darriwilian.

Lower Dapingian black shales occur in the northern Precordillera (Guandacol and Gualcamayo rivers area), after the inundation of the Lower Ordovician carbonate platform. Graptolite faunas of this area are characteristic of the deep-water biofacies in low latitudes, which are dominated by isograptids and dichograptids. In the late Dapingian the entrance of the first axonophorans (*Exigraptus*) is documented for the Precordillera. Graptolite-bearing sections of Darriwilian age are widespread in the Precordillera. During the early Darriwilian graptolite assemblages show a high diversity corresponding to low-paleolatitudinal regions. In the late Darriwilian, the first appearance of *Dicellograptus*, *Dicranograptus*, and *Nemagraptus* is recorded, preceding the entrance of the *Nemagraptus gracilis* fauna. Most part of the Sandbian graptolites are pandemics, achieving a widespread distribution in the Precordillera. They are also recorded in the Bloque de San Rafael and in the Cordillera Frontal. Graptolite faunas of Katian age are poorly known in the Precordillera. They are represented by low diversity assemblages and are usually recorded in offshore environments. In most cases, they are distributed worldwide. Some of them represent the cosmopolitan epipelagic biotope (e.g., *Diplacanthograptus spiniferus*, *Rectograptus amplexicaulis*) and the cosmopolitan mesopelagic biotope (*Styracograptus tubuliferus*) of the model proposed by Goldman et al. (1995). The low diversity graptolite assemblage of the *Normalograptus extraordinarius* Zone from the Alcaparrosa Formation and the absence of the low-paleolatitudinal taxa (e.g., *Climacograptus hastatus* species group) suggest a high-paleolatitude position of the Precordillera in the early Hirnantian (Brussa et al., 1999), which probably extended to the late Hirnantian *Metabolograptus persculptus* Zone.

3. BIOSTRATIGRAPHY

3.1 Tremadocian

3.1.1 Conodont Zones

3.1.1.1 *Iapetognathus* Zone (TS: Lower Tr1)

The *Iapetognathus* Zone was formally defined in the mostly shaly upper part of the Cardonal Formation, Amarilla Creek section at the Cajas Range, by the FAD of *Iapetognathus fluctivagus* Nicoll et al. (Fig. 4(18)), as the basal unit of the Ordovician System. *Cordylodus lindstromi* (Fig. 5(13)), the nominal species of the uppermost Cambrian biozone, extends its range through the Lower Ordovician. The range of other species of *Iapetognathus*, such as *I. jilinensis*, is restricted to this biozone, while the long-ranging *Iapetognathus aengensis* (Fig. 5(9)) continues up to the middle Tremadocian. According to recommendations of the ICS, IUGS (Cooper et al., 2001) *Iapetognathus fluctivagus* as index species in the study section, which received a number of previous investigations (Rao, 1999, and references therein), can be considered as the reference section for the Cambrian/Ordovician boundary in South America (Albanesi et al., 2015b). This biozone is also approached in the Volcancito Formation of the Sistema de Famatina (Albanesi et al., 1999b, 2005), although the index taxon was still not documented. The intersystemic boundary is expected to be located in the middle part of the La Silla Formation in the Central Precordillera, where the referred index species might be recorded (cf. Lehnert, 1995b; Albanesi et al., 2016).

3.1.1.2 *Cordylodus angulatus* Zone (TS: Upper Tr1)

This biozone was initially documented by Rao and Hünicken (1995) and Rao (1999) in the Cajas Range, Cordillera Oriental. The *Cordylodus angulatus* Zone extends through the uppermost strata of the Cardonal Formation in the Amarilla Creek of the referred locality, and other classical sections of the Cordillera Oriental of Jujuy, where the eponymous taxon (Fig. 4(19)) was recorded (Albanesi et al., 2015b; with references), or other key associated species such as in sections of the Zenta Range (Zeballo et al., 2013). Endemic species characterized the *C. angulatus* Zone in the Cordillera Oriental; e.g., *Acanthodus raqueli* Zeballo and Albanesi (Fig. 4(11)), *Kallidontus gondwanicus* Zeballo and Albanesi (Fig. 4(12)), and *Variabiloconus crassus* Zeballo and Albanesi (Fig. 4(15)). The appearance of the characteristic Mid-continent species *Rossodus manitouensis* (Fig. 7(4)) in northwestern Argentine basins (Alfarcito Member of the Santa Rosita Formation in the Alfarcito area, Zeballo et al., 2005), and the Sistema de Famatina (Bordo Atravesado Formation), reveals the upper part of the *C. angulatus* Zone (Albanesi et al., 2005). This biozone should extend through the middle to upper section of the La Silla Formation in the Precordillera, considering that the index species of the overlying biozone was recorded in the upper strata of the unit in the type locality (Lehnert, 1995a), as well as in the Yanso section at Cerro Potrerillo (Albanesi et al., 1998).

3.1.1.3 *Paltodus deltifer* Zone (TS: Tr2)

The *Paltodus deltifer* Zone was recognized by Keller et al. (1994) in the carbonate La Silla Formation at the eponymous locality in the Precordillera.

This biozone can be divided into two subzones by the presence of the genus *Cordylodus* in the lower part of the biostratigraphic interval, and the two eponymous subspecies, *Paltodus deltifer pristinus* (Viira) (Fig. 7(2)) and *P. d. deltifer* (Lindström) (Fig. 4(2)), in northwestern Argentine basins, following the Baltoscandian scheme (Löfgren, 1997). *Paltodus deltifer* n. ssp. Zeballo et al. (Fig. 4(17)), which is intepreted as an intermediate form of the lineage, was recorded in strata spanning both subzones of the Cordillera Oriental (Zeballo et al., 2008). A number of endemic taxa typical of the *P. deltifer* Zone were recorded in the Cordillera Oriental; i.e., *Utahconus tortibasis* Zeballo and Albanesi (Fig. 4(8)), *Utahconus purmamarcensis* Zeballo and Albanesi (Fig. 4(9)), *Utahconus scandodiformis* Zeballo and Albanesi (Fig. 4(10)), *Acodus primitivus* Zeballo and Albanesi (Fig. 4(13)), *Tilcarodus humahuacensis* (Albanesi and Aceñolaza) (Fig. 4(14)), and *Drepanoistodus chucaleznensis* Albanesi and Aceñolaza (Fig. 4(16)). This zone was documented in the Cordillera Oriental at several classical localities, e.g., Nazareno, El Aguilar, Alfarcito, Purmamarca, Santa Victoria, Zenta (Manca et al., 1995; Rao and Flores, 1998; Albanesi et al., 2001; Zeballo et al., 2005, 2008, 2013; Zeballo and Albanesi, 2013b; Voldman et al., 2013; Giuliano et al., 2013) and the Sistema de Famatina at Bordo Atravesado (Albanesi et al., 2005), including characteristic species like *Paroistodus numarcuatus* (Lindström) (Fig. 7(3)). Certain series of records of particular assemblages with *P. deltifer* are interpreted to correspond to this interval, in diverse successions of carbonate rocks from the Cuyania terrane, including those from the Central Precordillera (Lehnert, 1995a; Albanesi et al., 1998), the Bloque de San Rafael in Mendoza Province (Lehnert et al., 1998), and the San Jorge Formation outcropping in the northwestern region of La Pampa Province (Albanesi et al., 2003).

3.1.1.4 *Macerodus dianae* Zone (TS: Upper Tr2)

Macerodus dianae Fåhraeus and Nowlan (Fig. 5(12)) is a typical North American Midcontinent key taxon (Ross et al., 1997; Ji and Barnes, 1994) that was recognized in formations from different setting of the Precordillera; i.e., the shallow-water carbonates of the La Silla Formation and hallodapic calcilutites and calcarenites of the Los Sombreros Formation (Albanesi et al., 2016; Voldman et al., 2016b). It is most useful to introduce this biozone because it refers to a constrained chronostratigraphic interval of the middle Tremadocian (Tr2), correlative to the lower part of the upper subzone of the *Paltodus deltifer* Zone. Consequently, it allows for precise linkage between conodont faunas from provinces of low and high latitudes.

3.1.1.5 *Paroistodus proteus* Zone (TS: Tr3, Lower Fl1)

In the Precordillera, this biozone is divided into the *Paroistodus proteus*–*Stiptognathus borealis* and *Oelandodus elongatus*–*Acodus deltatus* subzones, in the basal limestones of the San Juan Formation. Albanesi et al. (1998) proposed the base of the *O. elongatus*–*A. deltatus* subzone as the boundary between the Tremadoc and Arenig British Series, which correlates with the base of the global Floian Stage of the Lower Ordovician Series. The presence of *P. proteus* and *A. deltatus*, and a correlation with the respective biostratigraphic interval, was documented from diverse localities of the Precordillera (Lehnert, 1995b) and the Bloque de San Rafael (Lehnert et al., 1998).

3.1.1.6 *Paroistodus proteus–Acodus apex* Zone (TS: Tr3, Lower Fl1)

The chronostratigraphic interval that covers this biozone is equivalent to that originally defined by Albanesi and Ortega (2002) as *Paroistodus proteus*–*Acodus deltatus*, and to that recently defined by Voldman et al. (2016a) as *Acodus apex* Zone for the upper but not uppermost levels of the Santa Rosita Formation in the Santa Victoria area. It correlates accurately with the *P. proteus* Zone of the Precordillera. We prefer to maintain the original conception of assemblage zone, which is characterized by the nominal species and *A. apex* Albanesi and Zeballo, for the reason that *P. proteus* is a cosmopolitan key species that allows for regional and intercontinental correlations, and *A. apex*, synonymous of *A. deltatus sensu lato*, as documented by Albanesi et al. (2001), and Ortega and Albanesi (2005), in the Parcha area, is the most abundant and typical species of the referred interval from the Cordillera Oriental, although it is apparently endemic of the basin as far as current knowledge denotes; e.g., it was also reported in the Zenta Range by Zeballo et al. (2013).

3.1.2 Graptolite Zones

3.1.2.1 *Rhabdinopora flabelliformis parabola* Zone (TS: Lower Tr1)

The presence of *Rhabdinopora flabelliformis* cf. *R. f. parabola* (Bulman) (Fig. 3(1), 3(2), and 3(6)) in the upper succession of the Cardonal Formation at Amarilla Creek, Cajas Range, allows for recognizing this biozone in the western margin of the Cordillera Oriental, Jujuy Province (Ortega and Rao, 1995; Albanesi et al., 2008, 2015a,b). Specimens referable to *R. f. parabola* were also recovered from the upper Filo Azul Member of the Volcancito Formation at Sistema de Famatina, La Rioja Province, co-occurring with *R. f. socialis* (Salter) and *R. f. norvegica* (Kjerulf) (Gutiérrez-Marco and

Esteban, 2005). Recently, Giuliano et al. (2013) and Giuliano (2016) described *R. f.* cf. *R. f. parabola* from the upper part of the Las Vicuñas Formation at Lari Creek, Puna Occidental of Salta.

3.1.2.2 *Anisograptus matanensis* Zone (TS: Lower Tr1)

The entrance of *Anisograptus matanensis* Ruedemann (Fig. 3(3) and 3(4)) is recorded in the upper sequence of the Cardonal Formation (Amarilla Creek, Cajas Range) in western Cordillera Oriental of Jujuy, some meters above the first record of *R. f.* cf. *R. f. parabola.* In the lower part of this biozone, the co-occurrence of *A. matanensis* and *R. f.* cf. *R. f. parabola* is verified. As indicated by Ortega and Albanesi (2005), specimens of *Rhabdinopora* present in this biozone could correspond to another species like *R. f. canadensis* (Lapworth). In the Angosto del Moreno, south of Cajas Range, the *A. matanensis* Zone is identified by the assemblage of the nominal taxon and *Rhabdinopora flabelliformis* ssp. at the Saladillo Formation (Moya et al., 1994, 2003b; Ortega and Albanesi, 2005). According to Gutiérrez-Marco and Esteban (2005) the specimens of *Rhabdinopora* from the Angosto del Moreno would be referred to *R. f. acenolazai.* The *A. matanensis* Zone was also recorded in the Sistema de Famatina, where Gutiérrez-Marco and Esteban (2005) described this biozone in the Peña Negra Member, Volcancito Formation at the Peña Negra section, La Rioja Province. The assemblage is composed by *A. matanensis, Rhabdinopora turneri* Gutiérrez-Marco and Esteban, *R.* cf. *R. enigma* (Cooper and Stewart), *R. f. flabelliformis* (Eichwald), and *R. f. acenolazai.* Other locality that records this biozone is the Las Angosturas, northwest Tinogasta, Catamarca Province (Ortega et al., 2005).

3.1.2.3 *Rhabdinopora flabelliformis anglica* Zone (TS: Upper Tr1)

This biozone is documented in the middle part of the Peña Negra Member (Volcancito Formation) at the Peña Negra section, Sistema de Famatina. It is marked by the entrance of the *Rhabdinopora flabelliformis anglica* (Bulman), which ranges with *Anisograptus matanensis* throughout its lower part (Gutiérrez-Marco and Esteban, 2005). It could also be present in the upper part of the Cardonal Formation at Cajas Range, Cordillera Oriental, where fragmentary tubaria of *R. f.* cf. *R. f. anglica* (Fig. 3(7)) were collected with *A. matanensis* (Ortega and Albanesi, 2005).

3.1.2.4 *Adelograptus* Zone (TS: Upper Tr1)

A graptolite assemblage composed by *Adelograptus tenellus* (Linnarsson), *A. cuerdai* Monteros and Moya, and some taxa referred to kiaerograptids allow

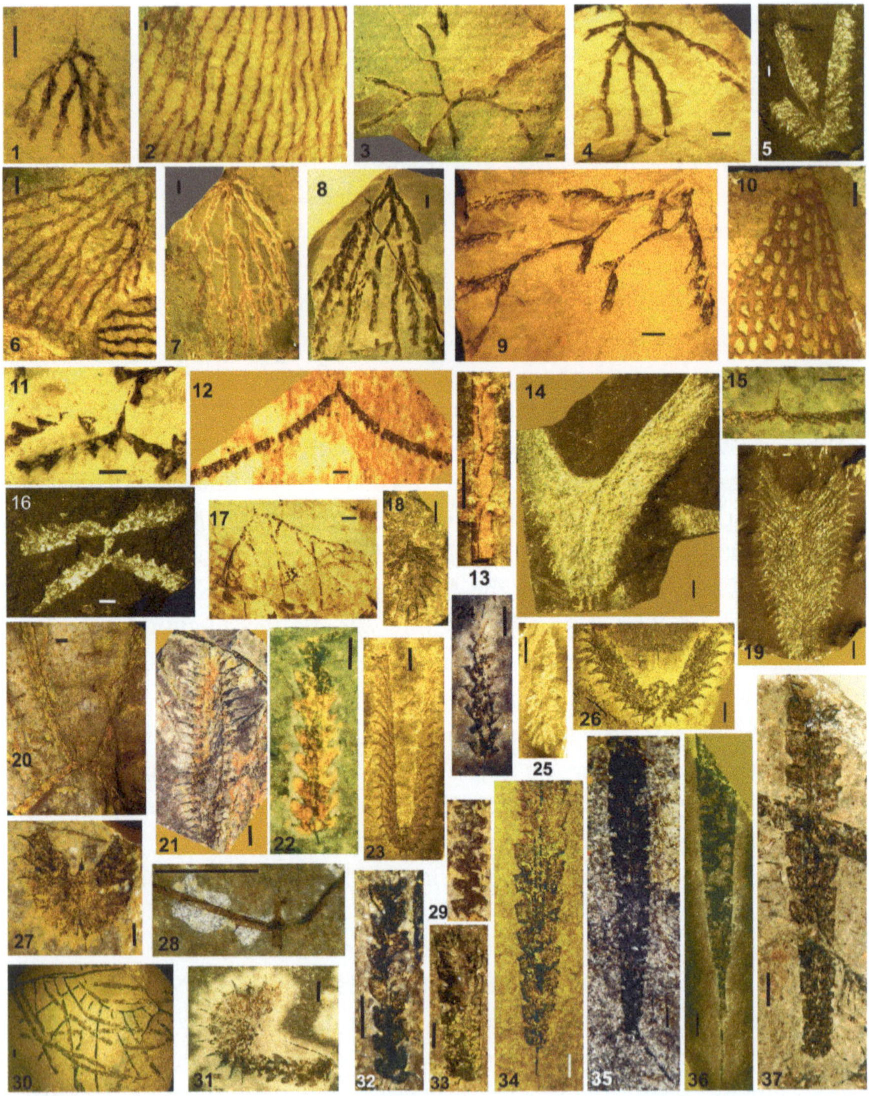

Figure 3

(1, 2, 6) *Rhabdinopora flabelliformis* cf. *R. f. parabola* (Bulman), *Rhabdinopora flabelliformis parabola* Zone, Cardonal Formation, Amarilla Creek, Cajas Range, Cordillera Oriental of Jujuy, 1, CORD-PZ 18,132B, 2, CORD-PZ 18,138, 6, CORD-PZ 23,836.

(3, 4) *Anisograptus matanensis* Ruedemann, *Anisograptus matanensis* Zone, Cardonal Formation, Amarilla Creek, Cajas Range, Cordillera Oriental of Jujuy, 3, CORD-PZ 18,136, 4, CORD-PZ23823.

(5) *Tetragraptus phyllograptoides* Strandmark, *Tetragraptus phyllograptoides* Zone, La Alumbrera Formation, La Alumbrera River, Sistema de Famatina, Catamarca Province, CORD-PZ 20,642.

(7) *Rhabdinopora flabelliformis* cf. *R. f. anglica* (Bulman), *Rhabdinopora flabelliformis anglica* Zone?, Cardonal Formation, Amarilla Creek, Cajas Range, Cordillera Oriental of Jujuy, CORD-PZ 23,827.

(8) *Bryograptus kjerulfi* Lapworth, *Bryograptus kjerulfi* Zone, Saladillo Formation, Angosto de Lampazar, Incamayo Creek, Cordillera Oriental of Salta, CORD-PZ 19,457.
(9) *Aorograptus victoriae* (T.S. Hall), *Aorograptus victoriae* Zone, Coquena Formation, Purmamarca, Cordillera Oriental of Jujuy, CORD-PZ 32,002.
(10) *Araneograptus murrayi* (J. Hall), *Araneograptus murrayi* Zone, Santa Rosita Formation, between Santa Victoria and Acoite villages, Cordillera Oriental of Salta, CORD-PZ 35,716.
(11) *Kiaerograptus kiaeri* (Monsen), *Aorograptus victoriae* Zone, Saladillo Formation, Angosto de Lampazar, Incamayo Creek, Cordillera Oriental of Salta, CORD-PZ 19,849.
(12) *Baltograptus* cf. *B. deflexus* (Elles and Wood), *Baltograptus* cf. *B. deflexus* Zone?, Acoite Formation, Cerro Siete Hermanos, Yavi, Cordillera Oriental of Jujuy, CORD-PZ 31,410.
(13) *Atopograptus woodwardi* Harris, *Holmograptus spinosus* Zone, Gualcamayo Formation, Villicum Range, Precordillera Oriental, San Juan Province, CORD-PZ 23,703.
(14) *Oncograptus upsilon* T.S. Hall, *Oncograptus upsilon* Zone, Gualcamayo Formation, Los Sapitos Creek, Guandacol, Precordillera Central, La Rioja Province, CORD-PZ 21,617.
(15) *Hunnegraptus copiosus* Lindholm, *Hunnegraptus copiosus* Zone, Parcha Formation, Abra de Sococha, Incamayo Creek, Cordillera Oriental of Salta, CORD-PZ 18,978.
(16) *Tetragraptus approximatus* Nicholson, *Tetragraptus phyllograptoides* Zone, Acoite Formation, Cristobal River, Cajas Range, Cordillera Oriental of Jujuy, CORD-PZ 20,890B.
(17) *Pterograptus elegans* Holm, *Pterograptus elegans* Zone, Los Azules Formation, Cerro Viejo of Huaco, Precordillera Central of San Juan, CORD-PZ 14,629.
(18) *Bergstroemograptus crawfordi* (Harris), *Holmograptus lentus* Zone (upper assemblage), Los Azules Formation, Cerro Viejo of Huaco, Precordillera Central of San Juan, CORD-PZ.
(19) *Cardiograptus morsus* Harris and Keble, *Cardiograptus morsus* Zone, Gualcamayo Formation, Los Sapitos Creek, Guandacol, Precordillera Central of La Rioja, CORD-PZ 21,462.
(20) *Tetragraptus akzharensis* Tzaj, *Tetragraptus akzharensis* Zone, La Alumbrera Formation, Sistema de Famatina, Catamarca Province, CORD-PZ 20,844A.
(21) *Paraglossograptus* cf. *P. tricornis* (Mu, Geh and Yin), *Undulograptus dentatus* Zone, Los Azules Formation, Cerro Viejo of Huaco, Precordillera Central of San Juan, CORD-PZ 13,979.
(22) *Levisograptus austrodentatus* (Harris and Keble), *Levisograptus dentatus* Zone, Los Azules Formation, Cerro Viejo of Huaco, Precordillera Central of San Juan, CORD-PZ 14,632.
(23) *Parisograptus caduceus* (Salter), *Oncograptus upsilon* Zone, Gualcamayo Formation, Los Sapitos Creek, Guandacol, Precordillera Central of La Rioja, CORD-PZ 21,615.
(24) *Undulograptus sinicus* (Mu and Lee), *Holmograptus lentus* Zone, Los Azules Formation, Cerro Viejo of Huaco, Precordillera Central of San Juan, CORD-PZ 13,984.
(25) *Skiagraptus gnomonicus* (Harris and Keble), *Cardiograptus morsus* Zone, Gualcamayo Formation, Los Sapitos Creek, Guandacol, Precordillera Central of La Rioja, CORD-PZ 21,462.

for identifying the *Adelograptus* Zone in the Floresta Formation at Mojotoro Range, Cordillera Oriental of Salta (Monteros and Moya, 2005).

3.1.2.5 *Bryograptus kjerulfi* Zone (TS: Lower Tr2)

The *Bryograptus kjerulfi* Zone, usually represented by a monospecific fauna, is documented in diverse localities of the Cordillera Oriental, Salta Province. It is present in the Saladillo Formation at Angosto de Lampazar (Fig. 3(8)) and in the Floresta Formation, Mojotoro Range (González Barri and Alonso, 1984; Moya et al., 1994; Ortega and Albanesi, 2005; Maletz et al., 2010). Recently, this biozone was recorded in the Santa Rosita Formation on

(26) *Pseudisograptus manubriatus* (T.S. Hall), *Cardiograptus morsus* Zone, Los Sapitos Creek, Guandacol, Precordillera Central of La Rioja, CORD-PZ 21,593.
(27) *Arienigraptus angulatus* (Mu), *Holmograptus lentus* Zone, Los Azules Formation, Cerro Viejo de Huaco, Precordillera Central of San Juan, CORD-PZ 23,357-B.
(28) *Nemagraptus gracilis* (J. Hall), *Climacograptus bicornis* Zone, Los Azules Formation, Cerro Viejo de Huaco, Precordillera Central of San Juan, CORD-PZ 24,651.
(29) *Haddingograptus* cf. *H. oliveri* (Bouček), *Holmograptus lentus* Zone (upper assemblage), Los Azules Formation, Cerro Viejo de Huaco, Precordillera Central of San Juan, CORD-PZ 23,575.
(30) *Nemagraptus gracilis* (J. Hall), *Climacograptus bicornis* Zone, Los Azules Formation, Cerro Viejo de Huaco, Precordillera Central of San Juan, CORD-PZ 13,441.
(31) *Levisograptus austrodentatus* (Harris and Keble) and *Arienigraptus zhejiangensis* (Yu and Fang), *Levisograptus dentatus* Zone, Los Azules Formation, Cerro Viejo de Huaco, Precordillera Central of San Juan, CORD-PZ 13,976-A.
(32) *Archiclimacograptus? ambiguus* Maletz, *Holmograptus lentus* Zone, Los Azules Formation, Cerro Viejo de Huaco, Precordillera Central of San Juan, CORD-PZ 23,262-A.
(33) *Levisograptus primus* (Legg), *Holmograptus lentus* Zone, Los Azules Formation, Cerro Viejo de Huaco, Precordillera Central of San Juan, CORD-PZ 23,398.
(34) *Styracograptus tubuliferus* (Lapworth), *S. tubuliferus* Zone, Sierra de La Invernada Formation, La Invernada Range, Precordillera of San Juan, CORD-PZ 24,739.
(35) *Diplacanthograptus spiniferus* (Ruedemann), *Diplacanthograptus caudatus* Zone, Sierra de La Invernada Formation, La Invernada Range, Precordillera of San Juan, CORD-PZ 24,854.
(36) *Diplacanthograptus caudatus* (Lapworth), *Diplacanthograptus caudatus* Zone, Sierra de La Invernada Formation, La Invernada Range, Precordillera of San Juan, CORD-PZ 24,733.
(37) *Climacograptus bicornis* J. Hall., *Climacograptus bicornis* Zone, Los Azules Formation, Cerro Viejo de Huaco, Precordillera Central of San Juan, CORD-PZ 14,245.
Scale bar: 1 mm.

the eastern part of the Cordillera Oriental, nearby Nazareno and Santa Victoria villages (Giuliano et al., 2013; Giuliano, 2016).

3.1.2.6 *Aorograptus victoriae* Zone (TS: Upper Tr2-Lower Tr3)

The *Aorograptus victoriae* Zone is a widespread graptolite assemblage in the Cordillera Oriental. It was localized in the San Bernardo Formation, Mojotoro Range (Monteros and Moya, 2002, 2003; Moya and Monteros, 2011), in the Saladillo Formation, Incamayo Creek (Ortega and Albanesi, 2002, 2003, 2005; Toro and Carrera, 2013; Toro et al., 2015), in the Coquena Formation, Purmamarca (Zeballo et al., 2008) (Fig. 3(9)), and in the Humacha Member, Santa Rosita Formation in Tilcara Range (Zeballo et al., 2008). Newly, this graptolite fauna was discovered in the Santa Rosita Formation, nearby the Santa Victoria village, in the northeastern Cordillera Oriental (Toro et al., 2015; Giuliano, 2016). The graptolite assemblage is integrated by *Paradelograptus antiquus* (T.S. Hall), *Paradelograptus onubensis* Erdtmann, Maletz and Gutiérrez-Marco, *P. mosseboensis* Erdtmann, Maletz and Gutiérrez-Marco, *Paradelograptus* spp., *Paratemnograptus isolatus* Williams and Stevens, *Adelograptus* cf. *A. altus* Williams and Stevens, *Kiaerograptus kiaeri* (Monsen) (Fig. 3(11)), *Kiaerograptus* sp., *Ancoragraptus bulmani* (Spjeldnaes), and *Ao. victoriae* (T.S. Hall), among others. An assemblage composed by *Kiaerograptus* cf. K. *kiaeri*, *Paradelograptus* sp., *Clonograptus* sp. allows for recognizing the *Ao. victoriae* Zone in the Chiquero Formation at the Susques area, Puna of Jujuy (Benedetto et al., 2002). The cited references can be consulted in Ortega and Albanesi (2005), Moya and Monteros (2011), and Toro et al. (2015). The *Ao. victoriae* Zone is also represented in the Bordo Atravesado Formation of the Sistema de Famatina, La Rioja Province, where tubaria of kiaerograptids and *Paradelograptus* sp. were identified (Albanesi et al., 2005).

3.1.2.7 *Araneograptus murrayi* Zone (TS: Lower Tr3)

Araneograptus murrayi (J. Hall) (Fig. 3(10)) is present in numerous localities of Argentina including Cordillera Oriental, Puna, and Precordillera, ranging through the *A. murrayi* and *Hunnegraptus copiosus* zones, and probably the lower *Tetragraptus phyllograptoides* Zone. In the Cordillera Oriental, the *A. murrayi* Zone was identified in the Mojotoro and Santa Victoria ranges (Turner, 1960; Gutiérrez-Marco and Aceñolaza, 1987; Voldman et al., 2014; Toro et al., 2015; Giuliano, 2016), Parcha and Incahuasi localities (Incamayo Creek) (Gutiérrez-Marco and Aceñolaza, 1987; Aceñolaza et al., 1996; Ortega and Albanesi, 2003, 2005) and the Zenta Range

(Albanesi et al., 2011). *"Kiaerograptus" supremus* and *A. murrayi* co-occur in the Santa Rosita Formation at the Zenta Range (Albanesi et al., 2011), the Santa Victoria Range (Toro et al., 2015; Giuliano, 2016), and in the Parcha Formation at the Incamayo Creek area (Toro and Carrera, 2013; Toro et al., 2015), suggesting the presence of the lower part of the *A. murrayi* Zone. In the Abra de Sococha, Incamayo Creek area, Ortega and Albanesi (2003, 2005) remarked the presence of a ca. 30-m-thick interval with *K. supremus*, *Paradelograptus* spp., and indeterminate two stiped declined and deflexed graptolites preceding the entrance of *A. murrayi*, which they referred to the *supremus* Zone. The *A. murrayi* Zone could be represented in the volcaniclastic Tolillar Formation, southern Puna of Catamarca (Zimmermann et al., 1999), and in the Los Sombreros Formation at the El Tontal Range, Precordillera of San Juan (Ortega et al., 2014), where the index fossil was recovered.

3.1.2.8 *Hunnegraptus copiosus* Zone (TS: Upper Tr3)

The *Hunnegraptus copiosus* Zone is recognized by the first appearance of the nominal taxon in the lower part of the Parcha Formation at the Incamayo Creek, Cordillera Oriental of Salta (Ortega and Albanesi, 2002, 2003, 2005; Toro et al., 2010). In this biozone *H. copiosus* Lindholm (Fig. 3(15)) co-occurs with singular species, such as *P. onubensis*, *Paradelograptus* cf. *P. rallus* (Jackson and Lenz), and *A. murrayi*. It is also recorded in the San Bernardo Formation, Mojotoro Range (Moya and Monteros, 2011), and recently, in the Santa Rosita Formation at the Zenta Range (Albanesi et al., 2011) and at the Santa Victoria Range, eastern margin of the Cordillera Oriental (Toro et al., 2015; Giuliano, 2016). Findings of typical taxa from the *H. copiosus* Zone were also reported from the Chiquero Formation, Puna of Jujuy (Benedetto et al., 2002).

3.2 Floian

3.2.1 Conodont Zones

3.2.1.1 *Prioniodus elegans* Zone (TS: Fl1)

This biozone includes two subzones in the Precordilleran scheme; i.e., *Prioniodus elegans–Tropodus sweeti* and *P. elegans–Oepikodus communis* subzones (Albanesi et al., 1998). Hünicken and Sarmiento (1980) published the first record of *P. elegans* Pander (Fig. 5(14)) from the lower part of the San Juan Formation overlying the carbonate Los Sapitos Formation (=La Silla Formation) in the Guandacol River section, northern Precordillera. The species was reported by several authors from correlative strata of the

Precordillera (Lehnert, 1993, 1995a; Albanesi et al., 1998; Mango et al., 2016). The first appearance of *O. communis* in association with *P. elegans* verifies a subdivision of the biozone. Lehnert et al. (1997) recognized an interval correlative with the *P. elegans* Zone in the Suri Formation, Sistema de Famatina, and Lehnert et al. (1998) reported this unit in the Bloque de San Rafael of the Cuyania terrane.

3.2.1.2 *Acodus triangularis* Zone (TS: Lower Fl1)

This zone is defined by the FAD of *Acodus triangularis* (Ding) (Fig. 4(1)) and extends up to the FAD of *Gothodus vetus* Voldman and Albanesi (Fig. 4(3)). *A. triangularis* is present in the uppermost limolitic levels of the Santa Rosita Formation and the lower gray shales of the Acoite Formation at the Chulpíos Creek, Santa Victoria área, as described by Voldman et al. (2016a). The basal strata of the Acoite Formation could extend from the uppermost Tremadocian Stage since the lowest record of *A. triangularis* together with other key species; i.e., *Acodus apex*. *Paltodus deltifer*, and *Araneograptus murrayi*, in the type section does not verify a Floian age for the base of the Acoite Formation (the ranges of all associated index species span the Tremadocian–Floian boundary). The appearance of the Gondwanan genus *Zentagnathus* is recorded within this zone. Considering the stratigraphic position and homotaxial succession of index species, the *A. tringularis* Zone most probably correlates with uppermost levels of the *Paroistodus proteus* Zone, and the lower subzone of the *Prioniodus elegans* Zone from the Precordillera.

3.2.1.3 *Gothodus vetus* Zone (TS: Upper Fl1)

Gothodus vetus Voldman and Albanesi is recorded from the lower part of the Acoite Formation up to its replacement by the offspring species *Gothodus andinus* (Rao et al.) at the Chulpíos Creek type section (Voldman et al., 2016a). Within this interval zone the first record of the genus *Baltoniodus* occurs, as well as the first appearance datum of *Drepanoistodus basiovalis* (Sergeeva). The *G. vetus* Zone may correspond to the upper subzone of the *Prioniodus elegans* Zone of the Precordillera.

3.2.1.4 *Oepikodus evae* Zone (TS: Fl2)

This zone is divided into the *Oepikodus evae–Juanognathus variabilis* and *O. evae–Scolopodus oldstockensis* subzones in the Precordillera as it was originally defined by Albanesi et al. (1998). The interval characterized by *O. evae* (Lindström) (Fig. 5(15) and 5(18)) was recognized in numerous sections of

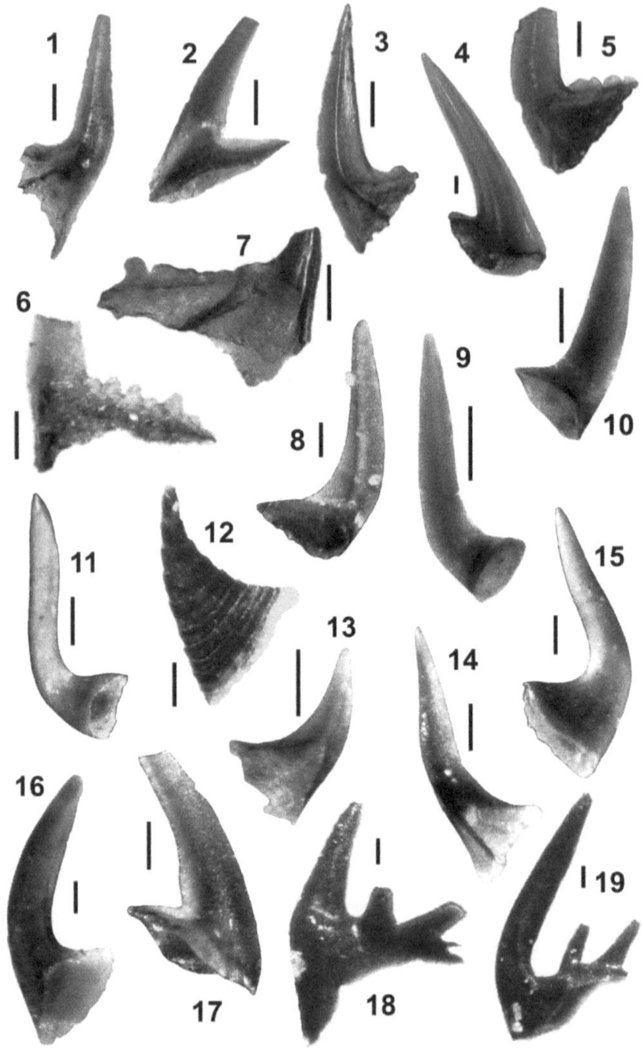

Figure 4

(1) *Acodus triangularis* (Ding). Pa element. Acoite Formation. *Acodus triangularis* Zone. Chulpíos, Santa Victoria. Cordillera Oriental. CORD-MP 30,003.
(2) *Paltodus deltifer deltifer* (Lindström). M element. Acoite Formation. *Acodus triangularis* Zone. Chulpíos, Santa Victoria. Cordillera Oriental. CORD-MP 30,024.
(3) *Gothodus vetus* (Voldman and Albanesi). Pb element. Acoite Formation. *Gothodus vetus* Zone. Chulpíos, Santa Victoria. Cordillera Oriental. CORD-MP 30,026.
(4) *Drepanoistodus costatus* (Abaimova). M element. Acoite Formation. *Gothodus andinus* Zone. Chulpíos, Santa Victoria. Cordillera Oriental. CORD-MP 30,061.
(5) *Gothodus andinus* (Rao et al.). Pb element. Acoite Formation. *Gothodus andinus* Zone. Chulpíos, Santa Victoria. Cordillera Oriental. CORD-MP 30,038.
(6) *Gothodus andinus* (Rao et al.). Pa element. Cieneguillas Formation. *Gothodus andinus* Zone. Lipán. Cordillera Oriental. CORD-MP 50,990.

the San Juan Formation in the Precordilleran region (Serpagli, 1974; Sarmiento, 1990; Mango et al., 2016, and references therein) through the stratigraphic interval of high stand sea level in the carbonate platform. Calcareous beds interbedded in the siliciclastic Cieneguillas Formation, cropping out in the Purmamarca area, and Lipán, Cordillera Oriental of Jujuy, yielded a low diversity conodont fauna dominated by *G. andinus* (Rao et al.) (Fig. 4(5) and 4(6)) and *D. costatus* (Abaimova) (Fig. 4(4)) that can be correlated with the *O. evae* Zone (Rao et al., 1994; Albanesi et al., 2014), as well as the stratigraphic interval of the Acoite Formation at Chulpíos Creek that is assigned to the *G. andinus* Zone (Voldman et al.,

(7) *Zentagnathus argentinensis* (Rao et al.). Pa element. Acoite Formation. *Gothodus andinus* Zone. Chulpíos, Santa Victoria. Cordillera Oriental. CORD-MP 30,053.
(8) *Utahconus tortibasis* Zeballo and Albanesi. M element. Santa Rosita Formation. *Paltodus deltifer* Zone. Huacalera, Alfarcito. Cordillera Oriental. CORD-MP 16,479/1.
(9) *Utahconus purmamarcensis* Zeballo and Albanesi. M element. Coquena Formation. *Paltodus deltifer* Zone. Purmamarca. Cordillera Oriental. CORD-MP 11,306/1.
(10) *Utahconus scandodiformis* Zeballo and Albanesi. M element. Santa Rosita Formation. *Paltodus deltifer* Zone. Huacalera, Alfarcito. Cordillera Oriental. CORD-MP 11,304/1.
(11) *Acanthodus raqueli* Zeballo and Albanesi. Sa element. Santa Rosita Formation. *Cordylodus angulatus* Zone. Huacalera, Alfarcito. Cordillera Oriental. CORD-MP 16,486/184.
(12) *Kallidontus gondwanicus* Zeballo and Albanesi. Sa element. Santa Rosita Formation. *Cordylodus angulatus* Zone. Huacalera, Alfarcito. Cordillera Oriental. CORD-MP 16,501/12.
(13) *Acodus primitivus* Zeballo and Albanesi. Pa element. Santa Rosita Formation. *Paltodus deltifer* Zone. Huacalera, Alfarcito. Cordillera Oriental. CORD-MP 16,635/7.
(14) *Tilcarodus humahuacensis* (Albanesi and Aceñolaza). Pa element. Santa Rosita Formation. *Paltodus deltifer* Zone. Alfarcito. CORD-MP 16,498/44.
(15) *Variabiloconus crassus* Zeballo and Albanesi. Sb element. Santa Rosita Formation. *Cordylodus angulatus* Zone. Huacalera, Alfarcito. Cordillera Oriental. CORD-MP 16,480/148.
(16) *Drepanoistodus chucaleznensis* Albanesi and Aceñolaza. P element. Santa Rosita Formation. *Paltodus deltifer* Zone. Rupasca, Alfarcito. Cordillera Oriental. CORD-MP 11,312.
(17) *Paltodus deltifer* n. ssp. Zeballo et al. M element. Coquena Formation. *Paltodus deltifer* Zone. Purmamarca. Cordillera Oriental. CORD-MP 11,294.
(18) *Iapetognathus fluctivagus* Nicoll et al. Sc element. Cardonal Formation. *Iapetognathus* Zone. Cajas. Cordillera Oriental. CORD-MP 29,730.
(19) *Cordylodus angulatus* Pander. Pa element. Cardonal Formation. *Cordylodus angulatus* Zone. Cajas. Cordillera Oriental. CORD-MP 29,764.
All pictures are optical photomicrographs. Scale bar: 0.1 mm.

2016a), as referred below. A characteristic association, including *O. evae* elements with endemic forms, was recovered from the Suri Formation of the Sistema de Famatina in the outcrops of the Saladillo River section (Lehnert et al., 1997; Albanesi and Astini, 2000a).

3.2.1.5 *Gothodus andinus* Zone (TS: Fl2)

The FAD of *G. andinus* (Rao et al.) defines the base of this biozone, which ends up to the FAD of *Trapezognathus diprion* (Lindström). *G. andinus* (Fig. 4(5) and 4(6)) was originally determined by Rao et al. (1994) as *Baltoniodus crassulus andinus* in the Cieneguillas Member of the Acoite Formation that crops out few kilometers west of Purmamarca in the Cordillera Oriental of Jujuy. In these strata, *G. andinus* is associated to *Cornuodus longibasis* (Lindström) (Fig. 7(14)), *Drepanodus arcuatus* Pander (Fig. 7(1)), *Drepanoistodus basiovalis* (Fig. 7(6)), *Drepanoistodus costatus* (Fig. 4(4)), and *Zentagnathus argentinensis* (*Trapezognathus argentinensis sensu* Rao et al.) (Fig. 4(7)), among other species. Albanesi et al. (2014) reexamined the exposure of the Cieneguillas Member at the Lipán section (Aguas Blancas area), Los Colorados Range of Cordillera Oriental, reporting graptolites assigned to the lower part of the *Baltograptus* cf. *B. deflexus* Zone (Toro, 1997a; Toro and Maletz, 2007) associated with a conodont assemblage similar to that of the type locality described by Rao et al. (1994). The record of *Erraticodon patu* Cooper, *Paroistodus parallelus* (Pander), *Protopanderodus gradatus* Serpagli, at this locality, as well as in the Chulpíos Creek section, Santa Victoria area, is consistent with the stratigraphic distribution of these taxa, whose FAD occurs in the worldwide recognized interval of *O. evae*, providing a feasible correlation with the lower part of this zone in the Precordillera (e.g., Albanesi et al., 1998, 1999a).

3.2.1.6 *Trapezognathus diprion* Zone (TS: Lower Fl3)

The *Trapezognathus diprion* Zone was recognized for the first time by Carlorosi (2011) in strata referred to as the "Sepulturas Formation" at the Espinazo del Diablo Range to the east of the Cajas Range in the Cordillera Oriental of Jujuy (Aceñolaza et al., 2008). Worth clarifying that this formation is actually exposed in the Los Colorados Range as defined by Harrington and Leanza (1957) and thoroughly described by Astini et al. (2004), where it is Darriwilian in age. A description of *T. diprion* Lindström is given by Carlorosi (2011) for its upper record in the Alto del Condor Formation, where it is associated to *Baltoniodus* cf. *B. triangularis* (Lindström), *Baltoniodus triangularis* (Lindström), *Erraticodon patu* Cooper, *Gothodus andinus*

(Rao et al.), and *Oepikodus intermedius* Serpagli (Fig. 5(17)) as index species of the Floian–Dapingian boundary interval. The geographic and stratigraphic distribution of *T. diprion* in the Cordillera Oriental and its correlation worldwide is discussed by Carlorosi and Heredia (2013).

3.2.1.7 *Oepikodus intermedius* Zone (TS: Fl3)

According to the definition of Albanesi et al. (1998) for the Yanso section, Cerro Potrerillo, Central Precordillera, *O. intermedius* (Fig. 5(17)) is diagnostic for the homonymous zone in the stratigraphic interval of the San Juan Formation where it is not associated to *Oepikodus evae* or *Tripodus laevis* Bradshaw sensu lato. This unit was also recognized in the same formation exposed in Niquivil, where Albanesi et al. (2006) reported a typical association that includes *Cooperignathus aranda* (Cooper) and *Texania heligma* Pohler (Fig. 5(16)), and the first appearances of *Juanognathus jaanussoni* Serpagli (Fig. 6(14)) and *Semiacontiodus potrerillensis* Albanesi (Fig. 6(9)) for this interval of the uppermost Floian Stage. The biostratigraphic interval of this zone at Talacasto, Central Precordillera, yielded other significant taxa for intercontinental correlation like *Jumudontus gananda* Cooper (cf. Ethington and Clark, 1981; Albanesi et al., 1998). This conodont association can also be recognized in the limestones succession of the Bloque de San Rafael (Lehnert et al., 1998), and it probably correlates to the *Baltoniodus* cf. *B. triangularis* Zone of the Cordillera Oriental, as discussed below.

3.2.1.8 *Baltoniodus* cf. *B. triangularis* Zone (TS: Upper Fl3)

Although it is not convenient to characterize a biostratigraphic interval by a taxon under the open nomenclature, because its identification could be ambiguous, we maintain this biozone as defined by Carlorosi and Heredia (2013) on the basis of the nominal species previously described by Carlorosi et al. (2011) for the type area, because its record characterizes the interval interspersed between the *Trapezonagnathus diprion* Zone and the *Baltoniodus triangularis* Zone in the Cordillera Oriental, correlative with the *Oepikodus intermedius* Zone of the Precordillera. The *B.* cf. *B. triangularis* Zone was recognized in the Laguna Verde section of the Zenta Range by Voldman et al. (2013). Within the homonymous biostratigraphic interval, *B.* cf. *B. triangularis* Lindström is accompanied by *Erraticodon patu* and *T. diprion* from the underlying biozone, which are valuable for worldwide correlation to refer the uppermost Floian Stage (Bagnoli and Stouge, 1997; Wang et al., 2005).

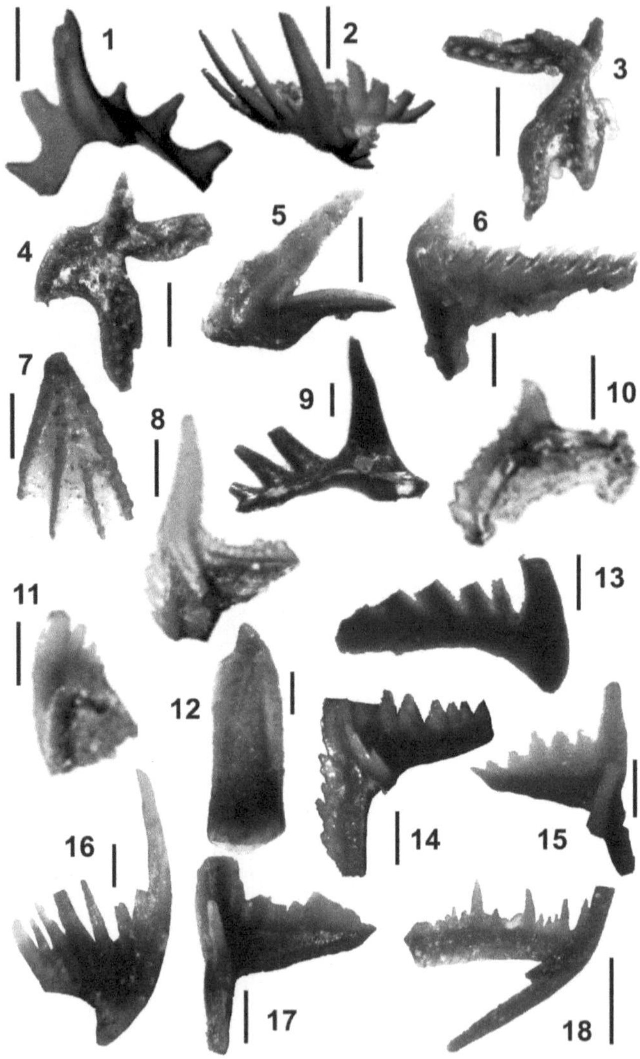

Figure 5

(1) *Erismodus* sp. Sa element. Capillas Formation. *Erismodus* Zone. Capillas, Sierras Subandinas. CORD-MP 11,324.

(2) *Erraticodon* sp. Pa element. Capillas Formation. *Erismodus* Zone. Capillas, Sierras Subandinas. CORD-MP 50,984.

(3) *Eoplacognathus pseudoplanus* (Viira). Pa element. Sierra de La Invernada Formation. *Eoplacognathus pseudoplanus* Zone. La Invernada. Precordillera. CORD-MP 50,985.

(4) *Cahabagnathus sweeti* (Bergström). Pa element. Los Azules Formation. *Pygodus anserinus* Zone. La Chilca. Precordillera. CORD-MP 50,986.

(5) *Periodon aculeatus* Hadding. M element. Los Azules Formation. *Pygodus anserinus* Zone. La Chilca. Precordillera. CORD-MP 50,987.

3.2.2 *Graptolite Zones*

3.2.2.1 *Tetragraptus phyllograptoides* Zone (TS: Lower Fl1)

The *Tetragraptus phyllograptoides* Zone was erected by Toro (1994, reference in Toro and Brussa, 2003) for the lower part of the Acoite Formation at Agua Blanca Creek, Cordillera Oriental of Jujuy. It can also be recognized by the record of *Tetragraptus approximatus* Nicholson (Fig. 3(16)), *T. phyllograptoides* Strandmark (Fig. 3(5)), *Baltograptus geometricus* (Törnquist) and *Cymatograptus* cf. *C. rigoletto* (Maletz, Rushton and Lindholm) in the Aguilar Range (Martín et al., 1987; Vento et al., 2012; Toro and Vento, 2013; references in Toro et al., 2015). It is also present in the Cajas Range, and the Angosto del Moreno in the western margin of the Cordillera Oriental

(6) *Pygodus anserinus* Lamont and Lindström. Pb element. Los Azules Formation. *Pygodus anserinus* Zone. La Chilca. Precordillera. CORD-MP 50,988.
(7) *Pygodus anserinus* Lamont and Lindström. Pa element. Los Azules Formation. *Pygodus anserinus* Zone. La Chilca. Precordillera. CORD-MP 50,989.
(8) *Baltoniodus variabilis* (Bergström). Pa element. Los Azules Formation. *Amorphognathus tvaerensis* Zone. Cerro Viejo de Huaco. Precordillera. CORD-MP 50,993.
(9) *Iapetognathus aengensis* (Lindström). Sb element. Cardonal Formation. *Iapetognathus* Zone. Cajas. Cordillera Oriental. CORD-MP 29,728.
(10) *Amorphognathus tvaerensis* Bergström. Pb element. Los Azules Formation. *Amorphognathus tvaerensis* Zone. Cerro Viejo de Huaco. Precordillera. CORD-MP 50,991.
(11) *Amorphognathus tvaerensis* Bergström. M element. Los Azules Formation. *Amorphognathus tvaerensis* Zone. Cerro Viejo de Huaco. Precordillera. CORD-MP 50,992.
(12) *Macerodus dianae* Fåhraeus and Nowlan. S element. La Silla Formation. *Macerodus dianae* Zone. Portezuelo Jáchal. Precordillera. CORD-MP 34,000.
(13) *Cordylodus lindstromi* Druce and Jones. M element. Cardonal Formation. *Iapetognathus* Zone. Cajas. Cordillera Oriental. CORD-MP 29,717.
(14) *Prioniodus elegans* Pander. Pb element. San Juan Formation. *Prioniodus elegans* Zone. Niquivil. Precordillera. CORD-MP 50,996.
(15) *Oepikodus evae* (Lindström). Pa element. San Juan Formation. *evae* Zone. Niquivil. Precordillera. CORD-MP 50,995.
(16) *Texania heligma* Pohler. Sc element. San Juan Formation. *O. intermedius* Zone. Niquivil. Precordillera. CORD-MP 50,997.
(17) *Oepikodus intermedius* Serpagli. Pa element. San Juan Formation. *intermedius* Zone. Peña Sombría, Los Sapitos. Precordillera. CORD-MP 50,994.
(18) *evae* (Lindström). Sc element. San Juan Formation. *evae* Zone. Niquivil. Precordillera. CORD-MP 50,994.
All pictures are optical photomicrographs. Scale bar: 0.1 mm.

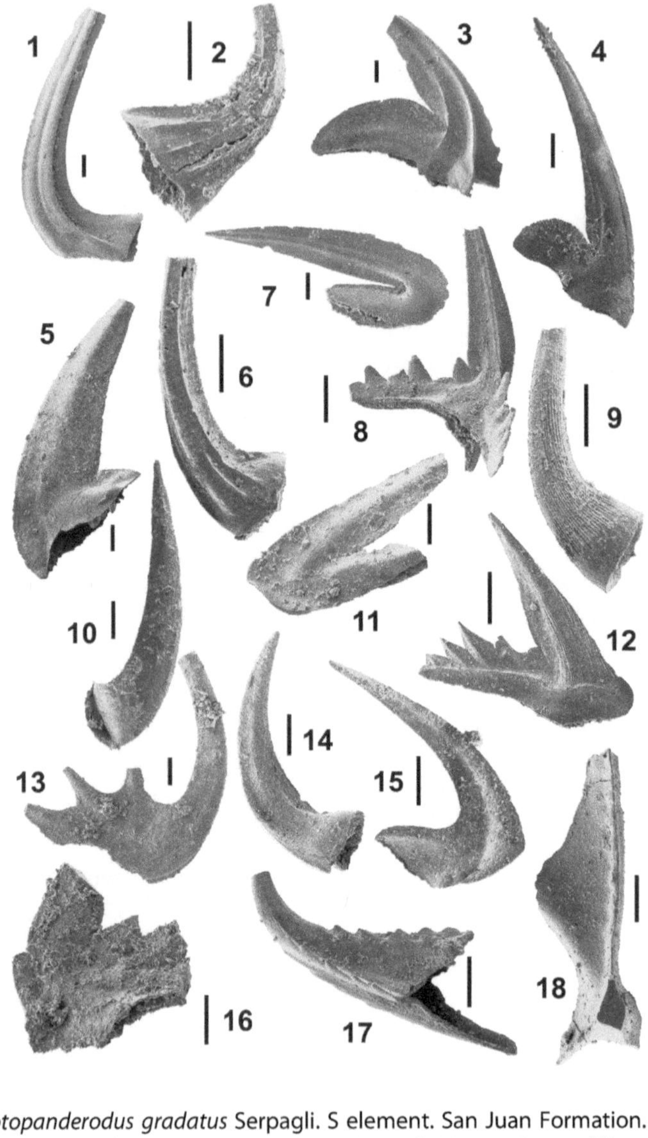

Figure 6

(1) *Protopanderodus gradatus* Serpagli. S element. San Juan Formation. *Yangtzeplacognathus crassus* Zone. Cerro Viejo de Huaco. Precordillera. CORD-MP 50,967.
(2) *Costiconus costatus* (Dzik). S element. Gualcamayo Formation. *Baltoniodus triangularis-Tripodus laevis* Zone. Los Sapitos. Precordillera. CORD-MP 50,964.
(3) *Oistodus lanceolatus* Pander. Sb element. San Juan Formation. *Yangtzeplacognathus crassus* Zone. Cerro Viejo de Huaco. Precordillera. CORD-MP 50,970.

(Moya et al., 1998; Ortega et al., 1998), and in the Santa Victoria Range on the eastern part of the Cordillera Oriental (Toro, 1999). In the Sistema de Famatina, this biozone was documented for the middle part of the La Alumbrera Formation, Catamarca Province (Aceñolaza et al., 1976; Toro, 1997b). In the Precordillera, the *T. approximatus* fauna is present in black shales of the Los Sombreros Formation at the Ojos de Agua Creek, El Tontal Range (Banchig and Moya, 2002, Ortega et al., 2014).

(4) *Pteracontiodus cryptodens* (Mound). S element. San Juan Formation. *Yangtzeplacognathus crassus* Zone. Cerro Viejo de Huaco. Precordillera. CORD-MP 50,968.
(5) *Rossodus barnesi* Albanesi. Pb element. Gualcamayo Formation. *Baltoniodus navis* Zone. Puesto Los Alamos. Precordillera CORD-MP 50,979.
(6) *Scolopodus rex* Lindström. S element. Gualcamayo Formation. *Baltoniodus navis* Zone. Puesto Los Alamos. Precordillera CORD-MP 50,973.
(7) *Paroistodus horridus* (Barnes and Poplawski). M element. San Juan Formation. *Lenodus variabilis* Zone. Cerro Viejo de Huaco. Precordillera. CORD-MP 50,971.
(8) *Gothodus* n. sp. Pa element. Gualcamayo Formation. *Baltoniodus triangularis-Tripodus laevis* Zone. Los Sapitos. Precordillera. CORD-MP 50,972.
(9) *Semiacontiodus potrerillensis* Albanesi. S element. Gualcamayo Formation. *Baltoniodus triangularis-Tripodus laevis* Zone. Los Sapitos. Precordillera. CORD-MP 50,978.
(10) *Parapaltodus simplicissimus* Stouge. M element. San Juan Formation. *Yangtzeplacognathus crassus* Zone. Cerro Viejo de Huaco. Precordillera. CORD-MP 50,969.
(11) *Drepanoistodus forceps* (Lindström). M element. Gualcamayo Formation. *Baltoniodus triangularis-Tripodus laevis* Zone. Los Sapitos. Precordillera. CORD-MP 50,975.
(12) *Protoprioniodus* n. sp. P element. Gualcamayo Formation. *Baltoniodus triangularis-Tripodus laevis* Zone. Los Sapitos. Precordillera CORD-MP 50,974.
(13) *Paroistodus horridus* (Barnes and Poplawski). S element. San Juan Formation. *Yangtzeplacognathus crassus* Zone. Los Azules. Precordillera. CORD-MP 50,976.
(14) *Juanognathus jaanussoni* Serpagli. S element. Gualcamayo Formation. *Baltoniodus navis* Zone. Puesto Los Alamos. Precordillera CORD-MP 50,981.
(15) *Paroistodus originalis* (Sergeeva). S element. Gualcamayo Formation. *Baltoniodus navis* Zone. Puesto Los Alamos. Precordillera CORD-MP 50,982.
(16) *Microzarkodina parva* Lindström. P element. Gualcamayo Formation. *Microzarkodina parva* Zone. Puesto Los Alamos. Precordillera CORD-MP 50,977.
(17) *Baltoniodus navis* (Lindström). Sd element. Gualcamayo Formation. *Baltoniodus navis* Zone. Puesto Los Alamos. Precordillera CORD-MP 50,980.
(18) *Baltoniodus navis* (Lindström). Pa element. Gualcamayo Formation. *Baltoniodus navis* Zone. Puesto Los Alamos. Precordillera CORD-MP 50,983.
All pictures are SEM photomicrographs. Scale bar: 0.1 mm.

3.2.2.2 *Tetragraptus akzharensis* Zone (TS: Upper Fl1)

A graptolite assemblage composed by *Tetragraptus akzharensis* Tzaj (Fig. 3(20)), *Expansograptus constrictus* (J. Hall), *Baltograptus vacillans* (Tullberg), *Baltograptus geometricus* (Törnquist) and *Cymatograptus protobalticus* (Monsen) characterize the T. *akzharensis* Zone in the Cordillera Oriental (Toro, 1994; Toro and Maletz, 2007). It can be identified at the Los Colorados Creek, Aguilar, Cajas, and Santa Victoria ranges (Toro and Vento, 2013; Toro et al., 2015, and references therein). The *T. akzharensis* Zone was documented in the Portezuelo de las Minitas and the La Alumbrera River, Sistema de Famatina (Aceñolaza and Gutiérrez-Marco, 2000; Toro, 1997b). A graptolite assemblage integrated by *Pendeograptus fruticosus* (J. Hall), *Tetragraptus approximatus*, *T. quadribrachiatus* (J. Hall), and *T.* aff. *T. bigsbyi* (J. Hall), among others, was recorded in allochthonous blocks with black shales in the lower part of the Empozada Formation, Precordillera of Mendoza (Bordonaro and Peralta, 1987). It represents the only record of the *T. akzharensis* Zone in the Precordillera at the present (Ortega et al., 2007b).

3.2.2.3 *Baltograptus* cf. *B. deflexus* Zone (TS: Lower Fl2)

The *Baltograptus* cf. *B. deflexus* Zone occupies the middle and upper part of the Acoite Formation in the Cordillera Oriental, where it is present in the Los Colorados Creek, and Aguilar and Santa Victoria ranges (Toro, 1994; Toro and Maletz, 2007, 2008; Vento et al., 2012; Toro et al., 2015). The biozone bears *B.* cf. *B. deflexus* (=*Baltograptus* sp. nov. *sensu* Toro and Maletz, 2007; =*Baltograptus jacksoni* Rushton) (Fig. 3(12)), *Baltograptus turgidus* (Lee), and *Acrograptus filiformis* (Tullberg), as frequent forms. A graptolite fauna with *Corymbograptus v-fractus* (Salter) and *Didymograptellus nitidus* (J. Hall) recovered in the San Bernardo Formation, Mojotoro Range, was referred to this biozone by Monteros (2005). In the Sistema de Famatina the *B.* cf. *B. deflexus* Zone was described for the Suri Formation, in the Salado River, Famatina Range (Toro and Brussa, 1997).

3.2.2.4 *Didymograptellus bifidus* Zone (TS: Lower Fl3)

Toro (1994) defined the *Didymograptellus bifidus* Zone in the Chamarra Creek, Cordillera Oriental of Jujuy. The assemblage of *D. bifidus* (J. Hall), *Baltograptus minutus* (Törnquist), *B. deflexus* (Elles and Wood), *Baltograptus* cf. *B. deflexus*, and *Pseudophyllograptus* sp. is present in the upper part of the Acoite Formation in different sections of the Cordillera Oriental. First mention of *D. bifidus* correspond to Loss (1949), who collected the species nearby Yavi village, to the east of La Quiaca city. This biozone was later recognized in the Santa Victoria Range (Toro et al., 2015). In the northeastern Puna, Jujuy Province, the *D. bifidus* Zone was recorded in the volcanic sedimentary succession of the Quichagua Range (Martínez et al., 1999). Another identification of this biozone corresponds to the Suri Formation in the Salado River section, Sistema de Famatina (Toro and Brussa, 1997).

3.3 Dapingian

3.3.1 Conodont Zones

3.3.1.1 *Baltoniodus triangularis* Zone (TS: Dp1)

The *Baltoniodus triangularis* Zone was defined by Carlorosi et al. (2011) in the shaly Alto del Condor Formation, Los Colorados Range of the Cordillera Oriental of Jujuy, as the basal biostratigraphic unit of the Dapingian Stage established for global correlation (Wang et al., 2005). In the type section, *B. triangularis* is associated to the taxa ranging from the underlying biozone and *Trapezognathus quadrangulum* Lindström, as significant species of the lowest Dapingian in the temperate to cold water domains.

3.3.1.2 *Baltoniodus triangularis* —*Tripodus laevis* Zone (TS: Dp1)

In the Argentine Precordillera, the joint record of *Baltoniodus triangularis* and *T. laevis* Bradshaw in correlative strata through the basin, makes it useful for defining an assemblage zone with taxa from the cold and warm waters domains (Albanesi et al., 1998, 1999a). This unique combination of species that typifies different provincial domains, as recorded in diverse areas of the Precordillera; e.g., Peña Sombría, Los Sapitos, and Cerro Potrerillo sections, provides a most precise datum for accurate regional and global correlations (Ethington and Clark, 1981; Wang et al., 2005; Albanesi et al., 2006; Della Costa et al., 2016). New short ranging taxa, such *Gothodus* n. sp. (Fig. 6(8)), and *Protoprioniodus* n. sp. (Fig. 6(12)), that span de Floian-Dapingian boundary in the Precordillera, represent useful index species for a high resolution correlation of this boundary interval throughout the basin (Albanesi, 1999). Other species associated within this zone are *Costiconus costatus* (Dzik) (Fig. 6(2)), *Drepanoistodus forceps* (Lindström) (Fig. 6(11)), *Semiacontiodus potrerillensis* Albanesi (Fig. 6(9)), and *Texania heligma* Pohler (Fig. 5(16)).

3.3.1.3 *Baltoniodus navis* Zone (TS: Dp2)

The *Baltoniodus navis* Zone was first documented in the Precordillera of San Juan by Lemos (1981), in the Buenaventura Luna section at Huaco,

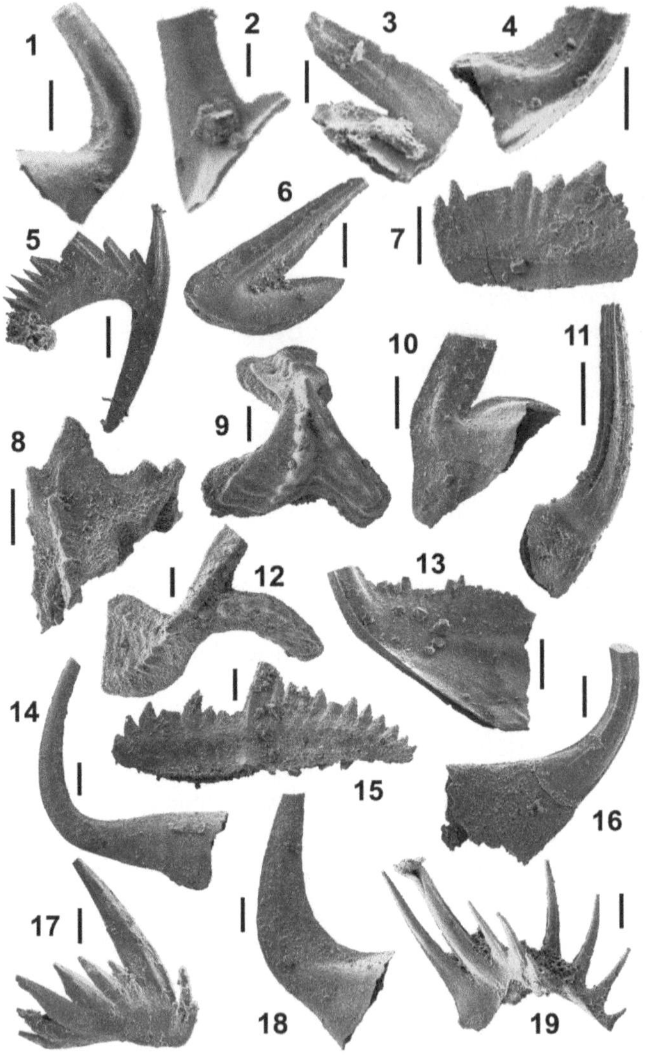

Figure 7

(1) *Drepanodus arcuatus* Pander. S element. Bordo Atravesado Formation. *Paltodus deltifer* Zone. Bordo Atravesado. Sistema de Famatina. CORD-MP 50,950.

(2) *Paltodus deltifer pristinus* (Viira). M element. Bordo Atravesado Formation. *Paltodus deltifer* Zone. Bordo Atravesado. Sistema de Famatina. CORD-MP 10,132.

(3) *Paroistodus numarcuatus* (Lindström). M element. Bordo Atravesado Formation. *Paltodus deltifer* Zone. Bordo Atravesado. Sistema de Famatina. CORD-MP 10,131.

(4) *Rossodus manitouensis* Repetski and Ethington. Sb element. Bordo Atravesado Formation. *Paltodus deltifer* Zone. Bordo Atravesado. Sistema de Famatina. CORD-MP 10,034.

(5) *Fahraeusodus* n. sp. Sc element. San Juan Formation. *Lenodus variabilis* Zone. Las Aguaditas. Precordillera. CORD-MP 50,952.

(6) *Drepanoistodus basiovalis* (Sergeeva). M element. San Juan Formation. *Lenodus variabilis* Zone. Las Aguaditas. Precordillera. CORD-MP 50,951.

although this unit could not be verified for the section in later works (Mestre, 2008; Mango et al., 2016). Albanesi et al. (1998) established formally this biozone in the Yanso section at Cerro Potrerillo by the FAD of *B. navis* (Lindström) (Fig. 6(17) and 6(18)), and the presence of key taxa, such as *Trigonodus brevibasis* (Sergeeva) and *Histiodella altifrons* Harris in the Niquivil section (Lehnert, 1995a; Albanesi et al., 2006), which allows for intercontinental linkages of the referred interval. Among species that range through this zone are mentioned *Scolopodus rex* Lindström (Fig. 6(6)), *Rossodus barnesi* Albanesi (Fig. 6(5)), *Juanognathus jaanussoni* Serpagli (Fig. 6(14)), and *Paroistodus originalis* (Sergeeva) (Fig. 6(15)). Strata that can be correlated with this biozones were identified in peripheral environments of the Bloque de San Rafael (Lehnert et al., 1998). The "*Baltoniodus navis*" Zone was recognized from the upper sequence of the Suri Formation at Chaschuil River section, Sistema de Famatina (Albanesi

(7) *Histiodella holodentata* Ethington and Clark. Pa element. San Juan Formation. *Lenodus variabilis* Zone. Las Aguaditas. Precordillera. CORD-MP 40,953.
(8) *Lenodus variabilis* (Sergeeva). Pa element. San Juan Formation. *Lenodus variabilis* Zone. Las Aguaditas. Precordillera. CORD-MP 50,956.
(9) *Pygodus* n. sp. Pa element. San Juan Formation. *Lenodus variabilis* Zone. Las Chacritas. Precordillera. CORD-MP 50,954.
(10) *Ansella jemtlandica* (Löfgren). M element. San Juan Formation. *Lenodus variabilis* Zone. Las Chacritas. Precordillera. CORD-MP 50,961.
(11) *Decoriconus peselephantis* (Lindström). S element. San Juan Formation. *Lenodus variabilis* Zone. Las Chacritas. Precordillera. CORD-MP 50,957.
(12) *Yangtzeplacognathus crassus* (Chen and Zhang). Pa element. San Juan Formation. *Yangtzeplacognathus crassus* Zone. Las Aguaditas. Precordillera. CORD-MP 50,955.
(13) *Ansella jemtlandica* (Löfgren). Sc element. San Juan Formation. *Lenodus variabilis* Zone. Las Chacritas. Precordillera. CORD-MP 50,960.
(14) *Cornuodus longibasis* (Lindström). S element. San Juan Formation. *Lenodus variabilis* Zone. Las Chacritas. Precordillera. CORD-MP 50,958.
(15) Gen. et sp. nov. A (ex *"Bryantodina"* aff. *"B." typicalis*). P element. San Juan Formation. *Lenodus variabilis* Zone. Las Chacritas. Precordillera. CORD-MP 50,959.
(16) *Protopanderodus rectus* (Lindström). S element. San Juan Formation. *Microzarkodina parva* Zone. Las Chacritas. Precordillera. CORD-MP 50,962.
(17) *Periodon macrodentatus* (Graves and Ellison). Pa element. San Juan Formation. *Yangtzeplacognathus crassus* Zone. Cerro Viejo de Huaco. Precordillera. CORD-MP 50,965.
(18) *Paltodus? jemtlandicus* Löfgren. S element. San Juan Formation. *Microzarkodina parva* Zone. Las Chacritas. Precordillera. CORD-MP 50,963.
(19) *Erraticodon alternans* (Hadding). Pa element. San Juan Formation. *Yangtzeplacognathus crassus* Zone. Cerro Viejo de Huaco. Precordillera. CORD-MP 50,966.
All pictures are SEM photomicrographs. Scale bar: 0.1 mm.

and Vaccari, 1994), by means of a diagnostic association of conodonts that included the closely related form *Baltoniodus navis sensu* Cooper.

3.3.2 Graptolite Zones

3.3.2.1 *Azygograptus* Zone (TS: Lower Dp1?)

A graptolite fauna with *Azygograptus eivionicus* Elles, *Dichograptus octobrachiatus* (J. Hall), *Xiphograptus svalbardensi* (Archer and Fortey), and *Pseudotrigonograptus minor* (Mu and Lee), was described by Monteros et al. (1996) for the Coquena Formation (Aguada de la Perdiz Formation *sensu* Brussa et al., 2008) in the Huaytiquina area, western Puna, Salta Province. It was referred to as the *Az. eivionicus* Zone by Brussa et al. (2008), which suggests an early Dapingian age. It was also documented in the Muñayoc section at Quichagua Range in the northeastern Puna, Jujuy Province (Martínez et al., 1999). Specimens of isograptids of the *victoriae* group were mentioned by Toro and Brussa (2000) for the Coquena Formation at the Tafna area, northeastern Puna, assigning the bearer strata to upper Dapingian or lower Darriwilian.

3.3.2.2 *Isograptus maximus* Zone (TS: Dp2)

The *I. maximus* Zone was identified in the calcareous-pelitic lower succession of the Gualcamayo Formation in the Guandacol River area at the Peña Sombría and Los Sapitos sections, northern Precordillera (Ortega et al., 1985; Ortega and Albanesi, 1999). It is the older graptolite assemblage deposited above the carbonate platform of the San Juan Formation. The association of *I. maximus* (Harris), *Xiphograptus lofuensis* (Lee), *X. v-deflexus* (Harris), species of *Arienigraptus*, dichograptids and sigmagraptids characterizes this biozone.

3.3.2.3 *Oncograptus upsilon* Zone (TS: Lower Dp3)

The FAD of *Oncograptus upsilon* T.S. Hall (Fig. 3(14)) permits the identification of this biozone at the Los Sapitos Creek, Guandacol River (Ortega and Albanesi, 1999), which was also identified in the Potrerillo Creek an affluent of the Los Piojos River in the northern Precordillera (Ortega et al., 1993). The index species ranges throughout the lower part of the Gualcamayo Formation with *Arienigraptus dumosus* (Harris), *Isograptus maximus*, *Xiphograptus lofuensis*, *X. v-deflexus*, *Skiagraptus gnomonicus* (Harris and Keble) (Fig. 3(25)), *Parisograptus caduceus* (Salter) (Fig. 3(23)), *Parisograptus horridus* (Harris), *Glossograptus acanthus* Elles and Wood, manubriate species and other taxa.

3.3.2.4 *Cardiograptus morsus* Zone (TS: Upper Dp3)

The first appearance of *Cardiograptus morsus* Harris and Keble (Fig. 3(19)) marks the base of the homonymous biozone in the lower member of the Gualcamayo Formation at the Los Sapitos Creek, northern Precordillera (Ortega et al., 1999). The graptolite assemblage is integrated by *C. morsus* and most part of the previous taxa, and it is characterized by the entrance of *Exigraptus clavus* Mu and *Paraglossograptus* sp.

3.4 Darriwilian

3.4.1 Conodont Zones

3.4.1.1 *Microzarkodina parva* Zone (TS: Dp3 – Dw1)

The biostratigraphic interval corresponding to this unit received a limited attention because of the particularly restricted facies for conodont preservation in the Central Precordillera, where it was formally defined in the Yanso section, Cerro Potrerillo (Albanesi et al., 1998). According to the Baltoscandian zonal scheme, the lower part of the *Microzarkodina parva* Zone spans the Dapingian/Darriwilian boundary defined by the FAD of *Levisograptus autrodentatus* (Harris and Keble) (Bergström et al., 2009). The Precordilleran *M. parva* Zone (Fig. 6 (16)) correlates with the *M. parva, Baltoniodus norrlandicus* and *Lenodus antivariabilis* zones (Bagnoli and Stouge, 1997). In the northern Precordillera, the *M. parva* Zone has been recognized at the Puesto Los Alamos section, through the lower member of the Gualcamayo Formation, Precordillera of San Juan (Albanesi et al., 1999a), including frequent specimens of *Protopanderodus rectus* (Lindström) (Fig. 7(16)), and *Paltodus? jemtlandicus* Löfgren (Fig. 7(18)). This interval is apparently present in the Pachaco section of the San Juan River (Serpagli, 1974), and the Niquivil section, where the key species *Trigonodus brevibasis* occurs (Lehnert, 1995a; Albanesi et al., 2006). *H. altifrons* Harris that also extends up to the *M. parva* Zone, was recorded in the La Silla, Niquivil, and puesto Los Potrerillos sections (Lehnert, 1995a). This unit is probably present in the Ponón Trehué Formation of the Bloque de San Rafael (Lehnert et al., 1998).

3.4.1.2 *Lenodus variabilis* Zone (TS: Dw1)

The *L. variabilis* Zone was formally defined in the San Juan and Gualcamayo formations at the Yanso section of the Cerro Potrerillo by Albanesi et al. (1998), where is divided into the *Periodon gladysae* and the *Paroistodus horridus* subzones. Nevertheless, this interval zone was first recognized in the Eastern Precordillera at the Don Braulio creek section in the Villicum Range by Sarmiento (1985, 1991), later verified by Mestre (2013), and the Rinconada section of the Chica de Zonda Range by Sarmiento et al. (1988), after de record of *Lenodus variabilis* (Sergeeva). The upper strata of the San Juan Formation were examined for conodonts in diverse localities of the Central Precordillera, where the zone typified by *L. variabilis* (Fig. 7(8)) was identified (e.g., Cerro Potrerillo, Albanesi et al., 1995a, 1998; Cerro Viejo, Ottone et al., 1999; Ortega et al., 2007a; Mango and Albanesi, 2014; Las Aguaditas, Albanesi et al., 2013; Feltes et al., 2016; La Chilca, Carrera et al., 2013), accompanied by a number of characteristic species, many of them of Baltoscandian affinity (cf. Löfgren, 2003); such as, *Ansella jemtlandica* (Löfgren) (Fig. 7(10) and 7(13)), *Cornuodus longibasis* (Lindström) (Fig. 7(14)), *Decoriconus peselephantis* (Lindström) (Fig. 7(11)), *Pygodus* n. sp. (Fig. 7(9)), as well as Midcontinent type taxa, namely *Fahraeusodus* n. sp. (Fig. 7(5)), *Histiodella sinuosa* (Graves and Ellison), Gen. et sp. nov. A (ex "*Bryantodina*" aff. "*B.*" *typicalis*) (Fig. 7(15)), and *Paroistodus horridus* (Barnes and Poplawski) (Fig. 6(7,13)). The subspecies *Paroistodus horridus primus* Albanesi and *P. horridus secundus* Albanesi characterize the *P. horridus* Subzone of the *L. variabilis* Zone (Albanesi and Barnes, 2000). This unit was recognized as well in the siliciclastic lower section of the Yerba Loca Formation at Ancaucha Creek in the western Precordillera (Albanesi et al., 1995b).

3.4.1.3 *Yangtzeplacognathus crassus* Zone (TS: Dw1 – Dw2)

This biozone was recognized by Albanesi et al. (2013), Feltes et al. (2013), Mestre and Heredia (2013), Serra et al. (2013), and Voldman et al. (2013) from diverse localities of the Central Precordillera, namely Las Aguaditas, Las Chacritas, La Chilca, and Oculta creeks, in the transitional interval between the San Juan Formation and ovelying units. *Yangtzeplacognathus crassus* (Chen and Zhang) (Fig. 7(12)) was also documented from sections of the western flank of the Cerro Viejo de Huaco, illustrated as *Lenodus variabilis* by Ortega et al. (2007a). Mestre (2012, 2014) reported the presence of the index species *Y. crassus* Zhang from the uppermost San Juan Formation and lower Gualcamayo Formation in the eastern flank sections of the Villicum Range, Eastern Precordillera. A cohort of frequent species from this zone are *Periodon macrodentatus* (Graves and Ellison) (Fig. 7(17)), *Erraticodon alternans* (Hadding) (Fig. 7(19)), *Protopanderodus gradatus* Serpagli (Fig. 6(1)), *Pteracontiodus cryptodens* (Mound) (Fig. 6(4)), *Parapaltodus simplicissimus* Stouge (Fig. 6(10)), *Oistodus lanceolatus* Pander (Fig. 6(3)), *Histiodella holodentata* Ethington and Clark (Fig. 7(7)) and *Paroistodus horridus* (Fig. 6(13)). A similar conodont association from calcareous debris deposits from the Los Sombreros Formation in the Westen Precordillera, was assigned by Voldman et al. (2009) to the *L. variabilis* Zone, but probably represents the *Y. crassus* Zone. A thorough review of the *Y. crassus* Zone considering its conodont fauna richness and biofacial analyzes, regional and intercontinental correlations, in particular with reference schemes of North America, Baltoscandia and China is provided by Serra et al. (2015) and Feltes et al. (2016).

3.4.1.4 *Eoplacognathus pseudoplanus* Zone (TS: Dw2)

The *E. pseudoplanus* Zone was recognized at the Las Chacritas Formation, which is exposed in the La Trampa Range, Central Precordillera (Heredia et al., 2005), after a prelimary report on the conodont faunas of the section provided by Albanesi and Astini (1994, 2000b). Afterwards, subsequent papers dealt with the conodont biostratigraphy of this section, with recurrent misintrepretations of the conodont biozonation, although the index species *E. pseudoplanus* was confirmed (Heredia et al., 2011; Heredia and Mestre, 2013). Recent works by Serra et al. (2013, 2015) revised the previous contributions on conodont biostratigraphy from the Las Chacritas River sections, demonstrating the homotaxial succession of the *Yangtzeplacognathus crassus* Zone (across the boundary between the San Juan and the Las Chacritas formations), and the *E. pseudoplanus* and *Eoplacognathus suecicus* zones through the Las Chacritas Formation. The Darriwilian zones reveal a high richness in conodonts, with several tens of species determined for each unit, elsewhere in the Central Precordillera. The *E. pseudoplanus* Zone was reported in the Las Aguaditas Creek section (Feltes et al., 2016), where the succession of index taxa from the neighboring Las Chacritas River section is reproduced. The latter works, from the Las Chacritas and Las Aguaditas sections, define the *Microzarkodina hagetiana* and *M. ozarkodella* subzones, in similar arrangement as defined originally in the Baltoscandian region (Bagnoli and Stouge, 1997; Löfgren, 1978; Löfgren and Zhang, 2003). Furthermore, this zonation scheme is complemented by the records of other contemporaneous key taxa of the genera *Histiodella, Dzikodus, Polonodus,* and *Periodon* (cf. Mestre and Heredia, 2012; Serra et al., 2015; Feltes et al., 2016), which provide a precise adjustment for correlation with the Darriwilian biozones of North America (Ethington and Clark, 1981; Sweet, 1984; Stouge, 1984, 2012; Johnson and Barnes, 1999, 2000; Bauer, 2010). Considering the appearance of *E. pseudoplanus* (Viira) (Fig. 5(3)), *Dzikodus tablepointensis* (Stouge) or *Polonodus magnus* Albanesi (Fig. 8(1)) associated in strata of the Argentine Precordillera, the *E. pseudoplanus*

Figure 8 Conodont elements preserved as casts on bedding plane surfaces: (1) *Polonodus magnus* Albanesi. Pa element. Gualcamayo Formation. *Eoplacognathus suecicus* Zone. Cerro Potrerillo. Precordillera. CORD-MP 50,998. (2) *Pygodus anitae* Bergström. Pa element. Gualcamayo Formation. *E. suecicus* Zone. Cerro Potrerillo. Precordillera. CORD-MP 50,999. (3) *Amorphognathus superbus* Rhodes. Pa element. Sierra de La Invernada Formation. *A. superbus* Zone. Sierra de La Invernada. Precordillera. CORD-MP 51,000. All pictures are optical photomicrographs. Scale bar: 1 mm.

Zone was also documented in siliciclastic environments, such as interspersed calcarenites of the Sierra de La Invernada Formation (Albanesi et al., 1999a) and interbedded calcarenites in the Yerba Loca Formation (Voldman et al., 2008).

3.4.1.5 *Eoplacognathus suecicus* Zone (TS: Dw2)

This zone was divided into the *Histiodella kristinae* and *Pygodus anitae* assemblage subzones by Albanesi and Ortega (2002). Although the FAD of *H. kristinae* Bergström occurs in uppermost levels of the *Eoplacognathus pseudoplanus* Zone, it is well represented in the lower *Eoplacognathus suecicus* Zone and is a diagnostic form of the homonymous subzone when its record is associated with *E. suecicus*. The derived species *H. bellburnensis* Stouge makes its entrance within the lower interval of the *E. suecicus* Zone (Serra et al., 2015). Other posible index species for the lower part of the *E. suecicus* Zone is *Polonodus magnus* Albanesi (Fig. 8(1)), although its FAD occurs like that of *H. kristinae* in the uppermost *E. pseudoplanus* Zone (Feltes et al., 2016). The upper subzone of the *E. suecicus* Zone is characterized by the presence of *Pygodus anitae* Bergström (Fig. 8(2)) (Albanesi et al., 1995a). The range of *P. magnus* extends within this interval in the Cerro Potrerillo (Albanesi et al., 1998), as well as that of *Dzikodus tablepointensis*, which is preserved on bedding plane surfaces of black shales, together with *P. anitae* and *P. magnus*, at Cerro Viejo de Huaco (Hünicken and Ortega, 1987; Ottone et al., 1999), Consequently, we maintain the original zonation proposed by Albanesi and Ortega (2002) for the present revised biostratigraphic chart. A regional stratotype section for the definition of the base of the *E. suecicus* Zone is located at Las Chacritas River section (Albanesi et al., 1998; Albanesi and Astini, 2000b; Serra et al., 2015). Uppermost strata of the San Juan Formation at Talacasto, Central Precordillera, bear a conodont association that could be referred to this zone (Lehnert, 1995a). Albanesi et al. (2013) demonstrated the occurrence of the middle Darriwilian Isotope Carbon Excursion Event (MDICE) at the *E. suecicus* Zone in the Las Chacritas Formation. The record of this event linked to the conodont zone in the Precordillera follows accurate global correlation with the chronostratigraphic Dw2 time slice as stated by Bergström et al. (2009).

3.4.1.6 "*Erismodus*" Zone (TS: Dw2, Dw3)

The Santa Gertrudis Formation, outcropping in the Gallinato Creek, eastern flank of the Cordillera Oriental, Salta Province, yielded a conodont fauna characterized by a number of new taxa associated to species of *Erraticodon* (Fig. 5(2)), *Erismodus* (Fig. 5(1)), and the new taxon illustrated as *Bryantodina typicalis* (Albanesi and Rao, 1996; Albanesi in Moya et al., 2003a). The overlapping range of these taxa probably refers to the Darriwilian two (Dw 2) time slice of Bergström et al. (2009) taking into account their co-occurrence in correlative strata of other continents (cf. Bauer, 1987, 2010; Löfgren and Zhang, 2003). Determining a Dapingian age for this unit by Heredia et al. (2015) is considered incorrect due to mis-indentification of the conodont specimens illustrated (new taxonomy in preparation by Albanesi). A similar conodont fauna was recovered from the Capillas Formation, Sierras Subandinas, Jujuy Province, where the age for the bearer strata is constrained to the middle-late Darriwilian according to the Sr isotope analysis (Albanesi et al., 2007b).The presence of *Erraticodon* and *Erismodus* in the Sepulturas Formation *sensu stricto*, Cordillera Oriental, Jujuy Province (Albanesi and Astini, 2002) suggest a correspondence of this formation with the units mentioned, or slighter younger strata considering the finding of remains of the oldest pteraspidomorph of South America, *Sacabambaspis janvieri* Gagnier et al., associated with the conodont fauna (see *Pygodus serra* Zone). Because of the limited biostratigraphical information of the "*Erismodus*" Zone, as interpreted herein it is subjected to further studies and precise delimitation while new conodont collections permits.

3.4.1.7 *Pygodus serra* Zone (TS: Dw3)

The first report of *P. serra* (Hadding) that allowed for the recognition of the Baltoscandian *P. serra* Zone was published by Hünicken and Ortega (1987) from the classic section of the shaly Los Azules Formation at the Cerro Viejo de Huaco, Central Precordillera of San Juan Province, and later discussed in detail and illustrated by Ottone et al. (1999). The *P. serra* Zone was identified in the Yerba Loca Formation, western Precordillera, with the *Eoplacognathus foliaceus* and *E. robustus* subzones (Albanesi et al., 1995b). The *E. robustus* Zone was also recognized by Heredia (1998) in the Ponón Trehué Formation (= Lindero Formation *sensu* Bordonaro et al., 1996), Bloque de San Rafael, Mendoza Province. This unit was thoroughly studied by Lehnert et al. (1999), Heredia (2001), Heredia and Rosales (2006) and Heredia et al. (2015), who documented the eponymous index species of subzones from the *Pygodus serra* Zone in the latter referred geological province. The *Eoplacognathus lindstroemi* Subzone was recorded from allochthonous clasts at the base of the La Cantera Formation, Villicum range, Eastern Precordillera of San Juan, by Albanesi et al. (1995c), associated to the pteraspidomorph *Sacabambaspis janvieri* Gagnier et al.

3.4.1.8 *Pygodus anserinus* Zone (TS: Dw3 – Sa1)

Heredia (1982) reported *P. anserinus* (Lamont and Lindström) (Fig. 5(6,7)) from the Bloque de San Rafael, Mendoza Province, for the first time. The basal level of this zone was determined in the carbonate sequence of the Bloque de San Rafael by the FAD of *P. anserinus* (Lehnert et al., 1999; Heredia, 2001; Heredia and Rosales, 2006), as well as that of the lower subzone, and the upper boundary is recorded in the Las Aguaditas Formation, Central Precordillera, by the FAD of *Amorphognathus tvaerensis* (Lehnert, 1995a; Albanesi and Ortega, 1998). The FAD of characteristic taxa; i.e., *Baltoniodus variabilis* (Bergström) (Fig. 5(8)) and *Cahabagnathus sweeti* (Bergström) (Fig. 5(4)) represent the basal upper subzone of the *P. anserinus* Zone in the Precordillera (cf. Lehnert et al., 1999). This zone is well represented in the Sierra de La Invernada Formation (Ortega et al., 2008), as well as, in the Las Aguaditas Formation in its type section (Albanesi and Ortega, 1998; Feltes et al., 2016), and Los Azules Formation at the La Chilca section (Albanesi in Peralta et al., 2003), and the Las Aguaditas Formation at the Las Chacritas section (Serra et al., 2015), with *Periodon aculeatus* Hadding as a common form (Fig. 5(5)).

3.4.2 Graptolite Zones

3.4.2.1 *Levisograptus austrodentatus* Zone (TS: Lower Dw1)

The FAD of *Levisograptus austrodentatus* (Harris and Keble) (Fig. 3(22) and 3(31)) in the uppermost lower member of the Gualcamayo Formation at the Los Sapitos Creek, allows for recognizing the homonymous biozone in the northern Precordillera (Ortega and Albanesi, 1999). The biozone ranges through the lower middle member of the Gualcamayo

Formation, with a rich graptolite assemblage. In that section, the *Arienigraptus zhejiangensis* and *Levisograptus sinicus* subzones are also recognized. To the south, in the Nazareno Creek near Las Plantas Creek, and Cerro Potrerillo section, the *L. austrodentatus* Zone is identified according to the record of the index species (Brussa et al., 1998; Ortega and Albanesi, 2000). The graptolites described by Brussa (1997) and Toro and Brussa (2003) from the lower part of the Sierra de La Invernada Formation, San Juan Precordillera, also correspond to the *L. austrodentatus* Zone.

The graptolite fauna composed by *Pendeograptus pendens liber* (Carter), *Aulograptus cucullus* (Bulman), *Tetragraptus* cf. *T. reclinatus* Elles and Wood, *T. quadribrachiatus*, *Etagraptus zhejiangensis* (Ge), *H.* cf. *H. bovis*, *Xiphograptus* sp., and diplograptids referable to the *Levisograptus austrodentatus* Zone were collected in the Cuesta del Toquero, Escaya Range, Puna of Jujuy (Aceñolaza, 1980; Gutiérrez-Marco et al., 1996; Brussa et al., 2008). Although, species like *A. cucullus* could correspond to a slightly younger assemblage.

3.4.2.2 *Levisograptus dentatus* Zone (TS: Upper Dw1)

The *Levisograptus dentatus* Zone was identified in the lower member of the Los Azules Formation, Cerro Viejo de Huaco, San Juan Precordillera (Mitchell et al., 1998; Brussa et al., 2003; Ortega and Rickards, 2003; Ortega et al., 2007a). This shaly succession bears a rich graptolite fauna that mostly consists of taxa ranging from the *Levisograptus austrodentatus* Zone, and *L.* cf. *L. dentatus* (Brongniart), *Levisograptus primus* (Legg) (Fig. 3(33)), *Arienigraptus* cf. *A. geniculatus* (Skevington), and *Paraglossograptus* cf. *P. tricornis* (Mu, Geh and Yin) (Fig. 3(21)). The presence of *Arienigraptus angulatus* (Yu and Fang) (Fig. 3(27)) indicates the upper subzone of the *L. dentatus* Zone. Graptolites of this zone were recovered from the middle member of the Gualcamayo Formation at La Corridita Creek, Gualcamayo River area (Máspero Castro et al., 2003), and it could also be present at the Cerro Potrerillo section. Some localities as Olaroz Grande and Lina Range, of the western Puna, probably correspond to this biozone according to the presence of *Didymograptus artus* Elles and Wood and *Levisograptus* cf. *L. austrodentatus* in the Esquina Colorada locality in the northern Puna (Bahlburg et al., 1990). These authors also mentioned *Didymograptus* cf. *D. artus* and *L. dentatus* in Olaroz Grande, and similar graptolite assemblages at the Rosario River section and Lina Range.

3.4.2.3 *Holmograptus lentus* Zone (TS: lower Dw2)

This biozone was first recognized by the record of *Holmograptus lentus* (Törnquist), *Paraglossograptus tentaculatus*, and *Acrograptus* sp., in the

Precordillera (Máspero Castro et al., 2003), ranging through the lower part of the middle member of the Gualcamayo Formation at La Corridita Creek, Gualcamayo River, San Juan Precordillera. Later, it was identified in the lower member of the Los Azules Formation, Cerro Viejo de Huaco by the entrance of *Archiclimacograptus*, *Haddingograptus*, and *Hustedograptus* (Brussa et al., 2003; Ortega and Rickards, 2003). The assemblage of *Cryptograptus schaeferi* Lapworth, *Bergstroemograptus crawfordi* (Harris) (Fig. 3(18)), *Archiclimacograptus? ambiguus* Maletz (Fig. 3(32)), *A. marathonensis* (Clarkson), and *Haddingograptus oliveri* (Bouček) (Fig. 3(29)) was pointed out by Ortega et al. (2007a) for the upper part of the *H. lentus* Zone from the Cerro Viejo section. The specimens classified as *Holmograptus spinosus* (Ruedemann) by Brussa (1999) from the Sierra de La Invernada Formation would correspond to the *H. lentus* considering the absence of dorsal spines on the prothecal folds. This finding allows for the recognition the *H. lentus* Zone in the La Invernada Range, Precordillera of San Juan.

3.4.2.4 *Holmograptus spinosus* Zone (TS: Middle Dw2)

A graptolite assemblage composed by *Holmograptus spinosus* (Ruedemann), *Bergstroemograptus crawfordi*, *Atopograptus woodwardi* Harris (Fig. 3(13)), *Archiclimacograptus? ambiguus*, and *Haddingograptus oliveri* were recently recorded in the upper part of the Gualcamayo Formation at Villicum Range, San Juan Precordillera (Kaufmann and Ortega, 2015). It represents the first mention of the *H. spinosus* Zone in Argentina. Specimens of *H. spinosus* were localized in a calcareous block from the middle part of the Rinconada Formation, San Juan Precordillera (Ortega et al., 2016). They co-occur with *Pseudophyllograptus* sp., *B. crawfordi*, *Cryptograptus schaeferi*, *Glossograptus* sp., and *Archiclimacograptus* spp., suggesting the presence of the *H. spinosus* Zone.

3.4.2.5 *Pterograptus elegans* Zone (TS: Upper Dw2—Lower Dw3)

The first record of the *Pterograptus elegans* Zone in the Precordillera was originally identified in the lower part of the Los Azules Formation middle member at Cerro Viejo, San Juan Precordillera (Fig. 3(17)) (Ortega, 1987, 1995; Ottone et al., 1999; Brussa et al., 2003; Toro and Brussa, 2003). The nominal species is associated to *Kalpinograptus parallelus* (Ni), *Wuninograptus* sp., *Reteograptus geinitzianus* J. Hall, *R. speciosus* Harris, *Cryptograptus schaeferi*, *Glossograptus hincksii* (Hopkinson), *Hustedograptus teretiusculus* (Hisinger), *H. vikarbyensis* (Jaanusson), *Archiclimacograptus angulatus* (Bulman), and *A. caelatus* (Lapwoth), among others. The *P. elegans* Zone was also identified in the

Cerro Potrerillo (Ortega and Albanesi, 2000), at the La Corridita Creek section (Máspero Castro et al., 2003), and recently, it was reported from the basal part of the La Cantera Formation at Villicum Range, San Juan Precordillera (Heredia et al., 2014).

3.4.2.6 *Hustedograptus teretiusculus* Zone (TS: Upper Dw3)

In the upper part of the middle member of the Los Azules Formation at the Cerro Viejo de Huaco, San Juan Precordillera, above the last appearance of *Pterograptus elegans*, the graptolite assemblage is dominated by biserial graptolites, some of them ranging from the previous *P. elegans* Zone (Ortega, 1987; 1995; Ottone et al., 1999). The graptolite fauna includes *Xiphograptus* cf. *X. robustus* (Ekström), *Cryptograptus schaeferi*, *Glossograptus hinksii*, *Reteograptus geinitzianus*, *R. speciosus*, *Hustedograptus teretiusculus*, *Hustedograptus* sp., *Archiclimacograptus riddellensis* (Harris), *A. caelatus*, *Archiclimacograptus* sp., *Normalograptus* sp., and *Eoglyptograptus* sp. among others. In the upper part of this biozone, the appearance of *Dicellograptus*, *Dicranograptus*, and *Nemagraptus* is recorded. The *teretiusculus* fauna from the Cerro Viejo de Huaco section is associated with conodonts of the *Pygodus serra* Zone. A slightly younger assemblage of the *H. teretiusculus* Zone was also recovered from the middle levels of the Sierra de La Invernada Formation, La Invernada Range, San Juan Precordillera, where the graptolite fauna shares most taxa with the Los Azules Formation. These graptolite records are associated to conodonts of the *Pygodus serra* and *Pygodus anserinus* zones. The entrances of *Nemagraptus* cf. *N. subtilis* Hadding, *Dicellograptus* specimens, and *Pseudazygograptus incurvus* (Ekström) are also recorded (Ortega et al., 2008). This biozone would also be present in the Estancia San Isidro Formation, San Isidro Creek, Mendoza Precordillera (Ortega et al., 2007b). The records of *Dicellograptus salopiensis* Elles and Wood, *Pseudamplexograptus* cf. *P. distichus* (Eichwald), and *Pseudoclimacograptus* sp. in the Rinconada Range, and the specimens of *Dicellograptus* from the Cochinoca Range and Cerro Queta, northern Puna (Toro et al., 2003; Bahlburg et al., 1990) suggest the upper part of the *H. teretiusculus* Zone, both from the Precordillera and northwestern Argentine basins.

3.5 Sandbian

3.5.1 Conodont Zones

3.5.1.1 *Amorphognathus tvaerensis* Zone (TS: Sa1)

The *Baltoniodus variabilis* Subzone and probably the *Baltoniodus gerdae* Subzone of the *Amorphognathus tvaerensis* Zone are represented in the upper member of the Los Azules Formation (Ottone et al., 1999), as well as in the Las Plantas Formation at Cerro Potrerillo (Albanesi et al., 1998), and the Las Aguaditas Formation of the Central Precordillera of San Juan, where *Eoplacognathus elongatus* (Bergström) appears as a key taxon for the referred units (Lehnert, 1995a; Albanesi and Ortega, 1998). *A. tvaerensis* Bergström (Fig. 5(10) and 5(11)) together with *B. variabilis* (Fig. 5(8)) were recorded in the La Pola Formation in the Villicum Range, Eastern Precordillera, by Heredia and Milana (2010). The *Baltoniodus alobatus* Subzone of the *A. tvaerensis* Zone was identified through the upper section of the Sierra de La Invernada Formation, with a typical association of species from the genera *Rhodesognathus*, *Protopanderodus*, *Periodon*, *Plectodina*, *Icriodella*, and *Drepanoistodus* (Ortega et al., 2008). Midcontinent type species of *Phragmodus* are frequent in the zone as well. The conodont fauna from the Upper Ordovician carbonate sequence of the Precordillera, the Sassito Formation, was referred provisionally to the *A. tvaerensis* Zone (Keller and Lehnert, 1998); however, the subsequent biostratigraphic conodont zone could be represented because the absence of index species precludes a precise recognition.

3.5.2 Graptolite Zones

3.5.2.1 *Nemagraptus gracilis* Zone (TS: Sa1)

The *Nemagraptus gracilis* Zone was recognized in the middle and upper part of the Las Aguaditas Formation, Los Blanquitos Range (Brussa, 1996), and in the Los Azules Formation, Cerro La Chilca (Peralta, 1998; Peralta and Finney, 2000), San Juan Precordillera. In both the cases, the FAD and lower range of nominal species are absent because of a stratigraphic gap in the base of the mentioned successions. In the Sierra de La Invernada Formation, La Invernada Range, San Juan Precordillera, Ortega et al. (2008) described a rich graptolite assemblage composed of *Nemagraptus gracilis* (J. Hall), *Pseudazygograptus incurvus* (Ekström), *Cryptograptus schaeferi* Lapworth, dicellograptid, dicranograptids, and biserial axonophorans, associated with conodonts of the *Pygodus anserinus* Zone. Other sections from the Precordillera bearing *N. gracilis* that could correspond to this biozone are the Portezuelo del Tontal Formation, El Tontal Range (Cuerda et al., 1986; Peralta et al., 2003), La Cantera Formation, Villicum Range (Peralta, 1990; Sánchez et al., 1991; Brussa et al., 2003), and Yerba Loca Formation, Jáchal River (Blasco and Ramos, 1976; Ortega et al., 1991) (Fig. 3(28) and 3(30)).

3.5.2.2 *Climacograptus bicornis* Zone (TS: Sa2)

The *Climacograptus bicornis* Zone was recognized in many localities of the Argentine Precordillera, where the association of the nominal taxon with

Nemagraptus gracilis, glossograptids, dicellograptids, dicranograptids, and biserial graptolites is usually present (Ortega and Brussa, 1990). In the Cerro Viejo de Huaco, this biozone ranges throughout the upper member of the Los Azules Formation (Fig. 3(37)) (Ortega, 1987; Ortega and Albanesi, 1998). The *C. bicornis* Zone was also recognized in the Las Vacas Formation (Astini and Brussa, 1997), in the Las Plantas Formation exposed in the Cerro Potrerillo, Precordillera of San Juan (Benedetto et al., 1991; Ortega and Albanesi, 1998), and in the Empozada Formation from the Precordillera of Mendoza (Ortega et al., 2007b). Besides, this biozone was documented in the Pavón Formation, Cerro Bola, Bloque de San Rafael (Cuerda et al., 1998), and recently, in the Las Lagunitas Formation at Cordón del Carrizalito, which is the first record of graptolites in the Cordillera Frontal of Argentina (Tickyj et al., 2009).

3.6 Katian

3.6.1 Conodont Zones

3.6.1.1 *Amorphognathus superbus* Zone (TS: Sa1 – Sa2 – Ka1)

Specimens representative of the *Amorphognathus superbus* Zone were recovered from reworked clasts of the Empozada Formation (Heredia et al., 1990), as well as casts on bedding plane surfaces of black shales from this unit of the southern Precordillera, Mendoza Province (Ortega et al., 2007c). Also, specimens of *A. superbus* Bergström (Fig. 8(3)) from bedding plane surfaces of limolites were recorded from the Trapiche Formation at the Cerro Potrerillo (Albanesi et al., 1995d), and Sierra de La Invernada Formation, Precordillera of San Juan Province (Ortega et al., 2008).

3.6.2 Graptolite Zones

3.6.2.1 *Diplacanthograptus caudatus* Zone (TS: Lower Ka1)

This biozone was identified in the Sierra de La Invernada Formation, La Invernada Range, San Juan Precordillera, according to the presence of the eponymous species (Fig. 3(36)), *Orthoretiolites hami* Whittington, *Diplacanthograptus spiniferus* (Ruedemann) (Fig. 3(35)), *Rectograptus amplexicaulis* (J. Hall), *Neurograptus margaritatus* (Lapworth), and species of *Corynoides*, Cryptograptus, *Dicellograptus*, *Dicranograptus*, and *Leptograptus* (Ortega et al., 2007c, 2008). Graptolite faunas corresponding to this zonal interval are present in the Cántaro de Oro Formation, El Tigre Range of the western Precordillera (Brussa, 1995; Ortega et al., 1991; Brussa et al., 1999).

3.6.2.2 *Styracograptus tubuliferus* Zone (TS: Ka2-Ka3)

The entrance of *Styracograptus tubuliferus* (Lapworth) (Fig. 3(34)) was documented in the uppermost strata of the Sierra de La Invernada Formation, La Invernada Range, San Juan Precordillera (Ortega et al., 2007c, 2008). This biozone is represented by a monotonous graptolite fauna dominated by *S. tubuliferus* and scarce specimens of *Corynoides* sp., *Rectograptus amplexicaulis*, and *Orthograptus* sp. showing a marked impoverishment of the graptolite diversity in this succession. The uppermost graptolite assemblage (A3) of the Arroyo Pavón Formation, Bloque de San Rafael, Mendoza, studied by Cuerda et al. (1998) could correspond to the *S. tubuliferus* Zone.

3.6.2.3 *Dicellograptus complanatus* Zone (TS: Lower Ka4)

A graptolite assemblage that consists of *Dicellograpthus complanatus* Lapworth, *D. morrisi* Hopkinson, *D. flexuosus* Lapworth, *D. minor* Toghill, *S. tubuliferus*, and *Normalograptus miserabilis* (Elles y Wood) was referred to the *D. complanatus* Zone by Mitchell et al. (1998). These graptolites were collected from the middle member of the Empozada Formation, Precordillera of Mendoza.

3.6.2.4 *Dicellograptus ornatus* Zone (TS: Lower Ka4)

The *Dicellograptus ornatus* Zone was recognized by the entrance of the nominal species in the middle member of the Empozada Formation, Precordillera of Mendoza (Mitchell et al., 1998). It also contains *Dicranograptus ramosus* (J. Hall), *Cryptograptus* cf. *C. insectiformis* Ruedemann, and *Orthoretiolites* cf. *O. hami* Whittington, and shares some taxa ranging from the previous *D. complanatus* Zone (e.g., *S. tubuliferus*, *N. miserabilis*).

3.7 Hirnantian

3.7.1 Graptolite Zones

3.7.1.1 *Normalograptus extraordinarius* Zone (TS: Hi1)

The *Normalograptus extraordinarius* Zone is present in the Alcaparrosa Formation outcropping in the western Precordillera. The graptolite assemblage includes *N. extraordinarius* Sobolevskaya, *Normalograptus miserabilis* (Elles and Wood), *Normalograptus normalis* (Lapworth), *Amplexograptus* cf. *A. latus* (Elles and Wood), and specimens of *Dicellograptus* and *Plegmatograptus*. *S. tubuliferus* is the dominant species (Brussa et al., 1999).

3.7.1.2 *Metabolograptus persculptus* Zone (TS: Hi2)

This biozone is recognized in diverse sections of the San Juan Precordillera. It was originally recorded in the La Chilca Formation at Talacasto Range

(Cuerda et al., 1988), and later identified in the same formation at the Escondido River section, Cerro del Fuerte, Central Precordillera (Rickards et al., 1996). In the Villicum Range, the *M. persculptus* Zone occurs in the Don Braulio Formation (Peralta and Baldis, 1990; Sánchez et al., 1991). The *persculptus* assemblage comprises *Normalograptus angustus* (Perner), *Metabolograptus persculptus* (Elles and Wood), *Normalograptus* cf. *N. medius* (Tornquist), *Metaclimacograptus robustus* Cuerda, Rickards, and Cingolani, among others.

4. DISCUSSIONS AND CONCLUSIONS

The updated biostratigraphic chart (Fig. 2) presents seventeen conodont zones (with eight biozones divided into subzones) recognized in the Argentine Precordillera, most of them developed in carbonate facies. In northwestern Argentine basins twelve conodont zones (one divided into two subzones) are determined in siliciclastic sequences. Only two biostratigraphic intervals from both basins are identified by the same index species; i.e., *Paltodus deltifer* and *Paroistodus proteus*, although they are characterized by different conodont assemblages of the Tremadocian.

In particular, the interval referred to as "*Baltoniodus navis*" in northwestern Argentina (Fig. 2) does not include the nominal species, but the associated species are coeval with those that occur in strata correlative to the *B. navis* Zone in the Precordillera. *Baltoniodus triangularis*, the ancestor of *B. navis*, was recorded in the Cordillera Oriental (Carlorosi et al., 2013), and a closely related form, *B. navis sensu* Cooper was reported in the Sistema de Famatina (Albanesi and Vaccari, 1994). Considering these records, the correlative interval is named to as "*B. navis*", under the expectation that the index species could eventualy be found in northwestern Argentine basins as well.

The Darriwilian "*Erismodus*" interval from northwestern basins still needs clarification regarding its chronostratigraphic boundaries, somehow further studies are required even to subdivide its stratigraphic range into distinct conodont zones. Above this interval, the Ocloyic orogenic event eliminated the representative strata of most of the upper Middle and lower Upper Ordovician, which ends up with the diamictites of the Zapla Formation (Moya, 2015, and references therein). A similar situation is observed for the Precordillera, where the Upper Ordovician biozones are not represented probably because of the paleoclimatic control produced by the finiordovician glaciation event. The Guandacolic orogenic event (Furque, 1963) suppresses significant stratigraphic successions corresponding to parts of the *Eoplacognathus pseudoplanus*, *Eoplacognathus suecicus*, *Pygodus serra*, and *Pygodus anserinus* zones of the upper Middle Ordovician, as verified in the central sector of the Central Precordillera, although conodont bearer strata of the same interval are diachronics and irregularly distributed and exposed through the basin, and even elder biozones could be involved in the nothern extreme of the Central Precordillera. Regarding the graptolite records, the *Nemagraptus gracilis* Zone can be completely suppressed, such as it occurs in the Cerro Viejo de Huaco section (e.g., Ortega and Brussa, 1990; Ortega and Albanesi, 1998).

The biostratigraphic chart (Fig. 2) presents a succession of thirty graptolite zones (with two biozones divided into subzones) established for the Ordovician System of Argentina. There were recognized eight biozones for the Tremadocian Stage of the Cordillera Oriental, some of which are present in the Puna region (e.g., *Araneograptus murrayi*, *Hunnegraptus copiosus*), and four zones for the Floian Stage. The *Azygograptus eivionicus*, *Levisograptus austrodentatus*, *Levisograptus dentatus*, and *Hustedograptus teretiusculus* zones are represented in the Puna region (Bahlburg et al., 1990; Toro and Brussa, 2003). The graptolite faunas of the Upper Ordovician cannot be recorded in northwestern Argentina because the stratigraphic succession was removed by the Ocloyic orogenic event (Astini et al., 2004; Moya, 2015), and the diamictite deposits of the Zapla Formation (Hirnantian) were deposited unconformably over older strata.

In the Precordillera of western Argentina 17 graptolite zones are recognized in dominantly clastic successions that starts in the Dapingian, which overlies the Cambro-Ordovician carbonate platform. Within this biozonal succession, two hiatus are determined. The hiatus that spans the middle Darriwilian (Da2) corresponds to the *Nicholsonograptus fasciculatus* Zone of the North American and Baltoscandian schemes. A second hiatus is recorded in the upper Katian, between the *Dicellograptus ornatus* and *Normalograptus extraordinarius* zones, which is approximately equivalent to the *P. pacificus* Zone of the North American biozonation. Moreover, in the Precordillera, two biostratigraphic intervals from the Lower Ordovician were identified in autochthonous strata of the Los Sombreros Formation (with *Araneograptus murrayi* posibly belonging to the homonymous zone). *Tetragraptus approximatus* and *Pendeograptus fruticosus* are reported from allochthonous blocks of the basal strata of the Empozada Formation, which are referred to the lower Floian Stage (Bordonaro and Peralta, 1987; Ortega et al., 2007b).

A couple of well-constrained geochronologic data for the Ordovician System in the Precordillera is approached by means of radiogenic zircon analyses. The absolute dating for the top of the San Juan Formation at the Cerro Viejo de Huaco section indicates ca. 464 Ma for a level close to the basal Dw2 (Huff et al., 1997), in the transition between the *Lenodus variabilis* and the *Yangtzeplacognathus crassus* zones or the basal *Levisograptus dentatus* Zone. Another absolute dating from the Central Precordillera corresponds to lower levels in the upper part of the San Juan Formation exposed at the Talacasto Creek (Baldo et al., 2003). It gives ca. 470 Ma for a level close to the basal Dp2, in the transition between the *Baltoniodus navis–Tripodus laevis* Zone and the *Baltoniodus navis* Zone. For northwestern Argentine basins, a similar age published by the same authors corresponds to a level within the "*Baltoniodus navis*" Zone in the Chaschuil area, Sistema de Famatina. These types of conodont–graptolite absolute age linkages secure a valuable chronostratigraphic chart for confident global correlations.

ACKNOWLEDGMENTS

The authors are grateful for the careful review of this chapter by Dr. Jörg Maletz and an anonymous reviewer, as well as to the editors, whose helpful comments and assistance have contributed to improve its final presentation. The authors acknowledge the continued support by CONICET and Universidad Nacional de Córdoba, Argentina, for studies on conodonts and graptolites.

REFERENCES

Aceñolaza, F.G., 1980. Presencia de *Tetragraptus* (Graptolithina) en el Ordovícico de Jujuy. Implicancias cronológicas y paleobiogeográficas. Rev. Asoc. Geol. Argent. 35, 539–545.

Aceñolaza, F.G., 1992. El Sistema Ordovícico de Latinoamérica. In: Gutiérrez-Marco, J.C., Saavedra, J., Rábano, I. (Eds.), Paleozoico inferior de Ibero-América. Universidad de Extremadura, pp. 85–118.

Aceñolaza, F.G., Aceñolaza, G.F., Esteban, S.B., Gutiérrez-Marco, J.C., 1996. Estructuras nemales de *Araneograptus murrayi* (J. Hall) (graptolito del Ordovícico Inferior) y actualización del registro perigondwánico de la especie. In: Memorias XII Congreso Geológico de Bolivia, pp. 681–689.

Aceñolaza, F.G., Gutiérrez-Marco, J.C., 2000. Graptolitos de la Formación Portezuelo de Las Minitas (Ordovícico Inferior) del Sistema de Famatina, La Rioja, Argentina. Bol. Acad. Nac. Cienc. Córdoba 64, 17–26.

Aceñolaza, F.G., Gorustovich, S., Solís, J., 1976. El Ordovícico del río de La Alumbrera, departamento Tinogasta, provincia de Catamarca. Ameghiniana 13, 269–288.

Aceñolaza, F.G., Heredia, S., Carlorosi, J.M.T., 2008. La "*Sepulturas Limestones*" (Harrington *in* Harrington y Leanza, 1957) en su área tipo, fósiles y edad, provincia de Jujuy, Argentina. Acta Geol. Lilloana 20 (2), 147–158.

Albanesi, G.L., 1999. New *Gothodus* and *Protoprioniodus* species from lowest middle Ordovician strata, Argentine Precordillera. In: Pander Society Annual Meeting. University of Calgary, Canada. Abstracts: 1.
Albanesi, G.L., Astini, R.A., 1994. Conodontofauna de los niveles cuspidales de la Formación San Juan (Llanvirniano) en el perfil de Las Chacritas, Provincia de San Juan. In: VI Congreso Argentino de Paleontología y Biostratigrafía. Resúmenes Paleoinvertebrados, pp. 48–49.
Albanesi, G.L., Astini, R.A., 2000a. Nueva fauna de conodontes de la Formación Suri (Ordovícico Inferior–Medio), Sistema de Famatina, Argentina. In: Reunión de Comunicaciones de la Asociación Paleontológica Argentina, Mar del Plata. Ameghiniana, 37 (4), p. 68R.
Albanesi, G.L., Astini, R.A., 2000b. Bioestratigrafía de conodontes de la Formación Las Chacritas, Precordillera de San Juan, Argentina. In: Reunión de Comunicaciones de la Asociación Paleontológica Argentina, Mar del Plata. Ameghiniana, 37 (4), p. 68R.
Albanesi, G.L., Astini, R.A., 2002. Faunas de conodontes y *Sacabambaspis janvieri* (Vertebrata) en el Ordovícico Medio de la Cordillera Oriental Argentina: implicancias estratigráficas y paleobiogeográficas. In: VIII Congreso Argentino de Paleontología y Bioestratigrafía, Corrientes, Resumen, 17a.
Albanesi, G.L., Barnes, C.R., 2000. Subspeciation within a punctuated equilibrium evolutionary event: phylogenetic history of the lower-middle ordovician *Paroistodus originalis* – *P. horridus* complex (Conodonta). J. Paleontol. 74 (3), 492–502.
Albanesi, G.L., Bergström, S.M., 2004. Conodonts: Lower to Middle Ordovician record. In: Webby, B., et al. (Eds.), The Great Ordovician Biodiversification Event. Columbia University Press, pp. 312–326.
Albanesi, G.L., Bergström, S.M., 2010. Early-Middle Ordovician conodont paleobiogeography with special regard to the geographic origin of the Argentine Precordillera: a multivariate data analysis. In: Finney, S.C., Berry, W.B.N. (Eds.), The Ordovician Earth System, vol. 466. Geological Society of America, pp. 119–139. Special Paper.
Albanesi, G.L., Ortega, G., 1998. Conodont and graptolite faunas from the las Plantas Formation and equivalent units (Caradoc) in the Argentine Precordillera. In: Seventh International Conodont Symposium Held in Europe, Bologna-Modena, Abstracts, pp. 1–2.
Albanesi, G.L., Ortega, G., 2002. Advances on conodont-graptolite biostratigraphy of the Ordovician System of Argentina. In: Aceñolaza, F.G. (Ed.), Aspects of the Ordovician System of Argentina, Ser. Correl. Geol., 16, pp. 143–165.
Albanesi, G.L., Rao, R.I., 1996. Conodont fauna from the Santa Gertrudis Formation (middle-later Ordovician), Eastern Cordillera, northwestern Argentina. In: Sixth International Conodont Symposium, Warszawa, p. 3.
Albanesi, G.L., Vaccari, N.E., 1994. Conodontos del Arenig en la Formación Suri, Sistema del Famatina, Argentina. Rev. Esp. Micropaleontol. XXVI (2), 125–146.
Albanesi, G.L., Hünicken, M.A., Ortega, G., 1995a. Review of Ordovician conodont-graptolite Bioestratigraphy of the Argentine Precordillera. In: Cooper, J.D., Droser, M.L., Finney, S.C. (Eds.), Ordovician Odyssey: Short Papers for the Seventh International Symposium on the Ordovician System, Las Vegas, SEPM, Fullerton, pp. 31–36.
Albanesi, G.L., Ortega, G., Hünicken, M.A., 1995b. Conodontes y graptolitos de la Formación Yerba Loca (Arenigiano-Llandeiliano) en las quebradas de Ancaucha y El Divisadero, Precordillera de San Juan, Argentina. Bol. Acad. Nac. Cienc. Córdoba 60 (3–4), 365–400.
Albanesi, G.L., Benedetto, J.L., Gagnier, P.-Y., 1995c. *Sacabambaspis janvieri* (Vertebrata) y conodontes del Llandeiliano temprano en la Formación La Cantera, Precordillera de San Juan, Argentina. Bol. Acad. Nac. Cienc. Córdoba 60 (3–4), 519–544.

Albanesi, G.L., Hünicken, M.A., Ortega, G., 1995d. *Amorphognathus* aff. *superbus* (Conodonta) en la Formación Trapiche (Ordovícico Superior), cerro Potrerillo, Departamento Jáchal, Provincia de San Juan, Argentina. Bol. Acad. Nac. Cienc. Córdoba 60 (3–4), 461–468.

Albanesi, G.L., Hünicken, M.A., Barnes, C.R., 1998. Bioestratigrafía de conodontes de las secuencias ordovícicas del cerro Potrerillo, Precordillera Central de San Juan, R. Argentina. Actas XII Acad. Nac. Cienc. Córdoba 7–72.

Albanesi, G.L., Ortega, G., Barnes, C.R., Hünicken, M.A., 1999a. Conodont-graptolite biostratigraphy of the Gualcamayo Formation (Middle Ordovician) in the Gualcamayo-Guandacol rivers area, Argentina Precordillera. Acta Univ. Caroliane Geol. 43 (1/2), 45–48.

Albanesi, G.L., Esteban, S.B., Barnes, C.R., 1999b. Conodontes del intervalo del límite Cámbrico-Ordovícico en la Formación Volcancito, Sistema de Famatina, Argentina. Temas Geológico-Minéros ITGE, Madr. 26, 521–526.

Albanesi, G.L., Ortega, G., Zeballo, F., 2001. Late Tremadocian Conodont-Graptolite Biostratigraphy from NW Argentine Basins. In: The Guide Book, Abstracts & Ordovician Silurian Correlation Chart for the Joint Field Meeting of IGCP 410 and IGCP 421 in Mongolia, Ulan Bator; Extended Abstract, pp. 121–123.

Albanesi, G.L., Bergström, S.M., Melchor, R.N., 2003. The San Jorge Formation, La Pampa Province, Argentina, dated by means of conodonts. In: Reunión anual de comunicaciones de la Asociación Paleontológica Argentina, Santa Rosa. Resumen, pp. 77R–78R.

Albanesi, G.L., Esteban, S.B., Ortega, G., Hünicken, M.A., Barnes, C.R., 2005. Bioestratigrafía y ambientes sedimentarios de las formaciones Volcancito y Bordo Atravesado (Cámbrico Superior – Ordovícico Inferior), Sistema de Famatina, provincia de La Rioja, Argentina. In: Dahlquist, J.A., Baldo, E.G., Alasino, P.H. (Eds.), Geología de la provincia de La Rioja: Precámbrico – Paleozoico inferior, Asoc. Geol. Argentina, Serie D: Publ. Especial, 8, pp. 41–64.

Albanesi, G.L., Carrera, M.G., Cañas, F.L., Saltzman, M., 2006. A proposed global boundary stratotype section and point (GSSP) for the base of the Middle Ordovician Series: the Niquivil section, Precordillera of San Juan, Argentina. Episodes 29 (1), 1–15.

Albanesi, G.L., Zeballo, F.J., Bergström, S.M., 2007a. The *Paltodus deltifer* Zone (late Tremadocian; Early Ordovician) in Argentina: new conodont data for intercontinental correlation and paleobiogeographic analysis. In: 10° International Symposium on the Ordovician System, Nanjing, China. Acta Palaeontol. Sin., 46. Science Press, pp. 16–22.

Albanesi, G.L., Monaldi, C.R., Ortega, G., Trotter, J.A., 2007b. The Capillas Formation (late Darriwilian) of Subandean Ranges, northwestern Argentina; age, correlation and environmental constraints. In: 10th International Symposium on the Ordovician System, Nanjing, China. Acta Palaeontol. Sin., 46. Science Press, pp. 9–15.

Albanesi, G.L., Ortega, G., Zeballo, F.J., 2008. Faunas de conodontes y graptolitos del Paleozoico inferior de la Cordillera Oriental Argentina. In: Coira, B., Zappettini, E.O. (Eds.), Geología y Recursos Naturales de la Provincia de Jujuy, Relatorio XVII Congreso Geológico Argentino, Jujuy, pp. 98–118.

Albanesi, G.L., Ortega, G., Monaldi, C.R., Zeballo, F.J., 2011. Conodontes y graptolitos del Tremadociano tardío de la sierra de Zenta, Cordillera Oriental de Jujuy, Argentina. Ameghiniana 48 (2), 242–263.

Albanesi, G.L., Bergström, S.M., Schmitz, B., Serra, F., Feltes, N.A., Voldman, G.G., Ortega, G., 2013. Darriwilian (Middle Ordovician) $\delta^{13}C_{carb}$ chemostratigraphy in the Precordillera of Argentina: Documentation of the middle Darriwilian isotope Carbon Excursion (MDICE) and its use for intercontinental correlation. Palaeogeogr. Palaeoclimatol. Palaeoecol. Elsevier (SCI-ISI) 389, 48–63.

Albanesi, G.L., Ortega, G., Monaldi, C.R., 2014. Precisiones sobre la edad de la Formación Cieneguillas (Ordovícico) en el Alto de Lipán, sierra de Los Colorados, Cordillera Oriental de Jujuy. In: XIX Congreso Geológico Argentino, Córdoba, S2-1.

Albanesi, G.L., Barnes, C.R., Trotter, J.A., Williams, I.S., Bergström, S.M., 2015a. Deciphering the movement of the Argentine Precordillera from tropical to higher latitudes, Late Cambrian – Late Ordovician, through conodont $\delta^{18}O$ paleothermometry. In: Leslie, S.A., Goldman, D., Orndorff, R.C. (Eds.), The Ordovician Exposed: Proceedings of the 12 International Symposium on the Ordovician System, Harrisonburgh, pp. 89–90.

Albanesi, G.L., Giuliano, M.E., Pacheco, F., Ortega, G., Monaldi, C.R., 2015b. Conodonts from the Cambrian-Ordovician boundary in the Cordillera Oriental, NW Argentina. Stratigraphy 12 (3–4), 237–256.

Albanesi, G.L., Cañas, F., Mango, M., 2016. Fauna de conodontes tremadocianos del techo de la Formación La Silla en el Cerro Viejo de San Roque, Precordillera Central de San Juan. In: III Jornadas de Geología de Precordillera, San Juan. Acta Geol. Lilloana, 28 (Supl.), pp. 14–20.

Astini, R.A., Brussa, E.D., 1997. Dos nuevas localidades fosilíferas en el Conglomerado de Las Vacas (Caradociano) en la Precordillera Argentina: importancia cronoestratigráfica. Ameghiniana, Resúmenes 34, 114.

Astini, R., Waisfeld, B.G., Toro, B.A., Benedetto, J.L., 2004. El Paleozoico inferior y medio de la región de Los Colorados, borde occidental de la Cordillera Oriental (provincia de Jujuy). Rev. Asoc. Geol. Argent. 59, 243–260.

Bagnoli, G., Stouge, S., 1997. Lower Ordovician (Billingenian-Kunda) conodont zonation and provinces based on sections from Horns Udde, north Öland, Sweden. Boll. della Soc. Paleontol. Ital. 35, 109–163.

Bahlburg, H., Breitkreuz, Ch, Maletz, J., Moya, M.C., Salfity, J.A., 1990. The Ordovician sedimentary rocks in the northern Puna of Argentina and Chile: new stratigraphical data based on graptolites. Newsl. Stratigr. 23, 69–89.

Baldo, E.G., Fanning, C.M., Rapela, C.W., Pankhurst, R.J., Casquet, C., Galindo, C., 2003. U-Pb shrimp dating of rhyolite volcanism in the Famatinian belt and k-bentonites in the Precordillera. In: Albanesi, G.L., Beresi, M.S., Peralta, S.H. (Eds.), Ordovician from the Andes. Proceedings of the 9° International Symposium of the Ordovician System, Serie Correlación Geológica, vol. 17. INSUGEO, Tucumán, pp. 185–190.

Banchig, A.L., Moya, M.C., 2002. La Zona de *Tetragraptus approximatus* (Ordovícico Inferior) en la sierra del Tontal, Precordillera occidental Argentina. In: 8° Congreso Argentino de Paleontología y Bioestratigrafía, Corrientes, p. 83.

Barnes, C., Fåhræus, L., 1975. Provinces, communities and the proposed nektobenthic habit of Ordovician conodontophorids. Lethaia 8 (2), 133–149.

Bauer, J.A., 1987. Conodonts and conodont biostratigraphy of the McLish and Tulip Creek formations (Middle Ordovician) of south-central Oklahoma. Okla. Geol. Surv. Bull. 141, 1–53.

Bauer, J.A., 2010. Conodonts and conodont biostratigraphy of the Joins and Oil Creek formations, Arbuckle mountains, south-central Oklahoma. Okla. Geol. Surv. Bull. 150, 1–44.

Benedetto, J.L., Herrera, Z.A., Ortega, G., 1991. Bioestratigrafía del Ordovícico en el flanco occidental del Cerro Potrerillo, Provincia de San Juan. Rev. Asoc. Geol. Argent. 46 (1–2), 60–68.

Benedetto, J.L., Brussa, E.D., Pompei, J.F., 2002. El Ordovícico de la región de Susques-Huancar (Puna Oriental de Jujuy): Precisiones sobre su edad y significado estratigráfico. In: Actas XV Congreso Geológico Argentino, El Calafate, vol. 1, pp. 572–577.

Bergström, S.M., 1971. Conodont biostratigraphy of the Middle and Upper Ordovician of Europe and eastern North America. In: Sweet, W.C., Bergström, S.M. (Eds.),

Symposium on Conodont Biostratigraphy, Geological Society of America Memory, vol. 127, pp. 83–157.

Bergström, S.M., Chen, X., Gutiérrez-Marco, J.C., Dronov, A., 2009. The new chronostratigraphic classification of the Ordovician System and its relations to major regional series and stages and to $\delta^{13}C$ chemostratigraphy. Lethaia 42, 97–107.

Blasco, G., Ramos, V.A., 1976. Graptolitos caradocianos de la Formación Yerba Loca y del Cerro La Chilca, Depto. Jáchal, Provincia de San Juan. Ameghiniana 13 (3–4), 312–329.

Bordonaro, O., Keller, M., Lehnert, O., 1996. El Ordovícico de Ponón Trehué en la provincia de Mendoza (Argentina): redefiniciones estratigráficas. In: XIII Congreso Geológico Argentino y III Congreso Exploración Hidrocarburos, Buenos Aires, Actas, vol. 1, pp. 541–550.

Bordonaro, O., Peralta, S.H., 1987. El Arenigiano Inferior de la Formación Empozada en la localidad de San Isidro, Mendoza, Argentina. In: 10° Congreso Geológico Argentino, Tucumán, vol. III, pp. 81–84.

Brackebush, L., 1883. Estudio sobre la Formación Petrolífera de Jujuy. Bol. Acad. Nac. Cienc. Córdoba V, 137–252.

Brussa, E.D., 1995. Upper Middle Ordovician Graptolites from the Western Precordillera. Graptolite News, Argentina, pp. 15–16.

Brussa, E.D., 1996. Las graptofaunas ordovícicas de la Formación Las Aguaditas, Precordillera de San Juan, Argentina. Parte I: Familias Thamnograptidae, Dichograptidae, Abrograptidae y Glossograptidae. Ameghiniana 33, 421–434.

Brussa, E.D., 1997. Graptolitos del Arenigiano tardío-Llanvirniano de la Formación Sierra de La Invernada, Precordillera occidental sanjuanina, Argentina. Parte 1. Ameghiniana 34, 357–372.

Brussa, E.D., 1999. El género *Holmograptus* (Graptolithina) en el Ordovícico de la Precordillera occidental Argentina. Rev. Esp. Paleontol. 14, 183–190.

Brussa, E.D., Mitchell, C.E., Astini, R.A., 1998. Late Yapeenian to early Darriwilian graptolite faunas from Nazareno creek, Guandacol region, Precordillera, Argentina. In: Proceedings 6th International Graptolite Conference, Madrid. Temas Geológico-Mineros ITGE, 23, pp. 158–160.

Brussa, E.D., Mitchell, C.E., Astini, R.A., 1999. Ashgillian (Hirnantian?) graptolites from the western boundary of the Argentine Precordillera. Acta Univ. Caroliane Geol. 43, 199–202.

Brussa, E.D., Toro, B.A., Benedetto, J.L., 2003. Bioestratigraphy. In: Benedetto, J.L. (Ed.), Ordovician Fossils of Argentina. Universidad Nacional de Córdoba, pp. 75–90.

Brussa, E.D., Toro, B.A., Vaccari, N.E., 2008. Bioestratigrafía del Paleozoico Inferior en el ámbito de la Puna. In: XVII Congreso Geológico Argentino. Geología y Recursos Naturales de La Provincia de Jujuy. Relatorio, pp. 93–97.

Carlorosi, J., 2011. La Zona de *Trapezognathus diprion* en la "Formación Sepulturas", Espinazo del Diablo, Cordillera Oriental Argentina. Ser. Correl. Geol. 27, 37–43.

Carlorosi, J., Heredia, S., 2013. The conodont Trapezognathus diprion (Lindström) in the Acoite Formation, Eastern Cordillera, northwestern Argentina. In: Albanesi, G.L., Ortega, G. (Eds.), Conodonts from the Andes, Proceedings of the 3° International Conodont Symposium, Paleontological Note PE APA, Mendoza, Argentina, pp. 1–4.

Carlorosi, J., Heredia, S., Aceñolaza, G., 2013. Middle Ordovician (early Dapingian) conodonts in the Central Andean Basin of NW Argentina. Alcheringa 37.

Código Argentino de Estratigrafía, 1992. Asociación Geológica Argentina, Serie B (Didáctica y Complementaria), vol. 20, pp. 1–64.

Carrera, M.G., Fenoglio, F., Albanesi, G.L., 2013. Conodonts, sequence stratigraphy and the drowning of the San Juan carbonate platform in the Ordovician of the Argentine Precordillera. In: Albanesi, G.L., Ortega, G. (Eds.), Conodonts from the Andes,

Proceedings of the 3° International Conodont Symposium. Paleontological Note PE APA, Mendoza, Argentina, pp. 5–12.
Cooper, R.A., 1999. The Ordovician time scale – calibration of graptolite and conodont zones. Acta Univ. Caroliane Geol. 43, 1–4.
Cooper, R.A., Nowlan, G.S., Williams, S.H., 2001. Global stratotype and point for the base of the Ordovician System. Episodes 24 (1), 19–28.
Cooper, R.A., Rigby, S., Loydell, D.K., Bates, D.E., 2012. Palaeoecology of the Graptoloidea. Earth Sci. Rev. 112, 23–41.
Cuerda, A.J., Cingolani, C.A., Schauer, O.C., Varela, R., 1986. Bioestratigrafía del Ordovícico (Llanvirniano-Llandeiliano) de la sierra del Tontal, Precordillera de San Juan. Descripción de su fauna graptolítica. Ameghiniana 23, 3–33.
Cuerda, A.J., Rickards, R.B., Cingolani, C., 1988. A new Ordovician- Silurian boundary section in San Juan Province, Argentina, and its definitive graptolite fauna. J. Geol. Soc. Lond. 145, 749–757.
Cuerda, A.J., Cingolani, C., Manassero, M., 1998. Caradoc age of the graptolite assemblages and facial relations from the cerro Bola section, San Rafael Block, Mendoza Province, Argentina. Temas Geológico-Mineros ITGE 23, 170–173.
Della Costa, G., Albanesi, G.L., Ortega, G., 2016. El establecimiento de los límites cronoestratigráficos del Ordovícico Medio en la Precordillera Argentina. In: Libro de Resúmenes del IX Congreso Latinoamericano de Paleotología, Lima, Perú, p. 43.
Dubinina, S.V., 2000. Conodonts and zonal stratigraphy of the Cambrian-Ordovician boundary deposits. Nauka, Trans. GIN RAS 517, 1–240.
Ethington, R.L., Clark, D.L., 1981. Lower and middle Ordovician conodonts from the Ibex area, western Millard County, Utah. Brigh. Young Univ. Geol. Stud. 28 (2), 1–160.
Feltes, N.A., Albanesi, G.L., Bergström, S.M., 2013. Middle Darriwilian conodont biozones from the lower member of the Las Aguaditas formation, central Precordillera of San Juan, Argentina. In: Albanesi, G.L., Ortega, G. (Eds.), Conodonts from the Andes, Proceedings of the 3° International Conodont Symposium. Paleontological Note PE APA, Mendoza, Argentina, pp. 25–32.
Feltes, N.A., Albanesi, G.L., Bergström, S.M., 2016. Conodont biostratigraphy and global correlation of the middle Darriwilian – lower Sandbian Las Aguaditas formation, Precordillera of San Juan, Argentina. Andean Geol. 43 (1), 60–85.
Fortey, R.A., Harper, D.A.T., Ingham, J.K., Owen, A.W., Parkes, M.A., Rushton, A.W.A., Woodcock, N.H., 2000. A revised correlation of Ordovician rocks in the British Isles. Geol. Soc. Special Rep. 24, 1–83.
Furque, G., 1963. Descripción geológica de la Hoja 17b – Guandacol (Provincia de La Rioja-Provincia de San Juan). Bol. Dirección Nac. Geol. Minería, Buenos Aires 92, 1–104.
Giuliano, M.E., Albanesi, G.L., Ortega, G.C., Zeballo, F.J., Monaldi, C.R., 2013. Conodonts and graptolites of the Santa Rosita Formation (Tremadocian) at the Nazareno area, Santa Victoria range, Cordillera Oriental of Salta, Argentina. In: Albanesi, G.L., Ortega, G. (Eds.), 3° International Conodont Symposium, Conodonts from the Andes, pp. 39–44.
Giuliano, M.E., 2016. Biozonas y biofacies de conodontes y graptolitos en comarcas ordovícicas del noroeste argentino. Tesis Doctoral. Universidad Nacional de Córdoba, 257 pp. (unpublished).
Goldman, D., Bergström, S.M., Mitchell, C.E., 1995. Revision of the Zone 13 graptolite biostratigraphy in the Marathon, Texas, standard succession and its bearing on Upper Ordovician graptolite biogeography. Lethaia 28, 115–128.
Goldman, D., Leslie, S.A., Nölvak, J., Young, S., 2007. The black knob ridge section, south-eastern Oklahoma, USA: the global stratotype-section and point (GSSP) for the base of the Katian Stage of the upper Ordovician series. Acta Palaeontol. Sin. 46 (Suppl.), 144–154.

Goldman, D., Maletz, J., Melchin, M.J., Junxuan, F., 2013. Graptolite palaeobiogeography. In: Harper, D.A.T., Servais, T. (Eds.), Early Palaeozoic Biogeography and Palaeogeography, vol. 38. Geological Society, London, Memoirs, pp. 415–428.
González Barri, C., Alonso, R., 1984. Nuevos graptolites del Tremadociano superior del norte de Argentina. In: 3° Congreso Latinoamericano de Paleontología, México, pp. 62–67.
Gradstein, F.M., Ogg, J.G., Hilgen, F.J., 2012. On the geologic time scale. Newsl. Stratigr. 45 (2), 171–188.
Gutiérrez Marco, J.C., Aceñolaza, F.G., 1987. *Araneograptus murrayi* (Hall, 1865) (Graptoloidea, Anisograptidae): su identidad como "*Dictyonema yaconense*" Turner, 1960 y distribución en España y Sudamérica. In: Actas 10° Congreso Geológico Argentino, Tucumán, vol. 1, pp. 321–334.
Gutiérrez-Marco, J.C., Esteban, S.B., 2005. Graptolitos del Tremadociense (Ordovícico Inferior) de la Formación Volcancito, Sistema de Famatina (La Rioja, Argentina). Rev. Esp. Paleontol. 20 (1), 65–118.
Gutiérrez-Marco, J.C., Aceñolaza, G.F., Esteban, S.B., 1996. Revisión de algunas localidades con graptolitos Ordovícicos en la Puna salto-jujeña (Noroeste de Argentina). In: Memorias 12° Congreso Geológico de Bolivia, vol. 2, pp. 725–731.
Harrington, H.J., Leanza, A.F., 1957. Ordovician Trilobites of Argentina. In: University of Kansas Special Publication, vol. 1. University of Kansas Press, Lawrence, pp. 1–276.
Heredia, S.E., 1982. *Pygodus anserinus* Lamont and Lindström (Conodonto) en el Llandeilian de la Formación Ponón Trehué, Provincia de Mendoza, Argentina. Ameghiniana 19 (3–4), 229–233.
Heredia, S.E., 1998. *Eoplacognathus robustus* (Conodonta) en la Formación Ponón Trehué (Ordovícico Inferior), Sierra Pintada, provincia de Mendoza, Argentina. Ameghiniana 35 (3), 337–344.
Heredia, S.E., 2001. Late Llanvirn conodonts from the Ponón Trehué Formation, Mendoza, Argentina. GAIA 16, 101–117.
Heredia, S.E., Mestre, A., 2013. Advances in the middle Darriwilian conodont biostratigraphy of the Argentine Precordillera. In: Albanesi, G.L., Ortega, G. (Eds.), 3° International Conodont Symposium, Conodonts from the Andes, pp. 45–47.
Heredia, S.E., Milana, J.P., 2010. Conodontes Sandbianos (Ordovícico Superior) en la Quebrada La Pola, Sierra de Villicum, Precordillera de San Juan. (Argentina). Ameghiniana 47 (4), 515–525.
Heredia, S., Rosales, C., 2006. Biofacies de conodontes de la Formación Ponón Trehué y la importancia bioestratigráfica como sección tipo para el límite del Ordovícico Medio-Ordovícico Superior de Cuyania (Argentina). Ser. Correl. Geol. 21, 7–16.
Heredia, S.E., Gallardo, G., Maldonado, A., 1990. Conodontes caradocianos en las calizas alóctonas del Miembro Superior de la Formación Empozada (Ordovícico medio y superior), San Isidro (Mendoza, Argentina). Ameghiniana 27 (3–4), 197–206.
Heredia, S., Rosales, C., Peralta, S., Beresi, M., 2005. Conodontes de la Formación Las Chacritas (Darriwiliano) y su significado tecto-sedimentario en el bloque Sierra de la Trampa-Los Blanquitos, Precordillera de San Juan. In: 16° Congreso Geológico Argentino, La Plata, Actas, vol. 3, pp. 445–450.
Heredia, S., Beresi, M., Mestre, A., 2011. La estratigrafía del Ordovícico Medio del río Las Chacritas, Precordillera Central de San Juan. Ser. Correl. Geol. 27, 18–26.
Heredia, S., Kaufmann, C., Mestre, A., Soria, T., Ortega, G., 2014. La edad de la base de la Formación La Cantera (Ordovícico) en la Precordillera oriental, sierra de Villicum, San Juan. In: 19 Congreso Geológico Argentino, Córdoba, vol. S2, p. 13.
Heredia, S., Carlorosi, J., Sarmiento, G., 2015. The presence of the lower-middle Ordovician genus *Baltoniodus* in the central Andean Basin, Argentina: the stratigraphic significance. Sediment. Rec. 13 (2), 33–35.

Huff, W.D., Davis, D., Bergström, S.M., Krekeler, M.P.S., Kolata, D.R., Cingolani, C.A., 1997. A biostratigraphically wellconstrained K-bentonite U–Pb zircon age of the lowermost Darriwilian stage (Middle Ordovician) from the Argentine Precordillera. Episodes 20, 29–33.

Hünicken, M.A., Gallino, E., 1970. Los Conodontes de la Formación San Juan (Llanvirniano). Revista Facultad de Ciencias Exactas, Físicas y Naturales Universidad Nacional de Córdoba. Ser. Cienc. Geol. 1, 5–12.

Hünicken, M.A., Ortega, G., 1987. Lower Llanvirn-lower Caradoc (Ordovician) conodonts and graptolites from the Argentine Central Precordillera. In: Austin, R.L. (Ed.), Conodonts: Investigative Techniques and Applications, vol. 7. Ellis Horwood Limited, Chichester, pp. 136–145.

Hünicken, M.A., Sarmiento, G.N., 1980. The Baltoscandian conodont *Prioniodus elegans* pander (lower Arenigian) from the San Juan Formation of the Precordillera, Guandacol, river, La Rioja, República Argentina. Bol. Acad. Nac. Cienc. Córdoba 53 (3–4), 293–306.

Iglesias, S., 1949. Consideraciones preliminares sobre la presencia de conodontes y otros microfósiles en el Paleozoico inferior de la provincia de Jujuy, vol. II (6). Publicaciones del Instituto de Geología y Minería, San Salvador de Jujuy, Argentina, pp. 77–78.

Ji, Z., Barnes, C.R., 1994. Lower Ordovician conodonts of the St. George Group, Port au Port Peninsula, western Newfoundland, Canada. Palaeontogr. Can. 11, 1–149.

Johnston, D.I., Barnes, C.R., 1999. Early and Middle Ordovician (Arenig) conodonts from St. Pauls inlet and Martin Point, Cow Head group, western Newfoundland, Canada. 1: Biostratigraphy and paleoecology. Geol. Palaeontol. 33, 21–70.

Johnston, D.I., Barnes, C.R., 2000. Early and Middle Ordovician (Arenig) conodonts from St. Pauls inlet and Martin Point, Cow Head Group, western Newfoundland, Canada. 2: Systematic and paleoecology. Geol. Palaeontol. 34, 11–87.

Kaufmann, C., Ortega, G., 2015. La Zona de *Holmograptus spinosus* (Darriwiliano medio) en la sierra de Villicum, Precordillera Oriental, provincia de San Juan. Acta Geol. Lilloana 28 (Suppl.), 101–105.

Kayser, E., 1897. Beitrage zur Kenntnis einiger palaeozischer Faunen Sud-Amerikas. Zeitschr. Dtsch. Geol. Gesselch., Band. XLIX, 274–317.

Keller, M., Cañas, F., Lehnert, O., Vaccari, N.E., 1994. The upper Cambrian and lower Ordovician of the Precordillera (western Argentina): some stratigraphic reconsiderations. Newsl. Stratigr. 31 (2), 115–132.

Keller, M., Lehnert, O., 1998. The Rio Sassito sedimentary succession (Ordovician): a pinpoint in the geodynamic evolution of the Argentine Precordillera. Geol. Rundsch 87, 326–344.

Lehnert, O., 1993. Bioestratigrafía de los conodontes arenigianos de la Formación San Juan en la localidad de Niquivil (Precordillera sanjuanina, Argentina) y su correlación intercontinental. Rev. Esp. Paleontol. 8 (2), 153–164.

Lehnert, O., 1995a. Ordovizische Conodonten aus der Präkordillere Westargentiniens: Ihre Bedeutung für Stratigraphie und Paläogeographie. Erlanger Geol. Abh. 125, 1–193.

Lehnert, O., 1995b. The Tremadoc/Arenig transition in the Argentine Precordillera. In: Cooper, J.D., Droser, M.L., Finney, S.C. (Eds.), Ordovician Odyssey: Short Papers for the Seventh International Symposium on the Ordovician System, Las Vegas, SEPM, Fullerton, pp. 145–148.

Lehnert, O., Bergström, S.M., Vaccari, N.E., 1997. Arenig conodonts from the Famatina Range, northwestern Argentina: faunal affinities and paleogeographic implications. In: 18° IAS Regional European Meeting of Sedimentology, Abst., Gaea Heidelbergensis, vol. 3, p. 216.

Lehnert, O., Keller, M., Bordonaro, O., 1998. Early Ordovician conodonts from the southern Cuyania terrane (Mendoza Province, Argentina). Palaeontol. Pol. 58, 47–65.

Lehnert, O., Bergström, S.M., Keller, M., Bordonaro, O., 1999. Ordovician (Darriwilian) conodonts from the San Rafael region, west-central Argentina: biostratigraphic, paleoecologic, and paleogeographic implications. Boll. della Soc. Paleontol. Ital. 37 (2–3), 199–214.

Lemos, V.B., 1981. Conodontes ordovicianos da Formação San Juan, Precordillera. Departamento Jáchal, Provincia de San Juan, República Argentina (Bioestratigrafía). In: Anais II Congresso Latinoamericano de Paleontologia, Porto Alegre, vol. 1, pp. 31–43.

Lindström, M., 1971. Lower Ordovician conodonts of Europe. In: Sweet, W.C., Bergström, S.M. (Eds.), Symposium on Conodont Biostratigraphy, Geological Society of America Memoir, vol. 127, pp. 21–61.

Löfgren, A., 1978. Arenigian and Llanvirnian conodonts from Jämtland, northern Sweden. Fossils Strata 13, 1–129.

Löfgren, A., 1997. Conodont faunas from the upper Tremadoc at Brattefors, south-central Sweden, and reconstruction of the *Paltodus* apparatus. GFF 119, 257–266.

Löfgren, A., 2003. Conodont faunas with *Lenodus variabilis* in the upper Arenigian to lower Llanvirnian of Sweden. Acta Palaeontol. Pol. 48 (3), 417–436.

Löfgren, A., Zhang, J., 2003. Element association and morphology in some middle Ordovician platformequipped conodonts. J. Paleontol. 77, 723–739.

Loss, R., 1949. Sobre la presencia de Phyllograpti y la estratigrafía del Ordovícico Inferior entre La Quiaca y Yavi (Prov. de Jujuy), vol. II (4). Instituto de Geología y Minería, Universidad Nacional de Tucumán, pp. 39–62.

Maletz, J., 2009. *Holmograptus spinosus* and the middle Ordovician (Darriwilian) graptolite biostratigraphy at Les Mèchins (Quebec, Canada). Can. J. Earth Sci. 46, 739–755.

Maletz, J., Egenhoff, S.O., 2001. Late Tremadoc to early Arenig graptolite faunas of southern Bolivia and their implications for a worldwide biozonation. Lethaia 34, 47–62.

Maletz, J., Egenhoff, S.O., Alonso, R., 2010. The upper Tremadocian (Ordovician) graptolite *Bryograptus* in Argentina and Norway: biostratigraphy and biogeography. Palaeontology 53, 59–75.

Manca, N., Heredia, S., Hünicken, M.A., Rubinstein, C., 1995. Macrofauna, Conodontes y Acritarcos de la Formación Santa Rosita (Tremadociano), Nazareno, Provincia de Salta, Argentina. Bol. Acad. Nac. Cienc. Córdoba 60 (3–4), 267–276.

Mango, M.J., Albanesi, G.L., 2014. Bioestratigrafía de conodontes de la Formación San Juan (Ordovícico Inferior-Medio) en el Cerro Viejo de Huaco, Precordillera, Argentina. In: Actas 19 Congreso Geológico Argentino, Córdoba. CD-ROM.

Mango, M.J., Varini, M.V., Albanesi, G.L., 2016. Biozonas de conodontes de la Formación San Juan (Ordovícico) y cronoestratigrafía de los cadilitos carbonáticos de la Formación Guandacol (Carbonífero) en el anticlinal de Huaco, Precordillera de San Juan. Acta Geol. Lilloana 28 (Supl.), 122–125.

Martín, J.L., Malanca, S., Sureda, R., 1987. La fauna graptolítica de la sierra de Aguilar, Jujuy, Argentina. Algunos comentarios sobre las formaciones ordovícicas. In: 4° Congreso Latinoamericano de Paleontología, Santa Cruz de la Sierra, vol. II, pp. 599–619.

Martínez, M., Brussa, E.D., Pérez, B., Coira, B., 1999. El Ordovícico de la sierra de Quichagua (Puna nororiental Argentina): litofacies volcanosedimentarias y graptofaunas. In: XIV Congreso Geológico Argentino, Salta, vol. I, pp. 347–350.

Máspero Castro, B., Ortega, G., Albanesi, G.L., 2003. Middle Ordovician graptolite faunas of the Gualcamayo Formation (middle member) in the Corridita creek section, northern Precordillera, Argentina. Ser. Correl. Geol. 18, 61–66.

Mestre, A., 2008. Revisión bioestratigráfica del techo de la Formación San Juan (Ordovícico Inferior), en la sección del Monumento a Buenaventura Luna, Precordillera Central, San Juan. Acta Geol. Lilloana 20 (2), 127–136.

Mestre, A., 2012. Bioestratigrafía de conodontes del techo de la Formación San Juan y el miembro inferior de la Formación Los Azules, Cerro La Chilca, Precordillera Central. Ameghiniana 49 (2), 185–197.

Mestre, A., 2013. Middle Darriwilian conodont biostratigraphy of the Villicum range, eastern Precordillera, Argentina. Asoc. Paleontol. Argent. Publ. Especial 13, 69–72.

Mestre, A., 2014. Bioestratigrafía de conodontos del Darriwiliense medio (Ordovícico) en el borde oriental de la Sierra de Villicum (Precordillera Oriental, Argentina). Bol. Geol. Min. 125 (1), 65–76.

Mestre, A., Heredia, S., 2012. The Darriwilian species of the genus *Histiodella* (Conodonta) in the Argentina Precordillera. Alcheringa 36, 141–150.

Mestre, A., Heredia, S.E., 2013. La zona de Yangtzeplacognathus crassus (conodonta), Darriwiliano de la Precordillera Central, San Juan, Argentina. Ameghiniana 50 (4), 407–417.

Mitchell, C.E., Brussa, E.D., Toro, B.A., Astini, R.A., 1998. Late Ordovician graptolites from the Empozada Formation, Argentine Precordillera, an outer shelf, cool water, peri-Gondwanan assemblage? Temas Geológico-Mineros ITGE 23, 224–226.

Monteros, J.A., 2005. Las graptofaunas del Ordovícico Inferior en la sierra de Mojotoro, Cordillera Oriental argentina. Tesis Doctoral, Universidad Nacional de Salta, 198 pp. (unpublished).

Monteros, J.A., Moya, M.C., 2002. La Zona de *Aorograptus victoriae* (Ordovícico Inferior) en la sierra de Mojotoro, Cordillera Oriental Argentina. In: 8° Congreso Argentino de Paleontología y Bioestratigrafía, Corrientes, p. 88.

Monteros, J.A., Moya, M.C., 2003. Late Tremadocian graptolites from Mojotoro range, Argentine eastern Cordillera. Ser. Correl. Geol. 18, 73–78.

Monteros, J.A., Moya, M.C., 2005. La Zona de *Adelograptus* en el Tremadociano de la Cordillera Oriental Argentina. In: XVI Congreso Geológico Argentino, La Plata, Actas, vol. 3, pp. 399–406.

Monteros, J.A., Moya, M.C., Monaldi, C.R., 1996. Graptofaunas arenigianas en el borde occidental de la Puna Argentina. Implicancias paleogeográficas. In: Memorias Del XII Congreso Geologico de Bolivia, pp. 733–746.

Moya, M.C., 2015. La "Fase Oclóyica" (Ordovícico Superior) en el noroeste argentino. Interpretación histórica y evidencias en contrario. Ser. Correl. Geol. 31 (1), 73–110.

Moya, M.C., Monteros, J.A., 2011. The Tremadocian deposits of the Argentinian eastern Cordillera: a Scandinavian signal in the central Andes. Cuad. del Mus. Geomin. 14, 379–390.

Moya, M.C., Malanca, S., Monteros, J.A., Cuerda, A.J., 1994. Bioestratigrafía del Ordovícico Inferior en la Cordillera Oriental argentina, basada en graptolitos. Rev. Esp. Paleontol. 9, 91–104.

Moya, M.C., Monteros, J.A., Monaldi, C.R., 1998. Graptolite dating of a lower Ordovician unconformity in the Argentinian Andes. Temas Geológico-Mineros ITGE 23, 227–230.

Moya, M.C., Ortega, G., Malanca, S., Monteros, J.L., Albanesi, G.L., Buatois, L.A., Zeballo, F.J., 2003a. Ordovician and Silurian of the Cordillera Oriental and Sierras Subandinas, NW Argentina. In: Field Trip Guide of the 9th International Symposium on the Ordovician System, 7th International Graptolite Conference and 2003 Field Meeting of the Subcommission on Silurian Stratigraphy, Miscelánea 11, INSUGEO, Tucumán, pp. 1–92.

Moya, M.C., Malanca, S., Monteros, J.A., Albanesi, G.L., Ortega, G., Buatois, L.A., 2003b. Late Cambrian-Tremadocian faunas and events from Angosto del Moreno section, Eastern Cordillera, Argentina. In: Albanesi, G.L., Beresi, M.S., Peralta, S.H. (Eds.), Ordovician from the Andes, Ser. Correl. Geol., 17, pp. 439–444.

Ortega, G., 1987. Las graptofaunas y los conodontes de la Formación Los Azules, cerro Viejo, zona de Huaco, Departamento Jáchal, San Juan. Tesis Doctoral, Universidad Nacional de Salta, 210 pp. (unpublished).

Ortega, G., 1995. Graptolite zones of the Los Azules formation (Middle Ordovician) from Precordillera, western Argentina. In: 5° International Graptolite Conference, Graptolite Working Group, International Palaeontological Association. Graptolite News, Long Beach, California, pp. 57–59.
Ortega, G., Albanesi, G.L., 1998. The record of the *Nemagraptus gracilis* Zone in the Argentine Precordillera. Temas Geológico-Mineros ITGE 23, 231–235.
Ortega, G., Albanesi, G.L., 1999. Graptolite biostratigraphy of the Gualcamayo Formation (Middle Ordovician) at the Los Sapitos creek section, Argentine Precordillera. Acta Univ. Caroliane Geol. 43, 49–52.
Ortega, G., Albanesi, G.L., 2000. Graptolitos de la Formación Gualcamayo (Ordovícico Medio) en el Cerro Potrerillo, Precordillera central de San Juan, Argentina. Bol. Acad. Nac. Cienc. Córdoba 64, 27–59.
Ortega, G., Albanesi, G.L., 2002. Bioestratigrafía de graptolitos y conodontes del Tremadociano tardío de la Cordillera Oriental Argentina. In: XV Congreso Geológico Argentino, El Calafate, vol. 1, pp. 542–547.
Ortega, G.C., Albanesi, G.L., 2003. Late Tremadocian graptolite sequence and conodonts from the Parcha area,Eastern Cordillera, Argentina. Ser. Correl. Geol. 18, 79–85.
Ortega, G.C., Albanesi, G.L., 2005. Tremadocian graptolite- conodont biostratigraphy of the south American Gondwana margin (Eastern Cordillera, NW Argentina). Geol. Acta 3, 355–371.
Ortega, G., Albanesi, G.L., Rao, R.I., 1998. Lower Ordovician graptolites and conodonts from Cajas range and Parcha area, Eastern Cordillera, northern Argentina. Temas Geológico-Mineros ITGE 23, 236–240.
Ortega, G., Brussa, E.D., 1990. La Subzona de *Climacograptus bicornis* (Caradociano temprano) en la Formación Las Plantas en su localidad tipo, Precordillera de San Juan, Argentina. Ameghiniana 27, 281–288.
Ortega, G., Brussa, E.D., Astini, R.A., 1991. Nuevos hallazgos de graptolitos en la Formación Yerba Loca y su implicancia estratigráfica, Precordillera de San Juan, Argentina. Ameghiniana 28 (1–2), 163–178.
Ortega, G., Cañas, F., Hunicken, M.A., 1985. The occurrence of *Isograptus victoriae* Harris in the Gualcamayo formation, Guandacol, La Rioja province, Argentina. Bol. Acad. Nac. Cienc. Córdoba 56, 127–133.
Ortega, G., Rao, R.I., 1995. Nuevos hallazgos de graptolitos (Ordovícico Inferior) en la Sierra de Cajas, departamento Humahuaca, provincia de Jujuy, Argentina. Bol. Acad. Nac. Cienc. Córdoba 60 (3–4), 293–316.
Ortega, G., Rickards, R.B., 2003. A Darriwilian (Middle Ordovician) graptolite fauna of the lower member of the los Azules Formation, cerro Viejo, san Juan Precordillera; Argentina. In: Ortega, G., Aceñolaza, F.G. (Eds.), Proceedings of the 7° International Graptolite Conference and Field Meeting of the International Subcommission on Silurian Stratigraphy, San Juan, Argentina, Ser. Correl. Geol., 18, pp. 87–92.
Ortega, G., Toro, B.A., Brussa, E.D., 1993. Las zonas de graptolitos de la Formación Gualcamayo (Arenigiano tardío – Llanvirniano temprano) en el norte de la Precordillera (provincias de La Rioja y San Juan). Rev. Esp. Paleontol. 8, 207–219.
Ortega, G., Albanesi, G.L., Collo, G., Astini, R.A., 2005. La Formación Volcancito en Las Angosturas (Ordovícico Inferior), Sistema de Famatina, Argentina. In: XVI Congreso Geológico Argentino, La Plata, Actas, vol. I, pp. 335–342.
Ortega, G., Albanesi, G.L., Frigerio, S.E., 2007a. Early Darriwilian graptolite and conodont biofacies in the los Azules Formation, cerro Viejo section, central Precordillera, Argentina. In: Munnecke, A., Servais, T. (Eds.), Early Palaeozoic Palaeogeography and Palaeoclimate, Palaeogeography, Palaeoclimatology Palaeoecology, 245, pp. 245–263 (1-2).

Ortega, G., Albanesi, G.L., Heredia, S.E., Beresi, M.S., 2007b. Nuevos registros de graptolitos y conodontes (Ordovícico) en las formaciones Estancia San Isidro y Empozada, quebrada San Isidro, Precordillera de Mendoza. Ameghiniana 44 (4), 697–718.

Ortega, G., Albanesi, G.L., Banchig, A.L., Peralta, G.L., 2007c. Early Katian graptolites and conodonts (Upper Ordovician) of the Sierra de La Invernada Formation, San Juan Precordillera, Argentina. In: Jun, L:, Junxuan, F., Percival, I. (Eds.), The Global Ordovician and Silurian, Acta Palaeontol. Sin., 46, pp. 357–363.

Ortega, G., Albanesi, G.L., Banchig, A.L., Peralta, G.L., 2008. High resolution conodont-graptolite biostratigraphy in the Middle-Upper Ordovician of the Sierra de La Invernada Formation (Central Precordillera, Argentina). Geol. Acta 6 (2), 161–180.

Ortega, G., Banchig, A.L., Voldman, G.G., Albanesi, G.L., Alonso, J.L., Festa, A., Cardó, R., 2014. Nuevos registros de graptolitos y conodontes en la Formación Los Sombreros (Ordovícico), Precordillera de San Juan, Argentina. In: 19° Congreso Geológico Argentino, Córdoba vol. S2, p. 16.

Ortega, G., Voldman, G.G., Banchig, A.L., Albanesi, G.L., Cardó, R., Alonso, J.L., Fernández, L.P., 2016. Primeros registros de la Zona de *Holmograptus spinosus* (Ordovícico Medio) en la sierra de Rinconada, Precordillera Oriental, provincia de San Juan, Argentina. In: XI Congreso de la Asociación Paleontológica Argentina, Río Negro (in press).

Ottone, E.G., Albanesi, G.L., Ortega, G., Holfeltz, G., 1999. Palynomorphs, conodonts and associated graptolites from the Ordovician los Azules Formation, central Precordillera, Argentina. Micropaleontology 45 (3), 225–250.

Peralta, S.H., 1990. Nuevas graptofaunas de la Formación La Cantera, Sierra de Villicum, Precordillera Oriental de San Juan, Argentina. In: V Congreso Argentino de Paleontología y Bioestratigrafía, Tucumán, vol. I, pp. 59–66.

Peralta, S.H., 1998. Graptolites of the *Nemagraptus gracilis* Zone in the black shale sequences of the San Juan Precordillera, Argentina: its biostratigraphic and paleoenvironmental significance. Temas Geológico-Mineros ITGE 23, 244–247.

Peralta, S.H., Baldis, B.A., 1990. *Glyptograptus persculptus* en la Formación Don Braulio (Ashgilliano tardío-Llandoveriano temprano) en la Precordillera Oriental de San Juan, Argentina. In: V Congreso Argentino de Paleontología y Bioestratigrafía, San Miguel de Tucumán, Actas, I, pp. 67–72.

Peralta, S.H., Albanesi, G.L., Ortega, G., 2003. Ordovician and Silurian of the Precordillera, San Juan Province, Argentina. In: Field Trip Guide of the 9th International Symposium on the Ordovician System, 7° International Graptolite Conference and 2003 Field Meeting of the Subcommission on Silurian Stratigraphy, Miscelánea, 10, INSUGEO, Tucumán, pp. 1–149.

Peralta, S.H., Finney, S.C., 2000. The Upper Ordovician graptolite faunas of the Cuyania Terrane: their biostratigraphic and paleogeographic significance in the western margin of Gondwana. In: Aceñolaza, F.G. (Ed.), Aspects of the Ordovician System in Argentina, Serie Correlación Geológica, vol. 16. INSUGEO, pp. 41–70.

Pohler, S.M.L., Barnes, C.R., 1990. Conceptual models in conodont paleoecology. Cour. Forschungsinstitut Senckenberg 118, 409–440.

Pyle, L.J., Barnes, C.R., 2002. Taxonomy, evolution and biostratigraphy of conodonts from the Kechika Formation, Skoki Formation and Road River Group (Upper Cambrian to Lower Silurian), northeastern British Columbia. Natl. Res. Counc. Can. Monogr. Ser. 44461, 227 pp.

Rao, R.I., 1999. Los conodonte Cambro-Ordovícicos de la Sierra de Cajas y del Espinazo del Diablo, Cordillera Oriental, República Argentina. Rev. Esp. Micropaleontol. 31 (1), 23–51.

Rao, R.I., Flores, F.J., 1998. Conodontes ordovícicos (Tremadoc superior) de la sierra de Aguilar, provincia de Jujuy, República Argentina. Bioestratigrafía y tafonomía. Rev. Esp. Micropaleontol. 30 (1), 5–20.

Rao, R.I., Hünicken, M.A., 1995. Conodontes del Cámbrico Superior-Ordovícico Inferior en el área de Purmamarca, Cordillera Oriental, Provincia de Jujuy, R. Argentina. Bol. Acad. Nac. Cienc. Córdoba 60 (3–4), 249–266.

Rao, R.I., Hünicken, M.A., Ortega, G., 1994. Conodontes y graptolitos del Ordovícico Inferior (Tremadociano-Arenigiano) en el área de Purmamarca, Provincia de Jujuy, Argentina. An. Acad. Bras. Ciênc. 66 (1), 59–83.

Rickards, R.B., Brussa, E.D., Toro, B.A., Ortega, G., 1996. Ordovician and Silurian graptolite assemblages from Cerro del Fuerte, San Juan Province, Argentina. Geol. J. 31, 101–122.

Ross Jr., R.J., Hintze, L.F., Ethington, R.L., Miller, J.F., Taylor, M.E., Repetski, J.E., 1997. The Ibexian, lowermost series in the North American Ordovician. U.S. Geol. Surv. Prof. Pap. 1579, 50 pp.

Sadler, P.M., Cooper, R.A., Melchin, M., 2009. High-resolution, early Paleozoic (Ordovician-Silurian) time scales. GSA Bull. 121 (5–6), 887–906.

Sánchez, T.M., Benedetto, J.L., Brussa, E.D., 1991. Late Ordovician stratigraphy, palaeoecology and sea level changes in the Argentine Precordillera. In: Barnes, C.R., Williams, S.H. (Eds.), Advances in Ordovician Geology, Geological Survey of Canada, 90–99, pp. 245–258.

Sarmiento, G.N., 1985. La Biozona de *Amorphognathus variabilis – Eoplacognathus pseudoplanus* (conodonta), Llanvirniano inferior, en el flanco oriental de la sierra de Villicum. Prim. Jorn. sobre Geol. la Precordillera, San Juan, Act. I, 119–123.

Sarmiento, G.N., 1990. Conodontos ordovícicos de Argentina. Treballs Mus. Geol. Barc. 1, 135–161.

Sarmiento, G.N., 1991. Conodontos de la Zona *suecicus* (Llanvirniano inferior) en la sierra de Villicum, Precordillera de San Juan, Argentina. Rev. Esp. Micropaleontol. 23 (3), 113–132.

Sarmiento, G.N., Vaccari, N.E., Peralta, S.H., 1998. Conodontes ordovícicos de La Rinconada, Precordillera de San Juan, Argentina. In: IV Congreso Argentino de Paleontología y Bioestratigrafia, Mendoza, Actas, vol. 3, pp. 219–224.

Serpagli, E., 1974. Lower Ordovician conodonts from Precordilleran Argentina (Province of San Juan). Boll. Soc. Paleontol. Ital. 13 (1–2), 17–98.

Serra, F., Albanesi, G.L., Bergström, S.M., 2013. Middle Darriwilian conodont biostratigraphy of the Las Chacritas Formation, central Precodrillera of San Juan, Argentina. In: Albanesi, G.L., Ortega, G. (Eds.), Conodonts from the Andes, 3° International Conodont Symposium, Asoc. Paleontol. Argent., Special Publ., 13, pp. 109–115.

Serra, F., Albanesi, G.L., Ortega, G., Bergström, S.M., 2015. Conodont-graptolite biostratigraphy and global correlation for the Mid-Late Ordovician in the Las Chacritas river section, Precordillera of San Juan, Argentina. Geol. Mag. 152 (5), 813–829.

Skevington, D., 1974. Controls influencing the composition and distribution of Ordovician graptolite faunal provinces. Special Pap. Palaeontol. 13, 59–73.

Stouge, S., 1984. Conodonts of the middle Ordovician table head formation, western Newfoundland. Foss. Strat. 16, 1–145.

Stouge, S., 2012. Middle Ordovician (late Dapingian–Darriwilian) conodonts from the Cow Head Group and Lower Head Formation, western Newfoundland, Canada. Can. J. Earth Sci. 49 (1), 59–90.

Sweet, W.C., 1984. Graphic correlation of upper middle and upper Ordovician rocks, North American Midcontinent Province, U.S.A. In: Bruton, D.L. (Ed.), Aspects of the Ordovician System, vol. 295. Palaeontological Contributions from the University of Oslo, Universitetsforlaget, Oslo, pp. 23–35.

Sweet, W.C., Bergström, S.M., 1974. Provincialism exhibited by Ordovician conodont faunas. Soc. Econ. Paleontol. Mineral. Special Publ. 21, 189–202.

Tickyj, H., Rodríguez Raising, M., Cingolani, C.A., Alfaro, M., Uriz, N., 2009. Graptolitos ordovícicos en el sur de la Cordillera Frontal de Mendoza. Rev. Asoc. Geol. Argent. 64 (2), 295–302.

Toro, B.A., 1994. Las zonas de *Didymograptus (Didymograptellus) bifidus* (Arenigiano Medio) y *Didymograptus (Corymbograptus) deflexus* (Arenigiano inferior) en la Formación Acoite, Cordillera Oriental, Argentina. Ameghiniana 31, 209–220.

Toro, B.A., 1997a. La fauna de graptolitos de la Formación Acoite, en el borde occidental de la Cordillera Oriental argentina. Análisis bioestratigráfico. Ameghiniana 34, 393–412.

Toro, B.A., 1997b. Asociaciones de graptolitos del Arenig de la localidad tipo de la Formación La Alumbrera, Sistema de Famatina, Argentina. Rev. Esp. Paleontol. 12, 43–51.

Toro, B.A., 1999. Early Ordovician (Arenig) graptolites of northwestern Argentina (Cordillera Oriental and Famatina): paleogeographic remarks. Acta Univ. Caroliane Geol. 43, 437–440.

Toro, B.A., Brussa, E.D., 1997. Graptolitos de la Formación Suri (Arenig) en el Sistema de Famatina, Argentina. Rev. Esp. Paleontol. 12, 175–184.

Toro, B.A., Brussa, E.D., 2000. Graptofaunas del Ordovícico Inferior (Arenigiano) del Cerro Tafna, Puna Oriental. Ameghiniana 37 (Suppl. 4), 79R–80R.

Toro, B.A., Brussa, E.D., 2003. Graptolites. In: Benedetto, J.L. (Ed.), Ordovician Fossils of Argentina. Universidad Nacional de Córdoba, Córdoba, pp. 441–505.

Toro, B.A., Carrera, M.G., 2013. Nuevos registros de graptolitos y poríferos del Tremadociano superior en la Cordillera Oriental de Argentina. Ameghiniana Supl. Resúmenes 50, R73.

Toro, B.A., Maletz, J., 2007. Deflexed *Baltograptus* species in the early to mid Arenig graptolite biostratigraphy on northwestern Argentina. Acta Palaeontol. Sin. 46, 489–496.

Toro, B.A., Maletz, J., 2008. The proximal development in *Cymatograptus* (Graptoloidea) from Argentina and its relevance for the early evolution of Dichograptacea. J. Paleontol. 82, 974–983.

Toro, B.A., Vento, B.A., 2013. Reevaluación de las biozonas de *Tetragraptus phyllograptoides* y *T. akzharensis* (Ordovícico Inferior, Floiano) de la Cordillera Oriental Argentina. Ameghiniana 50, 287–297.

Toro, B.A., Brussa, E.D., Rodríguez, G., 2003. Middle Ordovician graptolites from the Sierra de Rinconada, northern Puna, Argentina. In: Ortega, G., Aceñolaza, F.G. (Eds.), Proceedings of the 7° International Graptolite Conference and Field Meeting of the International Subcommission on Silurian Stratigraphy, San Juan, Argentina, Ser. Correl. Geol., 18, pp. 99–103.

Toro, B.A., de la Puente, G.S., Rubinstein, C.V., 2010. New graptolite, chitinozoan and acritarch records from the Pascha–Incamayo area, Cordillera Oriental, Argentina. Comptes Rendus Palevol 9, 23–30.

Toro, B.A., Meroi Arcerito, F.R., Muñoz, D.F., Waisfeld, B.G., de La Puente, G.S., 2015. Graptolite-trilobite biostratigraphy in the Santa Victoria area, northwestern Argentina. A key for Regional and worldwide correlation of the lower Ordovician (Tremadocian–Floian). Ameghiniana 52 (4), 535–557.

Turner, J.C.M., 1960. Faunas graptolíticas de América del Sur. Rev. Asoc. Geol. Argent. 14, 1–180.

Vento, B.A., Toro, B.A., Maletz, J., 2012. New insights into the paleobiogeography of the Early Ordovician graptolite fauna of northwestern Argentina. Comptes Rendus Palevol 11, 345–355.

Vento, B.A., Toro, B.A., Maletz, J., 2014. Paleoecological and paleobiogeographic considerations of Ordovician graptolites from the Cordillera Oriental, Jujuy province, Argentina. Hist. Biol. 26 (6), 765–774.

Voldman, G.G., Albanesi, G.L., Do Campo, M., 2008. Conodont palaeothermometry of contact metamorphism–metasomatism in middle Ordovician rocks from the Precordillera of western Argentina. Geol. Mag. 145 (3), 449–462. Figs. 1–7.

Voldman, G.G., Albanesi, G.L., Ramos, V.A., 2009. Ordovician metamorphism in the carbonate platform of the Argentine Precordillera: implications for the geotectonic evolution of the proto-Andean margin of Gondwana. Geology 37 (4), 311–314.

Voldman, G.G., Genge, M.J., Albanesi, G.L., Barnes, C.R., Ortega, G., 2012-13. Cosmic spherules from the Ordovician of Argentina. Geological Journal 48, 222–235. http://dx.doi.org/10.1002/gj.2418.

Voldman, G.G., Albanesi, G.L., Zeballo, F.J., Monaldi, C.R., 2013. Early Ordovician (late Floian) conodonts from the Zenta range, Cordillera Oriental, NW Argentina. In: Albanesi, G.L., Ortega, G. (Eds.), Conodonts from the Andes, Proceedings of the 3° International Conodont Symposium. Paleontological Note PE APA, Mendoza, Argentina, pp. 123–128.

Voldman, G.G., Ortega, G., Albanesi, G.L., Monaldi, C.R., Giuliano, M.E., 2014. Observaciones sobre la edad del grupo Santa Victoria (Ordovícico) en su área tipo, sierra de Santa Victoria, provincia de Salta. Actas. In: XIX Congreso Geológico Argentino, Córdoba. CD-ROM.

Voldman, G.G., Albanesi, G.L., Ortega, G., Giuliano, M.E., Monaldi, C.R., 2016a. New conodont taxa and biozones from the lower Ordovician of the Cordillera Oriental, NW Argentina. Geol. J. http://dx.doi.org/10.1002/gj.2766. On-line version.

Voldman, G.G., Alonso, J.L., Fernández, L.P., Banchig, L.A., Albanesi, G.L., Ortega, G., Cardó, R., Pini, G.A., 2016b. Cambrian-Ordovician conodont records from slump deposits of the Argentine Precordillera: new insights into its passive margin development. Geol. Mag. http://dx.doi.org/10.1017/S0016756816000832.

Webby, D.B., Cooper, R.A., Bergström, S.M., Paris, F., 2004. Stratigraphic framework and time slices. In: Webby, B., et al. (Eds.), The Great Ordovician Biodiversification Event. Columbia University Press, pp. 41–47.

Wang, X., Stouge, S., Erdtmann, B.-D., Chen, X., Li, Z., Wang, C., Zeng, Q., Zhou, Z., Chen, H., 2005. A proposed GSSP for the base of the middle Ordovician series: the Huanghuachang section, Yichang, China. Episodes 28, 105–117.

Youngquist, W., Iglesias, S.J., 1951. Ordovician conodonts from south America. J. Paleontol. 25, 408.

Zeballo, F.J., Albanesi, G.L., 2013a. Conodont biofacies and palaeoenvironments of Cambrian-Ordovician sequences from the Quebrada de Humahuaca, Cordillera Oriental of Jujuy, Argentina. Geol. J. 48, 194–201.

Zeballo, F.J., Albanesi, G.L., 2013b. New conodont species and biostratigraphy of the Santa Rosita Formation (late Furongian – Tremadocian) in the Tilcara range, Cordillera Orienta of Jujuy, Argentina. Geol. J. 48, 170–193.

Zeballo, F.J., Albanesi, G.L., Ortega, G., 2005. Conodontes y graptolitos de las formaciones Alfarcito y Rupasca (Tremadociano) en el área de Alfarcito, Tilcara, Cordillera Oriental de Jujuy, Argentina, Parte 2: Paleontología sistemática. Ameghiniana 42, 47–66.

Zeballo, F.J., Albanesi, G.L., Ortega, G., 2008. New records of late Tremadocian (early Ordovician) conodonts and graptolites from the eastern Cordillera, Jujuy province, Argentina. Geol. Acta 6 (2)), 131–145.

Zeballo, F.J., Albanesi, G.L., Voldman, G.G., Monaldi, C.R., 2013. New records of Tremadocian conodonts (lower Ordovician) from the Zenta Range, Jujuy Province, Argentina. In: Albanesi, G.L., Ortega, G. (Eds.), Conodonts from the Andes, Proceedings of the 3° International Conodont Symposium. Paleontological Note PE APA, Mendoza, Argentina, pp. 129–133.

Zhen, Y.-Y., Percival, I.G., 2003. Ordovician conodont biogeography—reconsidered. Lethaia 36, 357–370.

Zimmermann, U., Moya, M.C., Bahlburg, H., 1999. First finds of Lower Ordovician graptolites propose a new stratigraphic subdivision for the southern Puna (NW Argentina). In: XIV Congreso Geológico Argentino, Salta, vol. I, pp. 343–346.

CHAPTER THREE

Chemostratigraphy and Chemofacies of Source Rock Analogues: A High-Resolution Analysis of Black Shale Successions from the Lower Silurian Formigoso Formation (Cantabrian Mountains, NW Spain)

T. Ferriday[*,1] **and M. Montenari**[§]
[*]ERCL Ltd — Exploration Reservoir Consultants, Oxon, United Kingdom
[§]Keele University, Newcastle, United Kingdom
[1]Corresponding author: E-mail: tim.ferriday@ercl.com

Contents

Stratigraphy & Timescales, Volume 1
ISSN 2468-5178
http://dx.doi.org/10.1016/bs.sats.2016.10.004

Abstract

During the early Silurian, widespread anoxic graptolitic black shales developed predominantly on the northern margin of Gondwana. These black shales are attracting a considerable amount of scientific interest, as it is nowadays hypothesized that their genesis might be closely related to major changes within the paleogeographical configuration of continents, the oceanic circulation patterns, and also drastic global climatic perturbations. Furthermore, these black shale successions can be of highest economic significance, as some form the source rocks of the world's major oil reserves. For the present study, high-resolution whole rock geochemical data obtained from an early Silurian black shale source rock analogue (Formigoso Formation, Cantabrian Mountains, NW, Spain) are used to develop a chemostratigraphic framework and to reconstruct and further constrain the paleoenvironmental conditions that prevailed at the time of deposition. These chemostratigraphic interpretations and paleoenvironmental reconstructions are used, together with the results of previously published studies, to develop a detailed generic depositional model for the black shale successions of the Cantabrian Basin. This analogue model is compared and contrasted to previously proposed models. The new high-resolution chemostratigraphic data are combined with detailed biostratigraphic information from the published literature. The present study demonstrates that chemostratigraphic methods provide not only complementary but even vital information, especially where biostratigraphic data are absent or scarce, for larger scale intra- and interbasinal correlations.

1. INTRODUCTION

The lower Silurian period is characterized by the presence of widespread, anoxic graptolitic black shales, which were mainly deposited along the northern margin of Gondwana. In the present day, these deposits can be mainly found in northern Africa, Jordan, Saudi Arabia, and northern Spain (Iberia), although they have also been documented elsewhere. These black shales are attracting a considerable amount of scientific interest, as it is nowadays hypothesized that their genesis might be closely related to major changes within the paleogeographical configuration of continents, the oceanic circulation patterns, and also drastic global climatic perturbations. Furthermore, these black shale successions can be of highest economic significance, as some form the source rocks of the world's major oil reserves. For example, it is the lower Silurian black shales that source approximately 80–90% of the hydrocarbons derived from the Paleozoic in North Africa (Lüning et al., 2000, 2005).

Previous work encompasses much debate as to the controls on the deposition of the Silurian mudstone successions within the peri-Gondwanan region. Various models for the formation of these organically enriched deposits have been postulated. Lüning et al. (2000) suggest that these black shales formed during the onset of the lower Silurian eustatic transgression, basically following the previously developed basal transgressive "puddle model" of Wignall and Hallam (1991). Within this model, the patchy, discontinuous lateral distribution of the black shales is mainly controlled by the presence of paleotopographical features. However, in developing an alternative model, Armstrong et al. (2005) argue that the black shales were forming during the maximum extent of the Silurian marine transgression, thereby drawing on the "expanding puddle model" of Wignall and

Hallam (1991), and that the major factor controlling the formation of the lower Silurian black shales was not necessarily the presence or absence of a paleorelief on the continental shelves, but a stratified shelfal water column, which was generated by a prominent and partly persistent freshwater cap.

The main objectives of this study is to utilize high-resolution whole rock geochemical data obtained from the Silurian black shales of the Cantabrian Basin to construct a chemostratigraphic framework and to reconstruct the paleoenvironmental conditions that prevailed at the time of deposition. These paleoenvironmental data and reconstructions are used, together with the results of previously published studies, to develop a new generic depositional model for the black shale successions of the Cantabrian Basin. This model is then compared and contrasted to the previously proposed models. The new high-resolution chemostratigraphic data are combined with detailed biostratigraphic information from the published literature. Graptolites and palynomorphs, such as acritarchs and chitinozoans, have been successfully used in the past as important biostratigraphic markers to precisely constrain the age of the Silurian "hot" shales of northern Gondwana (e.g., Butcher, 2009; Marshall, 2000; Paris et al., 2012; with further references therein). However, their use for biostratigraphic purposes during field-based exploration campaigns has proved to be problematic (e.g., Lüning et al., 2000), as the biostratigraphic potential of these fossils is often hampered by their poor preservation or even absence in surface or near-surface samples. The organic material from such samples is commonly oxidized and hence destroyed. Under such circumstances, the use of chemostratigraphic methods can provide complementary and even vital information, especially where biostratigraphic data are absent or scarce, thus potentially enabling larger scale intra- and interbasinal correlations. For this purpose, the whole rock geochemical signatures of the Cantabrian black shales, with particular emphasis on redox-sensitive but conservative elements, like uranium or vanadium, were analyzed, interpreted, and correlated.

2. GEOLOGICAL SETTING

The Variscan Belt (or Hercynian Orogen) extends for over 3000 km from the Appalachian Mountains via southern Iberia to northeastern Bohemia and has a diameter between 700 and 800 km (e.g., Paris and Robardet, 1990; Weil et al., 2001). The Variscan Belt formed during the Late Paleozoic (Late Devonian to Carboniferous, approximately 370–290 Ma ago) as the result of the convergence and subsequent collision of the two major paleocontinents

Laurasia and Gondwana, including a number of smaller microplates and fragments of magmatic arcs (Abalos et al., 2002; Dias and Ribeiro, 1995; Martínez Catalán et al., 2009a; Weil et al., 2001). In the Iberian Peninsula, Upper Proterozoic to Carboniferous rocks outcrop prominently in its western part and form together the Iberian or Hesperian Massif.

The Iberian Massif provides the largest outcrops of Variscan and pre-Variscan rocks in Europe (Valladares et al., 2002). It formed between the Upper Devonian and the Late Carboniferous (Abalos et al., 2002; Fernández-Suárez et al., 2000) and is traditionally subdivided into five major domains based upon the different sedimentological development of the Lower Paleozoic sedimentary successions and the presence/absence of distinctive igneous suites (Abalos et al., 2002; Julivert et al., 1972): (1) The Cantabrian Zone (CZ), the (2) West Asturian-Leonese (WALZ) Zone, (3) Central Iberian Zone (CIZ), (4) Ossa-Morena Zone (OMZ), and (5) South Portuguese Zone (SPZ) (see Fig. 1). The boundaries between these zones are major Variscan thrust- and reverse faults, which were in some instances reactivated by extensional tectonics in post-Variscan time (Nance et al., 2010).

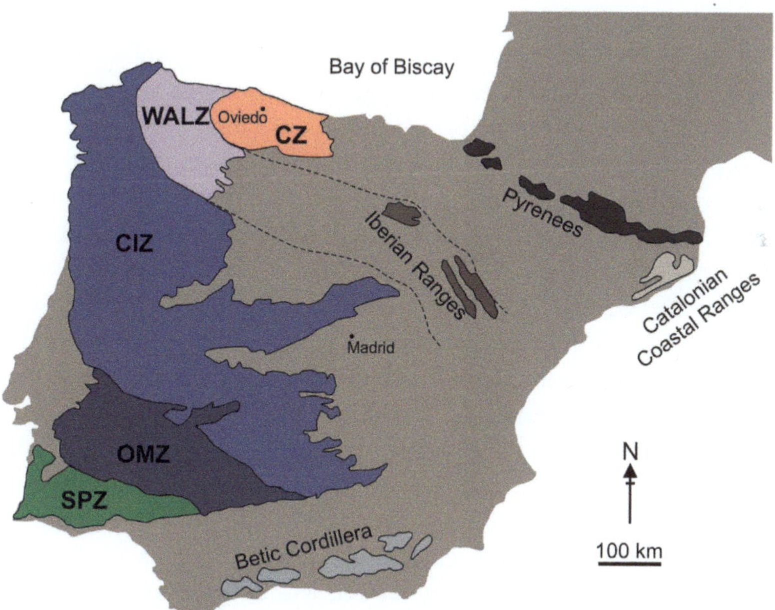

Figure 1 The Iberian Peninsula and the major Variscan units of the Iberian Massif according to Lotze (1945) and Julivert (1971). *CZ*, Cantabrian Zone; *CIZ*, Central Iberian Zone; *OMZ*, Ossa Morena Zone; *SPZ*, Southern Portuguese Zone; *WALZ*, West Asturian-Leonese Zone. *Modified after Garcia-Alcalde, J., Carls, P., Alonso, M.V.P., Lopez, J.S., Soto, F., Truyols-Massoni, M., Valenzuela-Rios, J.I., 2002. Devonian. In: Gibbons, W., Moreno, T. (Eds.), The Geology of Spain. Geological Society, London, pp. 67–91.*

The Cantabrian Mountains are located in northern Spain within the province of Castilla y Leon. These mountains form part of the Cantabrian Zone (CZ) and the Cantabrian Arc (Fig. 1). The CZ represents the external zone of the Iberian Massif and the external zone of the Variscan belt and is located near the core of the so-called Cantabrian Orocline (Aller et al., 2005; Garciá-López and Bastida, 2002; García-López et al., 2007; Paris and Robardet, 1990; Pastor-Galán et al., 2013). Oroclinal bending caused by Late Hercynian to post-Hercynian east—west compression is considered to have formed the arc shape of the CZ (Gutiérrez-Alonso et al., 2012; Keller et al., 2008; Weil et al., 2000, 2001, 2010).

The CZ comprises relatively thin preorogenic Paleozoic sedimentary successions (Martínez Catalán et al., 2009a), and owes its overall structural grain to thin-skinned tectonics, which developed from a combined Variscan and Alpine tectonic overprint. The general structural pattern of the CZ is dominated by a principle decollement which is commonly (but not exclusively) located within the carbonate-successions of the Lower-Middle Cambrian Lancara Formation (Aller et al., 2005; Julivert, 1971). The resultant tectonic shortening is estimated to be between 50% and 75% (Keller et al., 2008). The intricate relationship between the thrusts and the folds grants the Cantabrian Arc its distinctive geological configuration (Abalos et al., 2002; Julivert and Marcos, 1973; Pérez-Estaún et al., 1988).

In the CZ, the Variscan deformation took place without any significant metamorphism. Very low- to low-grade metamorphic overprints only appear in the north and northwestern parts of the Zone, close to the boundary with the WALZ (Aller et al., 2005; Garciá-López and Bastida, 2002; García-López et al., 2007; Pastor-Galán et al., 2013).

During the Cambro-Ordovician the CZ was one of many graben structures within the northern margins of the Gondwanan paleocontinent. It comprises a stratigraphic and sedimentological succession that is similar to other areas surrounding the Mediterranean, like the Anti-Atlas, the southern Montagne Noire and southwest Sardinia (Álvaro et al., 2000a,b; Aramburu-Zabala Higuera and García Ramos, 1993). The Cambrian to Devonian sediments of the CZ are considered to have been deposited in a passive margin setting, which gradually developed into a Carboniferous foreland basin. The sediments preserved within this foreland basin are interpreted to be supplied by the erosion of the internal zones of the Variscan Orogen (Garciá-López and Bastida, 2002).

The sedimentological development, paleontological content, and zircon populations of the Iberian Paleozoic succession show strong similarities to those found on the Saharan Craton and zircon-spectra documented from Libya. Based on these evidences it is commonly suggested that NW Iberia occupied a paleogeographical position close to present-day northern Africa during this time interval (Keller et al., 2008; Martínez Catalán et al., 2009b).

The CZ comprises two major tectonostratigraphic units (García-López et al., 2007; Julivert, 1978; Marcos and Pulgar, 1982): (1) a preorogenic (pre-Variscan) unit, which contains pre-Carboniferous successions (Abalos et al., 2002; Aller et al., 2005). It is located at the western margin of the CZ (Garciá-López and Bastida, 2002) and comprises Cambrian to Devonian mixed siliciclastic-calcareous series. Paleocurrent data indicate a sediment source to the east, yet the precise provenance area remains still controversial (Pastor-Galán et al., 2013). The preorogenic unit overlies unconformably the Proterozoic rocks of the Narcea Antiform (see Fig. 2) (Abalos et al., 2002). (2) A synorogenic unit formed by Carboniferous successions that appear throughout the CZ (García-López et al., 2007; Sarmiento et al., 1994). These successions are interpreted to have been deposited in a foreland basin setting and are commonly subdivided into several clastic series, which are thought to have been generated by the successive filling of various foredeep basins that formed in front of the major thrust units (Aller et al., 2005; Julivert, 1978). The clastic influx was derived from the western and southern internal zones of the Variscan orogen (Aramburu-Zabala Higuera and García Ramos, 1993). The CZ can be further subdivided into a number of smaller thrust units. Julivert (1971) and Pérez-Estaún et al. (1988) separated the CZ into five distinct units: (1) Folds and Nappes (typified by a near complete Paleozoic succession), (2) Central Asturian coal-field (mainly formed by Carboniferous series with an overall thickness of up to 6000 m and Cambro-Ordovician strata bound to the basal thrust system (Garciá-López and Bastida, 2002), (3) Ponga Nappe (characterized by the absence or only very thin developed Silurian and Devonian series and the presence of several large thrust units), (4) Picos de Europa (an imbricate system of thrust sheets made predominantly of Carboniferous successions), and the (5) Pisuerga-Carrion province (a para-autochthonous unit that consists predominantly of Carboniferous strata, although Silurian and Devonian successions are also present to a subordinate extent) (Abalos et al., 2002).

Based on differences in the stratigraphic development, the Folds and Nappes province is further subdivided (Abalos et al., 2002; Julivert, 1971; Pérez-Estaún et al., 1988) into: (1) the Somiedo-Correcilla unit, (2) the

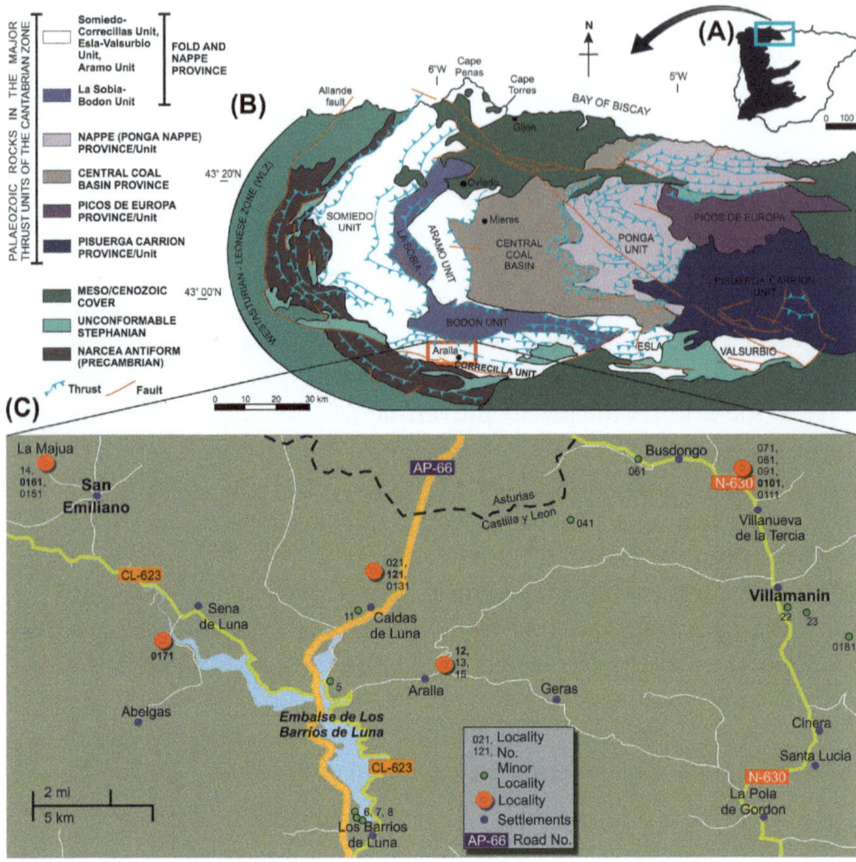

Figure 2 Overview, tectonostratigraphic units and sampled localities; (A) The Iberian Peninsula, the gray shaded area represents the Palaeozoic Massif; (B) tectonostratigraphic units of the Cantabrian Zone. (C) Area investigated for the present study: the locality numbers are listed next to points; the bold numbers represent the number assigned to a major locality. The results for the locality "Aralla" will be detailed within this chapter. *(B) Modified after Aller, J., Valín, M.L., García-López, S., Brime, C., Bastida, F., 2005. Superposition of tectono-thermal episodes in the southern Cantabrian Zone (foreland thrust and fold belt of the Iberian Variscides, NW Spain). Bull. Soc. Geol. Fr. 176 (6), 487–497; Julivert, M., 1971. Decollement tectonics in the Hercynian Cordillera of northwest Spain. Am. J. Sci. 270 (1), 1–29; Pérez-Estaún, A., Bastida, F., Alonso, J.L., Marquínez, J., Aller, J., Alvarez-Marrón, J., Marcos, A., Pulgar, J.A., 1988. A thin-skinned tectonics model for an arcuate fold and thrust belt: the Cantabrian Zone (Variscan Ibero-Armorican Arc). Tectonics. 7 (3), 517–537.*

La Sobia-Bodon unit, (3) the Aramo unit, (4) the Esla nappe unit, and finally (5) the Valsurbio unit. Occasionally, the last two units were unified into the Esla-Valsurbio unit (Abalos et al., 2002; Garciá-López and Bastida, 2002). The Somiedo-Correcillas unit (Fig. 2), located near the center of the

Cantabrian Zone, is the area of interest for the present study as it contains well-preserved and nearly complete black shale successions of the Silurian Formigoso Formation.

3. SILURIAN STRATIGRAPHY OF THE CANTABRIAN MOUNTAINS

The Silurian succession of Iberia consists predominantly of siliciclastic sedimentary rocks with pelagic fauna and is very similar to other northern Gondwanan "hot" shale deposits (Lüning et al., 2000). It is developed in all Iberian regions except the South Portuguese Zone, which possibly belonged to the Avalonian microplate (Gutiérrez-Marco and Rábano, 1998; Robardet and Gutierrez-Marco, 2002) (Fig. 3).

After the Late Ordovician glaciation the Early Silurian sea level fluctuated greatly. These fluctuations are interpreted to be caused by the waxing and

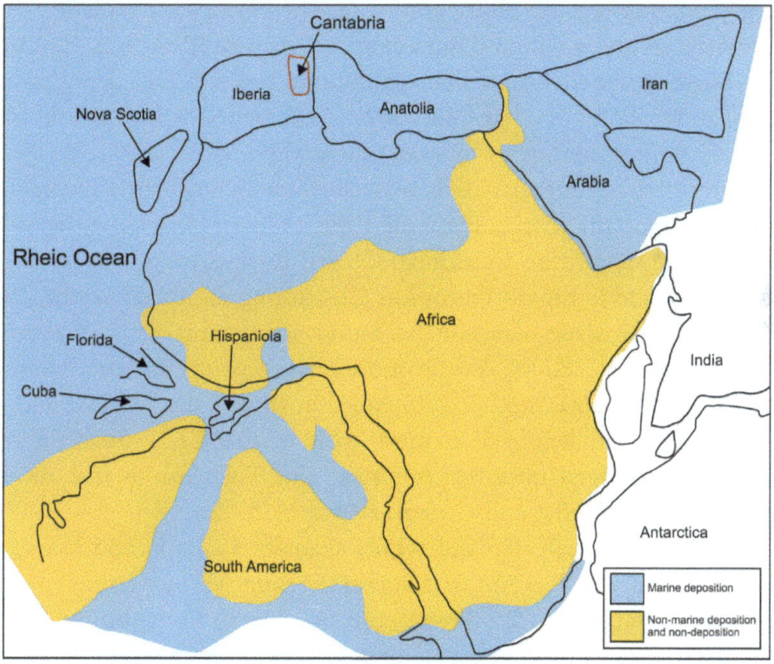

Figure 3 Early Silurian Paleogeography for Gondwana. Large parts of the Gondwanan shelf are flooded by the sea enabling the deposition of marine developed sediments. The position of "Cantabria" is highlighted. *Modified after Lüning, S., Craig, J., Loydell, D.K., Štorch, P., Fitches, B., 2000. Lower Silurian 'hot shales' in North Africa and Arabia: regional distribution and depositional model. Earth-Science Rev. 49 (1–4), 121–200.*

waning of ice sheets in NW Gondwana (Sachanski et al., 2010), which in turn is considered to be related to orbital forcing (Armstrong et al., 2009; Le Heron and Craig, 2008; Le Heron et al., 2005; Le Heron et al., 2009; Sutcliffe et al., 2000). Zhang et al. (2006, with previous references therein), concluded that the Latest Ordovician-Early Silurian sea level was not only affected by the melting of the Hirnantian ice sheets, but also by three short-lived early Silurian glacial episodes during the (1) early Aeronian, (2) latest Aeronian to early Telychian, and (3) latest Telychian to earliest Wenlock. The relative sea-level lowstand during the Hirnantian was followed by a major Silurian eustatic transgression (Rubino, 2003). The onset of this relative sea-level rise is dated to be within the late Crispus to early Griestoniensis Biozone and it reached its maximum during the Telychian Spiralis Biozone (Sachanski et al., 2010). As a consequence, large parts of the continental shelf areas of Gondwana were invaded by marine waters (Paris and Robardet, 1990; Tait et al., 1997). Anoxic conditions prevailed in the north Gondwanan area most probably as a result of water column stratification caused by freshwater capping and subsequent shut down of the oceanic circulation as a consequence of the melting Hirnantian ice masses (Paris and Robardet, 1990). These paleoenvironmental conditions led to the deposition of predominantly fine grained siliciclastic sediments (now black shale successions) during the Lower Silurian within the CZ.

It is important to note that the base of the Silurian black shale succession is diachronous across the Iberian Peninsula. The onset of black shale deposition occurred in the OMZ, the Catalonian Coastal Ranges, certain isolated areas of the WALZ, and in the northernmost CIZ during the Rhuddanian. These Rhuddanian black shales can also be found across large parts of northern Africa (Lüning et al., 2000). Yet in most regions of the Iberian Peninsula, like within the Penalba and Mondonedo domains of the WALZ, and also in the Iberian Cordillera and in most areas of the CIZ, the black shale deposition was delayed until the Aeronian and Telychian as it is the case within the CZ (Robardet and Gutierrez-Marco, 2002).

Rapid deepening affected the whole Cantabrian Basin and led to the sedimentation of the lowermost Silurian Formigoso Formation above a transgressive surface (Fig. 4). Overall, the Formigoso Formation exhibits a general shallowing and coarsening upward trend. It is suggested within this study that the sedimentary succession was generated by repeated high-frequency transgressive—regressive cycles.

The overall thickness of the Silurian Formigoso Formation is highly variable and ranges from <50 m in the southern Cantabrian Zone to

>400 m in the northern areas. The lateral variation in thickness seems to be controlled by a combination of the extensional tectonics related to the opening and widening of the Rheic Ocean and the Hirnantian glacial processes as mentioned earlier.

Within the CZ, the Silurian is developed in the "folds and nappes region" and the Pisuerga-Carrion unit of the Palentian region (Gutiérrez-Marco and Rábano, 1998; Robardet and Gutierrez-Marco, 2002). The thickest and most complete Silurian succession is within the "Fold and Nappe Province." The approximately 400-m-thick series is composed in its lower part of the mud and sandstones of the Formigoso Formation and in its upper part by the coarser siliciclastics of the lower San Pedro Formation (Garciá-López and Bastida, 2002).

The questionably earliest Silurian rocks occur at the Cabo Peñas area within the "folds and nappes region." The Viodo Member of the El Castro Formation has yielded a fossil assemblage, which probably indicates a pre-Telychian age (Robardet and Gutierrez-Marco, 2002).

Early Silurian (?Rhuddanian) conodonts were reported from the southern slope of the Cantabrian Mountains at the village of Getino. The conodonts were recovered from a layer of dolomitic limestone at the top of the so-called Transitional Beds (Bosch, 1969; García-Ramos et al., 1989). Within the more recent literature these "Transitional Beds" have also been termed "Getino Beds" and have sometimes even been assigned formation status despite their relatively thin development (Aramburu-Zabala Higuera and García Ramos, 1993; García-Ramos et al., 1989; Sarmiento et al., 1994). These Getino Beds overlie the Cambro-Ordovician Barrios Formation in most parts of the southern Correcillas unit and precede the lower Silurian Formigoso Formation. The Getino Beds have been described as heterolithic tidal deposits, consisting of a few meters of alternating sandstones, siltstones, mudstones, and carbonates, the latter now preserved predominantly as dolomites and siderites. All beds are highly bioturbated and show frequent enrichments with glauconite and Fe-oxides. The thickness and lithological development are extremely variable (Garciá-López and Bastida, 2002). The transition from the Barrios Formation to the Formigoso Formation is considered to be normal and gradual (Sarmiento et al., 1994). However, field-based observations made in the course of the present study demonstrated the contact to be always sharp and discrete, suggesting the existence of a potential stratigraphic gap and hence an unconformable boundary. The Ordovician Luarca, Sueve, and El Castro Formations are not developed within the southern CZ.

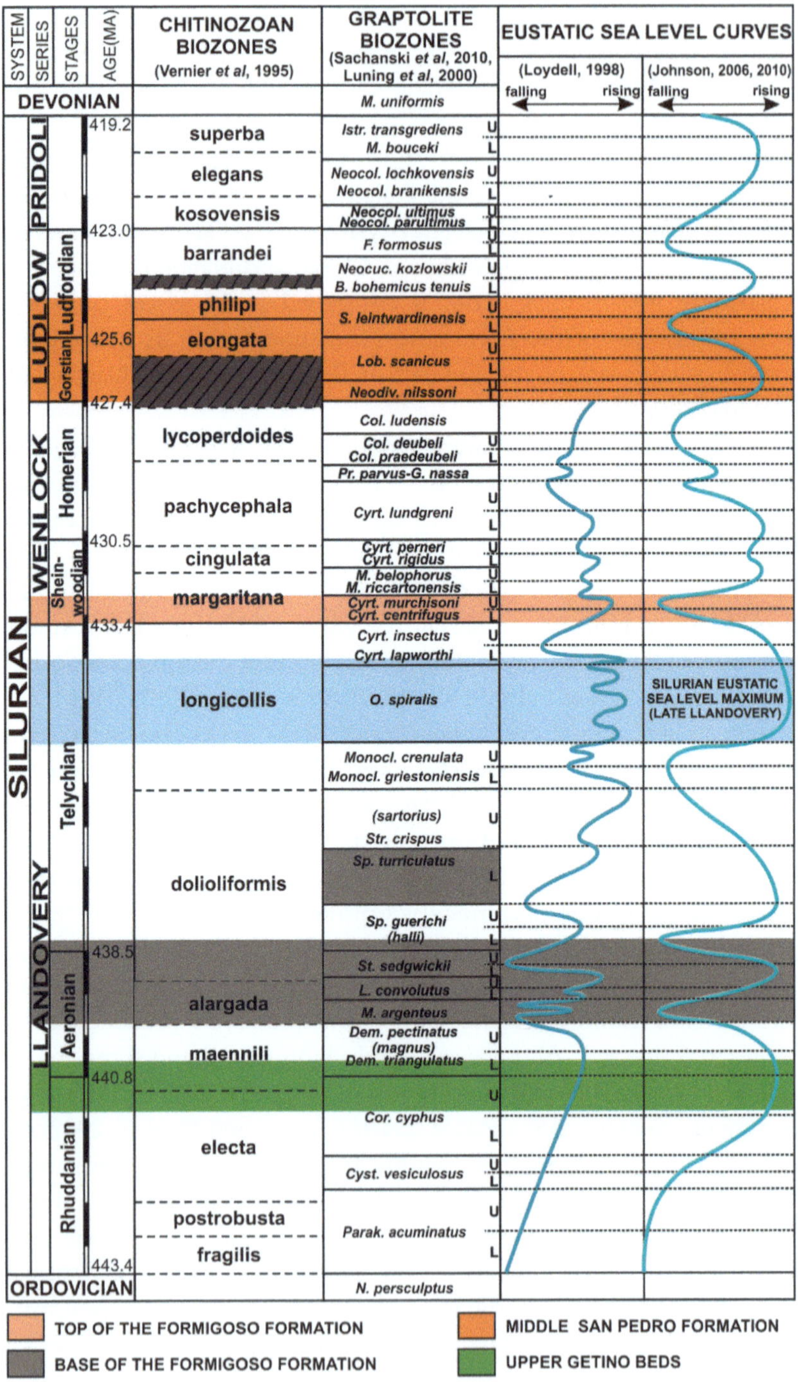
SYSTEM
SERIES
STAGES
AGE(MA)
CHITINOZOAN BIOZONES
(Vernier et al, 1995)
GRAPTOLITE BIOZONES
(Sachanski et al, 2010, Luning et al, 2000)
EUSTATIC SEA LEVEL CURVES
(Loydell, 1998)
(Johnson, 2006, 2010)
falling
rising
DEVONIAN
SILURIAN
ORDOVICIAN
PRIDOLI
LUDLOW
WENLOCK
LLANDOVERY
Ludfordian
Gorstian
Homerian
Shein-woodian
Telychian
Aeronian
Rhuddanian
419.2
423.0
425.6
427.4
430.5
433.4
438.5
440.8
443.4
superba
elegans
kosovensis
barrandei
philipi
elongata
lycoperdoides
pachycephala
cingulata
margaritana
longicollis
dolioliformis
alargada
maennili
electa
postrobusta
fragilis
M. uniformis
Istr. transgrediens
M. bouceki
Neocol. lochkovensis
Neocol. branikensis
Neocol. ultimus
Neocol. parultimus
F. formosus
Neocuc. kozlowskii
B. bohemicus tenuis
S. leintwardinensis
Lob. scanicus
Neodiv. nilssoni
Col. ludensis
Col. deubeli
Col. praedeubeli
Pr. parvus-G. nassa
Cyrt. lundgreni
Cyrt. perneri
Cyrt. rigidus
M. belophorus
M. riccartonensis
Cyrt. murchisoni
Cyrt. centrifugus
Cyrt. insectus
Cyrt. lapworthi
O. spiralis
Monocl. crenulata
Monocl. griestoniensis
(sartorius)
Str. crispus
Sp. turriculatus
Sp. guerichi
(halli)
St. sedgwickii
L. convolutus
M. argenteus
Dem. pectinatus
(magnus)
Dem. triangulatus
Cor. cyphus
Cyst. vesiculosus
Parak. acuminatus
N. persculptus
SILURIAN EUSTATIC SEA LEVEL MAXIMUM (LATE LLANDOVERY)
TOP OF THE FORMIGOSO FORMATION
BASE OF THE FORMIGOSO FORMATION
MIDDLE SAN PEDRO FORMATION
UPPER GETINO BEDS

Previously, the Getino Beds were interpreted to form the base of the Formigoso Formation (Comte, 1959; Evers, 1967; Kegel, 1929; Sarmiento et al., 1994). Biostratigraphic evidence supporting this assumption was provided by fossil assemblages from near the top of the succession at two localities (Pontedo and Getino). These fossil assemblages include graptolites (*Pristiograptus* sp., *Metaclimacograptus* aff. *hughesi*), trilobites (*Leonaspis* sp., *Calymene* sp.), and bivalves, crinoids, conodonts (*Panderodus unicostatus, Walliserodus sanctidairi*) and point to a late Rhuddanian to early Aeronian age (Garciá-López and Bastida, 2002). Furthermore, Sarmiento et al. (1994) confirmed the late Rhuddanian to early Aeronian age by comparing the conodonts from the upper parts of the Getino Beds to the fauna reported from the Viodo Member of the Castro Formation. They concluded that the top of the Getino Beds could tentatively be assigned to either the Aspelundia petila Zone or the Aspelundia fluegeli Zone. Based upon lithological features such as the degree and development of the bioturbation, the fossil content, and the presence of glauconite, it is generally assumed that the Getino Beds were at least partly deposited within a shallow marine environment (Garciá-López and Bastida, 2002).

The Formigoso Formation (Comte, 1959; Kegel, 1929) consists of up to 400 m of black/gray silt and mudstones with an increasing number of sandstone intercalations toward the top. The sandstones typically exhibit strong bioturbation and display ripple marks as well as hummocky cross-stratification (Robardet and Gutierrez-Marco, 2002; Suárez de Centi et al., 1989). In the Cabo Peñas area the Formigoso Formation overlies the Castro Formation, yet in most regions of the Fold and Nappe Province it is overlying the basal Silurian series (Getino Beds) or the Floian Barrios Formation with a paraconformity or disconformity (Comte, 1959; Garciá-López and Bastida, 2002; Rupke, 1965; Schneider et al., 2004). Kegel (1929) distinguished two members: the (1) Pizarras del Bernesga Member consists of graptolite-rich black shales with the graptolites *Monograptus convolutus* and *Monograptus turriculatus* indicating an age of late-mid Aeronian to early Telychian (Comte, 1959; Garciá-López and Bastida, 2002; Gutiérrez-Marco and Rábano, 1998;

Figure 4 Biostratigraphy of the Silurian formations of Cantabria. Highlighted are the stratigraphic sections for which reliable biostratigraphic data are available.Biostratigraphy compiled after Sachanski et al. (2010); Verniers et al. (1995); Lüning et al. (2000); Gutierrez-Marco and Štorch (1998). Relative sea-level curves are compiled after Loydell (1998); Johnson (2006); Johnson (2010).

Paris et al., 1995; Sarmiento et al., 1994; Truyols et al., 1974; Truyols and Julivert, 1983). The age constraints for the lower Bernesga Member were largely confirmed by organic walled microfossils, mainly chitinozoans (Cramer and Diez, 1978). However, the palynological data need to be reinvestigated and reevaluated.

The (2) Capas de Villasimpliz Member is characterized by the intercalation of silt and sandstones. The sandstones become more prominent toward the top of the Member. Fossils are less abundant within the member, but graptolites (*Monoclimacis* cf. *vomerina, Monograptus priodon*), brachiopods, and palynomorphs, which together suggest a Late Llandovery to Early Wenlock age (Upper Telychian to Lower Sheinwoodian, Centrifuges-Murchisoni Zone), have been described (Aramburu et al., 1992; Garciá-López and Bastida, 2002; Gutiérrez-Marco and Rábano, 1998; Schweineberg, 1987).

The strata of the Bernesga Member are interpreted to have been deposited in an open marine to off-shore environment within a shallow epeiric sea with little circulation and a lack of oxygen (Evers, 1967; Garciá-López and Bastida, 2002; García-Ramos et al., 1989; Suarez de Centi, 1988). The shallowing upward trend of the Formigoso Formation is evidenced by the increased amount of coarser siliciclastic intercalations within the overlying Villasimpliz Member. Many of these intercalations have previously been interpreted to be of tempestitic origin, indicating a deposition above the storm wave base (Barba and Fernández, 1991; Robardet and Gutierrez-Marco, 2002; Suárez de Centi et al., 1989; Veselovsky, 2004). The low abundance of graptolites within the Villasimpliz Member and the following San Pedro Formation is probably due to environmental stress (e.g., changes in salinity) and not postdepositional destruction, as moderately well-preserved organic-walled microfossils have been reported from these series (e.g., Richardson et al., 2001; Robardet and Gutierrez-Marco, 2002).

The Furada-San Pedro Formation is developed throughout the entire Fold and Nappes Province. The formation was originally described as the so-called Ferruginous Sandstones of Furada by Barrois (1882) and as the "San Pedro Formation" by Comte (1937). The contact with the underlying Formigoso Formation is sharp and erosive in the areas closer to the Central Coal Basin and gradational in the areas near the Narcea Antiform. Its thickness ranges from 5 to 250 m (Garciá-López and Bastida, 2002) and it consists of gray to reddish ferruginous sandstones with mudstone intercalations, thin oolitic ironstone beds, and at the top some sandy limestone lenses. Its overall thickness increases from the

Central Coal Basin in the east toward the Narcea Antiform in the west (Bosch, 1969; Garciá-López and Bastida, 2002). The San Pedro Formation has been subdivided into three informal members:

The (1) Lower Member consists of red ferruginous, coarse-grained sandstones with minor oolitic ironstone intercalations that show trough cross-stratification. Occasionally, conglomeratic beds with phosphatic and sideritic siltstone clasts as well as intraclasts can be observed. Some of these clasts have been described to be of volcanic origin (Bosch, 1969; Richardson et al., 2001; Suarez de Centi, 1988). Late Wenlockian palynomorphs have been described from the lowermost part of the San Pedro Formation (Richardson et al., 2001).

The (2) Middle Member is characterized by medium- to fine-grained sandstones with intercalations of gray to green colored, intensively bioturbated mudstone- and siltstone horizons (Garciá-López and Bastida, 2002). Ludlowian and Pridolian brachiopods such as *Orbiculoidea striata* and "*Spirifer*" *vulcani* have been described from its middle part (Gutiérrez-Marco and Rábano, 1998).

The (3) Upper Member, contains fine-grained, white to gray colored quartzitic sandstones with cross-stratification, interbedded with dark gray to black shales. Around 50 m below the top of the formation graptolites from the Nilssoni- and Scanicus Zones and probably the Leintwardinensis Zone (*Saetograptus fritschi fritschi*, *Saetograptus chimaera saandweyi*) have been described, indicating a Gorstian to lowermost Ludfordian age. The upper member has also so far yielded early Devonian (Lochkovian) conodonts, trilobites (*Acastella tiro*, *Acastella heberti*), brachiopods (*Platyorthis verneuili*, *Mesodouvillina triculta*, *Howellella mercuri*, *Platyorthis* ex gr. *monnieri*), molluscs, bryozoans, echinoderms, conulariids, acritarchs, and chitinozoans (Aramburu et al., 1992; Garciá-López and Bastida, 2002; Robardet and Gutierrez-Marco, 2002; Truyols et al., 1974, 1990). According to these data, the Silurian–Devonian boundary is located within the upper member of the San Pedro Formation, and most studies suggest that it is situated between approximately 13–15 m below the Formation's top (Richardson et al., 2001; Robardet and Gutierrez-Marco, 2002; Suárez de Centi et al., 1989).

The depositional environment of the San Pedro Formation is interpreted to have been shallow marine (littoral to very shallow sublittoral), dominated by the supply of coarser clastic detritus (Garciá-López and Bastida, 2002). A transgressive–regressive cycle is suggested, with the middle member representing the deepest conditions (Richardson et al., 2001). The grain size is seen to decrease from "an emerged land area," located in the core of the Asturian Arc in the east, toward the west. Most of the iron preserved within the ferruginous sandstones is considered to be derived from the subaerial weathering of basic volcanic rocks (Garciá-López and Bastida, 2002). The transition to the overlying succession (Raneces or La Vid Group) is gradual and characterized by an increase in dolomitic cements. The boundary between the San Pedro Formation and the overlying La Vid Group is highly diachronous and systematically younging toward the Narcea Antiform in the west (González, 1983).

4. PALEOGEOGRAPHIC REMARKS ON THE SILURIAN OF CANTABRIA

The precise determination of the paleogeographic position of the Iberian Peninsula for the Silurian based on paleomagnetic data is problematic. Strong overprints such as, for example, at Almaden, southern CIZ, where basaltic lavas interbedded with Llandovery mudstones were remagnetized during the Late Devonian/Early Carboniferous (Parés and Van der Voo, 1992; Perroud et al., 1991; Robardet and Gutierrez-Marco, 2002) obstacle the confident interpretation of the data. Also the San Pedro Formation has been affected by a prefolding demagnetization during the Early Carboniferous (Perroud and Bonhommet, 2007; Robardet and Gutierrez-marco, 2002; Van der Voo, 2005). Therefore the Silurian paleogeographic position of the Iberian Peninsula is inferred based on the high/cold temperature latitudes for the latest Ordovician Hirnantian glacial deposits (approximately 50°S, Paris and Robardet, 1990; Robardet and Gutierrez-Marco, 2002) and the warm temperate to subtropical latitudes for the Lower Devonian carbonate facies successions (approximately 35°S, Robardet and Gutierrez-Marco, 2002). Gutiérrez-Marco and Rábano (1998) suggest that during the Early Silurian Iberia was positioned between 40 and 50°S as to 35–40°S during the Late Silurian.

5. BLACK SHALE GENESIS

Black shales are generally dark-colored mudrocks, which consist of 40–60% clay minerals, up to 30% of silt- to sand-sized mineral grains (carbonates, cherts, and phosphates) and varying amounts of organic matter (Boggs, 2009; Cox and Lowe, 1995; Sethi and Schieber, 1998). The dark color is mainly caused by two constituents: (1) the contained organic matter and/or (2) finely disseminated pyrite (Schieber, 2003). Black shales are usually deposited under oxygen-depleted bottom water conditions and predominantly, but not exclusively, accumulate in deep-water basinal environments (Ketris and Yudovich, 2009; Wignall and Newton, 2001). The low oxygen content of the bottom water is maintained either by (1) a stratified water column, in which the well-oxygenated photic zone is separated from the anoxic nutrient-rich bottom waters (Preservation Model), or (2) high productivity rates, where the increased flux of organic matter (OM) and in turn the demand for dissolved O_2 exceeds its supply (Productivity Model). These oxygen-poor conditions lead to the preservation of organic material (mostly planktonic algae) within black shales. Presently, most black shales (Fig. 5) form in marine environments over areal extents of thousands of square kilometers (Arthur and Sageman, 1994). The black shales or "oil shales" may contain "shale oil," which can be extracted by pyrolysis (Ketris and Yudovich, 2009). The amount and the nature of the organic matter determine the type of hydrocarbons produced (Tourtelot, 1979; Tyson, 1995). The global "oil shale" reserves are considered to exceed the petroleum reserves (Ketris and Yudovich, 2009).

The majority of shales that meet the aforementioned classification tend to contain a total organic carbon (TOC) percentage of one or more; a few shales can reach in excess of 20% TOC as, e.g., the Lower Silurian transgressive "hot" shales of Northern Africa (Lüning et al., 2000, 2003a,b; Lüning and Kolonic, 2006). Many black shale successions are severalfold enriched in metals compared to ordinary mudstones (Oschmann, 2000; Piper and Calvert, 2009). Some black shale horizons are reported to have even formed host rocks for economical viable syngenetic metal deposits

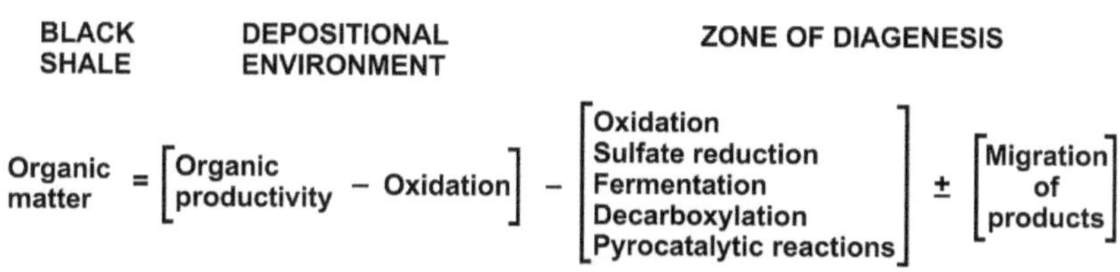

Figure 5 Chemical processes and relationships between the original depositional environment, diagenetic alterations, and resulting black shales. Note the high number of processes that finally lead to the destruction or the diminishing of the organic matter. Highlighted is the importance of the high original organic productivity for black shale genesis and subsequent preservation within the geological record. *Slightly modified after Tourtelot, H.A., 1979. Black shale; its deposition and diagenesis. Clays Clay Minerals 27 (5), 313–321.*

(Swanson, 1961; Tourtelot, 1979). Vine and Tourtelot (1970) described black shale successions in which certain horizons are over 50 times enriched in the elements Ag, Mo, Zn, Ni, Cu, Cr, and V and to a lesser extent enriched in the elements Co, Se, and U compared to a standard mudstone. These elements form together part of the redox-sensitive element suite and are predominantly adsorbed to organic matter and Fe-Mn-hydroxides. Additionally, they may act as sulfide-forming trace metals, which are widely utilized to reconstruct the prevailing environmental conditions during black shale deposition. For example, as U can be adsorbed to organic matter under chemically reducing conditions, it can be used not only to determine the redox state of the water column for the time of deposition, but also as a proxy to estimate the amounts of organic carbon present (Fertl and Chilingar, 1988; Lüning and Kolonic, 2003, 2006).

Black shales occur throughout Earth's geological history (Fig. 6). They are formed at times when specific geological settings and paleoenvironmental controls are suitable for the deposition of organic-rich sediments. The basic concept for generating organic-rich sediments is that they form when and where organic material accumulates at a faster rate than it can be destroyed (Tourtelot, 1979).

5.1 Depositional Mechanisms

Different models involving different processes and mechanisms have been proposed to explain the generation of black shale deposits. Two of these models attracted the most attention within the previously published literature: the productivity model (Pedersen and Calvert, 1990) and the preservation model (Demaison and Moore, 1980). Both have led to and still do stimulate much debate over the formation of organic-rich sediments. The productivity model suggests an enrichment of organic matter within the sediment as a consequence of high settling rates of OM related to high primary productivity in the surface waters (with the nutrients being supplied by upwelling or terrestrial influxes). The preservation model argues that the physicochemical conditions of the water column, such as depletion of dissolved O_2 (dysoxia/anoxia), stratification (presence of a pycnocline), and low sedimentation rates, lead to the formation of organic-rich sediments (Arthur and Sageman, 1994; Harris et al., 2004; Killops and Killops, 2009; Meyers, 2006; Murphy et al., 2000; Nijenhuis et al., 1999; Tyson, 1995; Wei et al., 2012). Tyson (1987) and Wignall (1994) pointed out the incompatibility of these two models, because high productivity and enhanced preservation do not coincide and may even inhibit each other. High productivity rates suggest a sustained

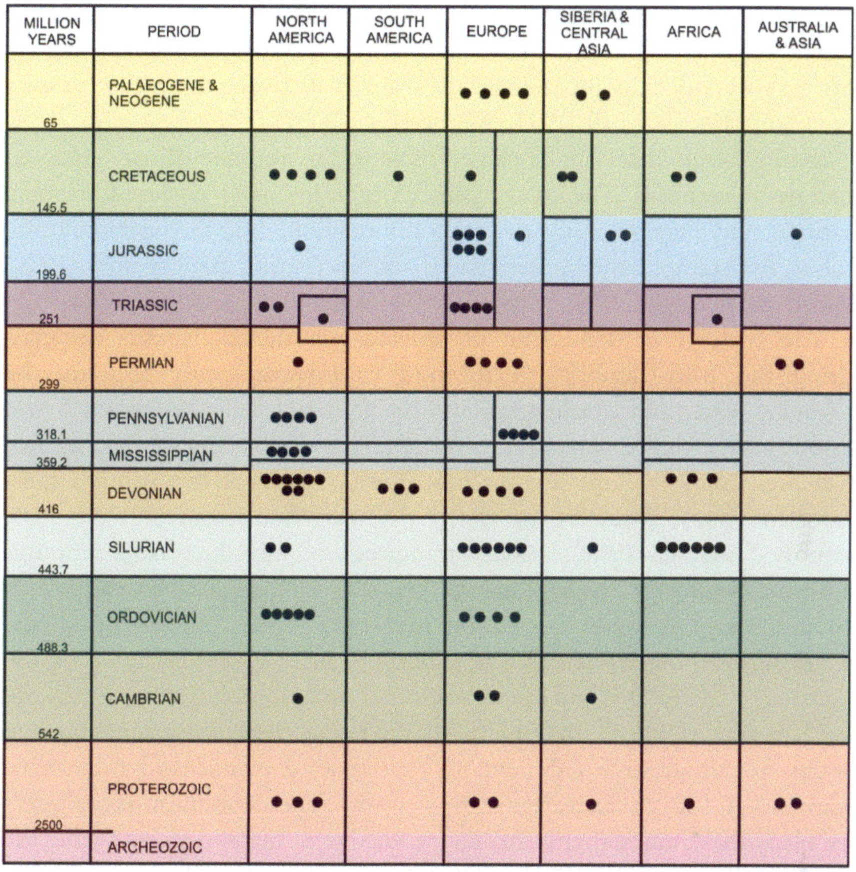

Figure 6 Temporal and spatial distribution of significant black shale deposits. The filled black circles represent occurrences. The occurrences differ greatly in thickness, areal extent, and other characteristics. Background color code after the International Chronostratigraphic Chart (Gradstein et al., 2012). *Modified after Tourtelot, H.A., 1979. Black shale; its deposition and diagenesis. Clays Clay Minerals 27 (5), 313–321; Lüning, S., Craig, J., Loydell, D.K., Štorch, P., Fitches, B., 2000. Lower Silurian 'hot shales' in North Africa and Arabia: regional distribution and depositional model. Earth-Science Rev. 49 (1–4), 121–200.*

nutrient supply and hence vigorous perturbation of the water column, whereas preservation of organic matter is typically facilitated by stratification, which intrinsically implies reduced circulation (Wignall, 1994). However, some studies (Calvert and Fontugne, 2001; Rimmer, 2004; Rinna et al., 2002; Wei et al., 2012; Werne et al., 2002) point out that anoxia and high productivity cannot occur completely independent, but instead may even induce and subsequently reinforce each other as coupled processes. The application of either the productivity model or the preservation model for ancient organic-rich sedimentary rocks is notoriously problematic due to poor paleoproductivity proxies and the inability to constrain the precise sedimentation rates (Wignall, 1994; Werne et al., 2002).

Marine black shales can be classed either as "marine shelfal" or "enclosed silled basin" deposits. The latter basins are typically deep, enclosed, strongly stratified, and characterized by a redox boundary high in the water column, such as, for example, the modern day Black Sea (Arthur and Sageman, 1994; Killops and Killops, 2009; Wignall, 1994; Wignall and Hallam, 1991).

The widespread deposition of organic-rich shelfal deposits occurred during the Mid-Cambrian, early mid Ordovician, early Silurian, the Devonian, early and late Jurassic, and the mid and late Cretaceous. These periods correspond to polar ice cap minima and consequently global sea-level highstands (Killops and Killops, 2009). Many of the World's most important source rocks are associated with this "transgressive" black shale category (Wignall, 1994). Marine transgressions introduce large amounts of nutrients into the shallow marine areas, causing high levels of primary productivity. The water circulation may be restricted due to postglacial stratification of the water column, leading to intense anoxia caused by the degradation of the OM and the formation of organic-rich sediments (Killops and Killops, 2009). Tourtelot (1979) suggested there are essentially two groups of transgressive sediments: (1) alternating marine and nonmarine sediments and (2) exclusively marine deposits. The deposition of transgressive black shales can either occur above erosive or hiatal surfaces during the initial stages of the relative sea-level rise under starved basin conditions ("puddle model" of Wignall and Hallam, 1991) or during the peak of transgression ("maximum flooding black shales") at the time of maximum shoreline retreat ("expanding puddle model" of Wignall and Hallam, 1991). The latter is essentially a variation of the "silled basin model" proposed and discussed by Stanley and Blanpied (1980), Wignall (1994) and Armstrong et al. (2005).

Tourtelot (1979) developed three models to describe the depositional environments for marine black shales based on oxygen content and water depth (Figs. 7 and 8). (1) Restricted circulation model (Fig. 7): This model is widely used to explain the processes that are apparent within the modern day enclosed stratified basin of the Black Sea. The oxygen content of the water column is not renewed by circulation, meaning the oxygen consumed by the oxidation of organic matter is not replenished fast enough or the amount of oxygen consumed is in constant equilibrium with the oxygen

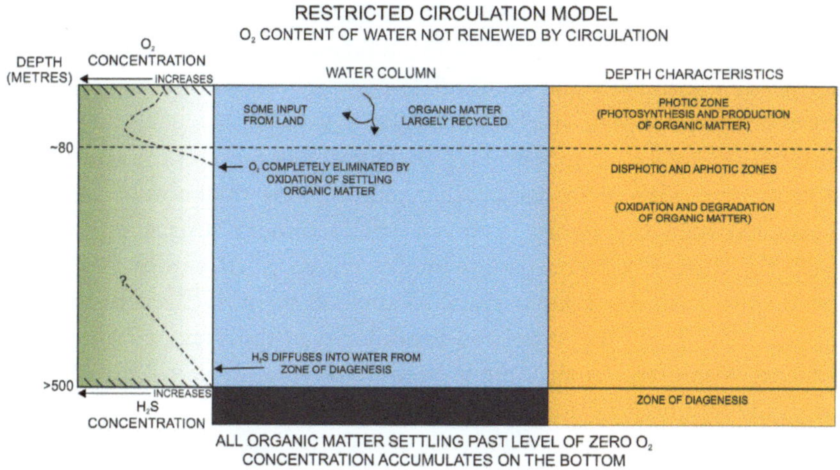

Figure 7 Restricted circulation model for the accumulation of organically rich black shales. *Modified after Tourtelot, H.A., 1979. Black shale; its deposition and diagenesis. Clays Clay Minerals 27 (5), 313–321.*

supplied. Under these conditions organic-rich sediments can accumulate and be preserved even if the bioproductivity rate is relatively low. (2) Open ocean model (Fig. 8): This model is commonly applied to explain the enrichment in OM within modern day deep marine environments. The oxygen content of the water column is controlled by thermohaline

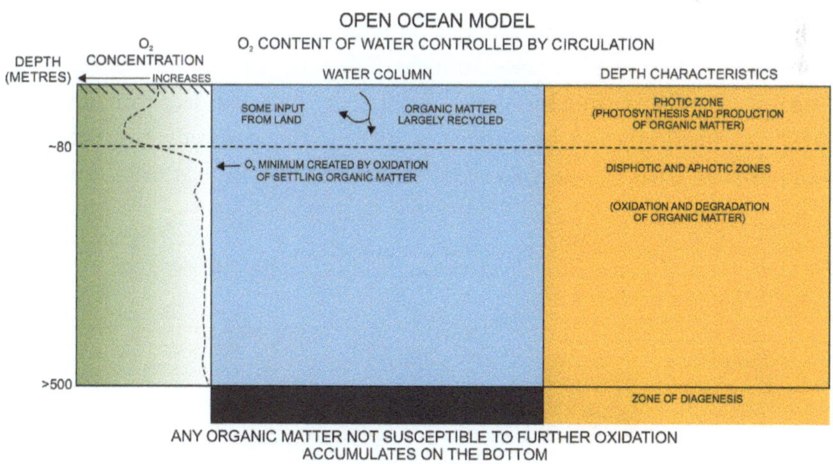

Figure 8 Open ocean model for the accumulation of organically rich black shales. *Modified after Tourtelot, H.A., 1979. Black shale; its deposition and diagenesis. Clays Clay Minerals 27 (5), 313–321.*

circulation and the oxidation rate of OM escaping the photic zone. The oxygen minimum is again created by the oxidation of the organic material. Organic-rich sediments can accumulate even if the sedimentary environment is completely oxic.

(3) Continental shelf model (Fig. 9): This model requires the same input factors as the models mentioned earlier, although the thickness of the water column between the photic zone and the sediment surface is reduced. Hence the organic material is less intensively recycled within the photic zone due to a more rapid transfer to the sediment surface. Because of bacterial activity and the associated breakdown of the organic content the sediments within this environment become oxygen depleted already very near to the sediment—water interface.

5.2 Organic Carbon, Bioproductivity, and Paleoproductivity

It is commonly accepted that the organic matter preserved within black shales is the result of the incompleteness of the organic carbon cycle as illustrated in Fig. 10 (Hedges and Keil, 1995; Oschmann, 2000; Walther, 2009). The initial fixation of carbon takes place during the oxygenic steps of photosynthesis. Müller and Suess (1979) and Meyers (2006) concluded that less than 1% of the organic matter produced in the photic zone reaches the seafloor. The organic matter accumulating on the seafloor is composed of the remains of the single-celled photosynthetic phytoplanktonic organisms that inhabit the photic zone (Meyers, 2006). A combination of these primary producers and both land- and marine-derived organic matter is more apparent in marine sediments close to river mouths (Meyers, 1997). The enrichment of organic carbon within sediments occurs, if the OM accumulates faster than it can be destroyed by processes such as the aerobic degradation in the uppermost parts of the oxic water column and the bacterial oxidation

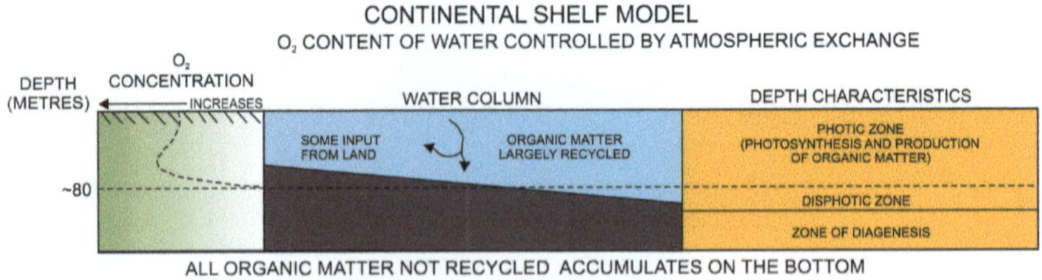

Figure 9 Continental shelf model for the accumulation of organically rich black shales. *Modified after Tourtelot, H.A., 1979. Black shale; its deposition and diagenesis. Clays Clay Minerals 27 (5), 313–321.*

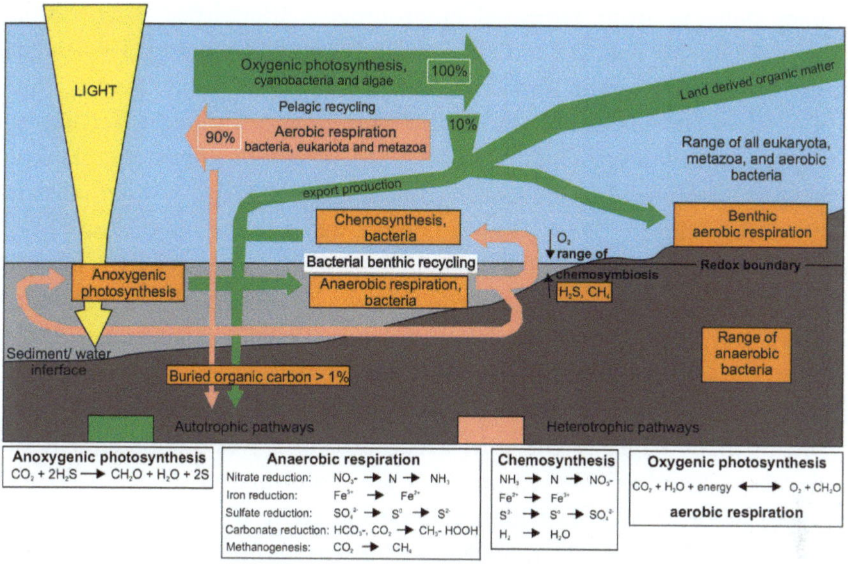

Figure 10 The organic carbon cycle and the main related redox reactions. *Modified after Oschmann, W., 2000. Microbes and black shales. In: Riding, R., and Awramik, S.M. (Eds.), Microbial Sediments, Springer, Berlin, pp.137–148.*

within the sediment. Arthur (1979), Bohacs et al. (2005), Meyers et al. (2005), Sageman et al. (2003), and Stow et al. (2001) highlight three fundamental processes controlling the burial of organic carbon: (1) dilution (caused by variations in sedimentation rates, terrigenous clastic and skeletal flux), (2) production (rates of marine and terrigenous OM flux), and (3) decomposition versus preservation.

Bioproductivity is a variable process. It is largely controlled and sustained by the levels of nutrients that are available in the photic zone of the water column. The level of nutrients in turn are controlled by and limited to depths of, seasonality (light and temperature), oceanic circulation, salinity, and variation in atmospheric CO_2 (Bonn et al., 1998; Hillaire-Marcel and De Vernal, 2007). The vast majority of bioproductivity takes place within the topmost 100 m of the water column (Meyers, 2006). Variations in atmospheric CO_2 are believed to result from changes in the productivity rates and the burial of carbon (Pfeifer et al., 2001). Hence, the varying degrees of paleoproductivity are used to infer and reconstruct paleoclimatological conditions. Nutrient recycling, mainly the replenishment of nitrate, phosphate, and silicate in the surface waters, is controlled and influenced by the vertical movement of the water masses, upwelling, vertical mixing, and nutrient delivery from terrestrial environments (Bernárdez et al., 2008; Chester, 1990; Meyers, 2006; Piper and Calvert, 2009; Tyson, 1995). The salinity levels also exert an impact on the rates of productivity. Fully marine developed organisms exhibit normally only a very limited tolerance to changes in salinity levels (Killops and Killops, 2009). Furthermore, brackish surface waters can lead to the salinity-driven establishment of water column stratification (Meyers, 2006; Tyson, 1995), which prevents the supply of nutrients via upwelling (Challands et al., 2009). The productivity is leading to a greater demand for oxygen and ultimately triggers the depletion of dissolved oxygen with increasing water depth (Prakash Babu et al., 2002). It could be suggested that high surface water productivity leads to high fluxes of OM to the seafloor sediment. However, this assumption proves to be invalid in many cases, as the OM accumulation is heavily dependent on the levels of dissolved O_2 within the water column and its depth-related decrease. In other words, a high paleoproductivity in the photic zone is not always and necessarily recorded within the sedimentary rocks (Bonn et al., 1998; Paytan and Gornitz, 2008; Wignall, 1994). Brumsack (1981) and Tyson (1995) demonstrated that a high productivity is not essential to produce marine sediments with TOC values up to 10%; it can be achieved by decreasing sediment dilution and enhancing the OM preservation by increased levels of anoxia. The concentration of OM within sediments represents only a fraction of the biological productivity in the surface waters. Under normal marine settings approximately 10% of the original productivity leaves the euphotic zone, and in upwelling areas this may increase to up to 30%; however, the amount of OM that actually reaches the sediment may only represent a few percent or less (Reitz et al., 2004; Tribovillard et al., 2006; Tyson, 1995). Phosphorus (P) and nitrogen (N) concentrations have been used as productivity indicators (Hetzel et al., 2011; Schmitz et al., 1997), yet these elements are intensively biologically recycled (biocycling) within the water column, and P in particular is known to be poorly correlated with surface water productivity (Wignall, 1994). Furthermore, P contents can also be severely affected by postburial redox conditions (Delaney, 1998; Paytan and Gornitz, 2008). Biogenic silica (Si_{bio}) concentrations produced by siliceous plankton and biogenic opal are also frequently used as an indicator for bioproductivity (Bonn et al., 1998; Schmitz et al., 1997). Over 90% of the organic silica is commonly preserved in sediments (Wignall, 1994). However, factors such as Si dissolution within the water column, at the sediment/water interface and within the sediment make it difficult to estimate original bioproduction rates for the surface waters (Bonn et al., 1998; Hillaire-Marcel and De Vernal, 2007). By far the most widely used paleoproductivity indicator is barium (Ba) or biogenic Ba (Ba_{bio}). Reducing conditions, together with the sulfate release from decaying OM within the oxygen minimum zone, remove Ba from solution as syngenetic barite ($BaSO_4$), which can be effectively incorporated into the siliceous phytodetritus (Bonn et al., 1998; Calvert and Pedersen, 2007; Chun and Delaney, 2006; Kiipli et al., 2004; Paytan and Gornitz, 2008; Pfeifer et al., 2001; Reitz et al., 2004; Schmitz et al., 1997; Scopelliti et al., 2006; Turgeon and Brumsack, 2006; Wignall, 1994). This process can generate a strong and significant positive correlation between Ba_{bio} and OM fluxes (Dymond et al., 1992; Paytan and Gornitz, 2008; Prakash Babu et al., 2002). However, the use of Ba as a paleoproductivity proxy is not unproblematic. The relationship between Ba and productivity as described earlier was demonstrated to work effectively in deep water environments (>1000 m), but only week correlations between Ba and productivity seem to exist in shallow water settings (Hetzel et al., 2011; Wignall, 1994). A further problem for the interpretation of Ba concentrations within the geological record arises from the highly mobile nature of Ba during diagenesis, specifically within the sulfate reduction zone (Calvert and Pedersen, 2007; Hetzel et al., 2006; Wignall, 1994; McManus et al., 1998; Prakash Babu et al., 2002; Turgeon and Brumsack, 2006). Ba associated with terrigenous silicates, like K-feldspars and micas, where it can substitute for K (Calvert and Pedersen, 2007), Fe-Mn oxides, and -hydroxides (Anderson and Winckler, 2005), as well as postdepositional hydrothermal activity, can mask the original biogenic Ba signal (Paytan and Gornitz, 2008; Wei et al., 2012).

5.3 Anoxia and its Relation to OM Preservation

Principally, oxygen deficiency (or anoxia) develops when the oxygen demand for the oxidation of OM is exceeding the rate of oxygen supply, which in turn is strongly controlled by climatic, hydrographic, or oceanographic factors (Dean et al., 1984; Tyson, 1995). The redox classification of sedimentary environments is based on the concentration of dissolved O_2 and comprises: (1) oxic (>2.0 mL O_2 L^{-1} H_2O), (2) dysoxic or suboxic (2.0–0.2 mL O_2 L^{-1} H_2O), and (3) the anoxic facies. The last facies can be further subdivided into the nonsulfidic (<0.2 mL O_2 L^{-1} H_2O, no free H_2S) and sulfidic or euxinic (0 mL O_2 L^{-1} H_2O, with free H_2S) subfacies (Algeo and Maynard, 2004; Algeo et al., 2004; Tyson, 1995; Wignall, 1994). There has been considerable confusion about the various terms

(Jorissen et al., 2007); however, the above-mentioned definitions will be used within the present study.

Anoxic conditions are commonly described as "euxinic," meaning "Black Sea-like," the technical term being derived from the Latin *Euxinus Pontus* and the ancient Greek εὔξεινος πόντος (King, 2004; Piper and Calvert, 2009). They can be broadly classified as either nonsulfidic or sulfidic. Sulfidic conditions prevail when free hydrogen sulfide (H_2S) percolates from the underlying sediments into the water column (Wignall, 1994). The H_2S is commonly regarded to be a by-product of the activity of sulfate-reducing bacteria in the sediments. Euxinic conditions may extend as far as the photic zone, such as in the modern day Black Sea. Suboxic conditions are characterized by low O_2 concentrations, where elevated H_2S levels are restricted to the pore waters below the sediment–water interface (Tribovillard et al., 2006).

Under oxic conditions, where surface waters contain between 4.5 and 8.0 mL O_2 L^{-1} H_2O (Algeo and Maynard, 2004; Wright and Colling, 1995), OM is degraded by the activities of aerobic bacteria and other organisms, decreasing the dissolved levels of O_2 within the water column. The oxygen consumption rates vary with the nature of the OM present (Tyson, 1995). Under depleted O_2 conditions (amount of OM exceeds the O_2 present) the OM is further broken down by anaerobic microbes. These anaerobia utilize secondary oxidant sources such as iron and manganese oxides/sulfates and dissolved species of Cr, Re, U, and V, which are metabolically reduced to less soluble states and accumulate on the seafloor (Piper and Calvert, 2009). Anoxic conditions are established when all oxidants have been depleted and sufficient amounts of H_2S are produced by sulfate-reducing bacteria. The presence of reduced species of Cr, Re, U, and V as well as FeS, CdS, and ZnS in sediments is generally assumed to demonstrate the activity of sulfate-reducing bacteria (Calvert and Pedersen, 2007; Piper and Calvert, 2009). With increasing depth below the sediment/water interface, methanogenic bacteria begin to break down the OM, mostly via the oxidative-reductive disproportionation of carbohydrates (Tribovillard et al., 2006). The boundaries between the differing oxygenation regimes within the sediment (Fig. 11) are largely defined and determined by microbial-driven oxidation reactions (Piper and Calvert, 2009; Riding and Awramik, 2000).

Anoxia can also develop within stratified water columns with pronounced vertical temperature and salinity gradients. The anoxia there is commonly exacerbated by the presence of OM and bacterial respiration,

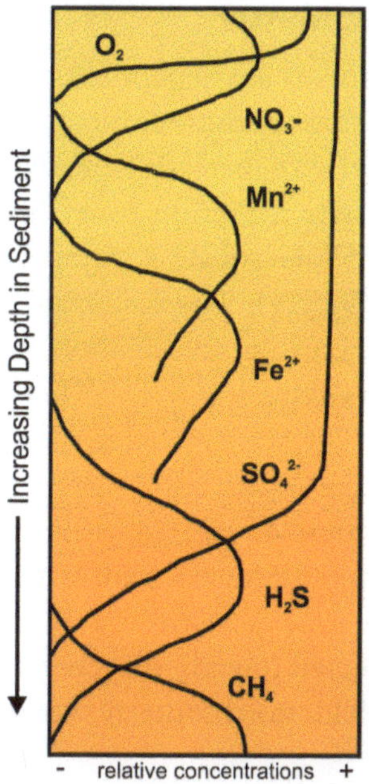

1. *Aerobic Respiration*

$$(CH_2O)_{106}(NH_3)_{16}(H_3PO_4) + 138O_2$$
$$106CO_2 + 16HNO_3 + H_3PO_4 + 122H_2O$$

2. *Nitrate Reduction*

$$(CH_2O)_{106}(NH_3)_{16}(H_3PO_4) + 94.4HNO_3$$
$$106CO_2 + 55.2N_2 + H_3PO_4 + 177.2H_2O$$

3. *Manganese Reduction*

$$(CH_2O)_{106}(NH_3)_{16}(H_3PO_4) + 236MnO_2 + 472H^+$$
$$236Mn^{2+} + 106CO_2 + 8N_2 + H_3PO_4 + 366H_2O$$

4. *Iron Reduction*

$$(CH_2O)_{106}(NH_3)_{16}(H_3PO_4) + 424FeOOH + 848H^+$$
$$424Fe^{2+} + 106CO_2 + 16NH_3 + H_3PO_4 + 742H_2O$$

5. *Sulphate Reduction*

$$(CH_2O)_{106}(NH_3)_{16}(H_3PO_4) + 53SO_4^{2-}$$
$$106CO_2 + 16NH_3 + H_3PO_4 + 53S^{2-}$$

6. *Methanogenesis*

$$(CH_2O)_{106}(NH_3)_{16}(H_3PO_4)$$
$$53CO_2 + 16NH_3 + H_3PO_4 + 53CH_4$$

Figure 11 Relationship between increasing depth and the chemical reactions facilitating the breakdown of organic matter. *Modified after Calvert, S.E., Pedersen, T.F., 2007. Elemental proxies for palaeoclimatic and palaeoceanographic variability in marine sediments: interpretation and application. In: Hillaire-Marcel, C., De Vernal, A. (Eds.), Proxies in Late Cenozoic Paleoceanography. Elsevier Science, pp. 518.*

which effectively restricts the circulation of O_2 (Piper and Calvert, 2009). Under these circumstances a well-oxygenated surface layer is separated from the underlying anoxic waters by the establishment of a pycnocline (Wignall, 1994; Wignall and Hallam, 1991; Wignall and Maynard, 1993), enhancing the development of anoxic/euxinic conditions toward the sediment/water interface. Tyson (1995) suggested that the oceanic thermohaline circulation, which is providing O_2 to the deep ocean (at least under present day conditions), is responsible for the generally low potential of oxygen depletion in the deep sea compared to shallower shelfal environments. The extent of the anoxic conditions and O_2 replenishment within the pore waters below the sediment–water interface is considered to be controlled by a wide variety of factors, such as the intensity of bioturbation (leading to O_2 replenishment), the mineralogical composition and texture of the sediments, the presence of turbiditic flows (which introduce oxygen-rich sediments into the system), and the strength of bottom currents (Dean et al., 1984; Tribovillard et al., 2006; Wignall, 1994).

5.4 Diagenesis and Degradation of Organic Matter

The early diagenetic processes affecting marine sediments are dependent on a series of chemical redox reactions, which are characteristic for major diagenetic zones. These zones are commonly termed (with increasing depth) (1) the aerobic oxidation zone (oxic zone), (2) the nitrate reduction zone, (3) the manganese zone, (4) the iron reduction zone, (5) the sulfate reduction zone, and finally (6) the zone of methanogenesis (Fig. 11). The systematic appearance of these zones is controlled by the decreasing availability of oxygen, as in OM-rich sediments O_2 is the primary oxidant, and the free energy yields of the zone-specific reactions (Bloch et al., 1999; Froelich et al., 1979; Martens et al., 1992).

Under anoxic bottom waters the same order of reactions and consequently the same succession of diagenetic zones are developed, with the important difference that the aerobic oxidation zone is absent together with intense bioturbation. The organic matter subjected to early diagenetic overprints is considered to have an overall Redfield stoichiometry with $(CH_2O)_{106}(NH_3)_{16}(H_3PO_4)$ (Chester, 1990).

Most marine environments are oxic. Therefore approximately 90% or more of the OM reaching the seafloor via normal particle flux is oxidized by O_2 close to the sediment/water interface. Only <10% of the OM flux is oxidized by secondary oxidants (Chester, 1990). Under oxic conditions and the presence of aerobic organisms (Fig. 11), remineralization and degradation of the OM occurs rapidly, and most of the OM is destroyed at this stage. The CO_2 released from the oxidation of the OM can lead to carbonate dissolution, and ammonia (NH_3) can be oxidized (nitrification) to nitrate (Chester, 1990). At the sediment/water interface under oxic bottom water conditions, the rate of oxidation can be significantly accelerated by bioturbation and the consequent mixing of sediments. Levels of oxygen within the sediment are partly controlled by the grain size. The small pore space within silts and muds is commonly leading to a rapid reduction of water circulation with increasing depth, thereby limiting the dissolved O_2 replenishment via diffusion (Killops and Killops, 2009). Open marine pelagic sediments are often oxygenated to depths of 0.5 m, whereas the oxygenation of shelfal, fine-grained deposits, in the absence of bioturbation, is limited to the upper 1–2 mm. Even with intense bioturbation the diffusion of O_2 will only extend as far as 20–30 mm (Killops and Killops, 2009; Tyson, 1995).

Aerobic degradation of OM is ubiquitous within marine sediments until the establishment of suboxic conditions with oxygen concentrations of

<0.2 mL L^{-1} H_2O (Tyson, 1995). With concentrations of dissolved O_2 falling below this level, anoxic conditions are reached. This prompts the degradation of OM via anaerobic processes. Most anaerobes are facultative, which enables them to use oxygen (if in sufficient supply), but with low oxygen levels or in the complete absence of O_2 they have the capability to utilize other inorganic oxidizing agents. These inorganic oxidizing agents are in order of decreasing energy return (1) manganese (IV) to (II), (2) nitrate, (3) iron (III) to (II), (4) sulfate, and (5) bicarbonate (Chester, 1990; Hedges and Keil, 1995; Killops and Killops, 2009). Nitrate reduction, a process typical for the denitrification zone, occurs under depleted O_2 conditions, when the dissolved O_2 levels reach approximately 5% of the respective concentration in oxic waters. The oxygen made available from the reduction of nitrate (or nitric acid, HNO_3) is used to facilitate further oxidation and hence degradation of the OM (Chester, 1990; Tyson, 1995). Due to low nitrate concentrations, the vertical extent of the denitrification zone is usually very limited in marine sediments (Killops and Killops, 2009). Only approximately 3% of the OM in shelf sediments is considered to be oxidized by denitrifying bacteria (Tyson, 1995). Examples for the use of other secondary oxidants, like manganese oxides or iron oxides, are illustrated and explained in Fig. 11.

Sulfate is reduced when dissolved O_2 and nitrate concentrations are further diminished (Tyson, 1995). Most of the OM is broken down within the sulfate reduction zone (SRZ). The depth of the SRZ is dependent on the relative concentration of the OM present (Killops and Killops, 2009). The vertical extent of the SRZ zone is normally two to three times larger than the extent of the oxic zone, as sulfate concentrations in seawater (typically between 20 and 30 mM) exceed those of O_2, which is generally less than or equal to 0.25–0.30 mM (Tyson, 1995). The degradation of OM causes the generation of hydrogen sulfide (H_2S) as a by-product of the activity of sulfate-reducing bacteria (Riding and Awramik, 2000). High concentrations of H_2S can ultimately inhibit the bacterial sulfate reduction (Tyson, 1995; Wignall, 1994). Pyrite (FeS_2) formation is dependent on the availability of free reactive iron. Under euxinic conditions with reactive iron depletion, H_2S can accumulate (sulfurization) and act as a weak acid, leading to poor preservation of OM (Wignall, 1994). The production of methane (CH_4) is predominantly facilitated within the methanogenesis zone. The chemical processes causing the formation of methane are initiated when all available sulfates are depleted (Tourtelot, 1979). Anaerobes (methanogens) synthesize biogenic methane from the smallest fermentation products (mostly alcohols with >C1). CO_2 and H_2 are necessary, where CO_2 is used as an electron donor during the reaction with H_2. Certain methanogens possess the metabolic capabilities to synthesize methane from simple C1-compounds like methanol and formates or methylated amines (Killops and Killops, 2009).

Two important factors are known to exert a strong control on the systematic chain of specific chemical reactions within the diagenetic zone. First, the magnitude of the OM flux from the surface areas to the seafloor. This process is largely controlled by the intensity of the bioproduction. Second, the sediment accumulation rate. Both factors tend to decrease from the continental margins to the deep sea environments. Variations of these two factors determine the amount of OM preserved within the sediments, and the extent to which the various oxidants are consumed (Chester, 1990).

During the late diagenesis, mineral reactions take place in response to burial and the consequent increase in pressure and temperature. The most common reactions are the dissolution–reprecipitation of silicates and carbonates, whereas calcite is generally depleted with increasing depth. Quartz and feldspars undergo dissolution and authigenetic mixed-layered clay minerals and/or illite are formed. The most significant diagenetic changes of the clay minerals occur with the transformation of smectite (montmorillonite) layers into illite layers (a process also called "I/S transformation," which describes the substitution of Al^{3+} and K^+ for Si^{4+} in the clay mineral structure). Smectite transforms into illite at depths around 4 km and temperatures of 120–140°C (Walther, 2009). At around 2.5 km depth chlorite is formed. This autogenesis is believed to be controlled by the concentrations of Mg^{2+} and Fe^{2+} in solution. Chlorite formation below 3.4 km in depth is considered to be related to the reaction of kaolinite with quartz, below 4 km, however, to be the by-product of the aforementioned I/S transformation. At temperatures exceeding 300°C and depths of >5.5 km illite recrystallizes into a sericite and finally a muscovite structure (Walther, 2009).

5.5 Elemental Geochemistry

In the following the contribution of major and trace elements toward the formation of black shales and their importance and use as paleoenvironmental indicators are described and discussed. In particular the major elements magnesium (as MgO), aluminum (as Al_2O_3), silicon (as SiO_2), potassium (as K_2O), calcium (as CaO), titanium (as TiO_2), manganese (as MnO), and iron (as Fe_2O_3), as well as the trace elements phosphorous (P), sulfur (S), scandium (Sc), vanadium (V), chromium (Cr), cobalt (Co), nickel (Ni), copper (Cu), zinc (Zn), arsenic (As), rubidium (Rb), strontium (Sr), zirconium (Zr), molybdenum (Mo), silver (Ag), cadmium (Cd), tin (Sn), antimony (Sb), barium (Ba), mercury (Hg), lead (Pb), bismuth (Bi), thorium (Th), and uranium (U).

Shales (or more precisely "mudstones") are predominantly composed of clay minerals and act as "element dustbins" or sinks. The elemental geochemistry (specifically at the trace element level) can be vast and varied, depending on the composition of the detrital material derived from the hinterland and the overall seawater chemistry. Within marine settings, black shales tend to form because of restricted circulation and/or elevated bioproductivity. A product of these factors are anoxic or reducing conditions within the sediment and also, under certain circumstances, the water column. These changes in redox conditions can drastically effect the absolute concentrations of certain redox-sensitive trace elements. In the past, these redox-sensitive elements have proved to be a powerful tool in the reconstruction of the environmental conditions that prevailed during the time of sedimentation and diagenesis of black shales. In the following, the geochemical behavior and the characteristics of these elements are briefly described and discussed. The major elements will be discussed in ascending order of their relative atomic mass.

5.5.1 Major Elements

Magnesium (as MgO). MgO is commonly associated with light detrital silicates. It also forms an important component of the dissolved ions in seawater. It is prominently represented within dolomites due to cation exchanges (Burgan et al., 2008; de Barry Barnett and Wilson, 1960; März et al., 2009). Burgan et al. (2008) demonstrated that elevated values of MgO and depletions of CaO are related to weathering effects of the original source material.

Aluminum (as Al_2O_3). The bulk of Al_2O_3 in shales belongs to the clay mineral fraction (detrital terrigenous silicates), and derives from the weathering and breakdown of alumosilicates, most prominently feldspars (Cox et al., 1995; de Barry Barnett and Wilson, 1960; Djarnia and Fekirine, 1998; Galindo et al., 2007; Wei et al., 2003; März et al., 2009). High concentrations of SiO_2, Al_2O_3, K_2O, Fe_2O_3 combined with low CaO and MgO contents are commonly interpreted to reflect the dominance of silicate minerals (e.g., illite, kaolinite, chlorite, smectite) and only a subordinate contribution of carbonates, like calcite and dolomite (Burgan et al., 2008). Elevated Al_2O_3 concentrations are interpreted to indicate a relatively high intensity of weathering and increased maturity of the sediment, and can also be related to clay minerals settling from the water column and/or the biological scavenging of Fe-oxyhydroxides, where biogenic opal is acting as a vessel transferring most of the aluminum from the water column to the seafloor (Wei et al., 2003). Moreover, increased Al_2O_3 levels can also be associated with the organic particle flux via adsorption of the dissolved Al to organic matter, and Al has consequently been used as a proxy element to estimate changes in paleobioproductivity rates (Calvert and Pedersen, 2007; Murray and Leinen, 1996). Aeolian dust transported via storm events into the sedimentary basins can have a major impact on the absolute Al_2O_3 concentrations (Calvert and Pedersen, 2007; Wei et al., 2003). Long-term, multiple recycling of sediments results in the sediments being enriched in quartz (sand- and silt-size fractions) and illite (within the clay minerals). This recycling process increases the compositional maturity of the sediments. As a consequence, the coarser grained sediments will start to exhibit trends toward pure quartz arenites and the finer grained mudrocks will become more potassic and aluminous (Cox and Lowe, 1995).

Silicon (as SiO_2). In sedimentary rocks, silicon is generally associated with components such as clastic quartz, siliceous fossils, and alumosilicate minerals (de Barry Barnett and Wilson, 1960; Kuroda et al., 2005). Enrichments of silica can be related to an increased influx of terrigenous material and/or biogenic processes, such as the amplified biomineralization and sedimentation of siliceous skeletons and shells (Calvert and Pedersen, 2007).

Potassium (as K_2O). K_2O is considered to represent the detrital fraction in shales (Wei et al., 2003), originating from the weathering and breakdown of feldspars. It is typical for light silicates (März et al., 2009) such as feldspars, micas, and slightly hydrated clays (Serra, 1984). The relative proportions of Al, Si, K, and Ti in shales are mainly controlled by weathering, transport, and depositional processes (Burgan et al., 2008). High concentrations of K_2O are taken to indicate the presence of larger proportions of illite within a sediment, as illite is commonly formed by the addition of K^+ to kaolinite (Wei et al., 2003). This process is also known as "illitization" of kaolinite (Fedo et al., 1997; Pearce et al., 1991). Elevated K_2O levels in sediments may also be caused by a lack of extensive leaching during the erosion of the hinterland material (Burgan et al., 2008). A significant to highly significant statistical positive correlation of K_2O to Al_2O_3 is interpreted to indicate a detrital origin of these elements (Abdou and Shehata, 2007; Burgan et al., 2008). K_2O/Al_2O_3 ratios are also used to elucidate the provenance of mudrocks (Cox et al., 1995). Elevated K values are interpreted to be indicative for granitic source areas. Another process leading to the enrichment of K is the long-term recycling of sediments similar to the recycling-induced accumulation of Al_2O_3.

Calcium (as CaO). Calcium can occur as carbonates, phosphates, and within aluminosilicate minerals in sedimentary rocks (Kuroda et al., 2005). In black shales calcium coexists in various mobile forms (Galindo et al., 2007), predominantly as calcite and/or dolomite. However, it can also be present within silicate minerals such as anorthite. For black shales, typically low CaO values document the carbonate deficiency of the environment. This depletion in Ca can be the result of the paleogeographical position of the respective sedimentary basin and/or it can be caused by Sr substituting for Ca (Burgan et al., 2008). Furthermore, Lipinski et al. (2003) suggested that carbonate contents are lower in TOC-rich sediments due to dilution effects.

Titanium (as TiO_2). TiO_2 is considered to represent the detrital (terrigenous) fraction in shales (Murray and Leinen, 1996; Wei et al., 2003). It is commonly associated with the heavy mineral rutile, which can also contain considerable amounts of the elements Cr, V, Mn, and Sn (de Barry Barnett and Wilson, 1960). The source of excess titanium in pelagic marine sediments is interpreted to be predominately aeolian dust (Sageman et al., 2003). Positive correlations of Mg and Ti with Al indicate their association with the alumosilicate fraction (Burgan et al., 2008). Ti/Al and Ti versus Al plots are commonly used to discriminate between various source areas, in particular to distinguish between areas of predominantly granitic or basic composition (Abdou and Shehata, 2007). Ti is considered to be a better proxy for the terrestrial influx compared to Al, as the total Al concentrations can be partly of authigenic or biogenic origin (Wei et al., 2003).

Manganese (as MnO). Under oxic conditions, Mn tends to form highly insoluble Mn(IV)-oxyhydroxides, which can act as a substantial trace element sink, preventing the recycling of trace elements into the overlying water column (Calvert and Pedersen, 2007). In anoxic sediments at the sediment/water interface, Mn is reduced and transformed from insoluble Mn(IV)-oxide to highly soluble Mn(II)-sulfide. Mn(II) is not found in black shales, as it is not readily adsorbed to organic matter (Algeo and Maynard, 2004; Calvert and Pedersen, 1993). As Mn(II) is highly soluble it easily migrates within the sediment's pore space upward or downward. The downward migration commonly leads to $MnCO_3$ precipitation within concretions under oxic bottom waters conditions, the upward diffusion finally results in the dissolution of Mn(II) into the water column and hence its removal from the sediment. This usually occurs in anoxic basins (Tribovillard et al., 2006). Mn is commonly associated with other redox-sensitive elements and is hence widely used as a proxy for redox conditions (Chun and Delaney, 2006; Elbaz-Poulicheet et al., 1997; Elbaz-Poulichet et al., 2005; Schröder and Grotzinger, 2007; Wei et al., 2003). However, Tribovillard et al. (2006) cautioned the use of Mn as a redox proxy, due to its highly mobile nature under chemically reducing conditions.

Iron (as Fe_2O_3). Iron is generally associated with alumosilicates, iron oxides, and sulfides such as pyrite in sedimentary rocks (Kuroda et al., 2005). The main source of iron in sedimentary basins is the detrital influx. Under oxygen-depleted (anoxic—euxinic) conditions Fe-oxide and -oxyhydroxides are commonly reduced, either biologically or nonbiologically (März et al., 2009). In the presence of free hydrogen sulfide (H_2S) Fe-oxides representing the reactive iron are transformed into Fe-sulfides, most commonly framboidal pyrite (Werne et al., 2002). If no free Fe is available for the reaction with H_2S, the H_2S will start to accumulate at the sediment—water interface and react with the organic material present (März et al., 2009; Wignall, 1994). High Fe concentrations can also be related to the formation of Fe-hydroxides and/or the presence of detrital biotite (Burgan et al., 2008). Ciglenečki et al. (2006) noted that FeS_2 can adsorb trace elements to its reactive surface and therefore act as an important sink for Mo, Mn, Co, Cu, Ni, and Cr. The presence of H_2S and the thereby induced reactions result in the removal of Ag, Cd, Cu, Ni, and Zn out of solution, and the formation of generally insoluble sulfides (Calvert and Pedersen, 2007). The absence of a strong correlation with Al_2O_3 can be interpreted to indicate an authigenic origin of Fe-oxides and may hence be used as a proxy for the prevailing redox conditions during Fe-oxide formation (Abdou and Shehata, 2007). Fe concentrations (and speciation) as well as Fe/Al ratios are widely used as a proxy for pyrite formation and redox conditions using the "Degree of Pyritization" (Lyons et al., 2003; Wei et al., 2012). In the presence of sufficient amounts of detrital iron and during times of enhanced organic matter burial, the amount of diagenetic pyrite formed will show a highly significant, positive correlation with TOC due to H_2S formation at the sediment—water interface (Schultz, 2004). In the absence of such a strong correlation, the iron may reside within the residual fraction, which is commonly composed of iron-containing silicate minerals (Galindo et al., 2007). A negative correlation between iron and silicon is commonly regarded to indicate changes within the environment or the provenance area of the original source material (Burgan et al., 2008).

5.5.2 Trace Elements

In the following section the geochemistry and the chemical behavior of trace elements analyzed for this study will be briefly discussed. Like the major elements, they have been sorted ascending, according to their relative atomic mass.

Phosphorous (P). P is mainly present as HPO_4^{2-} in modern day oceans, where it is strongly coupled to the cycles of Fe and OM (März et al., 2009). Under oxic conditions Fe oxides and oxyhyroxides can adsorb substantial amounts of P along with OM, both acting as carriers of P to the seafloor (Calvert and Pedersen, 2007; Tribovillard et al., 2006). The formation of Fe oxides can also lead to the coprecipitation of phosphates. Within reducing environments P is discoupled from the Fe oxides and OM and subsequently released as PO_4^{3-} back into the water column. There it is reutilized as a readily available nutrient by the phytoplankton. It becomes hence reintroduced to the biological cycle resulting in low burial rates of P together with OM. High bioproductivity combined with a high accumulation rate of OM may therefore develop without any significant P burial. Because of the rather intricate processes described earlier, the use of P as a productivity proxy is not straight forward (März et al., 2009; Tribovillard et al., 2006). In detrital form P exists predominantly as the calcium phosphate mineral apatite (de Barry Barnett and Wilson, 1960).

Sulfur (S). Sulfur is present in anoxic, OM-rich sediments as a result of bacterial sulfate reduction. The bacterial sulfate reduction requires sulfate dissolved in pore waters as an electron donor for the oxidation of OM in the absence of oxygen (Dean et al., 1984). Sulfate reduction occurs when all other oxidants (such as O_2 > Mn-oxides > NO_3 > Fe-oxides) have been used up and are hence depleted (Dean et al., 1984). Excess S is locked within metal sulfides in the overlying water column, if there are free reactive trace metals available (Dean et al., 1984; Lipinski et al., 2003; März et al., 2009).

Scandium (Sc). Little is known about the chemical behavior of Sc in organic-rich shales. Sc concentrations in seawater are very low. It is associated mostly with detrital terrestrial material in marine sediments. It is a highly compatible chemical element in the continental crust, which means that it is not highly concentrated in any nonbiological mineral (Wei et al., 2003). However, Sc can be slightly enriched within biogenic apatite and certain ferromanganese oxide minerals in pelagic sediments (Wei et al., 2003). Sc behaves like rare earth elements in sedimentary rocks and is enriched in weathering residues (Wei et al., 2003).

Vanadium (V). Vanadium is a redox-sensitive element with a strong euxinic affinity like U, Mo, and Zn (Algeo et al., 2004). Its geochemical behavior is similar to that of Ni (Wignall, 1994). V is present in a wide variety of minerals, by far the most important being patronite (VS_4, Hillebrand, 1907) and vanadinite, $Pb_5(VO_4)_3Cl$ (Trotter and Barnes, 1958). Under oxic conditions V is normally present as an oxy-anion (HVO_4^{2-}, $H_2VO_4^-$) or as V(V). It can be incorporated into Mn + Fe-oxides and hydroxides, possibly also into kaolinite, which all act as vessels for V to the seafloor. When dissolved, vanadium tends to occur as oxovanadium(IV) cation, which is easily adsorbed to clay particles once reduced to V(III) during burial (Elbaz-Poulicheet et al., 1997; Schröder and Grotzinger, 2007; Jones and Manning, 1994; Wignall, 1994). The removal from the water column and the subsequent transfer to the sediment are facilitated by principally two processes: (1) Under anoxic conditions V(V) is reduced to V(IV) because of the simultaneous oxidation of organic compounds, like humic and fulvic acids. This chemical reduction process leads to the formation of the soluble $VO(OH)_3$-hydroxyl species, which remains dissolved within the water column, as well as insoluble $VO(OH)_2$-hydroxides, which precipitate out of solution and are adsorbed onto the reactive surface of clay minerals at the sediment—water interface. The direct participation of organic substances (humic and/or fulvic acids) during these chemical reactions generally results in a strong positive correlation between V and TOC (Abanda and Hannigan, 2006; Abdou and Shehata, 2007; Algeo and Maynard, 2004; Arthur and Sageman, 1994; Brumsack, 1986; Calvert and Pedersen, 1993; März et al., 2009; Pi et al., 2013; Schröder and Grotzinger, 2007; Tribovillard et al., 2006). (2) Under sulfidic conditions (characterized by the presence of H_2S), V(IV) is further reduced to V(III) and forms insoluble $V(OH)_3$-hydroxides and V_2O_3-oxides, which directly precipitate out of solution and are subsequently transferred to the sediment. As these reactions are carried out without any involvement of organic materials, V shows typically no correlation with TOC values (Arthur and Sageman, 1994; Calvert and Pedersen, 1993; März et al., 2009; Pi et al., 2013; Tribovillard et al., 2006). A third process is offered by the incorporation of V into FeS during sulfide formation. However, evidence in support of this process is currently scarce and controversial, leaving this model as a matter of still ongoing debate (Algeo and Maynard, 2004; Tribovillard et al., 2006). During the lithification and following diagenesis, V(III) readily substitutes Al at the octahedral sites of authigenic clay minerals, a reaction known also to take place extensively in the course of the recrystallization of clay minerals (Tribovillard et al., 2006).

Chromium (Cr). Cr belongs to the group of redox-sensitive elements (Lipinski et al., 2003). Cr can be fixed in high amounts in sediments under reducing conditions either by OM via biocycling (according to Brumsack (1986) Cr is an important nutrient for phytoplankton) or by the adsorption to Fe- and Mn- oxyhydroxides (Calvert and Pedersen, 1993). In oxic environments Cr is present as Cr(VI) within the anion (CrO_2^{4-}). Under anoxic conditions Cr(VI) is typically reduced to Cr(III) to form the aquahydroxyl or hydroxyl cations $Cr(OH)^{2+}$. Cr(III)-containing molecular complexes are being precipitated as insoluble $Cr(OH)_3$ or Cr_2O_3 under high pH levels and incorporated into humic or fulvic acids under low pH conditions. Cr(III) is not known to form an insoluble sulfide and is hence not incorporated into pyrite due to its structural and electronic incompatibilities (Tribovillard et al., 2006), unlike other trace elements such as Mo, Pb, Cu, Ni, and Zn (Algeo and Maynard, 2004). Consequently, when OM and/or oxyhydroxides are remineralized by sulfate-reducing bacteria, Cr may be lost to the overlying water column (Tribovillard et al., 2006). Cr is frequently used together with V (V/Cr ratio) to evaluate and quantify redox conditions (Abdou and Shehata, 2007; Wignall, 1994) within the diagenetic nitrate reduction zone. V/Cr values above two are interpreted to indicate anoxia, whereas values below two are thought to be characteristic for oxic conditions (Nagarajan et al., 2007; Schröder and Grotzinger, 2007). Aside the phases that undergo changes and alterations while being metabolized, Cr is mainly hosted within the detrital fraction of sediments by minerals such as chromite as well as clay minerals, where Cr substitutes Al, and a wide range of ferromagnesian minerals, where Cr substitutes Mg (Algeo and Maynard, 2004; Calvert and Pedersen, 1993; Nagarajan et al., 2007; Tribovillard et al., 2006; Wignall, 1994).

Cobalt (Co). Co belongs to the redox-sensitive/sulfide-forming trace elements (Lipinski et al., 2003). However, in sediments, Co is strongly tied to the detrital fraction, limiting its use as a reliable redox proxy (Algeo and Maynard, 2004; Brumsack, 1986; Tribovillard et al., 2006). In oxic environments Co is typically present as a dissolved cation (Co^{2+}), although it can also be associated with humic and fulvic acids (Algeo and Maynard,

2004; Charriau et al., 2011; Tribovillard et al., 2006). Co can be accumulated in high amounts within sediments under reducing conditions, either by being bound to OM, by forming autonomous sulfides (CoS) or being coprecipitated together with Fe-sulfides (Algeo and Maynard, 2004; Lipinski et al., 2003). The Co uptake is kinetically very slow, its concentrations in authigenic sulfides are therefore low (Algeo and Maynard, 2004). In unaltered sediments formed under euxinic conditions Co shows positive correlations with other redox-sensitive elements such as U, V, and Zn.

Nickel (Ni). Ni is considered a redox-sensitive and sulfide-forming element (Lipinski et al., 2003). It is present in both the organic and detrital fraction of sediments in form of Fe oxides, hydrous Fe and Mn oxides, and clay minerals (Algeo and Maynard, 2004; Charriau et al., 2011). In oxic to suboxic waters, Ni appears either as a soluble cation (Ni^{2+}, $NiCl^{+}$) or adsorbed to humic and fulvic acids (Algeo and Maynard, 2004; März et al., 2009; Tribovillard et al., 2006). As an important nutrient Ni is a strong biolimiting element and involved within the biological cycle during primary production (Calvert and Pedersen, 1993; März et al., 2009; Piper and Calvert, 2009; Tribovillard et al., 2006; Wignall, 1994). The resulting OM serves as a carrier for Ni to the seafloor. Under dysoxic conditions Ni tends to form molecular complexes with OM. Under anoxic conditions the presence of H_2S leads to the precipitation of NiS. However, it may also be incorporated into FeS (Algeo and Maynard, 2004; Tribovillard et al., 2006; Wignall, 1994). Secondary coupling of Ni to OM can form during the process of OM sulfurization, although in the course of the same process Ni can also be released (März et al., 2009). The overall geochemical behavior of Ni is very similar to that of V.

Copper (Cu). Cu is considered to be a redox-sensitive element (Lipinski et al., 2003). It is either fixed in sediments under reducing conditions by being bound to OM [with Cu acting as a nutrient within the biocycle (Algeo and Maynard, 2004; Calvert and Pedersen, 1993], is precipitated out of solution in the form of autonomous sulfides, or coprecipitated together with Fe-sulfides (Lipinski et al., 2003). The transfer of Cu from the water column to the sediment via the soluble $CuCl^{+}$ ions is either facilitated by forming organic metal ligands in humic acids (Algeo and Maynard, 2004; Tribovillard et al., 2006) or by the adsorption to Fe-Mn-oxyhydroxides. Cu may be lost to the overlying water column when the OM and/or oxyhydroxides are being remineralized by sulfate-reducing bacteria (Algeo and Maynard, 2004; Tribovillard et al., 2006). Under anoxic conditions Cu(II) is reduced to Cu(I) and subsequently precipitated as CuS or Cu_2S as an independent sulfide phase or it is incorporated into FeS (Algeo and Maynard, 2004; Tribovillard et al., 2006). In unaltered sediments Cu therefore shows strong associations to OM and other redox-sensitive elements (e.g., Ni, Zn, and U). In (hemi-) pelagic sediments with low accumulation rates, Cu may be fixed during the authigenic formation of nontronite or smectite minerals (Tribovillard et al., 2006). Cu/Zn ratios are being used as redox indicators, with high values indicating reducing conditions and low values specifying oxidizing environments (Nagarajan et al., 2007).

Zinc (Zn). Zn is a redox-sensitive, sulfide-forming element (Lipinski et al., 2003) of strong euxinic affinity comparable to Mo, U, and V (Algeo et al., 2004). It is present under oxic to suboxic conditions in the form of soluble Zn^{2+} and $ZnCl^{+}$-cations or, as an important biolimiting nutrient for phytoplanktonic organisms, directly incorporated into OM (Algeo and Maynard, 2004; Brumsack, 1986; Calvert and Pedersen, 1993; Tribovillard et al., 2006). The OM resulting from the planktonic biomass serves as the main carrier from the water column to the sediment (Brumsack, 1986). It can also be adsorbed to precipitating Fe-Mn oxides and -hydroxides, which will act as additional vessels for the transfer to the seafloor (Tribovillard et al., 2006). Zn in the absence of OM and Fe-Mn oxides and -hydroxides adsorbs readily to the reactive surface of clay minerals and is hence used as a proxy to estimate the clay fraction within siliciclastic sediment (Svendsen and Hartley, 2001). For this purpose it is regarded to be far more reliable than the obvious Al_2O_3, as Al_2O_3 is also present within other minerals than clays, such as feldspars (Svendsen and Hartley, 2001). Under reducing conditions and in the presence of H_2S, zinc sulfides such as sphalerite ((Zn, Fe)S) are more readily formed than FeS, yet Zn can also be incorporated into the FeS crystal lattice (Algeo and Maynard, 2004; Tribovillard et al., 2006). The correlation or noncorrelation of the elemental ratios Zn/Al to TOC/Al and S/Al is commonly used to determine whether the sedimentary Zn is localized within the sulfurized OM or directly in pyrite (März et al., 2009). The bacterial decay of OM can liberate Zn from organic-metal complexes, from which it is incorporated into authigenic FeS (Algeo and Maynard, 2004), thus leading to a systematic Zn enrichment of the sulfide phases and a consequent depletion of Zn within the OM. High amounts of Zn within coarser clastic sediments are sometimes related to syn- or postdepositional hydrothermal activity (Charriau et al., 2011; Lipinski et al., 2003).

Arsenic (As). As can be adsorbed to Fe/Mn hydroxides. The dissolution of these hydroxides under reducing conditions results in the release of As. The distribution and amounts of the inorganic As species (As(III) and

As(V)) are largely controlled by the redox state of the environment, with the reduction of As(V) leading to the formation of As(III). During the formation of pyrite As(III) is readily adsorbed to its reactive surface leading to so-called arsenical pyrite (de Barry Barnett and Wilson, 1960; Blanchard et al., 2007). The oxidation of arsenopyrite is also known to be a prominent mechanism responsible for the accumulation of As (Xie et al., 2009). The organic species, monomethylarsonate (MMAs) and dimethylarsonate (DMAs) are frequently found within natural waters, and contribute significantly to As speciation. It has been shown that macroalgae and phytoplankton have the capability to reduce or methylate As(V) within the water column. The subsequent bacterial degradation of the phytoplankton and its associated organic matter is known to generate a wide variety of organoarsenical species (Elbaz-Poulicheet et al., 1997). Brumsack (1986) suggested that the biocycling of As may not be significant, as modern day seawater exhibits little concentration changes with increasing depth, unlike, for example, Cd and Zn. As a redox-sensitive element As is used to evaluate the environmental redox conditions.

Rubidium (Rb). Rb is considered a detrital element that is normally enriched in shales and siltstones. In shales it resides within clays and micas and hence shows commonly positive correlations with Al_2O_3 and K_2O (Dypvik and Harris, 2001; Kampunzu et al., 2005; Xu et al., 2010). For the geochemical characterization of siliciclastic sediments, Rb is often used as a grain size indicator. Furthermore, K/Rb ratios have also been demonstrated to be diagnostic for different salinity levels (Burgan et al., 2008), with K/Rb ratios between 250 and 300 indicating nonmarine to brackish environments and 150–200 being characteristic for marine shales (Abdou and Shehata, 2007; Campbell and Williams, 1965).

Strontium (Sr). The chemical behavior and systematics of Sr in sedimentary rocks is complicated (Veizer and Demovič, 1974). With a concentration of approximately 8 mg/L, Sr is a major component of modern day seawater (Turekian, 1964). The oceanic Sr concentrations are mostly controlled by the weathering of limestones (adding Sr to the oceanic Sr pool) and diagenetic processes (responsible for the removal of Sr). Sr is considered to be a nonessential biological element; however, because of its intermitting ion radius it can substitute for Ca^{2+} and Ba^{2+} in living organisms (De Vos et al., 2006; Turekian, 1964). Furthermore, Sr concentrations in sediments can also be controlled and changed by its adsorption to various clay minerals. However, Turekian (1964) suggested that clay minerals are not an efficient device for the removal of Sr in deep sea environments, as clay minerals exhibit only extremely low exchange-adsorption rates as defined by the respective mass–action equilibrium equations (Wahlberg et al., 1965). The major sink for Sr is considered to be deep-sea carbonate deposits via the extensive substitution of Sr^{2+} for Ca^{2+} ions in carbonate minerals (such as aragonite), and to a lesser extent by the exchange of K^{+} ions in minerals such as K-feldspar and plagioclase (De Vos et al., 2006; Krauskopf and Burg, 1995). Nonetheless, feldspars are considered to be the principal carrier of Sr in siliciclastic sedimentary rocks. Ba^{2+} ions within baryte ($BaSO_4$) and witherite ($BaCO_3$) can also, under certain circumstances, be substituted by Sr (De Vos et al., 2006). Low-temperature paleoenvironmental conditions, as typical for higher latitudes, can significantly influence the distribution of sedimentary Sr concentrations (Dypvik and Harris, 2001; Kiipli et al., 2004). The generally low concentrations of Sr in black shales are considered to be related to the low concentrations of Ca within this type of sediment and the absence of certain Sr-containing rock-forming minerals such as celestite ($SrSO_4$), strontianite ($SrCO_3$), hornblende, and plagioclase. The very rare occurrence of celestite and strontianite within fine-grained siliciclastic deposits is most commonly interpreted to indicate postdepositional hydrothermal activity or evaporitic conditions within extremely shallow water environments (De Vos et al., 2006; Turekian, 1964). Burgan et al. (2008) suggested that low concentrations of Sr in sediments are related to deposition under chemically reducing environmental conditions. In aqueous environments with pH values below 4.5, Sr exists as the highly mobile Sr^{2+} ion. Under neutral (pH 5–7.5) conditions $SrSO_4$ forms, and finally, under alkaline (>pH 8) conditions $SrCO_3$ is dominant (De Vos et al., 2006). Within the published literature Rb/Sr ratios have been used to infer the intensity of the chemical weathering that affected the hinterland from which the sediments derived (Romer and Hahne, 2010; Taylor et al., 1983; Xu et al., 2010). Higher values are interpreted to indicate intense weathering, lower values are thought to point toward more moderate weathering conditions. Chemical weathering can ultimately lead to leaching of the Ca-Sr system, leaving a residue enriched in K-Rb and depleted in Ca-Sr behind (Romer and Hahne, 2010; Xu et al., 2010). Higher Sr concentrations can occur within granitic rocks, where Sr and Ba correlate with Ca, particularly within feldspar minerals (Burgan et al., 2008). The fractionation of these two elements can occur during the selective weathering of these feldspars. Weathering of plagioclase results in the depletion of Sr relative to Ba as plagioclase is easier to weather than K-feldspar; furthermore, Sr is more mobile relative to Ba (De Vos et al.,

2006; Nyakairu and Koeberl, 2001). Sr/Ca ratios have been utilized as a salinity proxy with low values indicating freshwater conditions and opposing high values relating to the development of marine saline conditions (Chang et al., 2004; Dodd and Crisp, 1982; Limburg, 1995; Turekian, 1955).

Zirconium (Zr). Zirconium resides in the heavy mineral fraction and is commonly associated with the coarser grained fraction of otherwise fine-grained sediments (Svendsen and Hartley, 2001). Hence, higher Zr values are found in sandstones than in shales. Zr is commonly used as a terrigenous (detrital) proxy element together with Ti (Zr/Ti ratios). Higher Zr/Ti ratios are considered to indicate an input of siliciclastic sediments sourced from shallower and more energetic environments (de Barry Barnett and Wilson, 1960; Lipinski et al., 2003; März et al., 2009).

Molybdenum (Mo). Mo is considered a redox-sensitive element with a strong euxinic affinity. It has been demonstrated to be a useful and important paleoredox indicator (Algeo et al., 2004; Calvert and Pedersen, 1993; Lipinski et al., 2003; Wilde et al., 2004). In seawater under oxic to suboxic conditions, Mo is present as Mo(VI), mostly in the form of the soluble molybdate oxyanion, MoO_4^{2-} (Calvert and Pedersen, 1993; März et al., 2009; Tribovillard et al., 2006). With concentrations up to 10 ppb, Mo is one of the most abundant trace metals in modern oceans (Algeo and Maynard, 2004; Brumsack, 1986; Piper and Calvert, 2009; Schröder and Grotzinger, 2007; Werne et al., 2002; Wignall, 1994) and it is also one of the most enriched trace metals in organic-rich black shales (Brumsack, 1986; Wignall, 1994). The concentrations of Mo can be correlated to the amount of OM (Vine and Tourtelot, 1970), as it is incorporated into OM via the adsorption to humic acids (Brumsack, 1986; Tribovillard et al., 2006). High concentrations of Mo have been regarded to be a powerful proxy for high paleobioproductivity. Mo in its oxic form is considered to play an important role during the biological processes of N fixation and it is rapidly scavenged by Mn oxides or oxyhydroxides (Calvert and Pedersen, 1993; März et al., 2009; Lipinski et al., 2003). In the presence of H_2S, which is an essential component to Mo fixation (Lyons et al., 2003; Meyers et al., 2005), the Mo oxyanion is converted to particle-reactive (MoS_4^{2-}) thiomolybdates (März et al., 2009; Pi et al., 2013). These are readily extracted by and added to OM (predominantly humic substances) and also incorporated into FeS (Algeo and Maynard, 2004; Brumsack, 1986; Calvert and Pedersen, 2007; Elbaz-Poulichet et al., 2005; März et al., 2009; Schröder and Grotzinger, 2007; Tribovillard et al., 2006). Element ratios such as Mo/Al, S/Al and the ratio of TOC/Al are used to determine whether

Mo is incorporated within the nonbiogenic particles of the sediment, sulfurized OM, or FeS (Hild and Brumsack, 1998; März et al., 2008; Warning and Brumsack, 2000).

Silver (Ag). Silver is regarded to belong to the group of redox-sensitive elements. It can be fixed in high amounts within sediments under reducing conditions as it can be adsorbed to organic matter and sometimes even act as a nutrient (Calvert and Pedersen, 2007). However, little is known about the exact chemical behavior of Ag under reducing conditions (Lipinski et al., 2003).

Cadmium (Cd). Cd is an important nutrient for phytoplankton (Cullen et al., 1999; Lane et al., 2005; Xu and Morel, 2013). It is present under oxic to suboxic conditions as a soluble cation ($CdCl^+$ or Cd(II)) within the water column and incorporated into OM, which serves as its main carrier to the seafloor (Calvert and Pedersen, 1993; Tribovillard et al., 2006). Cd forms cadmium sulfide (CdS) in the presence of H_2S far more readily than FeS, and generates separate sulfide phases rather than being incorporated in FeS (März et al., 2009; Tribovillard et al., 2006). As Cd may be bound to OM via the biological cycle, the Cd/Al ratio is used as a paleobioproductivity indicator within the geological record (Lipinski et al., 2003). Cd is enriched in both mildly and strongly reducing sediments (Tribovillard et al., 2006). FeS_2 may incorporate/adsorb certain amounts of Cd, which makes it necessary to determine to what extent the sedimentary Cd is related to OM or pyrite by analyzing the correlations between Cd/Al, TOC/Al, and S/Al (Charriau et al., 2011; März et al., 2009).

Tin (Sn). Sn is considered to be detrital in nature. Within the water column the surface concentrations are high due to fluvial/aeolian input, but Sn is depleted rapidly with increasing depth throughout the water column because of its highly reactive nature (Brumsack, 1986).

Antimony (Sb). Sb has a similar geochemical behavior to As (Elbaz-Poulicheet et al., 1997) and under oxic conditions occurs in two stable forms (Sb(III) and Sb(V)). Sb is considered a redox-sensitive element (Lipinski et al., 2003), which is fixed to the sediment under reducing conditions either by OM or as autonomous sulfides that are coprecipitated together with FeS. A reduction step is required at the redox boundary to immobilize Sb in reducing environments (Lipinski et al., 2003).

Barium (Ba). Ba within marine systems is most commonly associated with detrital plagioclase crystals and micas. Concentrations of Al, K, or Ti are used to correct for this detrital fraction (Tribovillard et al., 2006). A further prominent Ba-containing mineral phase is baryte ($BaSO_4$). The enrichment of Ba in pelagic sediments relative to crustal values is normally associated with baryte (Calvert and Pedersen, 2007). Ba concentrations are widely applied as an indicator for paleoproductivity levels (Arnaboldi and Meyers, 2006), using the concentrations of biogenic Ba or Ba_{excess}. Under reducing conditions, baryte is commonly adsorbed to decaying organic detritus (Chun and Delaney, 2006; Kiipli et al., 2004; Tribovillard et al., 2006). Ba is also present in environments without significant amounts of organic matter, yet in far lower concentrations (Bonn et al., 1998; Brumsack, 1986). It has to be noted that the Ba proxy should be used with caution, as baryte dissolution is known to occur within sulfate-depleted pore waters, similar to the chemical behavior of Mn (Brumsack, 1986; Lipinski et al., 2003). In other words, Ba may accumulate in sediments with high OM fluxes, but can migrate with the onset of early diagenesis (Tribovillard et al., 2006). Furthermore, hydrothermal fluids are commonly enriched with Ba, Sr, Pb, Zn, and Mn, which can affect/overprint the biogenic barium signal (Tribovillard et al., 2006).

Mercury (Hg). Its main occurrence is in cinnabar (HgS). Cinnabar is known to form post-depositionally and is related to hydrothermal activity.

Lead (Pb). A redox-sensitive/sulfide-forming element. In oxic marine environments Pb is present as both the Pb^{+}-cation and in form of the soluble Pb-carbonate (Algeo and Maynard, 2004). Under anoxic conditions Pb is rapidly removed from the water column, as it is more readily scavenged than it is released (Brumsack, 1986). As insoluble PbS it is commonly precipitated independent from FeS (Algeo and Maynard, 2004). Because of the rapid depletion, Pb exhibits only weak positive correlations with TOC, but it tends to correlate well with Mo and V, as Pb and Mo can together be incorporated in pyrite and V is coprecipitated as an oxyhydroxide (Algeo and Maynard, 2004; Charriau et al., 2011).

Bismuth (Bi). Little is known about the chemical behavior of Bi. It is considered to be a redox-sensitive and sulfide forming element. It is easily bound to organic matter and precipitated as autonomous sulfides or coprecipitated together with Fe-sulfides, along with other redox-sensitive elements such as commonly Co, Cu, Cr, and Ni and to a lesser extent Ag, Cd, Mo, Re, Sb, Ti, U, and V (Lipinski et al., 2003). Slightly elevated Bi values are regarded to indicate its fixation in sulfides under anoxic conditions. However, these slight enrichments may also simply be related to its overall low abundance and relatively short residence time in seawater.

Thorium (Th). Th concentrations mostly correspond to the detrital influx of Th-bearing clay minerals, which originate predominantly from acidic igneous rocks (Serra, 1984). Th is mainly found in clays of detrital origin, where it is adsorbed to the platelets. A considerable amount of Th was demonstrated to reside within in FeS fraction (Galindo et al., 2007). To a lesser extent, humic acids are known to act as sinks for Th. The ratios of Th and U concentrations have been widely used as a redox indicator (Dypvik and Harris, 2001; Jones and Manning, 1994; Nagarajan et al., 2007), with U/Th values below 1.25 suggesting oxic environments and values greater than 1.25 indicating suboxic to anoxic conditions (Arthur and Sageman, 1994; Nagarajan et al., 2007; Svendsen and Hartley, 2001).

Uranium (U). U preserved in sedimentary basins derives from two fundamentally different sources. (1) The detrital U (U_{detr}) stems from U-bearing minerals (e.g., zircon) that have been extracted from rocks exposed to weathering and erosion in the hinterland, whereas (2) the authigenic U (U_{auth}) results from redox-driven precipitation of U out of solution. The major source of U in seawater is the weathering of igneous (acidic), metamorphic, and sedimentary rocks (Serra, 1984). The U is dissolved within ground waters and transported to the sea (Swanson, 1961). U is a highly redox-sensitive element and strongly related to anoxic bottom waters (Lipinski et al., 2003). Therefore U can appear in relatively high concentrations (Lüning et al., 2000) of up to approximately 200 ppm (Fisher and Wignall, 2001) within organic-rich black shales, compared to just 2.5 ppm within the average upper crust (Wedepohl, 1991) or 3.1 ppm within the international PAAS (Post-Archaean Australian Shale) standard (Taylor and McLennan, 1985). Under oxic water conditions, U occurs predominantly as a soluble uranyl (U^{6+}) (tri-)carbonate complex ($UO_2(CO_3)_3^{4-}$), only near the redox boundary it appears in the form of the also highly soluble uranyl (UO^{2+}) ion (Algeo and Maynard, 2004; Calvert and Pedersen, 1993; Fisher and Wignall, 2001; Nagarajan et al., 2007; Tribovillard et al., 2006; Wignall, 1994). Under anoxic conditions, U is reduced to an insoluble state and will precipitate out of solution: in anoxic environments with active sulfate reduction the uranyl (U^{6+}) complexes are reduced to insoluble (U^{4+} or U^{5+}) fluoride complexes, which are strongly adsorbed to organic matter and particle surfaces (Fisher and Wignall, 2001; Elbaz-Poulichet et al., 2005; Lev and Filer, 2004; Nagarajan et al., 2007; Schröder and Grotzinger, 2007). In particular, the fixation of uranyl ions to the organometal ligands of humic acids forms one of the major OM-driven sinks for authigenic U (Elbaz-Poulicheet et al., 1997; Lüning et al., 2000; Wignall, 1994). It can also precipitate unrelated to biological matter as crystalline uraninite (UO_2), mainly in the presence of H_2S (Algeo and Maynard, 2004; Pi et al., 2013). It has been suggested that the U associated with settling OM and adsorption only equates to a minor component of the total U budget in sediments. The concentration of U in diagenetic phases (such as apatite and monazite) is the main U sink in OM-rich sediments (Lev and Filer, 2004). Permanent pore-water anoxia is required for the constant U fixation, as diagenetically induced reoxygenation of the sediment will remobilize U as it is converted back to soluble U^{6+}, resulting in its escape back into the overlying water column (Tribovillard et al., 2006; Wignall, 1994). Three important factors control the concentrations of authigenic U within sediments: (1) the intensity of the anoxia, (2) the abundance of fixing components (OM), and (3) the sedimentation rate (Wignall, 1994). The relationship between U and TOC in sediments is regarded to be strongly influenced by (1) the U content of the water column, (2) the carbonate content, (3) the sedimentation rate, and finally, (4) the productivity versus preservation of OM in sediments (Lüning et al., 2003b). The U-TOC proxy is considered to be useful for even strongly weathered black shale successions, where the original OM has been largely destroyed by oxidation and dissolution (Lüning et al., 2003b).

When interpreting the total U concentrations of sedimentary rocks, it is important to determine and separate the authigenic and detrital fractions of U, to identify, reconstruct, and quantify the extent of potential anoxia that may have prevailed during the time of deposition. Wignall and Myers (1988) introduced a formula, in which the total U is formed as the sum of the concentrations of the detrital and the authigenic U. They noted that the amount of Th present within detrital U-bearing minerals is always on average three times higher than the concentration of U within these minerals. This allowed them to express the detrital amount of U as the equivalent of Th/3 and to arrive at the equation: $U_{auth} = U_{tot} - (Th/3)$.

5.5.3 Element Associations and Environmental Control

The main constituents of marine sediments are detrital minerals (mainly clays and quartz), authigenic sulfides, authigenic phosphates, and finally OM (Chester, 1990; Wignall, 1994). The intensity of the chemical weathering of the original rocks greatly affects the major element composition of mudrocks (Cox et al., 1995), such as the weathering and breakdown of feldspars ultimately controls the major elemental abundances of, for example, K and the trace elements Sr and Ba. The concentrations of specific elements within any siliciclastic sediment are predominantly controlled by the presence or absence of the following mineral classes (Fig. 12).

Mineral classes	Mineral phases	Elements and oxides
Silicates (detrital component)	quartz, feldspars, mica, clay minerals	**SiO_2, Al_2O_3, TiO_2**, Fe_2O_3, MgO, CaO, **Na_2O, K_2O, Rb**, Sr, **Zr**, Ba and transition metals (adsorbed)
Carbonates	calcite, dolomite, siderite, etc	**CaO, MgO**, Fe_2O_3, MnO, Sr, Ba C, O
Phosphates	apatite	**P_2O, CaO**, Ce
Metal-oxides/Hydroxides/ Sulphides (authigenic component)	goethite, hematite, pyrite etc.	**Fe, Mn, transition metals (adsorbed)**, Zr, Ba
Organic Matter		**C, transition metals (absorbed)**, Ba

Figure 12 Main components of siliciclastic sediments divided into mineral classes, associated mineral phases, and their chemical composition. The most significant elements are highlighted in bold. *Modified after Scheffler, K., 2004. Reconstruction of Sedimentary Environment and Climate Conditions by Multigeochemical Investigations of Late Palaeozoic Glacial to Postglacial Sedimentary Sequences from SWGondwana (Ph.D. thesis), Rheinischen Friedrich-Wilhelms-Universität Bonn, p. 243.*

The silicates correspond mostly to the detrital fraction of the sediments, whereas the formation of Fe-/Mn hydroxides/oxides and sulfides depends on and is associated with the redox conditions of the water column. The hydroxides/oxides formed in the oxic top waters act as carriers for various trace elements toward the seafloor. The hydroxides/oxides are subsequently reduced in the presence of H_2S and the trace elements hereby released are adsorbed/incorporated into OM or authigenic sulfides (mainly FeS). The chemical composition of the waters (and trace element enrichment in the sediment) at the sediment/water interface relies on (1) the sedimentation of metal-rich particulate material (detrital component), (2) trace metal adsorption onto clays, OM, and Fe-/Mn-hydroxides/oxides, and (3) the reduction of the OM, Fe-/Mn-hydroxides/oxides and consequent capture of the trace elements. The enrichment of certain chemical elements within sediments is controlled by: (1) heterolithic sedimentary successions and associated partial element mobilization during diagenesis, (2) prevailing redox conditions within the water column and the sediment, (3) potential hydrothermal alteration, (4) eolian influx, (5) fluvial runoff, and (6) elemental composition of the seawater (Nijenhuis et al., 1999). Piper and Calvert (2009) suggest that the concentrations of elements within marine sediments are controlled by two fundamental parameters: (1) terrestrial-derived clastics and subsequent transport to the oceans via rivers and the atmosphere, and (2) syn- and postdepositional hydrothermal activity affecting the elemental concentrations, e.g., Ba, Sr, Pb, Zn, Mn, and Fe (Tribovillard et al., 2006).

Trace elements in marine settings are classified either as lithogenous (detrital) or seawater sourced (authigenic). Algeo and Maynard (2004) developed and applied the following formulas to determine whether a trace element is related to one of these two classes or to the organic fraction of the sediment.

$$X_{detr} = X/Al_{back} \times Al \tag{1}$$

where X_{detr} is the detrital concentration of trace element X (ppm); X/Al_{back} is the average Al-normalized "background" concentration of trace element X ($\times 10^{-4}$); and Al is the Al concentration of the sample (wt.%).

$$X_{org} = [b_{at} + m_{at} \times TOC] \times Al - X_{detr} \tag{2}$$

where X_{org} and X_{detr} are the "organic" and "detrital" concentrations of element X (ppm); b_{at} and m_{at} relate to the y-intercept and slope of the "anoxic trend" of element X, respectively, with the "anoxic trend" referring to sediments forming under nonsulfidic anoxic conditions. The y-intercept and slope are to be taken from the correlation coefficients of the analyzed redox-sensitive element pairs (see Table 2 of Algeo and Maynard, 2004); TOC and Al are the TOC and Al concentrations of the sample (wt.%).

$$X_{sulf} = X_{tot} - \left(X_{org} + X_{detr}\right) \tag{3}$$

where X_{sulf} and X_{tot} are the "sulfidic" and total concentrations of element X (ppm), and X_{org} and X_{detr} are the values calculated from Eqs. (1) and (2).

Detrital elements are the major and trace elements, whose concentrations within sediments are controlled by the amount of siliciclastic influxes from the hinterland (mainly Si, Al, K, Ti, and Zr). The sedimentary detrital fraction consists principally of silicate minerals such as quartz, feldspars, mica, and clay minerals (Fig. 12). For the important clay minerals illite and smectite, a strong positive correlation between Al_2O_3 and K_2O is assumed to indicate their detrital origin (Burgan et al., 2008). According to Tribovillard et al. (2006), the detrital fraction of any given element can be calculated using the formula:

$$X_{detr} = (X/Al)_{average\ shale} \times Al_{sample} \quad (4)$$

where X_{detr} is the calculated detrital amount of the element in question, $(X/Al)_{average\ shale}$ corresponds to the X/Al ratio of the average shale within the analyzed section in regards to the element in question, and Al_{sample} relates to the absolute Al concentrations of the sample.

The authigenic elemental fraction is calculated using:

$$X_{auth} = X_{tot} - X_{detr} \quad (5)$$

where X_{auth} is the calculated authigenic amount of the element in question, X_{tot} is the total concentration of the element in question, and X_{detr} is the calculated detrital component for the given element using Eq. (4).

The determination of the detrital fraction of the elements (or normalization to Al or Ti) is necessary to assess and quantify the enrichment of redox-sensitive elements relative to given international standards.

Redox-sensitive and sulfide-forming elements. Redox-sensitive elements are major and trace elements, whose sedimentary concentrations vary depending on the amount of detrital siliciclastic influx, levels of OM and productivity rates, oxygen levels within the water column, and ultimately on the nature of the sediment and the sedimentation rates (Arthur and Sageman, 1994; Killops and Killops, 2009). During anoxic conditions the trace elements Ag, As, Bi, Cd, Co, Cr, Cu, Fe, Mn, Mo, Ni, P, Pb, Sb, U, V, and Zn are fixed either by incorporation into Mn/Fe-oxides and hydroxides, or by adsorption onto OM and clay minerals within the oxic, uppermost parts of an otherwise anoxic water column (Algeo and Maynard, 2004; Charriau et al., 2011; Lipinski et al., 2003; März et al., 2009). They are subsequently transferred to the sediment–water interface, where they become incorporated into authigenic sulfides (März et al., 2009; Tribovillard et al., 2006). For euxinic and H_2S-enriched environmental conditions, the redox-sensitive trace elements can be divided into two major groups, depending on their chemical reactivity with H_2S: (1) Trace elements of "strong euxinic affinity" are Mo, U, V, Zn (Algeo and Maynard, 2004; Berry and Wilde, 1978; Dean et al., 1984; Pettijohn et al., 1987), which usually show strong correlations with OM and are taken up as solid solution in FeS or other sulfide phases. (2) Trace elements of "weak euxinic affinity" (Cu, Ni, Cr, Co), which are not strongly influenced by the presence of H_2S (Algeo and Maynard, 2004), yet reside within the OM or detrital fractions of the sediment. The redox-sensitive trace element concentrations and ratios can be used to reconstruct environmental conditions for the time of deposition (Fig. 13).

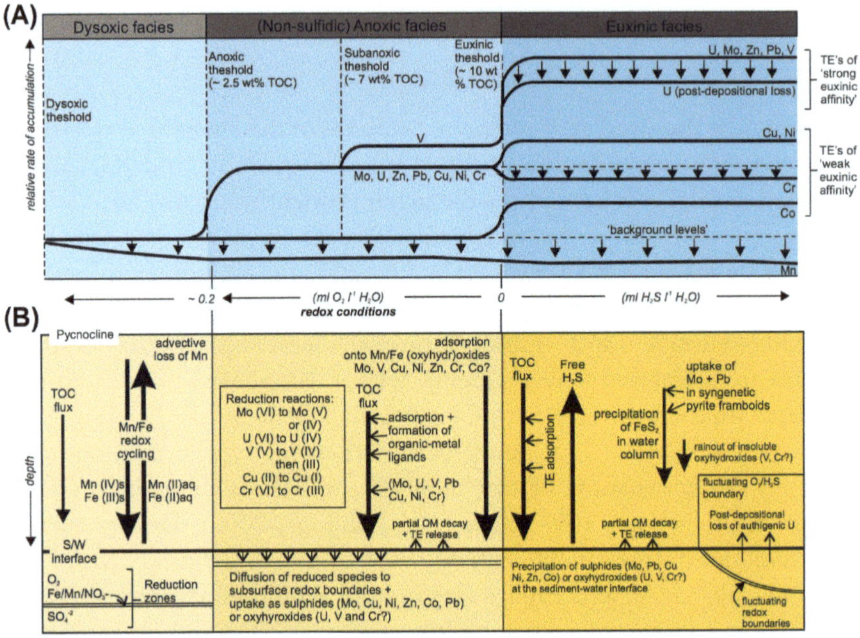

Figure 13 (A) Redox-dependent enrichments of trace elements (TE) and enrichment thresholds. The "anoxic threshold" is defined to be at 2.5 wt% TOC and permits only minor enrichments of trace elements (with the exception of Co, which can act as a chemical catalyst); the "subanoxic" threshold is set at 7 wt% TOC, where moderate enrichment takes place, and the "euxinic threshold" at 10 wt% TOC at which drastic and rapid enrichment of trace elements occurs. Downward pointing *arrows* indicate the rapid depletion of (1) Mn within low-oxygen environments followed by (2) Cr in euxinic facies and (3) U in euxinic facies through postdepositional remobilization. Benthic O_2 levels increase from the right to the left, whereas H_2S levels increase from the left to the right. (B) Process model showing the various chemical reaction types and reaction chains leading to systematic enrichment of shales by trace elements. Oxygen content decreases from left to right. Boundaries between "dysoxic"/"anoxic" facies and "anoxic"/"euxinic" facies are set at thresholds of 0.2 and 0 mL O_2 L^{-1} H_2O, respectively. *Modified from Algeo, T.J., Maynard, J.B., 2004. Trace-element behavior and redox facies in core shales of Upper Pennsylvanian Kansas-type cyclothems. Chem. Geol. 206 (3–4), 289–318.*

5.5.4 Mobile, Immobile, and Remobilized Elements

Most of the redox-sensitive elements described previously are highly mobile, and dependent on the oxygen concentration within the water column and the sediment, as well as on specific postdiagenetic conditions. When using trace element concentrations and ratios to reconstruct paleoenvironmental conditions (i.e., productivity or redox states), it is important to use multiple elements as proxies, as single element data can be unreliable and misleading (Rollinson and Rollinson, 1993). In the absence of postdepositional reoxygenation, metal sulfides are stable and not affected by diagenetic processes (Calvert and Pedersen, 2007; Tribovillard et al., 2006). Hence, redox-sensitive elements within the sediments are effectively recorded. However, relatively abrupt environmental changes caused by, e.g., not only bioturbation or larger scaled turbiditic flows, but also climatically induced glacial—interglacial transitions, can perturb the water column and resuspend sediments resulting in the replenishment of oxidizing agents (Rollinson and Rollinson, 1993). For example, only small amounts of O_2 are needed to affect the U contents of the uppermost parts of the sediments. Other elements such as V, Cd, and Mo may also be affected by these reoxygenation and remobilization processes (Calvert and Pedersen, 2007; Tribovillard et al., 2006), which have the potential to effectively mask if not completely obliterate the original paleoenvironmental signal archived within the geochemical signatures.

Elements considered to be relatively immobile in aqueous solutions are Ti, Zr, Y, Nb, and P (Rollinson and Rollinson, 1993). They are also relatively stable in the presence of hydrothermal fluids, during seafloor weathering, and remain immobile even up to medium metamorphic grades, making them ideal not only for paleoenvironmental reconstructions, but also for plate-tectonic provenance discriminations (Bhatia, 1983; Floyd and Winchester, 1978; Floyd et al., 1989; Rollinson and Rollinson, 1993).

5.5.5 Normalization of Major and Trace Elements and Enrichment Factors

Major and trace element concentrations of marine sediments are a combination of detrital and authigenic components. Marine sediments often contain varying amounts of mineral phases of biogenic origin, like $CaCO_3$ and opal. These additional mineral phases dilute and alter the trace element abundances of the sediment (Tribovillard et al., 2006). In fine-grained marine sediments Al is commonly used as a normalizing factor (Calvert and Pedersen, 2007) to eliminate the effects of this dilution and to smoothen the variability linked to the clay fraction, hence avoiding the influence of organic carbon, biogenic carbonate, and silica contents on the original elemental composition. Normalization to Al is applicable only if the Al resides within the detrital clay fraction of the sediments and is demonstrated to have remained immobile during diagenesis (Algeo and Maynard, 2004; Arnaboldi and Meyers, 2006; Calvert and Pedersen, 1993; Dellwig et al., 2000; Herut and Sandler, 2006; Scopelliti et al., 2006). Al normalized values of the trace elements are usually given as $\times 10^{-4}$. The elemental X/Al-normalized values of the samples are used in enrichment factor (EF) plots, with the element EFs (Dellwig et al., 2000; Rule, 1986; Scopelliti et al., 2006; Turgeon and Brumsack, 2006; Tribovillard et al., 2006) being calculated as

$$\mathrm{EF}_{\mathrm{element}\ \mathrm{X}} = \mathrm{X}/\mathrm{Al}_{\mathrm{sample}}/\mathrm{X}/\mathrm{Al}_{\mathrm{average\ shale}} \tag{6}$$

The $\mathrm{X}/\mathrm{Al}_{\mathrm{average\ shale}}$ component will vary depending on the selected standard for comparison. The EF plots compare the elemental concentrations of the analyzed sample in terms of enrichments (values greater than one) and depletions (values less than one) relative to a known standard. For the present study, the international standards for the "upper continental crust" (Taylor and McLennon, 1985) and various shales, such as the "Average Shale" (Turekian and Wedepohl, 1961; Wedepohl, 1971, 1991), "Average black shale" (Vine and Tourtelot, 1970), Post-Archean Australian Shale "PAAS" (Taylor and McLennon, 1985), and the North American Shale Composite "NASC" (Gromet et al., 1984), have been used.

Al normalization must be used with caution as highlighted by Van der Weijden (2002), as the concentrations of Al can also be affected biologically to a certain extent. It is necessary to determine and quantify this potential error source by carefully analyzing the correlation coefficients (r^2) of Al to relatively immobile elements of the detrital fraction such as Sc, Th, or Zr (Pi et al., 2013; Tribovillard et al., 2006). Ti is also used as a normalizing element, as it is associated with the heavy minerals of the detrital fraction (Kato et al., 2002). However, also the Ti-normalization is not without potential problems, as the postdepositional formation of authigenic rutile can compromise the significance of the total Ti readings and hence hamper the interpretation of the resulting Ti-normalized values (Calvert and Pedersen, 2007).

5.5.6 Sediment Recycling: Index of Compositional Variation

The index of compositional variation (ICV) is a parameter that can be used to determine the maturity of sedimentary rocks and the extent of sediment recycling (Cox et al., 1995). The ICV allows the discrimination between sedimentary rocks composed of first-cycle material and those that have been formed by reworking of this first-cycle material, which typically leads to increasing compositional maturity. Within tectonically inactive settings, such as cratons, siliciclastic sedimentary rocks will normally exhibit significantly higher calculated ICV values, indicating and reflecting multiple recycling processes and the consequent reworking of older sedimentary successions. In contrast, within tectonically active settings, the calculated ICV values will be typically low due to the relatively minor reworking and recycling of first-cycle sediments, and therefore exhibit a lower degree of compositional maturity (Cox et al., 1995). The ICV is commonly calculated by normalizing relatively mobile major elements to the relatively immobile aluminum:

$$ICV = [Fe_2O_3 + K_2O + Na_2O + CaO + MgO + MnO + TiO_2]/Al_2O_3 \tag{7}$$

The nonclay minerals show typically a higher ratio of the major elements to Al_2O_3 than the clay minerals (Cullers and Podkovyrov, 2000), thus giving a higher ICV index (Fig. 14). Immature shales may express ICV values >1, due to a high influx of nonclay minerals such as plagioclase, K-feldspar, amphiboles, and pyroxenes, which are generally found within tectonically active settings and/or within first-cycle deposits (Campos Alvarez and Roser, 2007; Cox et al., 1995; Cullers and Podkovyrov, 2000).

Mature shales have higher contents of clay minerals, such as kaolinite and illite as well as mica. As a consequence, they will normally display ICV values <1. These types of shales may form on cratons and other tectonically quiescent environments, e.g., within "passive margin settings," where sediment recycling is active. However, some first-cycle deposits have been documented to display high ICV values, as they were subjected to intense weathering (Cox et al., 1995; Cullers and Podkovyrov, 2000). The combination of ICV values with the ratio of certain alkali metals to aluminum (e.g., K_2O/Al_2O_3) is used to determine the alkali-feldspar to plagioclase and clay mineral ratio (Cox et al., 1995), which in turn is utilized to estimate the original mineralogy and the maturity of ancient mudrocks (Fig. 14). A K_2O/Al_2O_3 ratio of ≥ 0.5 indicates significant feldspar quantities relative to the clay mineral content and is typical for immature sediments. Values ≤ 0.4 suggest minimal concentrations of feldspar within the original shale and point toward highly mature sediments (Cox et al., 1995; Cullers

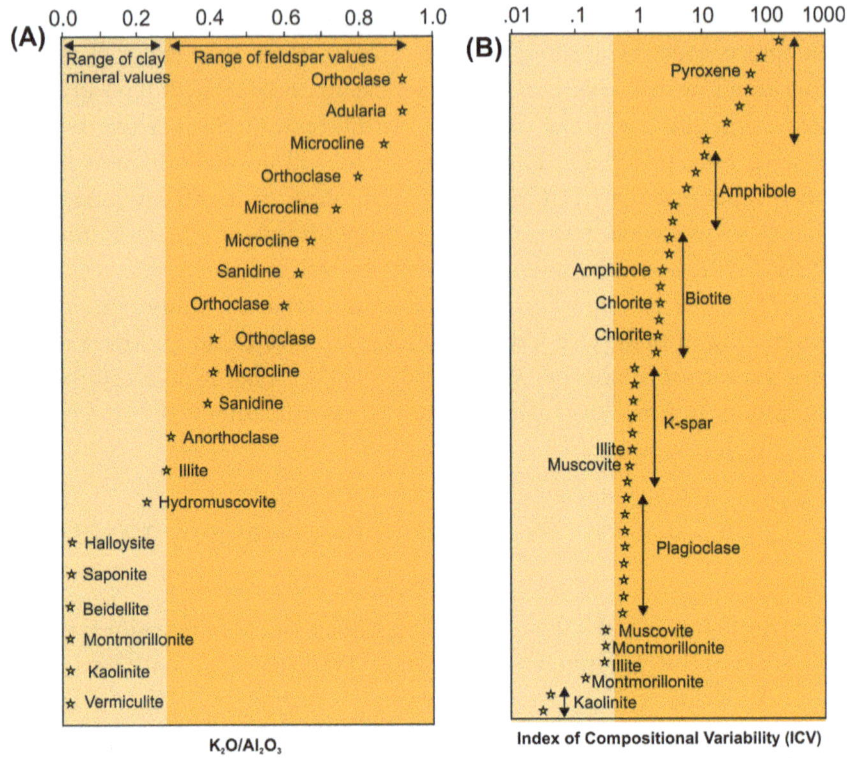

Figure 14 (A) K_2O/Al_2O_3 ratios for various K-feldspars and clay minerals. The stars represent the specific values for the minerals indicated. (B) Index of compositional variation (ICV), the stars locate typical values for the specific minerals. The *arrows* indicate the range of values for a particular mineral group. The background colors highlight unweathered minerals (darker coloration) and weathering products (lighter coloration). *Data from Deer, W.A., Howie, R.A., Zussman, J., 1966. An Introduction to the Rock-Forming Minerals. John Wiley and Sons, New York, pp. 528, modified after Cox, R., Lowe, D.R., Cullers, R.L., 1995. The influence of sediment recycling and basement composition on evolution of mudrock chemistry in the southwestern United States. Geochimica Cosmochimica Acta 59 (14), 2919–2940.*

and Podkovyrov, 2000). The ICV and K_2O/Al_2O_3 values are commonly described together with the weathering indices (chemical index of alteration and chemical index of weathering) to infer the intensity of the weathering, the degree of recycling, and the maturity of sediments (Barbera et al., 2006; Cullers and Podkovyrov, 2000; Okunlola et al., 2009).

5.5.7 Paleobioproductivity Calculations Using Barium

The quantification of paleobioproductivity using geochemical proxy data is under certain circumstances an intricate issue. Various calculation steps are needed to approximate the paleobioproductivity. The following methods were applied (Pfeifer et al., 2001).

A normative analysis is applied to correct for lithogenic Ba, as lithogenic Ba is not related to the bioproductivity within surface waters (Pfeifer et al., 2001). Barium in marine sediments occurs as biogenic barite or associated with alumosilicates (as terrigenous/lithogenic Ba), which is the main source of detrital Ba. The ratio of the detrital Ba/Al is the crucial factor (Bernárdez et al., 2008) to extract the proportion of the biogenic Ba (Ba_{bio}) from the total Ba (Ba_{total}). The terrigenous Ba (Ba_{terr}) is estimated by the ratio of the total Ba to the conservative elements Al or Ti ($Ba/Al_{aluminosilicate}$ or $Ba/Ti_{aluminosilicate}$). The Ba_{bio} is subsequently calculated by subtracting the Ba_{terr} from the Ba_{tot} concentrations. Ti can also be used as a proxy for the terrigenous fraction, as part of the Al concentrations may be of biogenic origin (Bonn et al., 1998; Murray et al., 1993). The following equations from Bonn et al. (1998) and Pfeifer et al. (2001) were applied to calculate Ba_{bio}. Ti and Al were used to normalize the Ba concentrations:

$$Ba_{bio}(Ba_{excess}) = Ba_{tot} - (Al \times Ba/Al_{aluminosilicate}) \tag{8}$$

$$Ba_{bio}(Ba_{excess}) = Ba_{tot} - (Ti \times Ba/Ti_{aluminosilicate}) \tag{9}$$

Under certain circumstances the applied formula (8) and (9) generated nonrealistic negative values. Therefore the approximation for Ba_{bio} developed by Prakash Babu et al. (2002) was also tested and subsequently used for the present study. The proxy of Prakash Babu et al. (2002) is based on the assumption that the lowest Ba/Al ratios are always equivalent to the local Ba background.

To calculate the accumulation rates of biogenic Ba (AR Ba_{bio}) the equations of Pfeifer et al. (2001) were applied:

$$AR\ Ba_{bio} = Ba_{bio} \times MAR/100 \tag{10}$$

where MAR is the mass accumulation rate in g cm^{-2} kyr^{-1}.

$$MAR = SR \times DBD \tag{11}$$

where SR is the sedimentation rate in cm/kyr and DBD is the dry bulk density in g/cm^3.

The sedimentation rate (SR) was calculated by using the available biostratigraphic age constraints for the base and the top of the lower part of the Formigoso Formation (Bernesga Member) and the number of samples taken. The Bernesga Member spans the time from the Mid-Late Aeronian to the Mid-Telychian. Using the numerical ages of Gradstein et al. (2012), a duration of approximately 4 Myr can be assigned to the Bernesga Member. A total of 181 shale samples were taken from the Bernesga Member at the Aralla locality at intervals of 25 cm. Applying the certainly simplistic assumption that the Bernesga Member does not include horizons affected by stratigraphic condensation and/or dilution, an average timespan of 22,099.44 yrs is represented by 25 cm of consolidated sediment, and therefore, 1 cm of the succession represents approximately 883.97 yrs. From this an approximate average sedimentation rate of 1.12 cm/kyr was calculated (11.2 m per million years). This estimate corresponds very well with previously published sedimentation rates for other black shales (Erbacher et al., 2005, 2004; Stein, 1990; Turgeon and Brumsack, 2006).

The Dry Bulk Density (DBD) was estimated by applying the average shale DBD ranges of 1.8–2.8 g cm^{-3} (Rieke and Chilingarian, 1974). Both the lower range value and upper range value for DBD together with the SR was used to calculate the minimum and maximum Mass Accumulation Rate (MAR):

$$\text{Minimum MAR} = 1.11603 \times 1.8 = 2.008854 \ \text{g cm}^{-2}\text{kyr}^{-1} \tag{12}$$

$$\text{Maximum MAR} = 1.11603 \times 2.8 = 3.124884 \ \text{g cm}^{-2}\text{kyr}^{-1} \tag{13}$$

To calculate the new production (P_{new}), the flux of Ba (F_{ba}) to the seafloor has to be estimated by first using Eq. (14) of Pfeifer et al. (2001). Dymond et al. (1992) related the amount of Ba_{bio} preserved to the MAR (expressed in mg cm^{-2} kyr^{-1}) of the sediment. To estimate F_{ba} on the basis of Ba accumulation rates the amount of preserved biogenic Ba ($\%Ba_{pres}$) is calculated using Eq. (15). Dissolved Ba at the sediment/water interface (F_{ba}) is calculated by dividing the AR Ba_{bio} by $\%Ba_{pres}$:

$$F_{ba} = \text{AR Ba}_{bio}/0.209 \log(\text{MAR}) - 0.213 \tag{14}$$

$$\%\text{Ba}_{pres} = 20.9 \log(\text{MAR}) - 21.3 \tag{15}$$

Consequently, the new production rate (P_{new}) is expressed by Eq. (16) in gC m^{-2} yr^{-1} (Francois et al., 1995; Pfeifer et al., 2001):

$$P_{new} = 1.95(F_{ba})^{1.41} \tag{16}$$

The primary production (or PP in gC m^{-2} yr^{-1}) is estimated by using Eq. (17) of Eppley and Peterson (1979), based on the P_{new} values calculated by using Eq. (16):

$$PP = 20(P_{new})^{0.5} \tag{17}$$

5.5.8 Sodium Proxy (Na_2O^{est})

The Niton XL3t XRF analyzer is unable to detect elements lighter than magnesium. Therefore the following formula was implemented to approximate the Na_2O values, using the observed relationship between Na_2O and K_2O in a suite of international black shale standards and the positive correlation of Na_2O to K_2O with $r^2 \geq 0.4$ described from floodplain sediments (De Vos et al., 2006). The international black shale standards selected were the (1) North American Shale Composite (NASC) of Gromet et al. (1984); the (2) three average shale composites of (i) Pettijohn (1975), (ii) Turekian and Wedepohl (1961), and (iii) Wedephol (1971, 1991); the (3) average black shale composite of Vine and Tourtelot (1970); the (4) SCO-1 shale standard of Abbey and Geological Survey of Canada (1983); and finally, the (5) Post-Archean Australian Shale (PAAS) of Taylor and McLennan (1985). The K_2O and Na_2O values for each of the aforementioned black shale standards were cross-plotted to determine the correlation ($r^2 = 0.437$). The equation of this curve was applied to total K_2O values to give a very rough approximation for sodium concentrations (Na_2O^{est}). The following equation was applied:

$$Na_2O^{est} = \left[0.277 \times \left(K_2O^{total}wt\%\right)\right] + 0.281 \tag{18}$$

Although this method gives only a very rough estimate for Na_2O^{est} values, it proved to be surprisingly robust and consistent. However, it has to be noted that the Na_2O^{est} values used and discussed within this study are not measured concentrations, but calculated estimates.

6. SILURIAN GONDWANA: PREREQUISITE FOR LLANDOVERY BLACK SHALES

During the Late Ordovician, Gondwana drifted over the paleo-South Pole (Cocks and Torsvik, 2002; Fortey and Cocks, 2003; Mac Niocaill et al., 1997; Scotese et al., 1999; Torsvik and Van der Voo, 2002). This together with a drop of global temperatures (Harland, 1972; Leton et al., 2012; Stanley, 1988) promoted the growth of ice sheets, which extended as far as 65°S paleolatitude (Scotese et al., 1999). The glaciation, almost entirely restricted to the Hirnantian, was of <0.5 My in duration (Cocks, 2007; Le Heron et al., 2009; Sutcliffe et al., 2000) and led to a global eustatic sea-level fall of 50–100 m, triggering the second largest global mass extinction event among life forms, during which 85% of all species were eliminated (Craig et al., 2002; Sheehan, 2001). The Silurian black shales were deposited along the northern margin of Gondwana during the Early Silurian transgression as a result of the melting Hirnantian ice masses (e.g., Fekirine and Abdallah, 1998; Lüning et al., 2000). These black shales form a 6000-km belt, extending from Mauritania to Saudi Arabia, and are composed of partly strongly organically enriched deposits, with TOC values locally exceeding >20% (Le Heron et al., 2013; Lüning et al., 2000). There are two horizons within the Silurian black shales of the nowadays Northern African region that are termed "hot shales," with American Petroleum Institute unit (API) values exceeding 200 API, due to a strong enrichment of U (Loydell et al., 2009; Lüning et al., 2000, 2003a, 2005). These "hot" shales are of highest economic importance as they form the source rock for 80–90% of the Paleozoic sourced hydrocarbons in North Africa (Lüning et al., 2000, 2005) (Fig. 15).

The spatial distribution of the lower Rhuddanian "hot" shale and the upper Telychian "hot" shale horizon varies. The Rhuddanian deposits are confined to paleolows during the initial transgression phase and seem to follow the relief and the tunnel valleys generated by the Hirnantian glaciation (Le Heron and Craig, 2008; Le Heron et al., 2004; Le Heron et al., 2007; Le Heron et al., 2009). The relief created by the glaciogenic processes was superimposed on and locally exaggerated the preexisting shelf topography that was caused by the Pan-African and Infracambrian tectonics (Le Heron et al., 2004; Lüning et al., 2000, 2005). These black shales are considered to be a large-scale synchronous anoxic event that is apparent across northern Gondwana (Lüning et al., 2000, 2005). Sea-level rise continued during the post-Rhuddanian leading to the deposition of "lean" shales with depleted TOC content and API values of approximately 120 (Fello et al., 2006; Lüning et al., 2000). These organically "lean" shales are difficult to date and correlate by means of biostratigraphy, as they contain only few fossils of biostratigraphic value (Butcher, 2009). The following Upper Telychian "hot" shale deposits were formed by a second large-scale synchronous anoxic event (Lüning et al., 2000, 2005). A second wave of "lean" shale deposits overlies the Telychian "hot" shales. Both lean shale deposits and the Telychian "hot" shales are laterally extensive (Figs. 16 and 17). It is important to note that only the basal

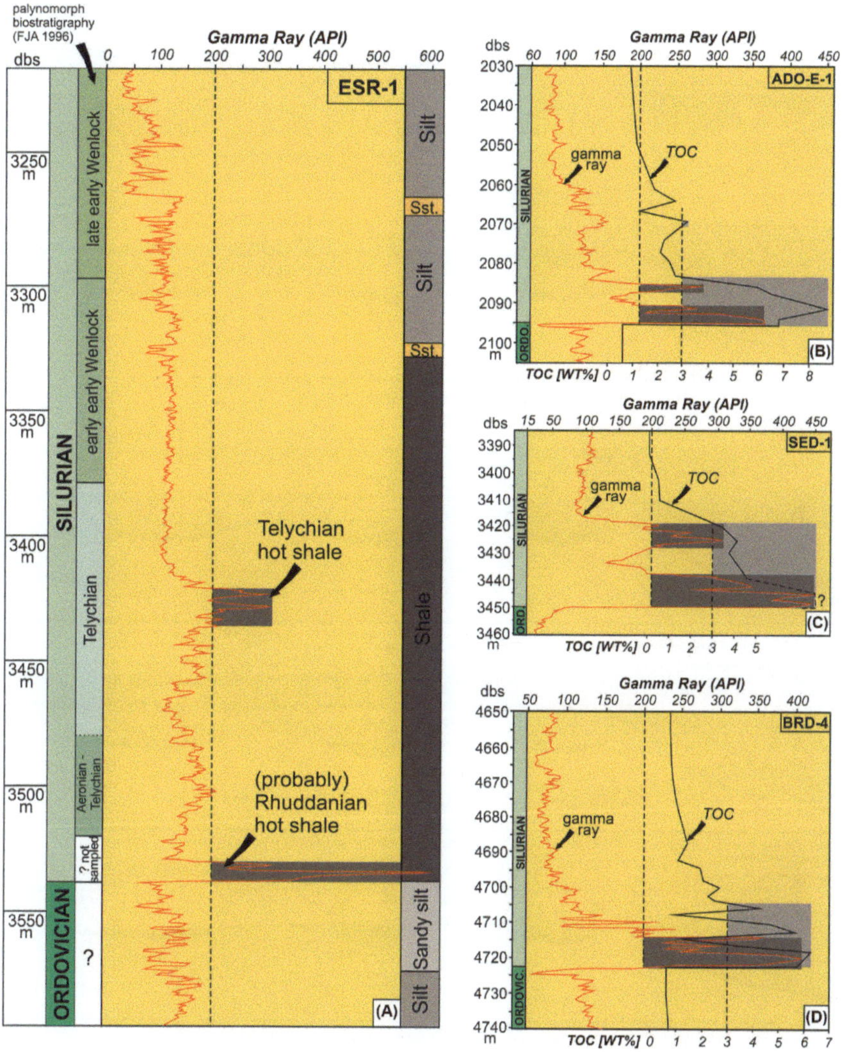

Figure 15 Gamma-ray and TOC signatures of Llandovery black shale successions from Algeria and Libya. (A) Gamma-ray signature of a black shale sequence, Ghadames Basin, Libya. Showing the two "hot" shale horizons. Biostratigraphy after Futyan, Jawzi, and Associates (FGA, 1996 in: Lüning et al., 2000). (B–D) Gamma-ray signals and TOC contents of coeval black shales from the Algerian part of the Ghadames basin and the Illizi basin of Algeria, showing the relationship between TOC and the gamma-ray signal (total API). The peaks in the gamma-ray signal correspond to the peaks in TOC. *API*, American Petroleum Institute units; *dbs*, depth below surface; *Sst.*, sandstone; *TOC*, total organic carbon. Modified after Lüning et al. (2000).

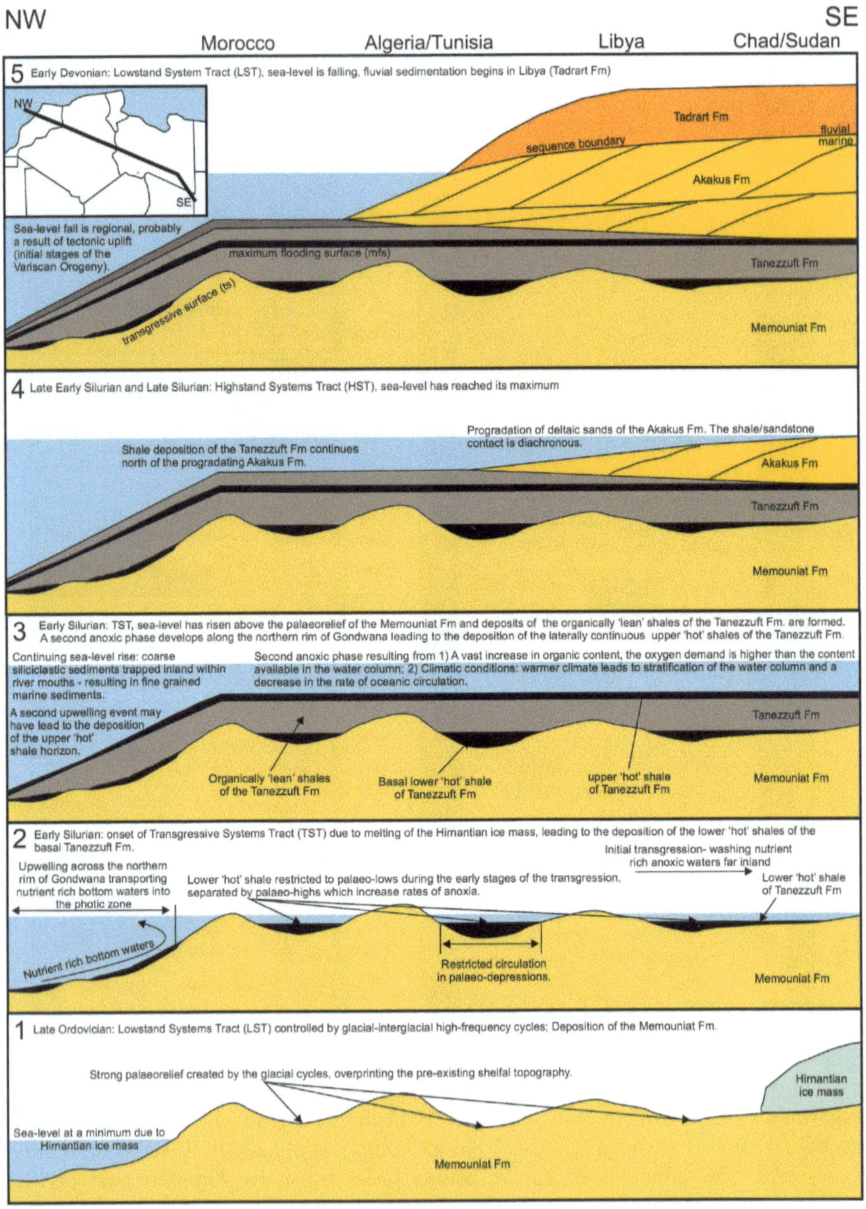

Figure 16 Depositional model for the lower and upper "hot" shales across northern Africa (Northern Gondwana, from the (1) Late Ordovician to the (5) Early Devonian). Cross sections 1–5 spanning from Morocco (North West) to the Chad/Sudan (South East) as indicated by inset in sketch number 5. *Modified after Lüning, S., Craig, J., Loydell, D.K., Štorch, P., Fitches, B., 2000. Lower Silurian 'hot shales' in North Africa and Arabia: regional distribution and depositional model. Earth-Science Rev. 49 (1–4), 121–200.*

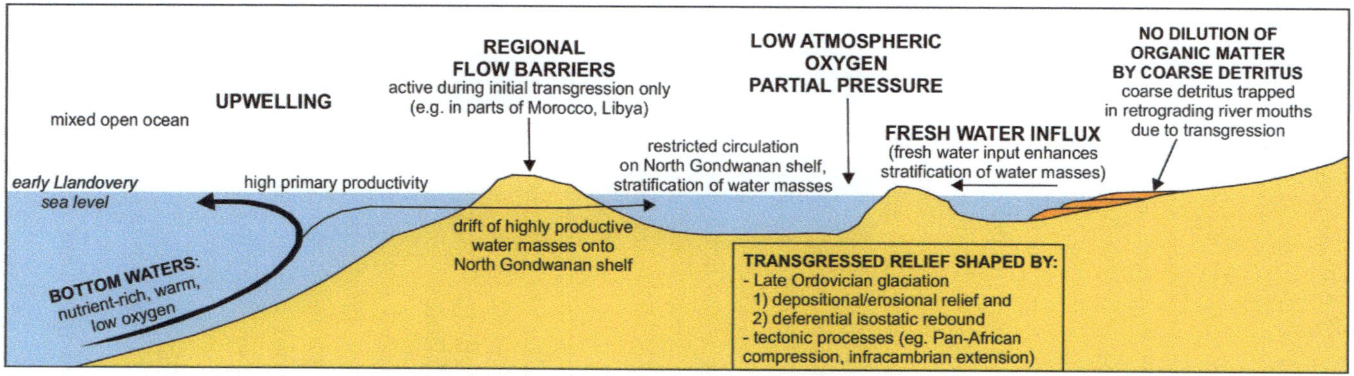

Figure 17 Paleo-oceanographic, sedimentological, and paleobiological processes that may have operated during the Early Silurian leading to the development of the Llandovery "hot" shales. *Modified after Lüning, S., Craig, J., Loydell, D.K., Štorch, P., Fitches, B., 2000. Lower Silurian 'hot shales' in North Africa and Arabia: regional distribution and depositional model. Earth-Science Rev. 49 (1–4), 121–200.*

Rhuddanian "hot" shales are restricted to the paleodepressions (Lüning et al., 2000, 2005). Le Heron et al. (2013) suggested that the patchy nature of the "hot" shales may also be related to lingering ice sheets that locally moderated the extent of anoxia.

The nowadays surface extent of the Rhuddanian "hot" shales is difficult to trace as they tend to lose their diagnostic features such as the elevated OM content and the corresponding dark color in the course of extensive weathering. However, the U distribution appears to remain largely unaltered by the weathering processes (Fello et al., 2006; Lüning and Kolonic, 2006). The U values are therefore being used as a proxy for the original OM content and hence to delineate the presence or absence of "hot" shales (Fello et al., 2006; Fertl and Chilingar, 1988; Lüning and Kolonic, 2006).

Using tectonic and biogeographical data it has been advocated that Iberia was, at least until the Early Silurian, part of the northern Gondwanan margin (Bassett, 2009; Cocks and Torsvik, 2006; Robardet, 2003; Scotese et al., 1999; Servais and Sinbutin, 2009). This assumption is further corroborated by the presence of Hirnantian glacial deposits and tunnel valleys in Iberia, comparable to those found across northern Africa (Gutiérrez-Marco et al., 2010). In the following the depositional models for the Silurian "hot" shales of northern Africa and the Middle East proposed by Lüning et al. (2000) and Armstrong et al. (2005) will be described and discussed in detail, as they provide valuable insights into the processes that may have also operated in Iberia and in particular within the Cantabrian basin.

6.1 Parameters Controlling Anoxia and the Deposition of Llandovery Black Shales

The processes and parameters that may have led to the development of the Silurian "hot" shales across Northern Gondwana are summarized and discussed. Many of the proposed models and theories addressing the problem of black shale formation invoke only one or just a very limited number of controlling factors. However, it has to be stressed that it may not have been necessarily just one or a few parameters that initiated and sustained the formation of black shales, but a combination and maybe even complex interaction of varied processes.

6.1.1 Upwelling at the Northern Margin of Gondwana

Upwelling transfers nutrient-rich waters from deep marine areas into the photic zone (Moore et al., 1993; Tyson, 1995). Demaison (1993) estimated that only 8% of the World's petroleum reserves are derived from "upwelling areas," in contrast to 78% being generated within dysoxic-anoxic "silled basins." Upwelling creates a favorable environment for the biological cycle, leading to an increase in plankton bioproductivity. However, it can also enhance the levels of anoxia by passing low oxygenated bottom waters onto the shallow marine shelf areas. Pedersen and Calvert (1990) suggest that the upwelling induced paleoproductivity contributes more toward the formation of organically enriched sediments than the water-column anoxia and that the key to black shale generation is the large-scale influx of organic carbon and not necessarily the redox status (see also Lüning et al., 2000). Moore et al. (1993) provided a numerical model for the extent of upwelling and the intensity of paleoproductivity for the early Silurian using modern day oceanographic circulation analogues and bioproductivity rates. The model highlighted large upwelling zones between 70 and 35°S, parallel to the extensive Northern Gondwanan margin, and assumed westerly winds to promote the intense upwelling. The high amount of organic matter in the water column and within sediment triggered the development of dysoxic to anoxic conditions, as large proportions of the dissolved oxygen were removed by the oxygenation and subsequent breakdown of OM. The results of the model also indicated that the prevailing winds and related upwelling were not only active during the Llandovery, but also persisted throughout the Wenlock, thus predicting a prolonged and continuous period of time for black shale formation. However, discrete black shale successions have only been documented from the Rhuddanian and the Telychian (Armstrong et al., 2005; Lüning et al., 2000). The confinement of these two black shale deposits to distinct stratigraphic levels emphasizes that additional processes, not considered within the numerical model, must have been responsible for the interruption and finally termination of the black shale formation (Lüning et al., 2000). The nature of these processes is currently a matter of ongoing debate, but the most widely discussed involve short-termed seasonal changes, geographical variations in nutrient sources, the effects of high-frequency sea level fluctuations, sedimentological condensation versus dilution due to under- or oversupply of terrestrial material and the punctuated, glacial–interglacial related development of a stratified water column (Armstrong et al., 2009; LaPorte et al., 2009; Le Heron et al., 2009; Lüning et al., 2000, 2003a; Małkowski and Racki, 2009; Nielsen, 2004; Page et al., 2007; Štorch, 1998; Walliser, 1996; with previous references therein).

Wilde et al. (1991) and Berry et al. (1998) proposed a monsoonal setting for northern Gondwana, with seasonal intense upwelling and the associated development of a pronounced oxygen minimum zone. Page et al. (2007) and Le Heron et al. (2009), on the other hand, argue that upwelling is unlikely to be the mechanism leading to black shale formation, as the necessary nutrient-laden waters stem, according to their model, from glacial melt waters derived from the hinterland, which reinforce the development of a shallow pycnocline and hence the development of a salinity-driven water column stratification.

6.1.2 Transgressions and Shelf Geometries

The deposition of the Silurian black shales across northern Gondwana correlates very well with eustatic transgressive phases, which reached their maximum during the Telychian Oktavites spiralis Biozone (Arthur and Sageman, 1994; Johnson, 2006, 2010; Loydell, 1998; Sachanski et al., 2010). The maximum extent of this transgression and the resulting land–sea relationship is illustrated in Fig. 3.16. It highlights that not only the majority of the Gondwanan shelf areas were submerged beneath the sea, but also a vast amount of the Gondwanan mainland, pushing the paleocoastline as far southward as present day Niger and Chad.

It is the initial stage of this eustatic sea level rise that is regarded to be of importance for the formation of the Rhuddanian black shale (Lüning et al., 2000). Most of the processes that were initiated during the early phase of this transgression and the geological features that were left behind by it can be explained by the basal transgressive "puddle model" of Wignall and Hallam (1991).

Next to the eustatic sea level rise, there are a number of additional factors that could have further controlled the development of the black shales: (1) Subbasins. Paleolows or topographical "depressions" are flooded during the initial stages of the marine transgression. These shelfal depressions are cut off from the ocean circulation by localized paleohighs, which act as flow barriers. Shelfal oxygen deficiency, created by the upwelling of oxygen-depleted bottom waters and high productivity rates, is therefore enhanced. These depressions are hence considered to be important for the localized concentration of OM (Fello et al., 2006; Le Heron et al., 2009; Lüning et al., 2000, 2005). The flooding of these paleodepressions and the environmental processes that operate within them can be interpreted by applying the "puddle model" of Wignall and Hallam (1991). Based on the observation that organically lean Hirnantian black shales have been found in Jordan, predating the basal Rhuddanian "hot" shales deposits of Lüning et al. (2000), Armstrong et al. (2005) concluded that a depositional model for the Rhuddanian "hot" shales could be closer related to the maximum transgression "expanding puddle model" of Wignall (1994). However, Lüning et al. (2006) stress that the age of the lean shales is only poorly constraint, and hence support their original idea of an initial transgressive "puddle model" for the Rhuddanian black shales. Furthermore, Lüning et al. (2006) pointed out that the basal "hot" shales are clearly Rhuddanian in age and that there is no evidence for Hirnantian "lean" shales. (2) Entrapment of coarse siliciclastics. The transgression during the latest Ordovician/earliest Silurian was extremely fast (Fekirine and Abdallah, 1998). The sea had moved more than 1000 km inland within a few hundred thousand years (Lüning et al., 2000). The rapid immersion left the shelfal areas to be starved of coarser grained siliciclastic sediments. The coarser sediments were trapped inland, in river mouths and around paleohighs, leading to relatively undiluted OM content in the more distal shelf areas (Lüning et al., 2000). (3) Incomplete recolonization after seasonal anoxia. Tyson and Pearson (1991) and Tyson (1995) proposed the development of seasonal anoxia: during the warm summer time, a thermocline/pycnocline formed, restricting the circulation and enabling the stratification of the water column. The water column stratification led to the establishment of a well-oxygenated, warm, low-density, mixed surface layer, which created effectively a film, capping the cooler, denser, oxygen-depleted waters below. This scenario explains well the termination of the aerobic benthic ecosystem and the introduction of ideal conditions for black shale formation (Tyson, 1995; Wignall, 1994). During winter the thermocline and/or pycnocline may not be developed, leading to the reestablishment of full water-column circulation. The circulation may be aided by vigorous wind-driven turbulence powered by winter storms, allowing nutrients from the deep bottom waters to reach the photic zone (Wignall, 1994). However, on wide continental shelf areas, the distances can be too vast for complete biological recolonization during the winter months. This could explain the rarity or absence of aerobic benthic faunas in most black shale sequences (Lüning et al., 2000; Tyson, 1995; Wignall, 1994).

6.1.3 Postglacial Freshwater Stratification

The melting of the Hirnantian ice masses led to a significant release of freshwater into the earliest Silurian sea. As freshwater has a lower density than saline water it will form a freshwater cap. This freshwater cap restricted the vertical exchange of shelfal water masses (Combaz, 1967, 1986; Lüning et al., 2000). Brackish surface waters are known to cause salinity driven water column stratification (Le Heron et al., 2009; Page et al., 2007; Tyson, 1995), effectively preventing the replenishment of nutrients from upwelling processes and enhancing the levels of anoxia. The majority of the ice masses had disappeared during the early Llandovery, resulting in the weakening and finally disappearance of the capping effect. However, this model strictly implies the existence of low salinity surface waters, which seems to be contradicted by the high abundance of graptolites found in the Rhuddanian "hot" shales (Lüning et al., 2000). Le Heron et al. (2013) interpreted the patchy nature of the Silurian "hot" shales to be related to lingering Hirnantian ice sheets, which in turn locally moderated the extent of anoxia.

6.1.4 Episodic Climatic Changes

The black "hot" shales of the northern Gondwanan region were deposited during the Rhuddanian and Telychian, respectively. The deposition of the basal Rhuddanian "hot" shales coincide with the "Spirodden Secundo episode," spanning from the latest, postglacial Ordovician to the Mid-Aeronian (Aldridge et al., 1993). Jeppsson (1990) describes "Secundo episodes" as being characteristic of warm climates in high latitudes, dry conditions in low latitudes, and warm, anoxic oceanic bottom waters. In contrast, the "Primo episodes" relate to cooler, more humid climates and cold, oxic oceanic bottom waters. The two episodes are controlled by climatic differences linked to the concentrations of atmospheric CO_2. The cold oceanic bottom waters of a Primo episode have a higher storage capacity for CO_2 than the warm bottom waters of a Secundo episode (Aldridge et al., 1993). As the northern margin of Gondwana was subject to upwelling conditions, warm, anoxic, nutrient-rich bottom waters of the Secundo episode could have been transported onto the Gondwanan shelf. These waters may have exacerbated the shelfal oxygen deficiency produced by the high influx of oxygen-consuming organic matter (Lüning et al., 2000). However, it has to be noted that the reported development of graptolite diversities throughout this time interval does not correspond to the Primo—Secundo episodes and this model should hence be used with caution (Lüning et al., 2000).

7. CHEMOSTRATIGRAPHY AND CHEMOFACIES OF CANTABRIAN BLACK SHALES

To establish the chemostratigraphy and characterize the chemofacies of the Lower Silurian black shales of Cantabria, a high-resolution study was conducted at Aralla (Castllia y Leon, NW Spain). The geochemical data were generated using a Niton XL3t XRF analyzer with helium purging and calibrated using international standards. The Aralla section hosts a complete stratigraphic succession with excellent exposure and preservation. It comprises the following formations and members: (1) the uppermost part of the Barrios Formation, (2) the Getino Beds and the Formigoso Formation, which is subdivided into the (3) Bernesga Member and (4) the Villasimpliz Member (see Aralla log in supplementary data).

At the base, the Aralla section shows the uppermost beds of the Barrios Formation, consisting of coarse-grained quartzarenite. The contact with the overlying Getino Beds is sharp. The Getino Beds are represented by 31 cm of interbedded fine-grained sand and siltstones. A distinct iron staining can be observed at the top of the succession. The boundary to the overlying Bernesga Member of the Formigoso Formation is again discrete. The Bernesga Member consists of 37 m of organically enriched black shales that are followed by 7 m of very fine-grained sandstone/shale alternations. At the base, the black shales are rich in monograptid graptolites. The relative abundance of the graptolites diminishes toward the top of the Bernesga Member, as the number of very fine-grained sandstone intercalations increases. Due to nonexposure, a gap of 27.3 m obscures the boundary to the overlying Villasimpliz Member. The Villasimpliz Member consists of 74 m of interbedded sand and siltstones. The thickness of these sandstone beds increases in a cyclic fashion. The vast majority of the thicker sandstone beds are heavily bioturbated at the base and symmetrical ripple marks are usually developed at the top. Channel deposits with planar cross-bedded sandstones occur toward the top of the succession. The contact of the Villasimpliz Member to the overlying San Pedro Formation is gradational.

7.1 Major Element Characteristics of the Bernesga Member

The Bernesga Member at Aralla is represented by 181 geochemical samples (see samples Aralla 4—184 in table "Aralla major and trace elements," in supplementary data). MgO concentrations are relatively low at the base and increase toward the top. The MgO concentrations vary from below detection limit (D.L.) to 2.12 wt%, with an average 1.15 wt% and a standard deviation (S.D.) of 0.4. Al_2O_3 values are relatively consistent throughout the Bernesga Member, ranging from 19.28 to 33.2 wt%, with an average concentration of 29.45 wt% and an S.D. of 1.66. SiO_2 concentrations vary between 36.4 and 70.81 wt% and increase systematically toward the top of the member. Sample "Aralla 170" (70.81 wt%) documents the onset of the increased input of terrigenous material. The SiO_2 values show an average of 53.59 wt% with an S.D. of 3.14. The K_2O values exhibit a depletion affecting the most basal units of the section and then increase consistently toward the upper most parts of the Bernesga Member with values ranging from 2.04 to 4.77 wt%, an average of 3.92 wt% and an S.D. of 0.42. TiO_2 values are relatively consistent, with a slight enrichment at the base of the section and systematic depletion toward the top. The values range from 0.65 to 1.3 wt%, averaging at 1.15 wt%, and show an S.D. of 0.08. The MnO concentrations are consistent throughout, varying between 0.05 and 0.2 wt%, averaging 0.05 wt% with an S.D. of 0.01. Some samples, e.g., Aralla 27 and 161 are highly enriched in MnO. This is most likely related to localized mineralization. Fe_2O_3 values show a general increase from the base to the top of the Member, varying from 3.47 to 37.86 wt%, have an average of 5.74 wt% and an S.D of 3.89. Three samples, Aralla 26, 38, and 43, show substantial enrichments, again most likely related to localized mineralization.

7.2 Major Element Characteristic of the Villasimpliz Member

The Villasimpliz Member at the Aralla section is represented by 57 geochemical samples (see samples Aralla 185–241 in table "Aralla major and trace elements," in supplementary data). MgO concentrations are highly erratic within the sand and siltstones, varying between below D.L. and 1.75 wt%, with an average 0.71 wt% and an S.D. of 0.66. Al_2O_3 concentrations are significantly lower than those of the Bernesga Member, ranging from 4.57 to 31.15 wt%, averaging at 17.4 wt% with an S.D. of 8.37. SiO_2 values show a marked increase varying from 48.73 to 82.57 wt%, averaging at 67.42 wt% with an S.D. of 8.86. The K_2O values show a wide range, from 0.85 to 6.2 wt%, averaging at 3.11 wt% and an S.D. of 1.58. TiO_2 concentrations vary between 0.28 and 1.29 wt%, averaging at 0.76 wt% with an S.D. of 0.29. The MnO values vary between 0.05 and 0.22, averaging at 0.06 wt% with an S.D. of 0.02. Fe_2O_3 concentrations are more consistent than in the lower member, ranging from 3.62 to 10.3 wt%, averaging at 5 wt% with an S.D. of 1.23 (Fig. 18).

7.3 Trace Element Characteristic of the Bernesga Member

The S concentrations are mostly below D.L., yet a few intervals toward the top of the Member show significant enrichments. The S values vary between below D.L. to 6148 ppm, with an average of 1743.9 ppm and an S.D. of 1340.4. Cl concentrations are relatively consistent with a few enrichments at the base and toward the top of the member. The values vary between 484 and 1213 ppm, with an average of 659.7 ppm and an S.D. of 96.6. The V values are again enriched at the base of the member. A rhythmic pattern can be observed in the basal parts of the section (three distinct peaks). The values decrease gradually toward the top of the member. The V values vary throughout from 91 to 394 ppm with an average of 208.6 ppm and an S.D. of 42.8. Cr values are enriched at the base of the member, decreasing toward the top, with a rhythmic pattern seen near the base. The Cr values range from 47 to 155 ppm, averaging at 120.2 ppm with an S.D. of 16.6. The Ni values increase from the base of the member toward the top, the values range from 34 to 236 ppm, averaging at 73.4 with an S.D. of 26.5. The Cu values are near D.L. at the base of the member. The values increase rapidly and begin to decrease gradually toward the top of the member. The Cu values range from below D.L. to 106 ppm, averaging at 30 ppm with an S.D. of 13.4. Zn values are mostly below D.L. at the base, the values increase from the middle of the member in what appears to be a rhythmic pattern. The values range from below D.L. to 292 ppm, averaging at 69.6 ppm with an S.D. of 47.7. The As values are relatively consistent throughout, ranging from below D.L. to 61 ppm, averaging at 18.4 ppm with an S.D. of 10.3. Rb values are consistent throughout the member, increasing slightly toward the top, the values range from 75 to 192 ppm, averaging at 157.7 ppm with an S.D. of 21.1. The Sr values are enriched at the base of the member, but decrease throughout in a logarithmic fashion, with values ranging from 163 to 300 ppm, averaging at 190.6 ppm with an S.D. of 21.9. Zr concentrations are low at the base and then gradually increase throughout, the values range from 88 to 258 ppm, averaging at 154.7 ppm with an S.D. of 26.9. Nb values are consistent, with values ranging from 11 to 25 ppm, averaging at 20.3 ppm with an S.D. of 2.2. Cs concentrations increase throughout the member, the values are below D.L. at the base followed by a series of enriched horizons before settling into a steady increase. Cs values range from below D.L. to 107 ppm, averaging at 26.1 ppm with an S.D. of 19. Ba values increase steadily from the base to the top with a few enriched horizons near the base, the values range from 38 to 1942 ppm, averaging at 398.7 with an S.D. of 215.8. The Hg concentrations are just above D.L., yet remain consistent from the base to the top, the values range from below D.L. to 10 ppm, averaging at 6.4 ppm with an S.D. of 3.2. Th values are enriched at the base of the member decreasing toward the top. A rhythmic pattern of the Th concentrations is again seen, with values ranging from below D.L. to 24 ppm, averaging at 15.6 ppm with an S.D. of 6.2. The

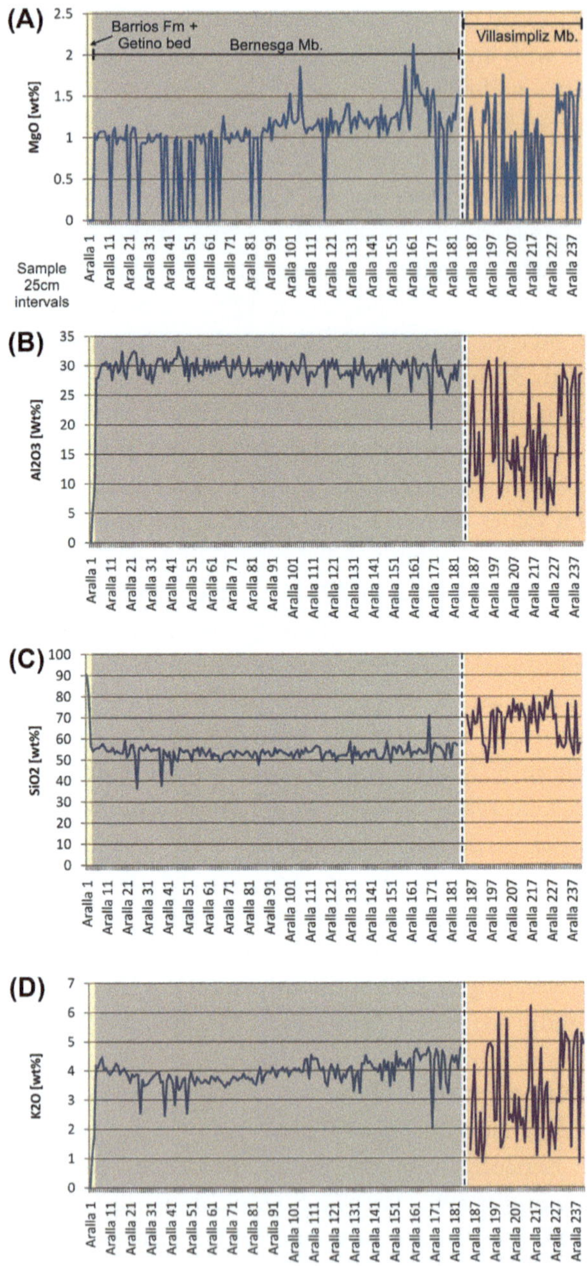

Figure 18 Major element variations for the Aralla section. Each sample number represents a 25-cm interval. The colored areas indicate the different lithostratigraphic units, starting with the Barrios Formation, followed by the Getino Beds (both in yellow), the Bernesga Member (gray), and finally the Villasimpliz Member (red). The *dashed line* represents a gap of 27.3 m. A single sample (Aralla 1) from the Barrios Fm. was analyzed

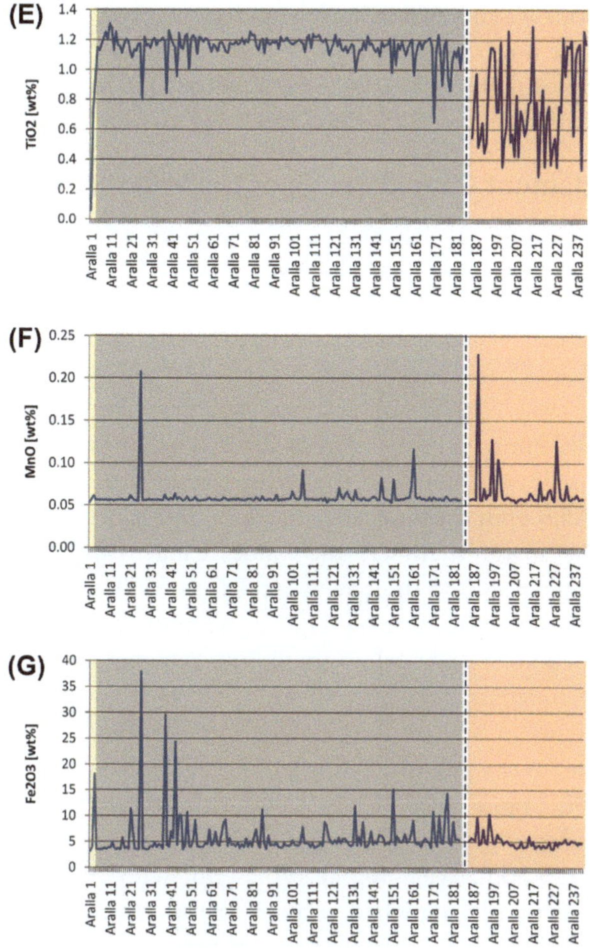

Figure 18 (*continued*).

as a control sample, the Getino Beds are represented by Aralla 2 and Aralla 3. The major elements of the Bernesga Member and Villasimpliz Member illustrated are described and discussed within the text. The absolute elemental values for the Barrios Formation and the Getino Beds are nonetheless documented in the geochemical data table (supplementary data). Due to limitations of the Niton XL3t XRF analyzer is the determination of lighter geochemical elements restricted. Hence, absolute Na values are not available for this data set. Ca values are mostly below detection limits (D.L.), and therefore are not included in this section; however, the data can be found in the geochemical data table (supplementary data).

U concentrations are enriched at the base of the member. A rhythmic pattern is seen at the base much like that of the V concentrations described previously (three distinct peaks); the values range from below D.L. to 18 ppm, averaging at 11.2 ppm with an S.D. of 3.4 (Fig. 19).

7.4 Trace Element Characteristic of the Villasimpliz Member

The S concentrations are below D.L. for this member and therefore not further discussed. Cl values are depleted relative to the lower member, with values ranging from 285 to 744 ppm, averaging 529.5 ppm with an S.D. of 89.6. The V values are again depleted relative to the lower member; values range from 29 to 215 ppm, averaging at 98 ppm with an S.D. of 48.8. Cr values are erratic. The values range from below D.L. to 120 ppm, averaging at 49 ppm with an S.D. of 36.9. The Ni values are depleted relative to the lower member, the values range from 29 to 110 ppm, averaging at 55.9 ppm with an S.D. of 18.8. The Cu values are depleted relative to the lower member, the values range from below D.L. to 35 ppm, averaging at 17.5 ppm with an S.D. of 7.9. Zn values are near D.L., the values range from below D.L. to 82 ppm, averaging at 15.8 ppm with an S.D. of 23.6. The As values are slightly depleted relative to the lower member. The concentrations range from below D.L. to 47 ppm, averaging at 12.5 ppm with an S.D. of 8. Rb values are depleted and highly erratic relative to the lower member, the values range from 26 to 195 ppm, averaging at 95.7 ppm with an S.D. of 55.1. Sr values are slightly erratic, yet average near to the values of the lower member. The values range from 156 to 185 ppm, averaging at 170.5 ppm with an S.D. of 8.5. The Zr values are drastically enriched relative to the lower member with the values ranging from 136 to 398 ppm, averaging at 257.4 ppm with an S.D. of 63.3. Nb values are depleted and highly erratic relative to the lower member with values ranging from 1 to 23 ppm, averaging at 12.5 ppm with an S.D. of 6.2. The Cs values are near D.L., however, they remain relatively consistent, ranging from below D.L. to 54 ppm, averaging at 17.4 ppm with an S.D. of 12.1. Ba values are lower relative to the lower member and are slightly more erratic, with values ranging from below D.L. to 816 ppm, averaging at 255.4 ppm with an S.D. of 229.7. The Hg concentrations are like those of the lower member, again close to the D.L., the values range from below D.L. to 8 ppm, averaging at 3.1 ppm with an S.D. of 3.1. The Th values are lower and more erratic than that of the lower member, ranging from below D.L. to 16 ppm, averaging at 7.4 ppm with an S.D. of 4.6. U concentrations are lower and more erratic in comparison to the lower member, with values ranging from below D.L. to 14 ppm, averaging at 4.8 ppm with an S.D. of 4.9 (Fig. 20).

8. INTERPRETATION OF THE CHEMOSTRATIGRAPHIC RESULTS

The interpretation of the geochemical data involves a number of proxies and elemental ratios that will be briefly described in this section. First, the geochemical data for all of the log locations were averaged and plotted in enrichment factor diagrams, to determine which elements are depleted or enriched relative to a selection of international standards. The geochemical data are then classified according to the schemes of Herron (1988), Roser and Korsch (1988), Floyd et al. (1989), and Pearce et al. (2010). The redox proxies of Fertl (1979), Hatch and Leventhal (1992), Jones and Manning (1994), Wignall (1994), and Kimura and Watanabe (2001) are applied systematically and described accordingly for each of the log localities. The paleosalinity proxy of Cambell and Williams (1965) is used to reconstruct salinity levels during the time of deposition for the individual log locations. The proxies of Akul'shina (1976) and Kiipli et al. (2012) are used to reconstruct the paleohumidity. The geochemical data for all of the log locations are systematically plotted in the clay discrimination diagrams of Cullers and Podkovyrov (2000), Eickmann et al. (2009), and Schlumberger (2009). The ICV of Cox et al. (1995) is used to reconstruct the extent of weathering and ultimately the level of sediment recycling. The salinity, redox, and grain-size proxies previously mentioned are systematically presented in 3D plots to determine and evaluate the relationships between the changing environmental conditions. The hydrothermal indicator of Boström (1973) is used to establish whether the sedimentary succession in question had been subject to any significant hydrothermal alteration. The quantification of any potential hydrothermal overprint is crucial to the development of the following paleobioproductivity reconstructions. The paleobioproductivity proxies of Pfeifer et al., 2001, Bonn et al., 1998, and Prakash Babu et al. (2002) are applied to the Ba concentrations.

8.1 Enrichment Factor Plots

The chemostratigraphic data of the Bernesga Member are used to characterize the black shale containing lower Formigoso Formation compared to various international standards. Only elemental values above the D.L. are introduced

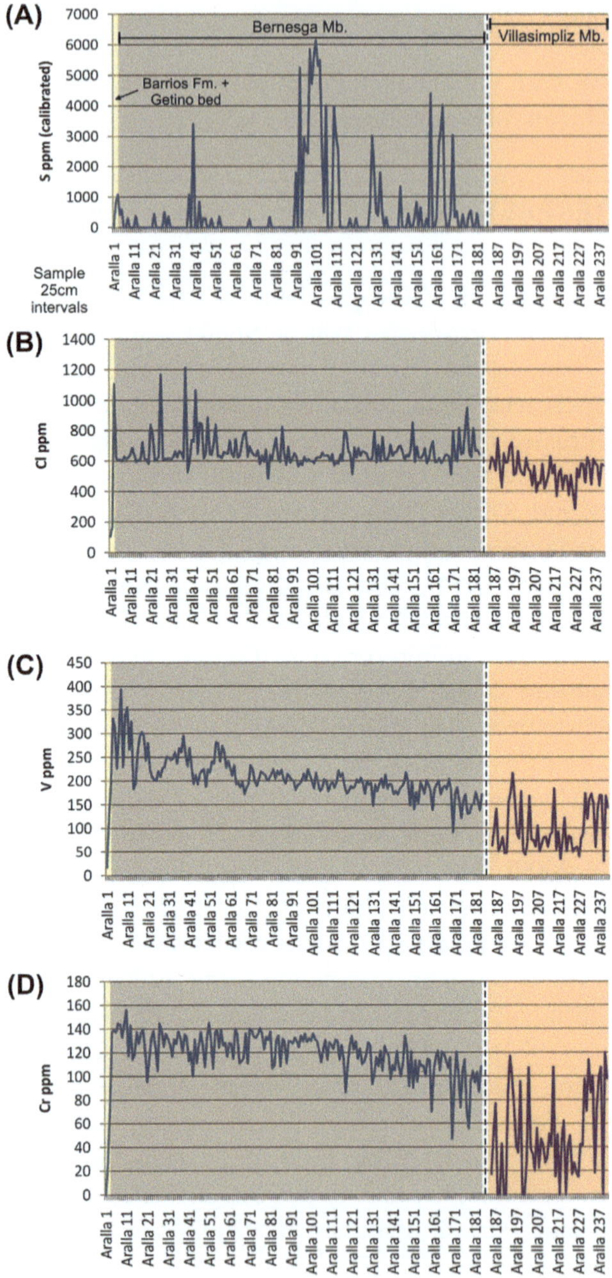

Figure 19 Trace element variations for the Aralla section, plotted against stratigraphy. Every sample number represents a 25-cm interval. The colored areas indicate the different lithostratigraphic units, starting with the Barrios Formation, followed by the Getino Beds (both in yellow), the Bernesga Member (gray), and finally the Villasimpliz Member (red). The *dashed line* represents a gap of 27.3 m. A single sample (Aralla 1) from the Barrios Fm. was analyzed as a control sample; the Getino Beds are represented by Aralla 2 and Aralla 3. Only the trace elements of the Bernesga Member and Villasimpliz Member will be described and discussed within the text. The absolute elemental values

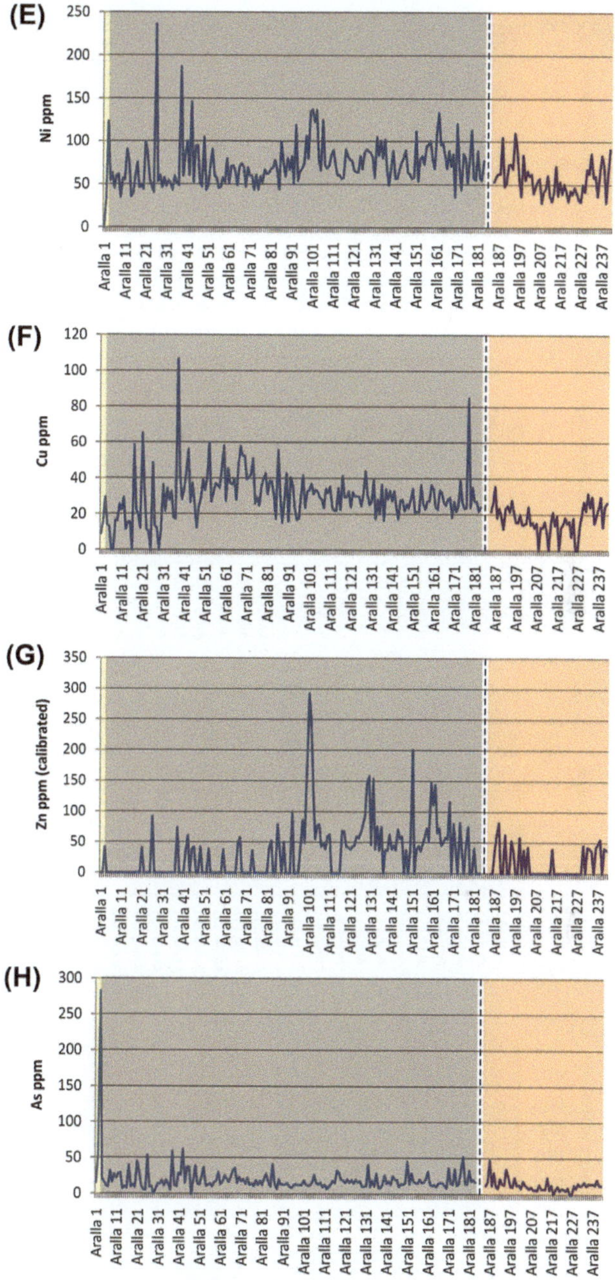

Figure 19 (*continued*).

for the Barrios Formation and the Getino Beds are documented in the geochemical data table (supplementary data). P and Bi concentrations are below D.L. and not further discussed. The elements Sc, Co, Se, Mo, Pd, Ag, Cd, Sn, Sb, Te, W, Au, and Pb are near to the D.L. and hence neither illustrated nor discussed; however, the data can be found in the geochemical data table (supplementary data).

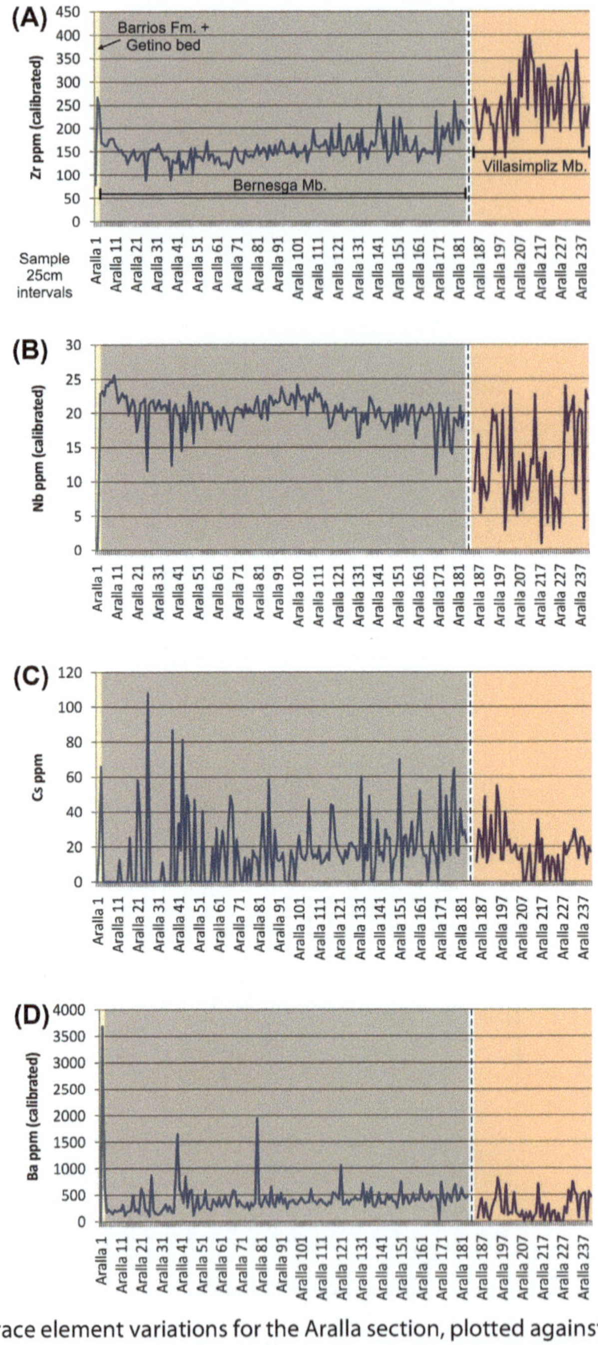

Figure 20 Trace element variations for the Aralla section, plotted against stratigraphy. Every sample number represents a 25-cm interval. The colored areas indicate the different lithostratigraphic units, starting with the Barrios Formation, followed by the Getino Beds (both in yellow), the Bernesga Member (gray), and finally the Villasimpliz Member (red). The *dashed line* represents a gap of 27.3 m. A single sample (Aralla 1) from the Barrios Fm. was analyzed as a control sample; the Getino Beds are

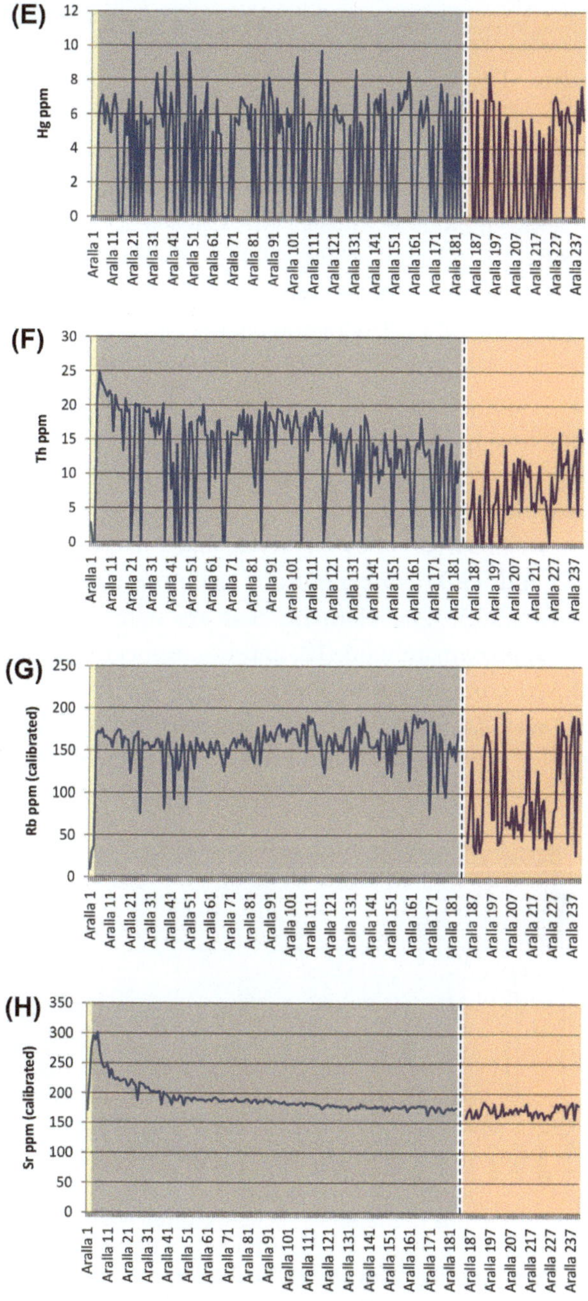

Figure 20 (*continued*).

represented by Aralla 2 and Aralla 3. Only the trace elements of the Bernesga Mb. and Villasimpliz Mb. are described and discussed within the text. The absolute elemental values for the Barrios Formation and the Getino Beds are documented in the geochemical data table (supplementary data) (Fig. 21).

to the data matrix. The values are compared to the following international standards: (1) Average Shale (Turekian and Wedepohl, 1961), (2) Average Shale (compiled from: Heinrichs et al., 1980; Wedepohl, 1971, 1991), (3) Average black shale (Vine and Tourtelot, 1970), (4) SCO-1 (compiled from: Abbey and Geological Survey of Canada (1983); Gladney and Roelandts, 1988; Govindaraju, 1989, 1994), (5) Post-Archean Australian Shale (PAAS, Taylor and McLennan, 1985), (6) Post-Archean Upper Continental Crust (PA-UCC, Taylor and McLennan, 1985, 1995), (7) Upper Crust (Shaw et al., 1967, 1976; Hans Wedepohl, 1995), and (8) North American Shale Composite (NASC, Gromet et al., 1984) (Fig. 22).

When normalized to the international standard "Average black shale" of Vine and Tourtelot (1970), the Bernesga Member shows similar elemental concentrations with regards to MgO, Cr, and Sr. A minor enrichment in the concentrations of Al_2O_3, Fe_2O_3, K_2O, Ba, Ni, Pb, Sc, V, and Zr is observed. MnO and TiO_2 concentrations are more strongly enriched (up to approximately 2.5×). The elements that are significantly enriched are Ag, Co, and U (approximately 60, 12, and 6×, respectively). On the other hand, Cu and Mo concentrations show a slight depletion (approximately 0.5×) and CaO and Zn are significantly depleted (approximately 0.95 and 0.9×, respectively). The normalization to the "Average shale" of Turekian

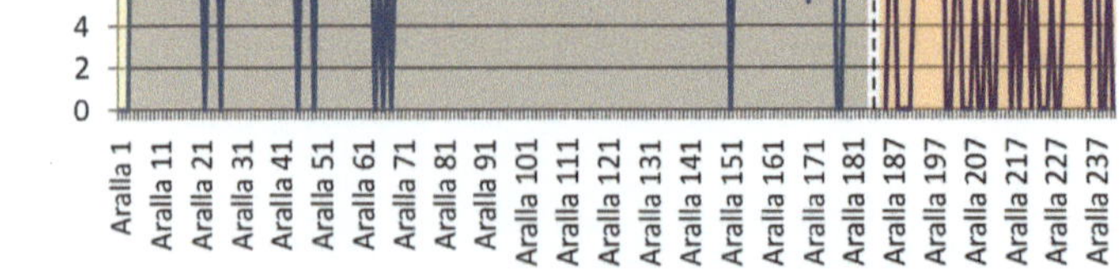

Figure 21 Uranium variations of the Aralla section, plotted against stratigraphy. Every sample number represents a 25-cm interval. The colored areas indicate the different lithostratigraphic units, starting with the Barrios Formation, followed by the Getino Beds (both in yellow), the Bernesga Member (gray), and finally the Villasimpliz Member (red). The *dashed line* represents a gap of 27.3 m. A single sample (Aralla 1) from the Barrios Formation was analyzed as a control sample; the Getino Beds are represented by Aralla 2 and Aralla 3. Only the uranium values of the Bernesga Member and Villasimpliz Member are described and discussed within the text. The uranium concentrations of the Barrios Formation and the Getino Beds are documented in the geochemical data table (supplementary data).

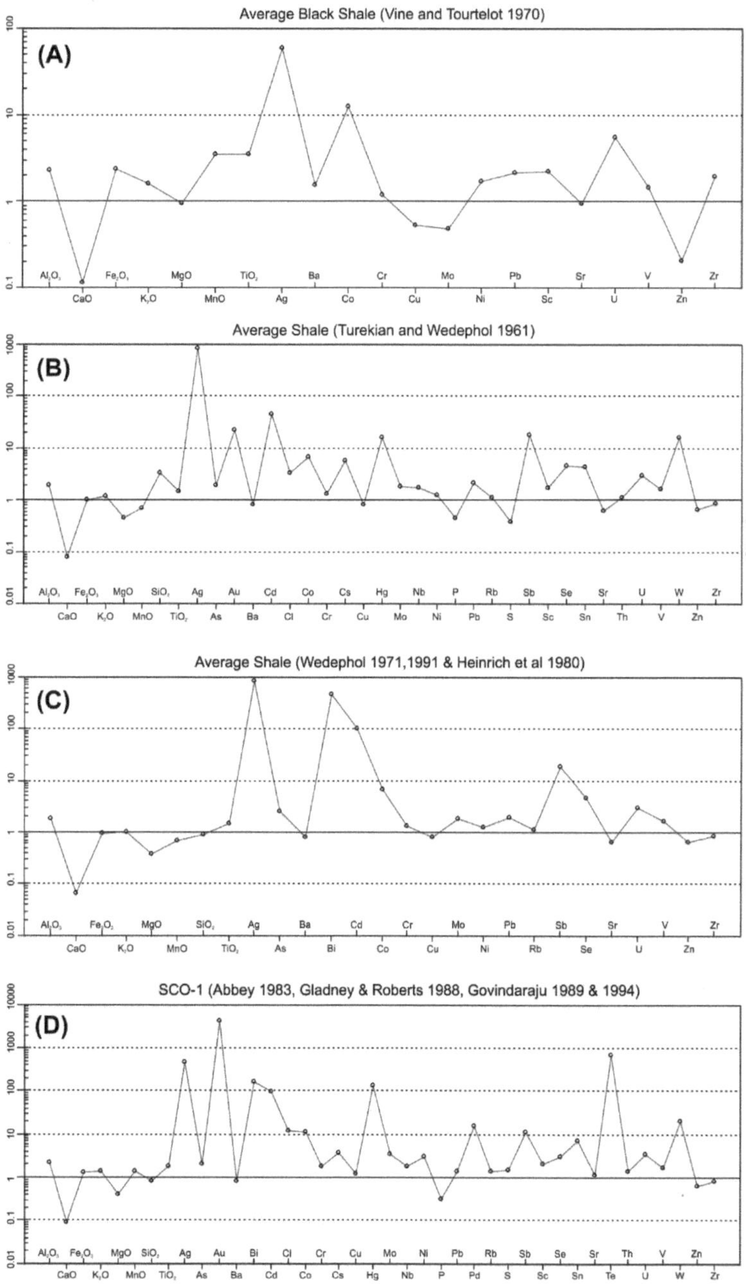

Figure 22 Enrichment factor diagrams for the Bernesga Member. Values below D.L. were excluded. Normalized to international standards. (A) Average black shale (Vine and Tourtelot, 1970); (B) Average shale (Turekian and Wedepohl, 1961); (C) Average shale (Heinrichs et al., 1980; Wedepohl, 1971, 1991); (D) SCO-1 (Abbey, 1983; Gladney and Roberts, 1988; Govindaraju, 1989, 1994).

and Wedepohl (1961) reveals that the concentrations of Fe_2O_3, K_2O, Ba, Cr, Cu, Ni, Rb, Th, and Zr are comparable. Al_2O_3, SiO_2, As, Cl, Cu, Hg, Pb, Se, Sn, U, and V are moderately enriched (approximately 2×), whereas the elemental concentrations of Ag, Au, Cd, Hg, Sb, and W show significant enrichments (approximately 900, 11, 50, 11, 11, and 11×, respectively). MgO, MnO, Sr, and Zn are very slightly (approximately 0.5×), CaO and S concentrations significantly depleted (approximately 0.09 and 0.5×, respectively). The Bernesga Member, when normalized to the "Average shale" of Wedepohl (1971, 1991) and Heinrichs et al. (1980), shows comparable Fe_2O_3, K_2O, SiO_2, Ba, Cr, Cu, Ni, Rb, and Zr concentrations. The enrichments of Al_2O_3, TiO_2, As, Mo, Pb, and V are insignificant (approximately 2×). On the other hand, the concentrations of Ag, Bi, Cd, Co, Sb, Se, and U show significant enrichments (approximately 900, 600, 100, 8, 11, 5, and 3×, respectively). CaO, MgO, Sr, and Zn are relatively depleted (0.08, 0.5, 0.7, and 0.7×, respectively). The normalization of the Bernesga Member to the "SCO-1" standard of Abbey (1983), Gladney and Roberts (1988), and Govindaraju (1989, 1994) highlights that Fe_2O_3, K_2O, MnO, Ba, Cu, Pb, Rb, Sr, Th, and Zr concentrations are comparable. The concentrations of Al_2O_3, TiO_2, As, Cr, Cs, Mo, Nb, Ni, S, Sc, Se, U, and V show mild enrichments (approximately 2×). Ag, Au, Bi, Cd, Cl, Co, Hg, Pd, Sb, Sn, Te, U, and W show significant enrichments (approximately 600, 5000, 200, 100, 10, 10, 150, 11, 10, 9, 700, 5, and 11×, respectively). CaO, MgO, P, and Zn are depleted (approximately 0.1, 0.4, 0.4, and 0.8×, respectively). The Bernesga Member concentrations of Al_2O_3, Fe_2O_3, K_2O, MnO, SiO_2, As, Ba, Cr, Sr, Th, and Zr are comparable to that of the "NASC" international standard of Gromet et al. (1984). Slight enrichments are observed for TiO_2, Au, Co, Cs, Nb, Ni, Pb, Rb, Sc, U, V, and W (approximately 1–2×). The concentrations of Ag, Bi, Cd, Sb, Se, and Te are significantly enriched (approximately 5000, 10,000, 6000, 10, 13, and 100,000×, respectively), whereas CaO and MgO are depleted (approximately 0.08 and 0.4×, respectively). When normalized to the "PAAS" international standard of Taylor and McLennon (1985) the elemental concentrations of K_2O, TiO_2, Cr, Nb, Rb, Sr, and Th are comparable. Al_2O_3, Cs, Ni, Sc, and V concentrations are mildly enriched (approximately 1.5×). The elements strongly enriched are Co, Mo, Pb, U, and W (5, 4, 2, and 3×, respectively). CaO, MgO, MnO, SiO_2, Ba, Cu, Zn, and Zr are depleted (0.2, 0.4, 0.6, 0.8, 0.7, 0.7, 0.8, and 0.7×, respectively). The Bernesga Member exhibits concentrations of K_2O, MnO, SiO_2, Ag, Ba, Cl, Nb, Th, and Zn comparable to that of the international standard

"Upper Crust" of Hans Wedepohl (1995). Mildly enriched are Al_2O_3, Fe_2O_3, TiO_2, Cr, Cs, Cu, Mo, Ni, Pb, Rb, S, Sc, U, and V (3×). As, Bi, Cd, Hg, Sb, Se, Sn, Te, and W concentrations are significantly enriched (approximately 11, 600, 150, 100, 100, 5, 10, 20,000, and 11×, respectively). CaO, MgO, Ba, Cl, Nb, P, Sr, and Zr concentrations are relatively depleted (0.07, 0.7, 0.8, 0.9, 0.8, 0.4, 0.7, and 0.6×, respectively). When normalized to the international standard "PA-UCC" of Taylor and McLennan (1985, 1995), comparable concentrations are observed for K_2O, MnO, SiO_2, and Ba. Elemental concentrations that are slightly enriched are Al_2O_3, Fe_2O_3, TiO_2, Ag, Cr, Cu, Nb, Ni, Pb, Rb, Sc, Th, and V (2×). As, Au, Bi, Cd, Co, Cs, Pd, Sb, Se, U, and W show significant enrichments (of up to 11, 7, 500, 150, 8, 8, 30, 150, 60, 5, and 11×, respectively). The elements Sr, Zr, and TiO_2 are comparable when normalized to the "Bulk continental crust" of Taylor and McLennon (1995). Ba, K_2O, and Nb concentrations are mildly enriched (2×). The concentrations of Cs, Rb, Th, and U are significantly enriched (30, 4, 4, and 10×, respectively). When normalized to the international "Chondrites" standard of Sun et al. (1982) the elemental concentrations of Rb, Ba, Th, U, and K_2O are significantly enriched (450, 150, 300, 900, and 250×, respectively), whereas the elements Nb, Sr, Zr, and TiO_2 are only mildly enriched (60, 20, 20, 10.5×, respectively). Finally the normalization to the "Chrondrites" of Thompson (1982) determines that the concentrations of Ba, Rb, Th, K_2O, and Nb are significantly enriched (80, 450, 400, 300, and 50×, respectively). Sr, P, Zr, and TiO_2 are enriched to a lesser extent (20, 8, 20, and 10×, respectively) (Fig. 23).

The enrichment of Al_2O_3 within the Formigoso Formation relative to the average shale composites is most likely the result of maturity effects. This interpretation is further corroborated by the consistently high weathering indices, which will be reviewed later. The Fe_2O_3 enrichments could be related to localized mineralization, as are the Ag and Pb concentrations. The TiO_2 and Zr enrichments could indicate an increased influx of detrital material, which is slightly elevated in contrast to the average shale composites. The U enrichment could be related to the organically rich basal shales of the Bernesga Member. These shales were most likely deposited under euxinic conditions and consequently enabled the accumulation of organic matter. Other elemental enrichments in relation to the upper continental crust composites are Bi, Cd, Hg, Sb, Te, and W. These are most likely related to the geochemical signature of the detrital source and/or the seawater chemistry during the time of deposition.

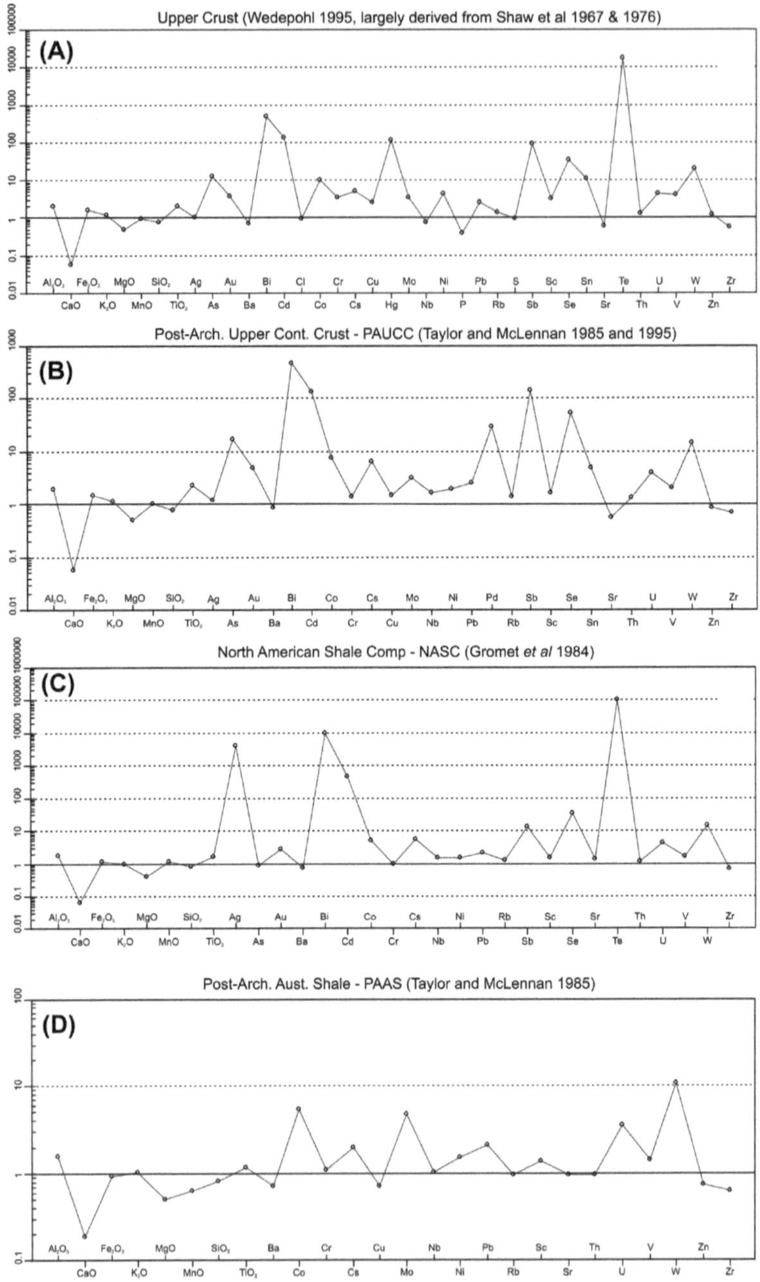

Figure 23 Enrichment factor diagrams for the Bernesga Member. Excluded are values below detection limits. Normalized to international standards. (A) Upper crust (Shaw et al., 1967, 1976; Hans Wedepohl, 1995); (B) Post-Archean upper continental crust, PAUCC (Taylor and McLennan, 1985, 1995); (C) North American shale composite, NASC (Gromet et al., 1984); (D) Post-Archean Australian shale, PAAS (Taylor and McLennan, 1985).

8.2 Geochemical Classification

All major and trace elemental concentrations obtained from the analyzed section were used to classify the sedimentary rocks geochemically. The (1) Ni versus TiO_2 plot of Floyd et al. (1989), (2) provenance discrimination diagram of Roser and Korsch (1988), (3) SiO_2-Al_2O_3-CaO ternary system of Ross and Bustin (2009), (4) Zr-15Al_2O_3-300TiO_2 diagram of Mongelli et al. (2006), and finally (5) the TOC-Fe(wt%)-TS10(wt%) all resulted in very similar signatures and patterns.

The data set is cross-plotted utilizing the log(Fe_2O_3/K_2O) versus log(SiO_2/Al_2O_3) of Herron (1988). This discrimination diagram was originally developed to separate aluminum-rich fine-grained sediments from coarser-grained immature sediments using the diagnostic differences within the ratios of the oxides for Fe and K and Si and Al, respectively. This diagram shows that the bulk of the Bernesga Member shales fall into the "shale" window, a couple of samples can be seen to push toward the "Fe-shale" category, fitting with the Fe_2O_3 enrichments seen in the elemental distribution plots. The majority of the sand and siltstones of the Villasimpliz Member of the Aralla section fall into the "Wacke" window, with a number of samples pushing toward the "Fe-sand." (Fig. 24).

The bivariate plot TiO_2 versus Ni of Floyd et al. (1989). This diagram was engineered for use with immature greywackes. It allows for the discrimination between sediments and sedimentary rocks composed of either a bulk acidic or basic composition when cross-plotting the concentrations of TiO_2 and Ni.

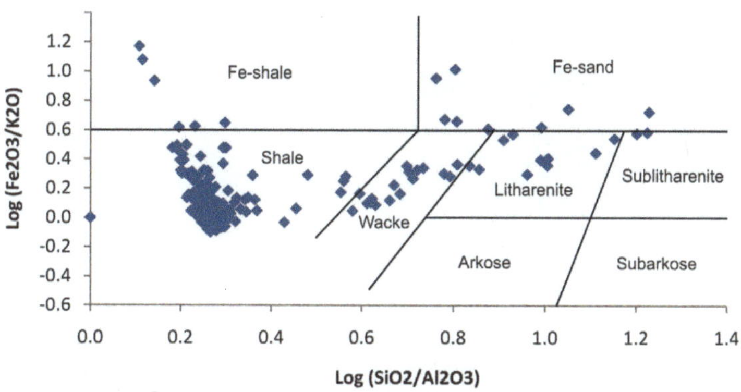

Figure 24 Chemical classification and discrimination of the sedimentary rocks analyzed using the log(SiO_2/Al_2O_3) versus log ($Fe_2O_3^t$/K_2O) system of Herron (1988). (A) Aralla, the shales of the Bernesga Member, are grouped within the "Shale" category, a few samples are plotting within the "Fe-shale, the sand and siltstones of the Villasimpliz Member trend toward the Fe-sand."

This diagram works effectively for discriminating between the mature shales of the Bernesga Member and the sand and siltstones of the Villasimpliz Member. A clear grain-size effect is observed. In terms of the provenance discrimination fields, however, the data do not discriminate within the predictable areas. For example, the shales of the Bernesga Member are not plotting near or at the trend line for mudstones, as they were expected to do. The sand and siltstones of the Villasimpliz Member do, however, fall within the proximity of the "acidic" composition field, which is not too farfetched, as the original sediments have certainly been sourced from a predominantly acidic basement of the hinterland (Fig. 25).

The provenance discrimination diagram of Roser and Korsch (1988). This diagram is widely used to distinguish between different provenance sources, such as the "quartzose sedimentary," "mafic igneous," intermediate igneous,' and finally "felsic igneous" provenances. Two different discriminant functions are used: (1) Discriminant function 1 (F1) = $(-1.773 \times TiO_2wt\%) + (0.607 \times Al_2O_3wt\%) + (0.76 \times Fe_2O_3^{tot}wt\%) + (-1.5 \times MgOwt\%) + (0.616 \times CaOwt\%) + (0.509 \times Na_2Owt\%) + (-1.22 \times K_2Owt\%) + (-9.09)$. (2) Discriminant function 2 (F2) = $(0.445 \times TiO_2wt\%) + (0.07 \times Al_2O_3wt\%) + (-0.25 \times Fe_2O_3^{tot}wt\%) + (-1.142 \times MgOwt\%) + (0.432 \times Na_2Owt\%) + (1.426 \times K_2Owt\%) + (-6.861)$. During the Silurian the Formigoso Formation is interpreted to have been deposited in a passive margin setting (see "Geological Setting").

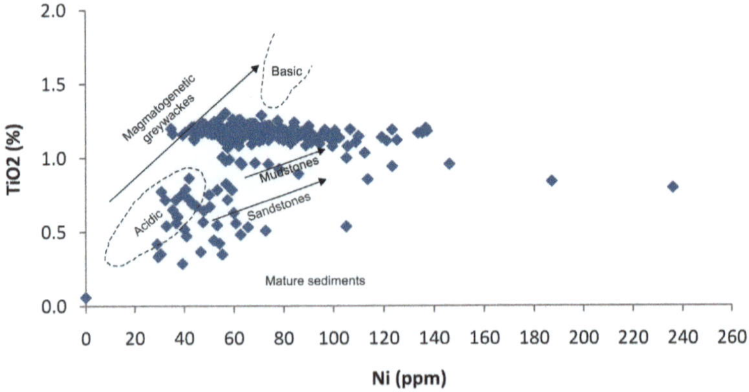

Figure 25 TiO_2 versus Ni after Floyd et al. (1989). Illustrated are data of the Aralla section (241 samples). The mudstones of the Bernesga Member display relatively uniform TiO_2 concentrations, with concentrations of Ni varying from approximately 35 ppm–140 ppm. The sand and siltstones of the Villasimpliz Mb. cluster toward the "Acidic" zone.

The hinterland and hence the source for the sediments is regarded to have been primarily a granitic basement. However, this is not shown in and cannot be deduced from the discrimination diagrams. However, the discriminant functions applied do show a clear systematic variation between the shales of the Bernesga Member and the sand and siltstones of the Villasimpliz Member. The maturity and grain-size effects are again clearly seen. These hinterland and tectonic discrimination diagrams have previously been widely used to elucidate the lithological composition of provenance areas and the very nature of their larger scale geodynamic setting (e.g., Arribas et al., 2007; Johnsson and Basu, 1993). However, a detailed critical evaluation of these diagrams and related discrimination functions by Armstrong-Altrin and Verma (2005) highlighted that they are not always reliable and consequently have to be applied with caution (Fig. 26).

Pearce et al. (2010) developed a geochemical classification scheme to discriminate between floodplain-, brackish to lacustrine-, marginal marine-, and fully marine developed mudstones. The parameters that are used to classify the sediments as well as sedimentary rocks are quantified by a scoring system: (1) U ppm values $\geq$6 are awarded a score of 4, concentrations $<$6 are given a value of 1; (2) a Th/U ratio of $\geq$3 is awarded a score of 1, values $<$3 are given a value of 4; (3) a P_2O/Al_2O_3 ratio $\geq$0.02 is given a score of 4, $<$0.02 is given a value of 1; (4) the ratios of Mo/Al_2O_3 are awarded a score of 4 if $\geq$0.08 and a value of 1 if $<$0.08; finally (5) a Zr/U ratio $>$65 is awarded a score of 2, a Zr/U ratio $<$65 gains 0. The sum of these parameters for each sample results in a geochemical classification value. These diagnostic values are used to identify the environmental setting in which the sediments were potentially deposited in. Geochemical classification values of $>$0 yet $\leq$4 are interpreted to be indicative of floodplain sediments, values of $>$4 and $\leq$7 point to a brackish to lacustrine environment of deposition, values of $>$7 and $\leq$10 infer marginal marine settings, and finally, values of $>$10 to 16 indicate fully marine developed sediments (Fig. 28).

When applying the scheme of Pearce et al. (2010), the depositional environment of most of the lower part of the Bernesga Member at Aralla can be classified as fully marine developed. The succession is, however, punctuated by three distinct phases, which exhibit values indicative for a strong floodplain influence. Toward the middle of the Bernesga Member a drop in the classification values can be observed, these shales are interpreted to be of marginal marine origin. Most of the sand and siltstones of the overlying Villasimpliz Member carry low values, diagnostic for marginal marine settings. Furthermore, there are a number of intercalated horizons,

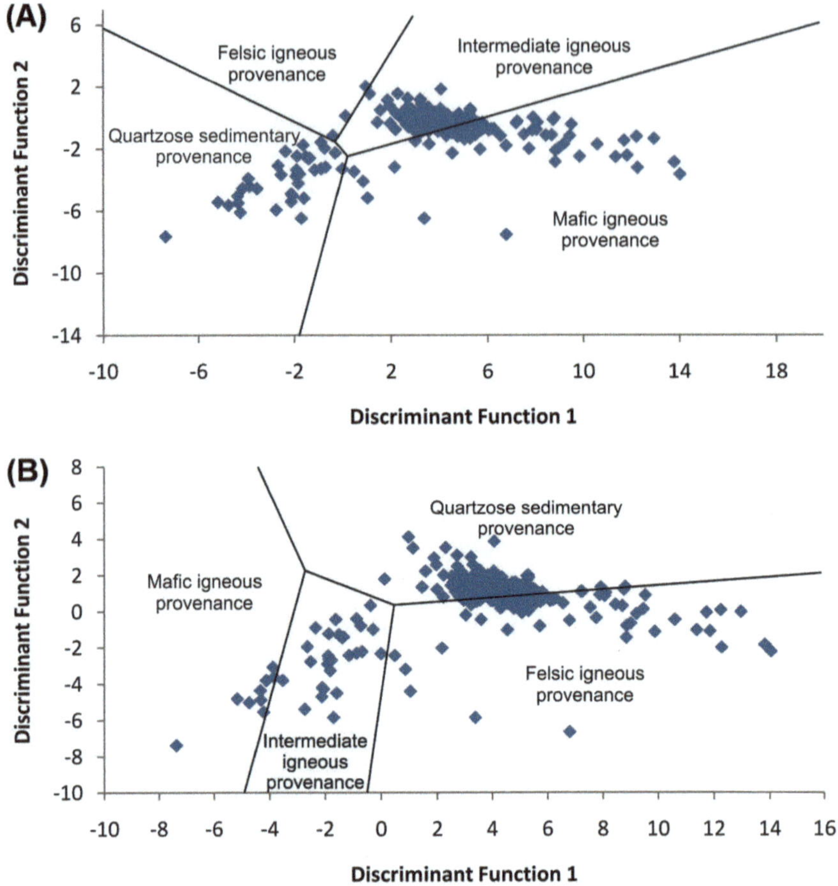

Figure 26 Chemostratigraphic data of the Aralla section (241 samples). (A) and (B): Provenance discrimination diagrams for shales. , after Roser and Korsch (1988). (A) Discriminant function 1 = ($-1.773 \times TiO_2$wt%) + ($0.607 \times Al_2O_3$wt%) + ($0.76 \times Fe_2O_3^t$wt%) + ($-1.5 \times$ MgOwt%) + ($0.616 \times$ CaOwt%) + ($0.509 \times Na_2O$wt%) + ($-1.22 \times K_2O$wt%) + (-9.09). Discriminant function 2 = ($0.445 \times TiO_2$wt%) + (0.07 $\times Al_2O_3$wt%) + ($-0.25 \times Fe_2O_3^t$wt%) + ($-1.142 \times$ MgOwt%) + ($0.432 \times Na_2O$wt%) + ($1.426 \times K_2O$wt%) + (-6.861). Applying these discriminant functions results in a clear compositional differentiation between the shales of the Bernesga Member and the sand and siltstones of the Villasimpliz Member. The sands are seen to trend toward the "Quartzose sedimentary provenance" field, whereas the shales of the Bernesga Member span from the "Intermediate igneous provenance" to the "Mafic igneous provenance." (B) Discriminant function 1 = ($-1.773 \times TiO_2$wt%) + ($0.607 \times Al_2O_3$wt%) + ($0.76 \times Fe_2O_3^t$wt%) + ($-1.5 \times$ MgOwt%) + ($0.616 \times$ CaOwt%) + ($0.509 \times Na_2O$wt%) + ($-1.22 \times K_2O$wt%) + (-9.09). Discriminant function 2 = ($0.445 \times TiO_2$wt%) + (0.07 $\times Al_2O_3$wt%) + ($-0.25 \times Fe_2O_3^{tot}$wt%) + ($-1.142 \times$ MgOwt%) + ($0.432 \times Na_2O$wt%) + ($1.426 \times K_2O$wt%) + (-6.861). These functions discriminate more effectively between the shales of the Bernesga Member and the sand and siltstones of the Villasimpliz Member. The Villasimpliz Member clusters within the "Mafic igneous provenance" and "Intermediate igneous provenance," whereas the shales of the Bernesga Member trend from the "Quartzose sedimentary provenance" to the "felsic igneous provenance." (Fig. 27).

which suggest a brackish to lacustrine environment of deposition. Overall, the Formigoso Formation at the Aralla section can be interpreted to display a decline in salinity levels from the fully marine developed conditions at the base over the marginal marine mud, silt, and sandstones of the upper Bernesga Member to the brackish—lacustrine environment inferred for parts of the Villasimpliz Member.

The SiO_2-Al_2O_3-CaO ternary plot (Fig. 29) of Ross and Bustin (2009) utilizes SiO_2 concentrations to determine the relative proportions of Quartz, Al_2O_3 to approximate the clay concentrations and CaO values to represent the carbonate fraction. The shales of the Bernesga Member display enrichment in Al_2O_3 relative to the "Average shale" of Wedepohl (1971), as also determined by the previously described elemental EF plots. The sand and siltstones of the Villasimpliz Member show a compositional maturity trend toward SiO_2 as expected. In general it is assumed that weathering and sedimentary recycling processes will lead to the demise of less resistant alumosilicates, and hence to a relative enrichment of the resistant heavy minerals and quartz. As a consequence, higher mature sediments will start plotting toward the SiO_2 end member. It is notable that all of the sedimentary rocks belonging to the Formigoso Formation are, compared to the international standards, strongly depleted with regards to CaO, indicating that no dilution of the siliciclastic sediments with a carbonate phase took place.

The $15Al_2O_3$-$300TiO_2$-Zr system (Fig. 30) after Mongelli et al. (2006) uses the oxides Al_2O_3 and TiO_2 and the element Zr to discriminate between different lithologies. The Al_2O_3 concentrations represent the clay content, whereas the TiO_2 and Zr concentrations are assumed to reside within the heavy mineral phases. The TiO_2 and Zr components are used to discriminate between (1) an acidic provenance, with the enrichment of Zr indicating the presence of zircon crystals originally formed within melts of granitic composition, and (2) a basic composition of the provenance, typified by the elevated concentrations of TiO_2 indicating the presence of Ti-oxides, for example, rutile. The increased Zr values are interpreted to be indicative for higher mature sandstones. The shales of the Bernesga Member plot within the "shale" category and are consequently trending toward the Al-rich shales. The shales display again an overall enrichment of Al_2O_3 in comparison to the international "PAAS" standard of Taylor and McLennon (1985). The sand and siltstones of the Villasimpliz Member are trending toward Zr, indicating a higher compositional maturity. The Getino Beds, represented by the blue triangles, have comparable

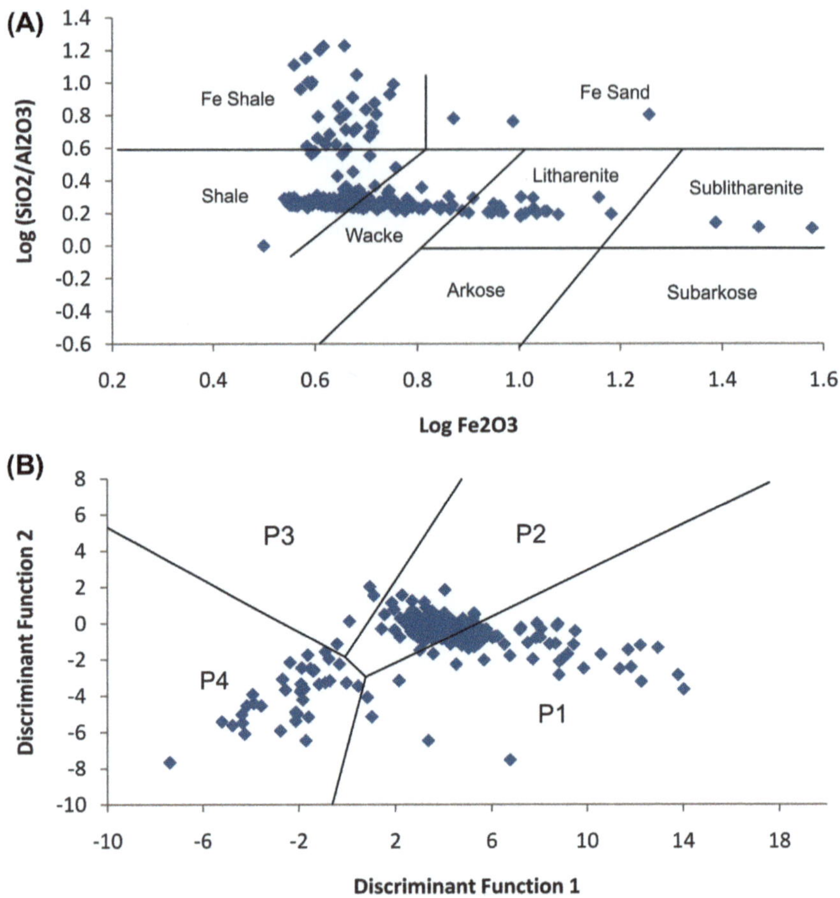

Figure 27 Chemostratigraphic discrimination of the Aralla section (241 samples). (A) Chemical classification, log(SiO_2/Al_2O_3) versus log(Fe_2O_3) after Herron (1988). The system discriminates well between the Bernesga Member and the Villasimpliz Member. The shales and siltstones of the Bernesga Member trend from the "Shale" category to the "Litharenite," whereas the sand and siltstones of the Villasimpliz Member are clustering toward the "Fe shale." (B) Plot of discriminant functions F1 and F2. Provenance fields are after Roser and Korsch (1988). P1 = mafic and lesser intermediate igneous provenance, P2 = intermediate igneous provenance, P3 = felsic igneous provenance, and finally P4 = recycled mature polycyclic quartzose detritus. The provenance discrimination diagram of Roser and Korsch (1988) clearly discriminates between the Bernesga Member and the Villasimpliz Member. The shales span from "P3" to "P1," whereas the sand and siltstones plot between "P4" and "P1."

concentrations of Al_2O_3 and Zr relative to the sand and siltstones of the Villasimpliz Member. However, they do exhibit a significant relative enrichment in TiO_2. The Getino Beds are interpreted to be the product of recycled sediments, with the predominant sediment source being the underlying Barrios Formation, combined with an influx of clay content from the

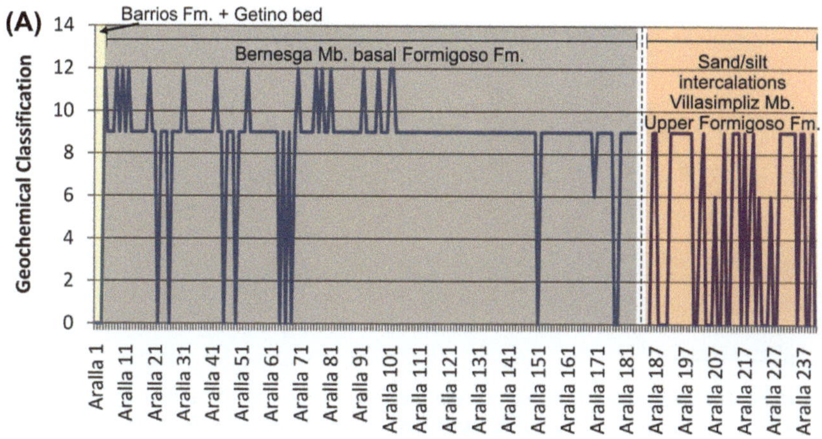

(B) **Criteria used for geochemical classification**

Parameter	Criteria	Score	Average values for each locality
U	> 6ppm < 6ppm	4 1	**a)** 9.1 **b)** 7.9 **c)** 4.8 **d)** 9.4 **e)** 9.5
Th/U	> 3.0 < 3.0	1 4	**a)** 1.2 **b)** 0.9 **c)** 1.1 **d)** 1.4 **e)** 1.0
P2O/Al2O3	> 0.02 < 0.02	4 1	**a)** n/a **b)** n/a **c)** n/a **d)** n/a **e)** n/a P2O values are below detection limit
Mo/Al2O3	> 0.08 < 0.08	4 1	**a)** 0.009 **b)** 0.04 **c)** 0 **d)** 0.02 **e)** 0.02
Zr/U	> 65	2	**a)** 17.5 **b)** 11.4 **c)** 16.6 **d)** 16.4 **e)** 11.5

Figure 28 (A) Geochemical classification values versus stratigraphy. (B) Criteria used for the calculation of the geochemical classification values. Geochemical classification based on Pearce et al. (2010). The geochemical classification values (0–14) indicate: >0 yet ≤ 4 = floodplain mudstone, >4 and ≤ 7 = brackish water or lacustrine, >7 and ≤ 10 = marginal marine mudstone, >10 to 16 = marine mudstone. If a sample has a very high Zr/U value (above 65), it is presumed to contain abundant heavy minerals and is awarded a final score of 2. If a sample is derived from coal, it is awarded a default score of 0. The P_2O and Mo values were near/below detection limits. The P_2O/Al_2O_3 ratio was excluded from the classification scheme calculation.

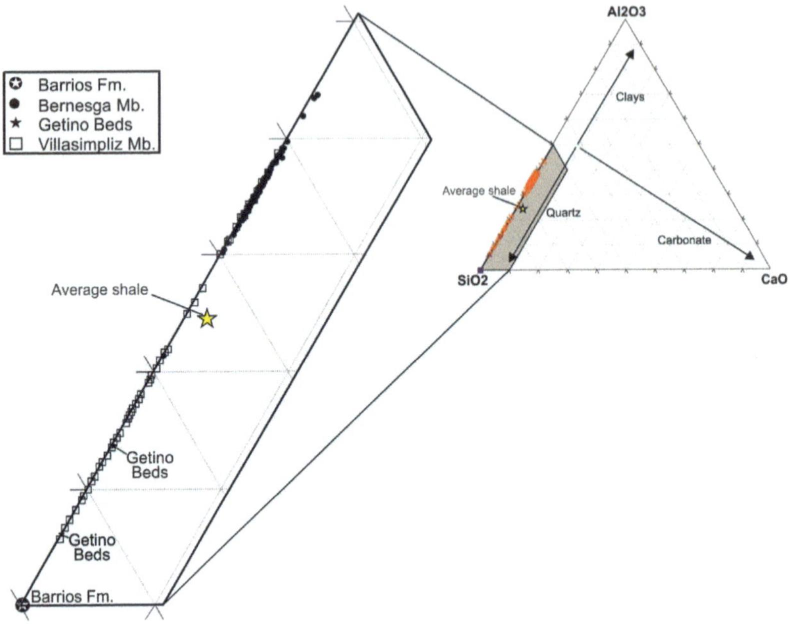

Figure 29 Ternary diagram showing relative proportions of the major shale/mudrock elements SiO_2 (quartz), Al_2O_3 (clays), and CaO (carbonates). The "Average shale" international standard after Wedepohl (1971) shown as star. The shales of the Bernesga Member (black circle) show an enrichment in Al_2O_3 relative to the "Average shale." The sand and siltstones of the Villasimpliz Member (square) exhibit a compositional maturity trend toward higher SiO_2 concentrations. The control sample of the highly mature Barrios Formation (black circle with inset star) as an example for an end member of such a maturity trend is plotting directly into the SiO_2 corner.

hinterland as a result of elevated chemical weathering rates. Most likely, the Getino beds represent transgression-related lag sediments with a relatively high content of heavy minerals such as zircon and rutile, with the latter causing the high TiO_2 concentrations (Fig. 31).

The TOC–TS10–Fe ternary plot after Dean and Arthur (1989) and Arthur and Sageman (1994). The differing concentrations of the parameters TOC, total sulfur (TS), and Fe are used to discriminate sediments formed under different redox conditions. The majority of the shales of the Bernesga Member trend toward high TOC values, whereas the Barrios Formation, Getino Beds, and Villasimpliz Member show no enrichment. As the TOC values for this study were approximated based on U ppm concentrations, there is a clear differentiation or grouping of the shales, (1) the U enriched basal shales or "hot shales," (2) the U lean shales, and finally (3)

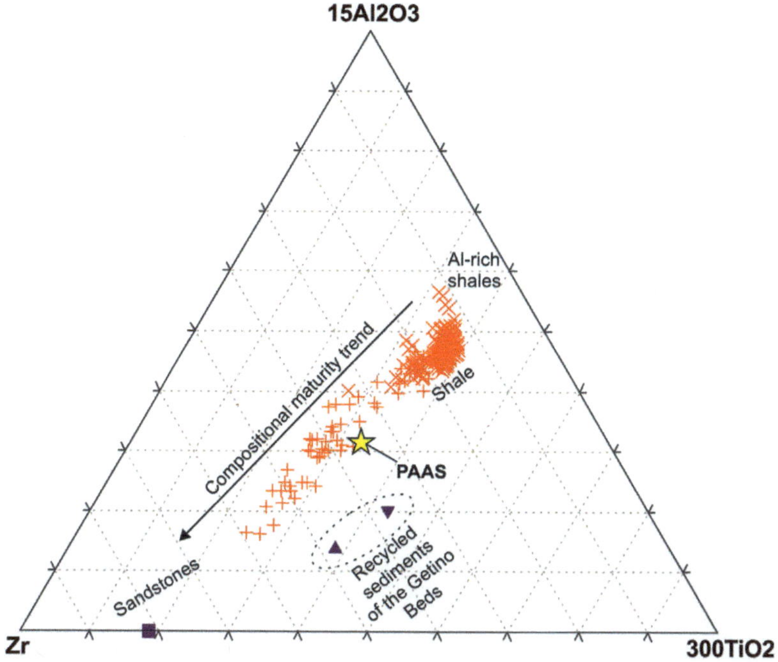

Figure 30 Ternary $15Al_2O_3$-$300TiO_2$-Zr plot of Mongelli et al. (2006); the international standard "PAAS" of Taylor and McLennon (1985) is plotted for comparison. The plot illustrates all geochemical data from the Aralla section (241 samples). The Barrios Formation is represented by a blue square, the Getino Beds by *blue triangles*, the Bernesga Member by the diagonal *red cross (x)* and finally the Villasimpliz Member by an upright cross (+). A compositional maturity trend is present within the sand and siltstones of the Villasimpliz Member toward increased Zr concentrations.

the shales or siltstones that are depleted in U relative to the very basal shales. A number of shale samples show elevated TS values, which do not coincide with Fe_2O_3 peaks on the variation plots, indicating that these S enrichments are not solely related to the formation of pyrite but also to other sulfide phases.

8.3 Environmental Reconstructions

For the environmental reconstruction a number of variables were approximated: (1) the paleoredox states, (2) paleosalinity variations, (3) paleohumidity fluctuations, (4) clay-typing, (5) chemical weathering conditions, (6) paleobioproductivity, and finally (7) the effects of potential hydrothermal

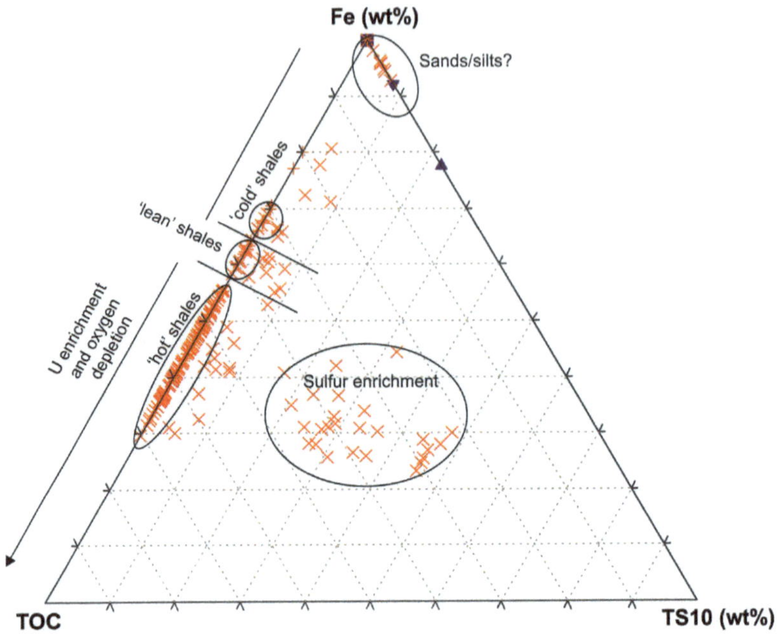

Figure 31 TOC–TS–Fe ternary diagram of Dean and Arthur (1989) and Arthur and Sageman (1994). *TOC*, total organic carbon; *TS*, total sulfur. Discrimination fields are indicated based on U concentrations: (1) the "hot" shales, (2) the "lean" shales, and (3) the "cold" shales. The Barrios Formation is represented by a *blue square*, the Getino Beds by *blue triangles*, the Bernesga Member by the diagonal *red cross (x)*, and finally the Villasimpliz Member by an upright cross (+).

overprinting on the concentrations of certain elements. The combination of these parameters was used to infer the paleoenvironmental conditions during the deposition of the (1) Getino beds, (2) Bernesga Member shales, and (3) the sand and siltstones of the Villasimpliz Member.

8.3.1 Anoxia Reconstructions

A number of geochemical proxies involving different chemical elements and elemental ratios were used to estimate the paleoredox conditions for the sedimentary succession of the Formigoso Formation, and also to critically test the value and robustness of the applied proxies compared to each other.

The Ni/V and V/Cr redox proxies of Jones and Manning (1994) rely on the sensitivity of V to changing redox conditions. Under oxic conditions V is relatively insoluble and accumulates in sediments, leading to low Ni/V and high V/Cr ratios. In contrast, under reducing conditions it is soluble and is incorporated into the water column (Harris et al., 2004). A V/Cr ratio ≤2 is interpreted to represent oxic environmental conditions, >2 to ≤4.25 dysoxic, and >4.25 suboxic to anoxic conditions. Kimura and Watanabe (2001) concluded that the degree of V-enrichment is best represented by the V/Sc ratio, as the reduced forms of V and Sc are equally insoluble. Generally, the V abundance in shales varies significantly relative to Sc, and not to other insoluble elements such as Ti and Al. A V/Sc ratio ≤9.1 is used to infer an oxic environment of deposition, and values above oxygen depletion. The V/(V + Ni) proxy of Hatch and Leventhal (1992) considers that both the V and Ni occur in highly stable organically derived tetrapyrrole structures (such as porphyrin), which are preserved under anaerobic conditions. Oxidized OM tends to exhibit a lower tetrapyrrole content and hence low Ni and V concentrations. The V may also be adsorbed onto the clay minerals, a process that is thought to occur postdepositionally (Rimmer, 2004). V/(V + Ni) ratios ≤0.46 are interpreted to represent oxic conditions, >0.46 to ≤0.60 dysoxic, >0.54 to ≤0.82 suboxic to anoxic, and >0.84 indicates a euxinic environment of deposition. Jones and Manning (1994) also engineered the U/Th redox proxy, using the redox behavior of the uranyl U^{+6} ion, which is highly soluble in oxic environments, but is reduced and precipitates under oxygen-depleted conditions. The Th resides in the detrital fraction that is not affected by the redox state. A U/Th ratio ≤0.75 indicates oxic conditions, >0.75 to ≤1.25 dysoxic, and >1.25 can be used to interpret a suboxic to anoxic environment of deposition. The Th/U proxy of Fertl (1979) uses the same principles highlighted above for the U/Th ratio. Values ≤2 relate to anoxic environmental conditions.

The values of the Ni/V and V/Cr redox ratios of Jones and Manning (1994) are used to support the interpretation that the basal shales of the Bernesga Member were deposited under highly reduced conditions (Fig. 32). The younger shales of the Bernesga Member (upper part) seem to have been deposited under progressively oxygenated conditions. The sand and siltstones of the Villasimpliz Member are interpreted as oxic when applying the Ni/V ratio. However, applying the V/Cr proxy, the Villasimpliz Member is displaying a higher state of anoxia than the basal shales of the Bernesga Member. This interpretation conflicts with the field observations of heavy bioturbation for the Villasimpliz Member. The redox proxies V/Cr, V/Sc, V/(V + Ni), of Jones and Manning (1994), Kimura and Watanabe (2001), Hatch and Leventhal (1992), respectively (Fig. 33), indicate that elevated redox conditions prevailed during the deposition of the basal shales of the Bernesga Member. In contrast, the U/Th ratio of Jones

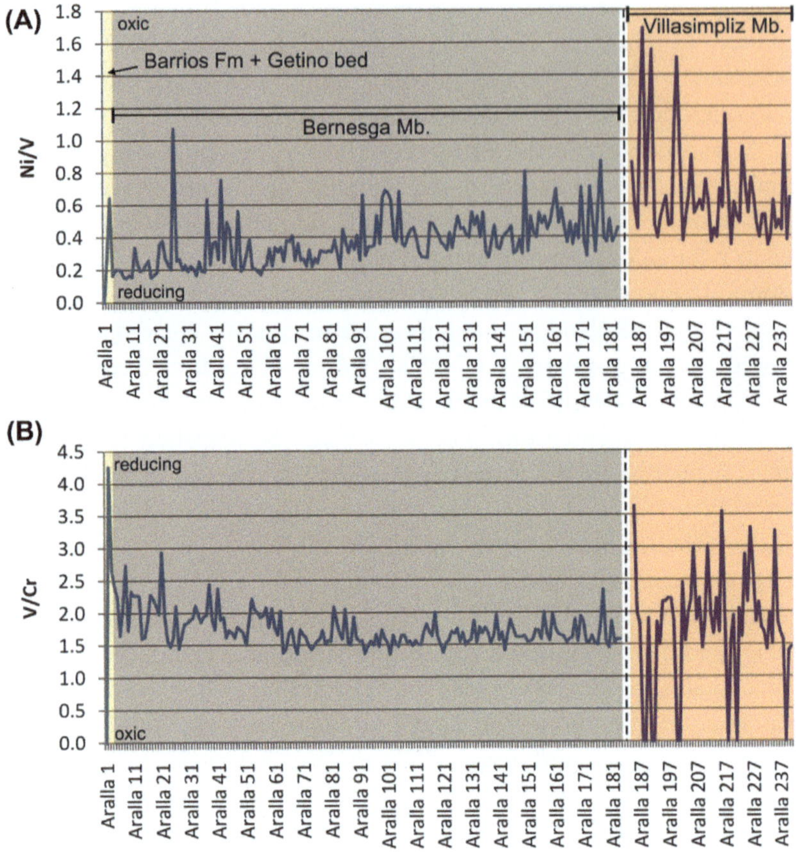

Figure 32 (A) Ni/V proxy of Jones and Manning (1994); the lowest values are observed at the base of the Bernesga Member indicating an oxygen-depleted environment of deposition, the values increase gradually throughout the Member. The sand and siltstones of the Villasimpliz Member show erratic and relatively higher values than the Bernesga Member, indicating the increased presence of free oxygen. (B) V/Cr proxy of Jones and Manning (1994); the values are elevated at the base of the Bernesga Member indicating reducing conditions. These values decrease throughout the section and become relatively stable. The sand and siltstones of the Villasimpliz Member have increased values, suggesting deposition under a reducing environment (this does not concur with the field observations); a few horizons display low values indicating sedimentation under oxic conditions.

and Manning (1994) results in values indicative for a predominantly oxic environment of deposition, hence demonstrating that this redox proxy has to be interpreted with caution. The Th/U ratio of Fertl (1979) may lead to the interpretation that increased reducing conditions were indeed

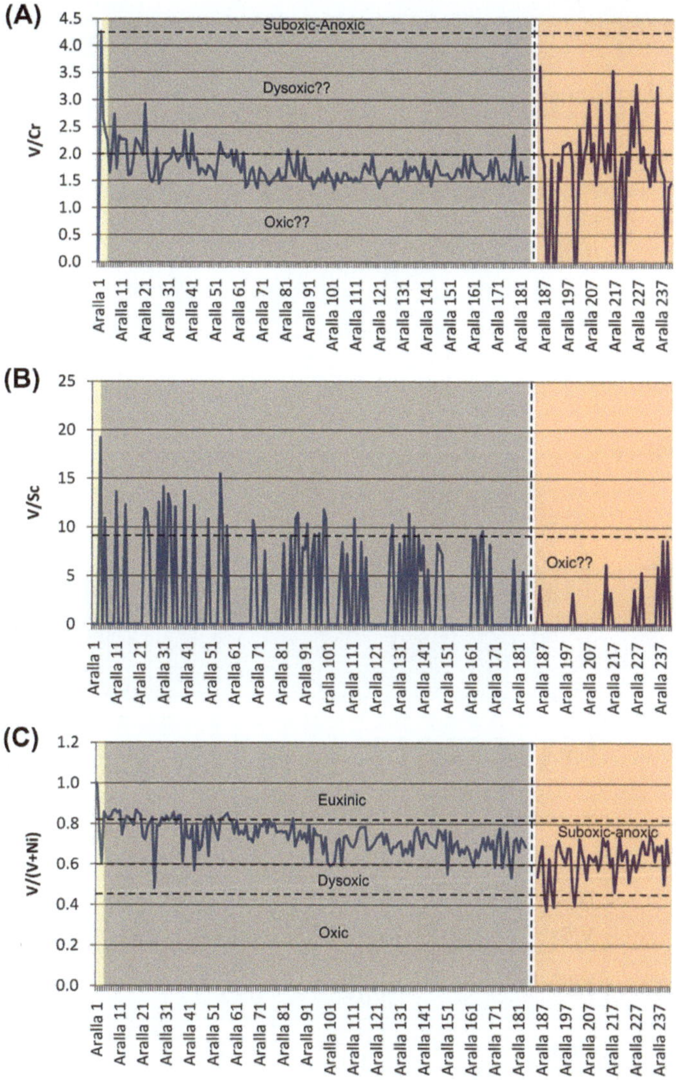

Figure 33 Proxies used to reconstruct paleoenvironmental redox conditions. (A) V/Cr ratios, illustrating the elevated reducing conditions at the base of the Bernesga Member; the sediments were progressively oxygenated toward the top of the Member. However, the Villasimpliz Member shows increased reducing conditions relative to the shales of the Bernesga Member. (B) V/Sc ratios indicating elevated reducing environments at the base of the Bernesga Member; the sediments are deposited under progressively oxygenated conditions toward the top of the Member including the overlying Villasimpliz Member. (C) V/(V + Ni) redox proxy, illustrating the progressive oxygenation of the environment from the "euxinic" base to the "dysoxic—suboxic" top. (D) U/Th values contradicting all other proxies, indicating that the vast majority of the Formigoso Formation was deposited under oxic environmental conditions. (E) Th/U redox proxy, interpreted to show reducing conditions at the base of the Bernesga Member, and an oxic environment toward its middle part, and progressively more reducing conditions toward the top, including the Villasimpliz Member. The table gives the various proxy values indicative for different environmental conditions.

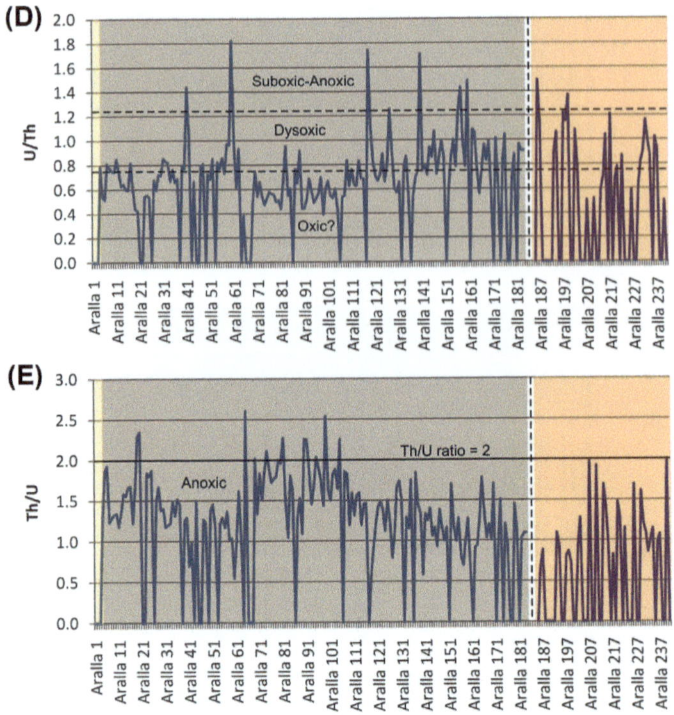

	Oxic	Dysoxic	Suboxic-Anoxic	Euxinic
Ni/Co[1]	< 5	5 - 7	> 7	
V/Cr[1]	< 2	2 - 4.25	> 4.25	
V/(V+Ni)[2]	< 0.46	0.46 - 0.60	0.54 - 0.82	> 0.84
U/Th[1]	< 0.75	0.75 - 1.25	> 1.25	
V/Sc[3]	< 9.1			

Figure 33 (***continued***).

established during the deposition of the basal shales. However, this proxy also indicates that oxic conditions prevailed during the sedimentation of the mid-Bernesga Member. The upper Bernesga Member and the overlying Villasimpliz Member have ratios typical for a more reducing environment of deposition than the basal shales of the Bernesga Member. When applying these proxies to the elemental concentrations of the Bernesga Member, it is evident that the ratios V/Sc, U/Th, and Th/U do not work as effectively as the V/(V + Ni) ratio. However, when the V/(V + Ni) ratio is applied to the elemental concentrations of the Getino Beds, the values point to an "euxinic" environment of deposition. This interpretation significantly contradicts the field observations, as the Getino Beds are again heavily bioturbated, indicating an oxic environment during the time of deposition.

The authigenic U (U_{auth}) concentrations were calculated following the method introduced by Wignall and Myers (1988). The U_{auth} (Fig. 34) contributes a significant proportion toward the absolute U concentrations. The average U_{auth} concentrations for the Bernesga Member equate to approximately 6 ppm, whereas the absolute U concentrations average at 11 ppm. Therefore the U_{auth} concentrations make up to 54% of the total U values analyzed. These high overall concentrations of the U_{auth} are related to the reducing conditions of the environment at the time of deposition. The high concentrations of U_{auth} within the Bernesga Member of the Aralla section correspond well to the "euxinic" conditions of the previously calculated V/(V + Ni) ratio. The cyclic signal observed in the U_{auth} concentrations is also present within the absolute U values. This seems to indicate that the cyclic U signal is the response to periodically changing environmental conditions at the time of the black shale deposition.

8.3.2 Paleosalinity Reconstruction

The paleosalinity reconstructions were calculated following the method of Campbell and Williams (1965). The ratio of the elements Rb and K are used to interpret varying levels of salinity. The use of this elemental ratio is based on the assumption that marine shales contain higher Rb

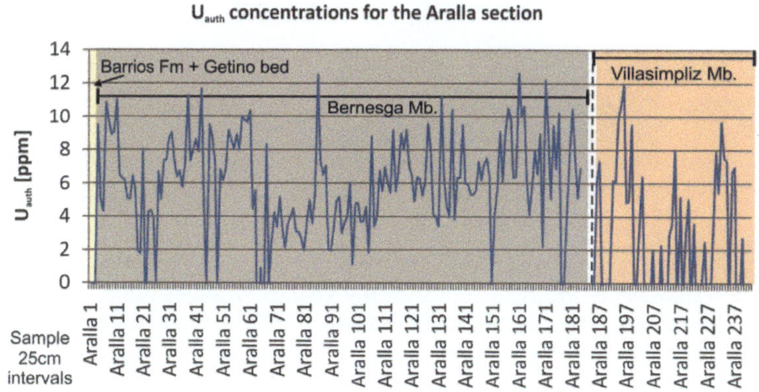

Figure 34 Authigenic uranium (U_{auth}) concentrations for the Aralla section. The U_{auth} shows the same similar cyclic signal as the absolute U concentrations.

concentrations due to the increased concentrations of Rb in seawater (0.12 ppm) relative to freshwater environments (0.0013 ppm). The K concentrations are used to represent the clay fraction. The ratio values are interpreted as the following: (1) Rb/K ratio of $\leq$0.004 indicates freshwater conditions, (2) >0.004 to $\leq$0.006 suggests freshwater to brackish conditions, and (3) >0.006 relates to fully marine developed environments (Fig. 35).

Following the Rb/K ratios, the Getino Beds are interpreted to be freshwater deposits. The black shales of the lower Bernesga Member seem to have been deposited under brackishwater conditions. This interpretation contradicts the previously discussed classification scheme of Pearce et al. (2010), where it was concluded that the most basal shales of the Bernesga Member are to be classified as "fully marine sediments." The Rb/K proxy suggests that the shales of the Formigoso Formation were never subject to full marine conditions. Under this assumption the shales were most likely confined to the "restricted shelfal margins," which relates well to the basal transgressive puddle model of Wignall (1994). The overlying Villasimpliz Mb. is typified by an increased influx of terrestrially derived material, interpreted to be the response to an overall relative sea level fall. The salinity levels reduce with the extensive freshwater input from the Gondwanan hinterland.

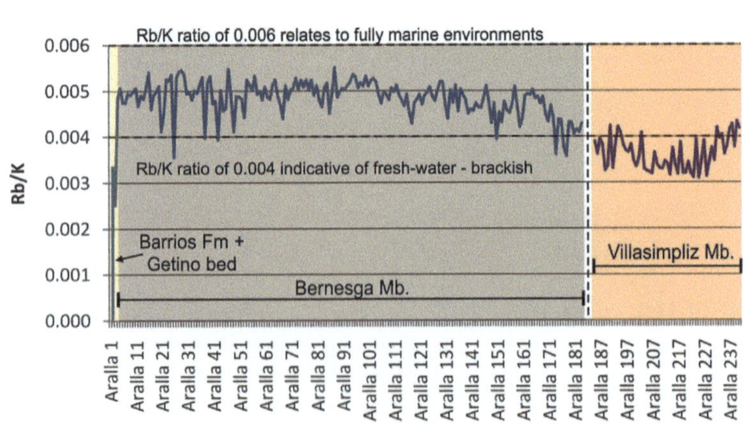

Figure 35 Paleosalinity reconstructions for the various localities using the Rb/K ratio, the *dashed line* values after Campbell and Williams (1965). Rb/K values above 0.004 indicate freshwater to brackish environments of deposition, whereas values above 0.006 are indicative of fully marine saline conditions. The Getino Beds are here interpreted to be freshwater deposits, the Bernesga Mb. deposited under brackish conditions and the Villasimpliz Mb. again under predominantly freshwater conditions.

8.3.3 Paleohumidity Reconstructions

The humidity proxy of SiO_2 versus ($Al_2O_3 + K_2O + Na_2O$) of Suttner and Dutta (1986) uses the relationship between the relative weathering resistant SiO_2 to the weathering prone oxides Al_2O_3, and K_2O. The Al_2O_3/TiO_2 proxy of Akul'shina (1976) is applied under the assumption that Al and Ti are relatively immobile in near-neutral (pH) environments. It is considered that under pH levels of <4 and >10 aluminum is mobilized, whereas Ti is mobile under pH conditions of <2. Ti is known to replace Al in the clay lattice depending on the prevailing intensity of chemical weathering. An Al_2O_3/TiO_2 ratio of ≤ 20 is used to interpret humid, >20 to ≤ 30 semihumid, and >30 semiarid environmental conditions.

Following the parameters introduced by Suttner and Dutta (1986) for humidity reconstructions, the Bernesga Member is seen to cluster toward the "arid" climatic conditions, whereas the Villasimpliz Member trends toward the "humid" environment (Fig. 36). This scenario is in agreement with the larger scale geotectonic setting. An "arid" interpretation for the Bernesga Member is in line with the high southern paleolatitude reconstruction of Scotese et al. (1999). A cool dry "arid" climate corresponds well with a polar to subpolar setting. The silt and sandstones of the upper Villasimpliz Member are trending toward "semiarid" to "humid" environments. During the Mid-Silurian the Iberian Peninsula, together with the northernmost margin of Gondwana, is interpreted to have drifted northward toward the paleoequator, thus leading to an overall increase in temperature and humid climatic conditions. The paleohumidity proxy of Akul'shina (1976) seems to indicate that the Getino Beds were deposited under a "humid" environment, the Bernesga Member under predominantly "semihumid," and the Villasimpliz Member under increasingly "humid" climatic conditions.

8.3.4 Clay Typing

In the following, a number of cross-plots and ternary systems are used to characterize the nature of the clay minerals present and the variation of them within the black shale succession: (1) Th versus K of Schlumberger (2009), (2) SiO_2 versus Al_2O_3 and Fe_2O_3 versus Al_2O_3 plots of Cullers and Podkovyrov (2000), and (3) SiO_2-Fe_2O_3-Al_2O_3 ternary plot of Konhauser (1998).

The Schlumberger (2009) system discriminates the clay mineralogy using the varying concentrations of Th and K_2O (Fig. 37). To analyze the nature and distribution of potentially different clay mineral associations throughout

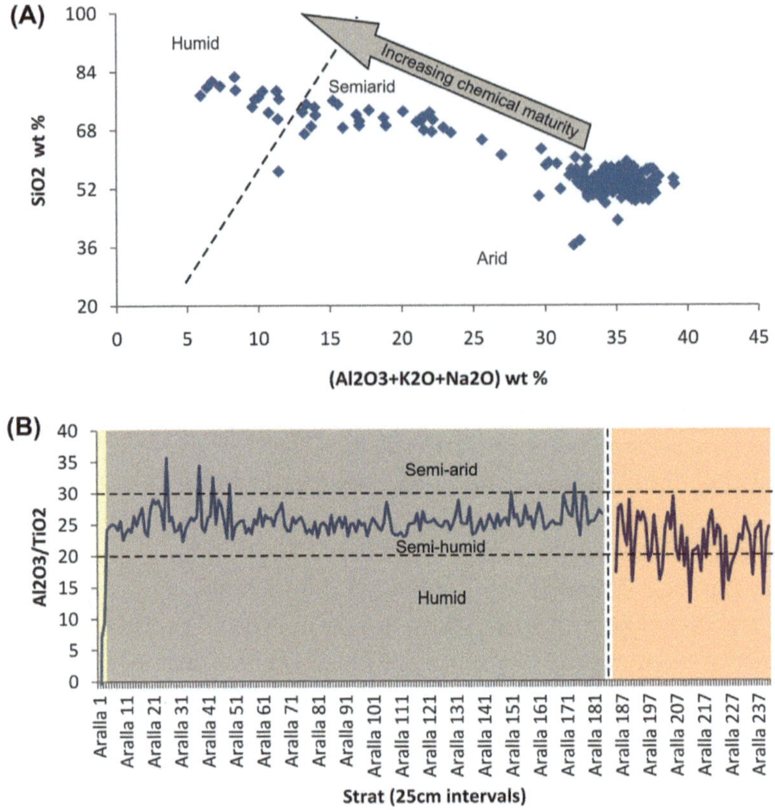

Figure 36 (A) Bivariate SiO_2 versus ($Al_2O_3 + K_2O + Na_2O$) paleoclimate discrimination diagram; fields after Suttner and Dutta (1986). The shales of the Bernesga Member cluster toward the "Arid" field, the Villasimpliz Member trends toward "Humid" conditions. (B) The Al/Ti proxy of Akul'shina (1976) for paleoclimatic conditions. (1) Ratios <20 represent humid environments, (2) 20–30 indicate semihumid, and (3) >30 are related to semiarid climatic settings. The Getino Beds are deposited under "Humid" conditions, the Bernesga Mb. within a predominantly semihumid environment, and the Villasimpliz Mb. displays an erratic decline toward a "Humid" paleoenvironment.

the Cantabrian black shale succession, only the data of the shales of the Bernesga Member were used. Variations in the clay mineralogy can be seen throughout the data set. A clear maturity trend toward the 100% kaolinite is evident, with the bulk of the shale data plotting within the illite/mixed layer clay mineral fields.

The majority of the shale data plot across the two zones "Illite" and "mixed layer clays" (Fig. 37). The presence of these clays can be used to infer the degree of weathering of the source rocks or maturity of the shales,

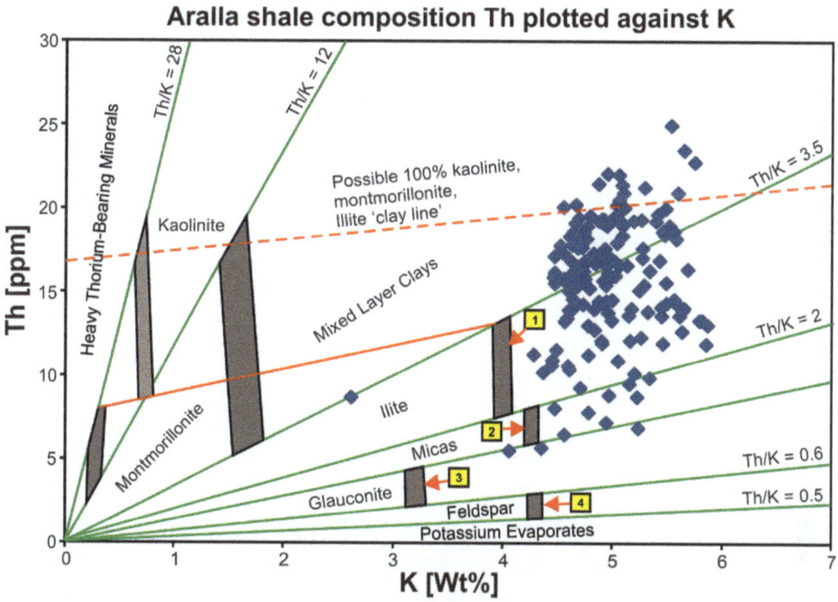

Figure 37 Cross-plot of K_2O (wt%) versus Th (ppm) after Schlumberger (2009), discriminating common clay minerals according to their thorium and potassium concentrations. Only the data of the Bernesga Member were used. The samples show a compositional maturity trend toward the 100% kaolinite, montmorillonite, and illite "clay line" with the bulk of the shale data residing within the illite/mixed layer clay zones.

respectively. The variation in the composition of clay mineral associations has also been used previously to infer paleolatitudes and corresponding climatic conditions. The presence of illite can indicate humid temperate climates (Einsele, 2000). However, Chamley (1989), Weaver (1989), and Diester-Haass et al. (1998) suggest chlorite and illite are being derived from areas subject to uplift (leading to increased active mechanical erosion) and cold-desert area conditions. Illites are the most common of the clay minerals in deep sea sediments and are present mostly at mid-latitudes. The presence of the "mixed-layer clays" suggests a weak development of chemical weathering (Chamley, 1989). Montmorillonite (smectite) clay types indicate temperate to subtropical climates (Einsele, 2000). Smectite commonly develops under warm, poorly drained continental areas (Chamley, 1989). Illite is the predominant clay mineral in marine shales as it is more stable than montmorillonite and kaolinite, the latter dominating freshwater sediments (Weaver, 1989). Montmorillonite forms under the lowest precipitation rates, illite and vermiculite at intermediate rates, and kaolinite/halloysite at the highest leaching intensities (Calvert and Pedersen, 2007). However, the diagram takes of course no diagenetic alterations, such as the transformation of various clay minerals into illite, into account.

The sedimentary rocks of the Aralla section display a wide variation of minerals present (Fig. 38). The Barrios Formation plots toward a pure "quartz" composition, the organically enriched shales of the Bernesga Member pool around the "illite" zone, but show a trend toward "kaolinite." The sand and siltstones of the overlying Villasimpliz Member also exhibit a compositional maturity trend toward the "quartz" end member.

When plotting the data of the Aralla section in the SiO_2-Fe_2O_3-Al_2O_3 discrimination diagram (Fig. 39), the Barrios Formation plots, as expected, at the SiO_2 corner. The Getino Beds, represented as black stars, show a wide compositional distribution, ranging from SiO_2 enrichment to an enrichment of Fe_2O_3, which may indicate the presence of the mineral glauconite. The black shales of the Bernesga Member (black circles) cluster around the illite zone, with a number of samples trending toward chamosite. The potential presence of chamosite could be interpreted to be the product of localized Fe enrichment. Fe_2O_3 peaks are always apparent at the base of the Bernesga Member. The sand and siltstones of the overlying Villasimpliz Member display a compositional maturity trend toward the SiO_2 component. Other ternary systems [MgO-CaO-Al_2O_3, Garnier et al. (2008); K_2O-Al_2O_3-SiO_2, Konhauser (1998) and Konhauser and Urrutia (1999); $Al_2O_3-(CaO + Na_2O + K_2O)-(Fe_2O_3+MgO)$, Moosavirad et al. (2011)] were also utilized for the clay mineral discrimination. These systems only reiterated the results of the SiO_2-Fe_2O_3-Al_2O_3 discrimination diagram and are therefore not further discussed here.

8.3.5 The Index of Compositional Variation

The K_2O/Al_2O_3 proxy of Cox (1995) was applied to the geochemical data of the analyzed sedimentary successions belonging to the Aralla section (Fig. 40). The Barrios Formation and Getino Beds fall within the uppermost limits of the "range of clay minerals." The shales of the Bernesga Member are located well within the "range of clay minerals." The overlying sand and siltstones of the Villasimpliz Member display elevated K_2O/Al_2O_3 ratios, toward the upper limits of the "range of clay minerals." The ICV calculation (Fig. 40C) details that the Barrios Formation and the Getino Beds exhibit values >1. The shales of the Bernesga Member plot within the "clay window," in proximity to the plagioclase category and between the muscovite/illite values. The silt and sandstones of the Villasimpliz Member

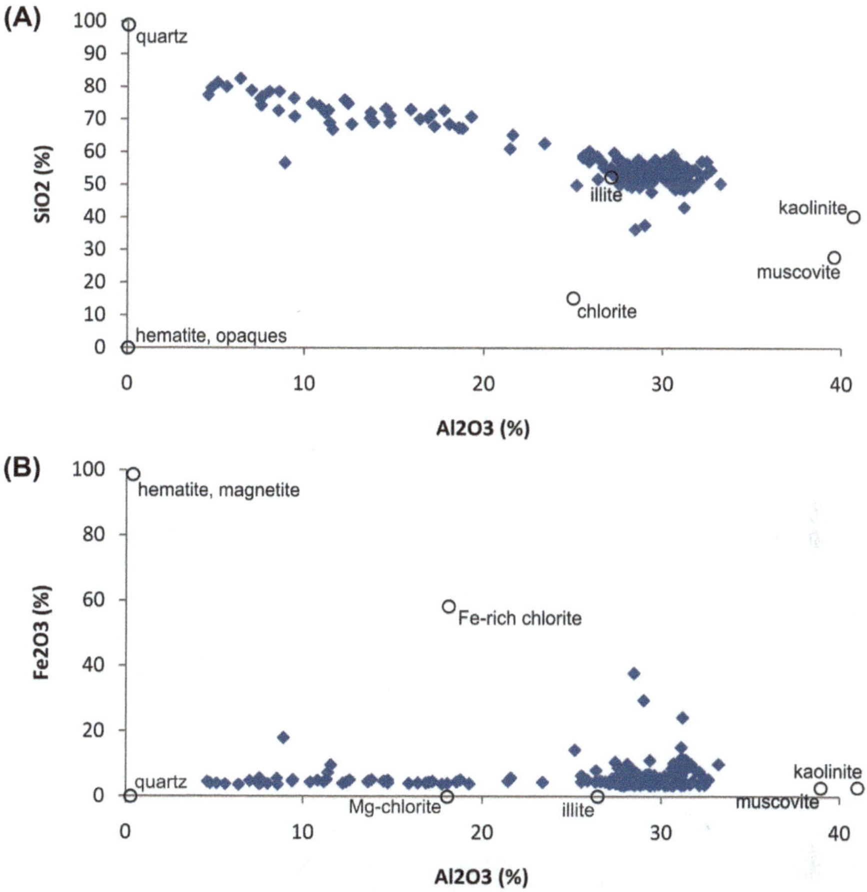

Figure 38 (A) SiO_2 versus Al_2O_3 of Cullers and Podkovyrov (2000), concentrations are plotted relative to the ideal composition of the observed minerals. Much of the variation in composition may be accounted for by different concentrations of the end members quartz and muscovite. The shales of the Bernesga Member are clustering around "illite," with the bulk of the samples trending toward "kaolinite." The silt and sandstones of the Villasimpliz Member trend toward the compositionally mature "quartz." (B) $Fe_2O_3^{tot}$ versus Al_2O_3 of Cullers and Podkovyrov (2000); concentrations plotted relative to the composition of the observed minerals. Again much of the variation in composition may be accounted for by the relative abundance of quartz and/or muscovite. The shales of the Bernesga Member cluster around the "illite" and are trending toward the "kaolinite" end member. The sedimentary successions of the Villasimpliz Member are seen trending toward "quartz."

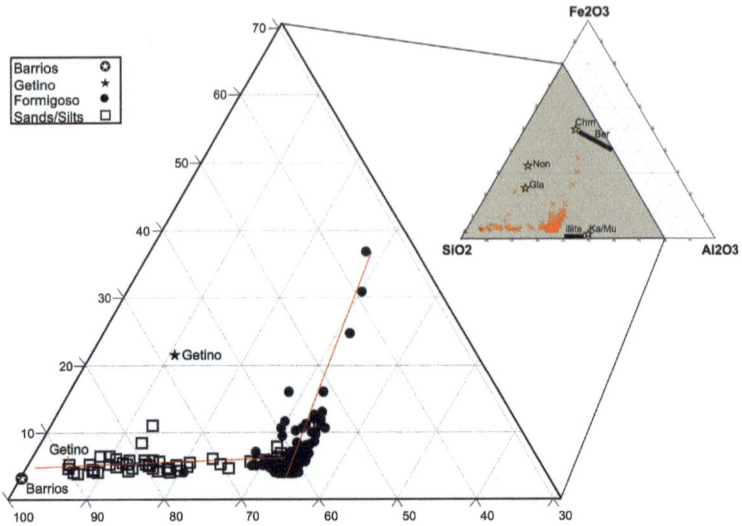

Figure 39 Distribution of the elements Fe, Al, and Si within the Aralla section, compared to several ideal clay minerals, including chamosite (Chm) and berthierine (Ber), kaolin and muscovite (Ka/Mu), nontronite (Non), illite and glauconite (Gla) after Konhauser (1998), Konhauser and Urrutia (1999), and Eickmann et al. (2009). *Black rectangles* indicate the existence of solid solutions between end members. The Barrios Formation plots at the SiO_2 corner, the shales of the Bernesga Member pool toward illite with a number of samples trending toward enrichment in chamosite. The sand and siltstones of the Villasimpliz Member display a compositional maturity trend toward SiO_2.

plot within and above the plagioclase category, with values greater than the illite category. The ICV values are low at the base of the Bernesga Member and increase gradually toward the top. The average ICV value of the Bernesga Member is 0.45. ICV values greater than one are interpreted to indicate the presence of immature shales, whereas values below one indicate the presence of compostionally mature shales. In this instance the Bernesga Member shales are classified as "highly mature shales." The average ICV value of the Villasimpliz Member is 0.69, suggesting that this Member is significantly less compositionally mature than the underlying shales of the Bernesga Member. The maturity trend decreases from the base to the top of the Formigoso Formation, with the introduction of detrital material from the hinterland.

8.3.6 3D Model Reconstructions: Synthesis of Environmental Factors

This section combines a selection of the previously interpreted environmental proxies. The factors were introduced into 3D plots to establish any potential relationship between the controlling environmental parameters. The factors combined were (1) the Rb/K salinity proxy of Campbell and Williams (1965); (2) the redox proxies, V/(V + Ni) of Hatch and Leventhal (1992), and Th/U of Fertl (1979); and (3) the grain-size proxies, Zr/Rb and Si/Al.

The 3D plots (Fig. 41) combine a salinity proxy (Rb/K), a redox state proxy (V/(V + Ni)), and a grain-size indicator (Zr/Rb). The shales of the Bernesga Member show high Rb/K ratios indicating elevated levels of salinity relative to the sand and siltstones of the Villasimpliz Member. Furthermore, the data of the Bernesga Member display a higher redox state relative to those of the Villasimpliz Member. The sand and siltstones of the Villasimpliz Member display an expected wide range of grain sizes, significantly wider than the underlying uniform shales of the Bernesga Member. The relation between these parameters can be used to infer a direct link between a reducing environment, higher salinity levels, and diminished siliciclastic (detrital) input. From this interpretation, the rising sea level combined with basin subsidence seems to be the main driving force for the establishment of the starved basin conditions and elevated reducing environmental conditions. This scenario is well suited for the development of the organically enriched black shale deposits. Furthermore, this interpretation is well in line with the different environmental proxies (such as the salinity proxy Rb/K, the redox proxy Th/U, and the grain-size indicator Si/Al).

The 3D plots prove to be powerful instruments, as they discriminate effectively the sedimentary successions of the Bernesga Member and the overlying Villasimpliz Member. The interpretation of the 3D plots offers also a straight explanation for the existence of small-sized benthic fauna within the Bernesga Member. Predominantly small sized representatives of the brachiopod genus *Lingula* have been previously described from its lower parts (Aramburu et al., 2002). The shales were forming under starved basin and reducing conditions, the small-sized benthic fauna may have also been affected by a severe lack of nutrients, as there were no elevated detrital influxes provided from the hinterland. The overall low-biodiversity graptolite assemblages documented from the Bernesga Member (Aramburu et al., 2002) may reinforce the interpretation of temporarily reduced salinity conditions. During seasonal phases with normal marine salinity (winter), the graptolites were able to invade the shelfal areas and to thrive, whereas during the summer, the increased freshwater influx caused by melting ice caps and glaciers would reduce the salinity levels leading to brackish water conditions and the establishment of a stratified water column. Both, the small-sized

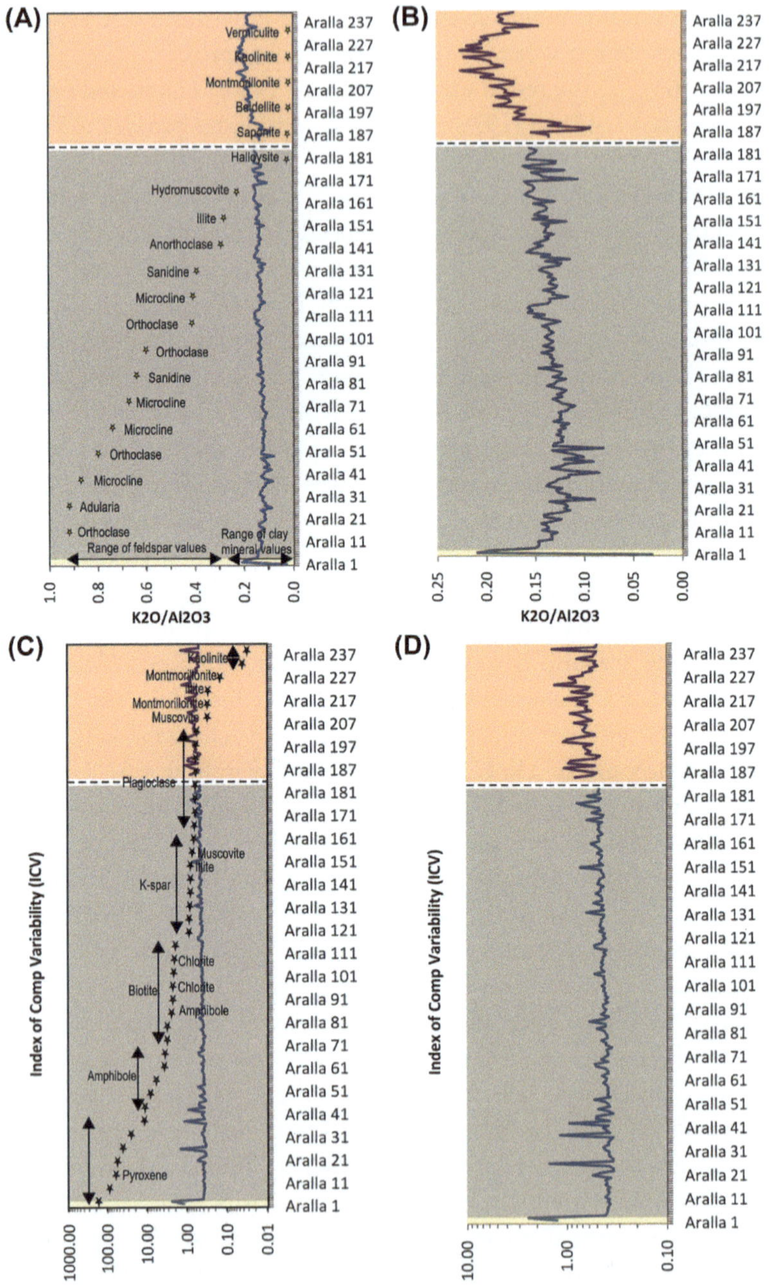

Figure 40 (A) K_2O/Al_2O_3 ratio of the Aralla section (discrimination between K-feldspars and clay minerals). The stars represent values for specific minerals, data from Deer and Howie (1966) overlay after Cox et al. (1995). (B) Scaled plot of (A), illustrating the K_2O/Al_2O_3 variability. (C) Index of Compositional Variation (ICV): $(Fe_2O_3 + K_2O + CaO + MgO + MnO + TiO_2)/Al_2O_3$, the stars represent values for specific minerals for comparison, the *arrows* show the range of values for the particular

benthic fauna and the low-biodiversity graptolite fauna point toward a high-stress environment.

The sand/silts horizons fall within the less reducing zones, most likely related to prograding deltas providing oxygen rich sediments into the predominantly anoxic black shale regime. The salinity levels and redox states are much lower within the sand and siltstone intercalations, which is in line with the high abundance of trace fossils (e.g., *Cruziana* isp. and *Planolites* isp.) of organisms thriving in the oxygen and nutrient-rich sediments (Fig. 42).

8.3.7 Paleobioproductivity Reconstructions

The use of Ba as a paleoproductivity proxy has its limitations. The Ba concentrations within sediments can be altered by hydrothermal overprinting, leading to false interpretations of the paleobioproductivity. However, as previously stated, the shales of the Bernesga Member do not show any significant hydrothermal overprint. In this instance, the critical application of Ba as a paleobioproductivity indicator seems to be viable.

Sedimentary Al concentrations can be effected by biogenic processes, leading to Ba/Al ratios of no particular value. The plot of Al_2O_3 versus TiO_2 ($r^2 = 0.87$) indicates that the Al_2O_3 was derived from the detrital clay fraction, suggesting that biogenic processes did not affect the relative Al_2O_3 concentrations. However, the Al_2O_3 concentrations are enriched relative to the international shale standards. The following ratios were implemented to determine the source of this enrichment. The Al_2O_3/TiO_2 ratios, when compared to those of Gordon et al. (1997) and Murray and Leinen (1996), indicate that the shales of the Bernesga Member seem to be derived from a predominantly granitic source. This interpretation is supported by Fe_2O_3/Al_2O_3 ratios, which also seem to point to a granitic source for the sediments. These ratios may indicate that chemical weathering of the predominantly granitic hinterland led to the elevated Al_2O_3 concentrations.

For the calculation of the paleobioproductivity data the following Ba/Al ratios were used: (1) 0.002913 (the average Ba/Al ratio value representative

mineral groups; data from Deer and Howie (1966), after Cox et al. (1995). (D) Scaled plot of (C). K_2O/Al_2O_3—range of the Bernesga Member: 0.0804–0.1620; K_2O/Al_2O_3—range of the Villasimpliz Member: 0.0933–0.2259; ICV—range of the Bernesga Member: 0.3423–0.7831 (1.5004 if Fe peak included), ICV—range of the Villasimpliz Member: 0.4635–1.3832. The ICV values increase gradually from the base to the top of the Formigoso Formation, indicating a decreasing maturity trend caused by the enhanced influx of terrigenous material.

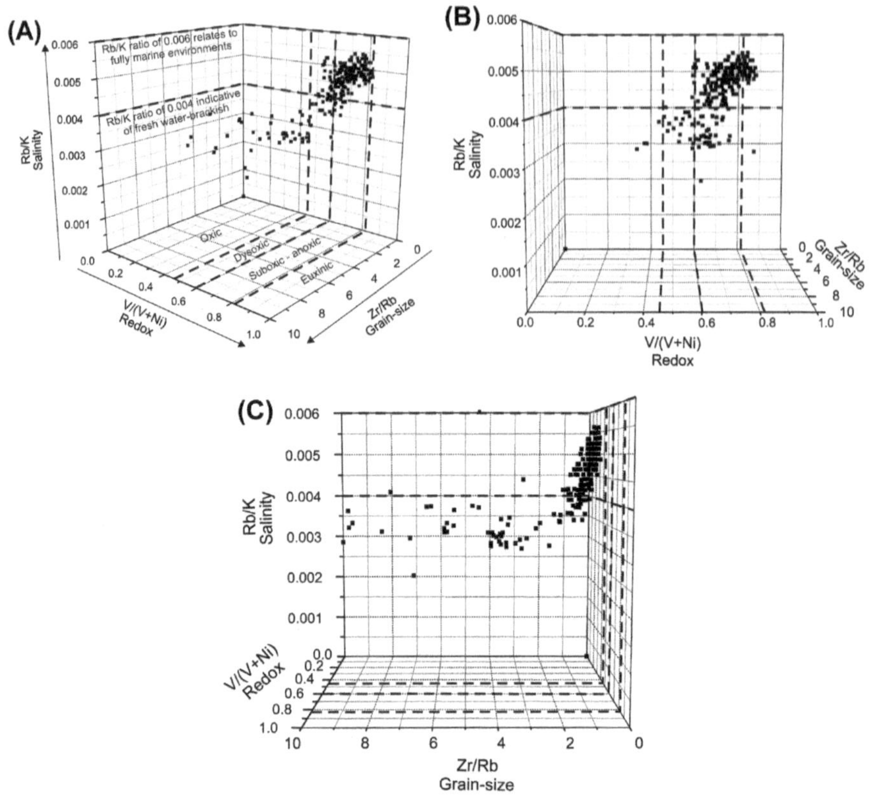

Figure 41 (A) 3D plot documenting the relationships between the redox (V/(V+Ni)), salinity (Rb/K) and grain-size (Zr/Rb) proxies. (B) The view of the 3D plot has been rotated to illustrate the relationship between the redox and salinity proxies. (C) Rotated view to illustrate the relationship between the grain-size and salinity proxies. Paleoenvironmental reconstruction, using the elemental ratios of Rb/K, V/(V + Ni), and Zr/Rb. The Rb/K ratio is used as a paleosalinity proxy, a value of ≤ 0.004 represents freshwater, values >0.004 to ≤ 0.006 indicate brackish waters, and values >0.006 suggest a fully marine developed environment (Campbell and Williams, 1965). The V/(V + Ni) is used as redox indicator, the values <0.46 "oxic," 0.46–0.60 "dysoxic," 0.54–0.82 "suboxic–anoxic," and >0.82 "euxinic" after Hatch and Leventhal (1992). The Zr/Rb ratio is used as a grain size indicator (Zr associated with quartz grains, Rb associated with Al_2O_3, TiO_2, and K_2O in the clay fraction), with high values reflecting coarse-grained successions and lower values mudstones.

of all black shale samples of the Bernesga Member) and (2) 0.0037 the average global Ba/Al crustal ratio from Reitz et al. (2004) and Bernárdez et al. (2008). The latter was used for comparison. It has to be noted that the average Ba/Al ratio for the Bernesga Member is lower than that of the average shale 0.0073 of Wedepohl (1971). This effect seems to be related

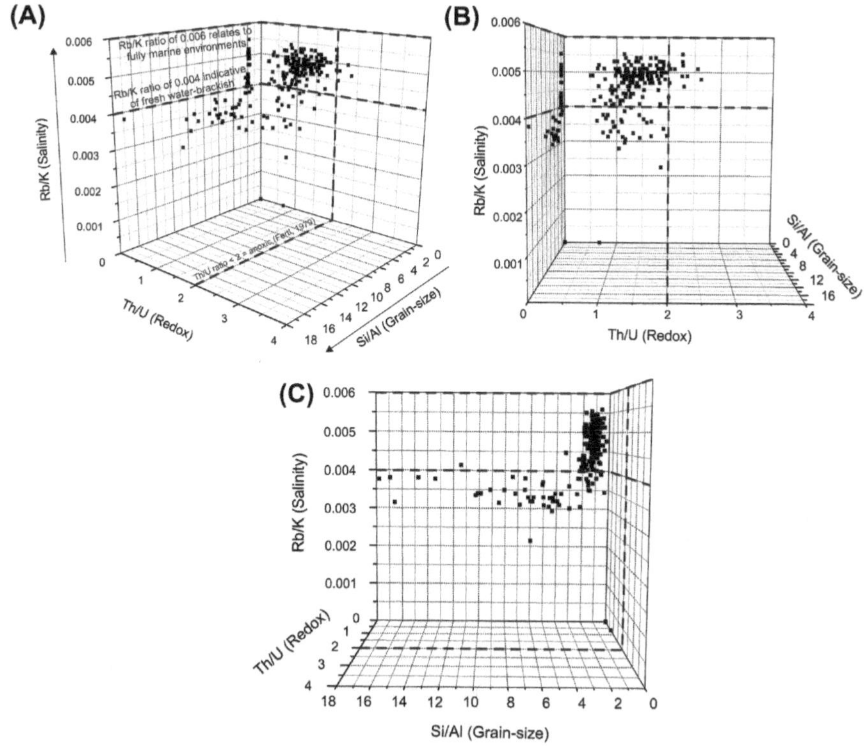

Figure 42 (A) 3D plot documenting the relationships between the redox (Th/U), salinity (Rb/K) and grain-size (Si/Al) proxies. (B) The view of the 3D plot has been rotated to illustrate the relationship between the redox and salinity proxies. (C) Rotated view to illustrate the relationship between the grain-size and salinity proxies. Paleoenvironmental reconstruction, using the ratios Rb/K, Th/U, and Si/Al (different grain size proxy compared to Fig. 41). The Rb/K ratio is used as a paleosalinity proxy, a value of ≤ 0.004 represents freshwater, values >0.004 to ≤ 0.006 indicate brackish waters, and values >0.006 suggest a fully marine developed environment (Campbell and Williams, 1965). Th/U is used as a redox indicator, the dashed black line = Th/U ratio of 2, less than 2 indicates anoxic conditions according to Fertl (1979). Uranium concentrations within the silt and sandstones of the Villasimpliz Member are all below D.L. (indicating an oxic environment). The Si/Al ratio is used as a grain-size indicator (Si associated with quartz grains, Al associated with Al_2O_3, TiO_2, and K_2O in the clay fraction) high values reflect coarse-grained units, lower values mudstones.

to the grain-size variations observable within the Bernesga and Villasimpliz Members, where the lowest Ba/Al values correspond systematically to the smallest grain-size fraction of the Bernesga Member and the highest Ba/Al values relate to the sand and siltstones of the Villasimpliz Member. To estimate the biogenic Ba (Ba_{bio}) concentrations the method proposed by Prakash Babu et al. (2002) is applied. The minimum Ba/Al ratio (Ba/Al_{min})

present at each of the analyzed section was used. It is assumed that the lowest Ba/Al value represents the regional detrital Ba background value. The determined Ba-background for the Aralla section is 0.001204. Ti concentrations were also used to calculate Ba_{bio} values. The $Ba/Ti_{aluminosilicate}$ values used were (1) 0.068156 (the average value for the Bernesga Member) and (2) 0.126 as the global Ba/Ti average from Turekian and Wedephol (1961). Again, the lowest Ba/Ti ratios (Ba/Ti_{min}) for the analyzed section were used, under the assumption that the lowest Ba/Ti values represent the regional detrital Ba background value. The lowest Ba/Ti ratio for the analyzed section is 0.01488309. The calculated paleobioproductivity (PP) values lead to the interpretation of elevated primary paleobioproductivity rates for the lower parts of the Bernesga Member (Fig. 43). Ba/Al and Ba/Ti ratios of the Bernesga Member (Figs. 43A and B, respectively) were used to calculate the concentrations of Ba_{bio}. The results were compared to calculations performed using the global Ba/Al ratio (Fig. 43C). When using the global average Ba/Ti, PP values diminish from the base to the top of the Formigoso Formation including the Villasimpliz Member (Fig. 43D). The elevated PP values calculated for the basal section of the Bernesga Member coincides with the high abundance of monograptid graptolites observed at the base of the section.

9. DISCUSSION AND CONCLUSIONS

9.1 Geochemical Classification

Various geochemical classification schemes were applied on the data set generated for this study. The geochemical discrimination system log(Fe_2O_3/K_2O) versus log(SiO_2/Al_2O_3) of Herron (1988) characterizes the vast majority of the sedimentary rocks of the Bernesga Member correctly as "shales." A number of samples from the Bernesga Member showed "Fe-shale" properties, which is here interpreted to be the effect of very localized, fracture-bound, postdepositional Fe_2O_3 mineralization. The majority of the sand and siltstones of the Villasimpliz Member plot within the "wacke" category. The bivariate plot of TiO_2 versus Ni (Floyd et al., 1989) discriminated well between the shales of the Bernesga Member and the overlying sand and siltstones of the Villasimpliz Member, with the latter clustering at the "acidic" provenance field. This is in good agreement with previous provenance studies, which concluded a bulk granitic composition of the hinterland. Other provenance discrimination diagrams applied

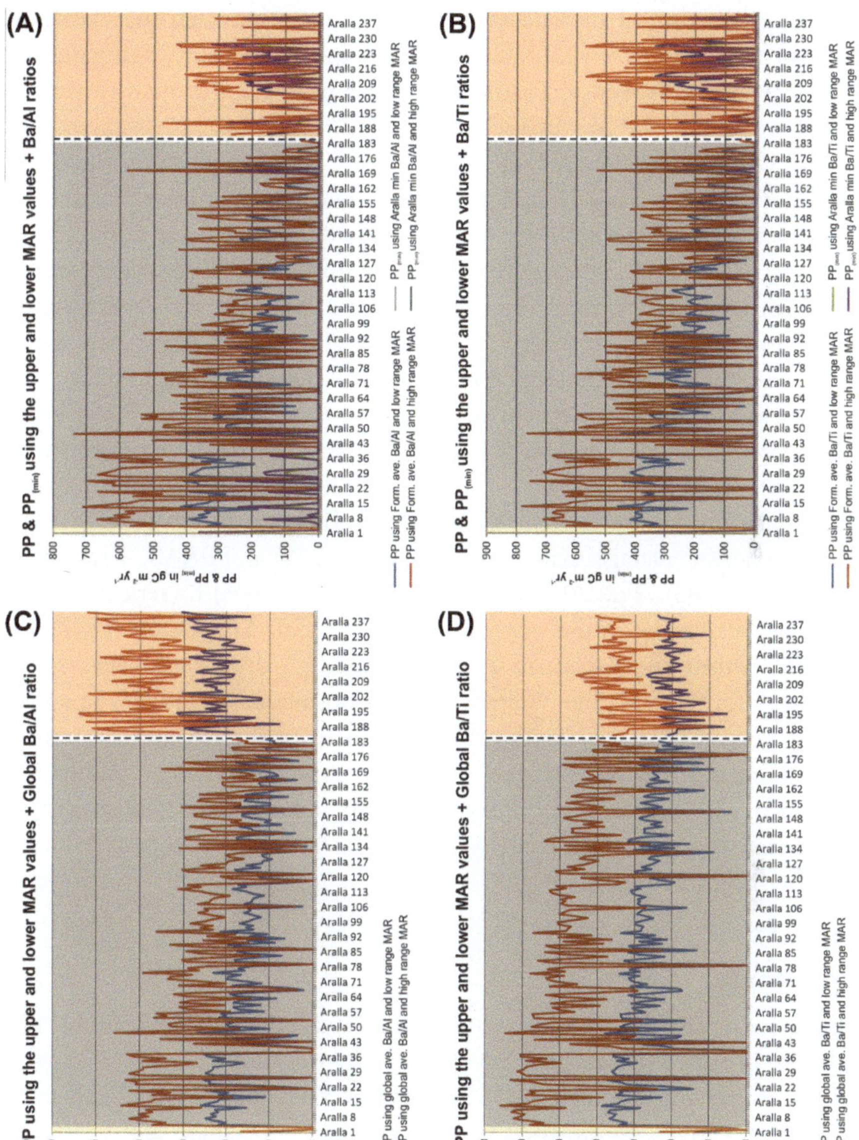

Figure 43 Paleobioproductivity (PP) for the Aralla section using the equation of Pfeifer et al. (2001) and Bonn et al. (1998): $PP = 20\ (P_{new})^{0.5}$ expressed in gC $m^{-2}\ yr^{-1}$. (A) Reconstructed PP values using the average Ba/Al ratio from the Bernesga Member. The productivity levels are significantly elevated at the base of the Bernesga Member relative to its top. (B) Approximated PP values for the average Ba/Ti ratio of the Bernesga Member. The productivity values are elevated for the basal shales. (C) PP values using the global Ba/Al ratio. In this instance the Villasimpliz Member displays higher productivity values relative to the underlying Bernesga Member. (D) The calculated PP values using the global Ba/Ti ratio. The PP values gradually fall throughout the Formigoso Formation including the sand and siltstones of the Villasimpliz Member.

(see earlier discussion) did successfully discriminate between the Bernesga Member from the overlying Villasimpliz Member; however, compared to each other, they proved to be inconclusive and did not allow any certain reconstructions of the provenance area.

The classification scheme of Pearce et al. (2010) successfully discriminated between the basal Bernesga Member and the deposits of its middle and upper parts. The basal Bernesga Member discriminated as fully marine developed mudstones, whereas the middle to upper parts were characterized as marginal marine deposits. This geochemical discrimination corresponds well to initial field-based sedimentological observations, which were tentatively interpreted to indicate an overall shallowing-upward succession with an increased terrestrial influx toward the top of the Formigoso Formation.

The analysis of the Cantabrian Formigoso data using the ternary systems of Mongelli et al. (2006) and Ross and Bustin (2009) revealed a pronounced Al_2O_3 enrichment of the Bernesga Member relative to international average shale standards ("average shale" of Wedepohl (1971) and PAAS, Taylor and McLennan, 1985). The Al_2O_3 enrichment is interpreted to reflect intense chemical weathering of the original source material. The TOC—TS10—Fe diagram (Arthur and Sageman, 1994; Dean and Arthur, 1989) enabled to distinguish for the first time different shale groups within the otherwise rather homogenous sedimentary rocks of the basal parts of the Bernesga Member: (1) basal organic-rich "hot" shales (with a high content of U) are conformably overlain by (2) organic carbon depleted "lean" shales (with diminished U values) and these are followed by organic-poor "cold" shales (exhibiting low U concentrations).

9.2 Anoxia

Various ratios of redox-sensitive elements (V/Cr, V/Sc, V/(V + Ni), U/Th, and Th/U, after Jones and Manning (1994), Kimura and Watanabe (2001), Hatch and Leventhal (1992), Jones and Manning (1994), and Fertl (1979), respectively) were used to estimate and reconstruct the extents of potential anoxia that prevailed during the time of deposition of the Formigoso Formation. The onset of the black shale deposition is characterized by highly anoxic conditions, as indicated by all of the applied redox proxies. The V/(V + Ni) ratio of Hatch and Leventhal (1992) appeared to be the most effective redox proxy for the black shales of the Bernesga Member. However, the informative value of the V/(V + Ni) ratio was seen to be limited for the sand and siltstones of the underlying Getino

Beds and overlying Villasimpliz Member, due to the relatively low Ni concentrations within this slightly coarser grained sedimentary rocks.

9.3 Paleosalinity

The salinity levels were reconstructed using the Rb/K ratio of Campbell and Williams (1965). This salinity proxy suggests that the Bernesga Member was deposited under brackish water conditions, thus contradicting the previously interpreted "fully marine developed" environment based on the classification scheme of Pearce et al. (2010). However, the sand and siltstones of the Getino Beds at Aralla discriminate clearly as brackish water deposits using the Rb/K proxy, which is supported by field observations. The glauconitic Getino Beds exhibit a high degree of bioturbation and have previously already been interpreted to represent very shallow marine, well-oxygenated deposits. The assumption of a sporadic freshwater input leading to brackish water conditions as indicated by the used proxy seems plausible.

9.4 Paleohumidity

The humidity proxies of Akul'shina (1976) and Suttner and Dutta (1986) indicate that the sand and siltstones of the Getino Beds were most likely derived from a hinterland subjected to "arid" to "humid" environmental conditions. The sudden changes in paleohumidity from "arid" to "humid" could be interpreted to represent glacial—interglacial periods, causing the "waxing" and "waning" of the Hirnantian ice masses. According to the proxy applied, the Bernesga Member is interpreted to have been sourced from areas under "arid" conditions. The interpretation of an "arid" hinterland is in good agreement with current paleogeographic reconstructions, which place the Iberian Peninsula in a high-latitude, polar setting for the time of the Early/Mid Silurian (Scotese et al., 1999). The sand and siltstones of the overlying Villasimpliz Member display a trend toward semiarid/humid conditions. This interpretation is supported by the northward drift of Iberia during the Mid/Late Silurian into lower latitudes (Scotese et al., 1999), leading to a consequent elevation of temperatures and humidity.

9.5 Clay Typing

The Schlumberger (2009) K_2O versus Th system, commonly used for the identification of different clay types shows that the majority of the data for the Bernesga Member plot across the "illite" and "mixed-layer clays" discrimination fields. The SiO_2 versus Al_2O_3 and Fe_2O_3 versus Al_2O_3 plots of Cullers and Podkovyrov (2000) illustrate the predominant nature of the clay minerals within the Bernesga Member to be illite, only a small number of samples, however, trend toward kaolinite. The sand and siltstones of the underlying Getino Beds exhibit a trend toward the SiO_2 end member as do the overlying sand and siltstones of the Villasimpliz Member. The Fe_2O_3 versus Al_2O_3 system revealed that a few samples of the Bernesga Member trend toward Fe-rich chlorite. This is interpreted to be the result of very localized Fe_2O_3 mineralization.

In the SiO_2-Al_2O_3-Fe_2O_3 ternary system of Konhauser (1998), Konhauser and Urrutia (1999), and Eickmann et al. (2009), the Bernesga Member pools near to the illite zone. The sand and siltstones of the underlying Getino Beds trend toward the SiO_2 end member, with a few samples indicating the presence of glauconite. The sand and siltstones of the overlying Villasimpliz Member display a compositional maturity trend toward the SiO_2 component.

9.6 Weathering Indices

The K_2O/Al_2O_3 ratio and ICV of Cox et al. (1995) allow for the interpretation that the basal shales of the Bernesga Mb. are of a higher compositional maturity than that of the middle and upper parts of the Bernesga Mb., which are diluted by an increased influx of terrestrial material. Additionally, the ICV values again indicate illite to be the predominant clay mineral within the Bernesga Mb. shales, reiterating the result derived from the clay-typing discrimination diagrams (see earlier discussion).

9.7 3D Environmental Reconstruction Models

The 3D environmental reconstructions combine three of the previously discussed parameters. (1) The redox state using the $V/(V + Ni)$ proxy of Hatch and Leventhal (1992) and Th/U ratio of Fertl (1979), (2) the paleosalinity indicator Rb/K of Campbell and Williams (1965), and (3) the detrital influx proxies Zr/Rb and Si/Al. The relation between these parameters can be used to infer a direct link between a reducing environment, higher salinity levels, and diminished siliciclastic (detrital) input.

9.8 Hydrothermal Activity

The diagram of Boström (1973) was utilized to determine any potential hydrothermal overprint of the sedimentary rocks analyzed. The shales of the Bernesga Mb. do not show any consistent and significant hydrothermal alteration. Additionally, the shales analyzed plot in very close proximity to the international shale standards NASC and PAAS, indicating that no major elemental changes relative to these standards have taken place. This may be a result of the low permeability of the shales.

However, the sands and silts of the underlying Getino Beds and overlying Villasimpliz Member may have been affected slightly by hydrothermal activity. A number of samples from these successions are trending toward the hydrothermal end member. This is interpreted to reflect the higher permeability of the sandstone intercalations, allowing the percolation of hydrothermal fluid.

9.9 Paleobioproductivity

As described earlier, the Bernesga Member seems to have not been affected by any significant hydrothermal alteration process. This permitted the application of barium-based geochemical proxies (Bernárdez et al., 2008; Bonn et al., 1998; Dymond et al., 1992; Francois et al., 1995; Peterson, 1979; Pfeifer et al., 2001; Prakash Babu et al., 2002) to estimate the paleoproductivity. The highest productivity rates coincide with the onset of the black shale deposition. The productivity rates are elevated within the basal, graptolite-rich parts of the Bernesga Member and decrease systematically toward the top of the succession. Overall the high productivity rates for the Bernesga Member at the Aralla section vary between 570 and 800gC m^2yr^{-1}, which is in good agreement with previously published productivity rates for upwelling-related ancient black shales (400−1200 gC m^2yr^{-1}, Einsele, 2000; Kuypers et al., 2002) and modern day analogues ($\geq$600 gC m^2yr^{-1}, Handoh et al., 2003).

The elevated productivity values during the onset of black shale deposition across the Cantabrian Basin correspond well to the Early Silurian upwelling model of Moore et al. (1993), which implies the introduction of nutrient-rich waters to the shelfal environments of northern Gondwana. This upwelling triggered a significant increase in primary productivity, which in turn is regarded to be one of the key factors responsible for the development of the widely distributed Silurian "hot" shales, not only within the Cantabrian Basin but throughout large parts of the northern Gondwanan margin.

ACKNOWLEDGMENTS

Financial support for the present study was provided by EPSAM-ACORN (grant to MM). Logistical support by ERCL Ltd (UK) is gratefully acknowledged. We thank Melissa Moore (SCHLUMBERGER), Peter J.A. Klinkenberg (Cospedal, Spain, and Amsterdam, The Netherlands) for discussions and comments during the fieldwork and the write-up phase, and two anonymous reviewers for their careful and constructive reviews, which helped to improve an earlier version of the manuscript. This is a contribution to UNESCO IGCP-503.

SUPPLEMENTARY DATA

Supplementary data related to this article can be found online at http://dx.doi.org/10.1016/bs.sats.2016.10.004.

REFERENCES

Abalos, B., Carreras, J., Druguet, E., Viruete, J.E., Pugnaire, M.T.G., Alvarez, S.L., Quesada, C., Fernandez, L.R.R., Gil-Ibarguchi, J.I., 2002. Variscan and pre-Variscan tectonics. In: Gibbons, W., Moreno, T. (Eds.), The Geology of Spain. Geological Society, London, pp. 155–183.

Abanda, P.A., Hannigan, R.E., 2006. Effect of diagenesis on trace element partitioning in shales. Chem. Geol. 230 (1–2), 42–59.

Abbey, S., Geological Survey of Canada, 1983. Studies in "standard samples" of silicate rocks and minerals, 1969–1982. Can. Geol. Surv. Pap. 83-15, 114.

Abdou, A., Shehata, M., 2007. Geochemical study of the shales of Gebel Ghorabi member, Bahariya Oasis, western Deser, Egypt. Aust. J. Basic Appl. Sci. 1 (4), 553–560.

Akul'shina, E., 1976. Methods for determining weathering conditions, sedimentation and post-sedimentary transformations according to clay minerals. In: Clay Minerals as Indicators of Rock Forming Conditions: Novosibirsk Nauka, pp. 9–37.

Aldridge, R.J., Jeppsson, L., Dorning, K.J., 1993. Early Silurian oceanic episodes and events. J. Geol. Soc. 150 (3), 501–513.

Algeo, T.J., Maynard, J.B., 2004. Trace-element behavior and redox facies in core shales of Upper Pennsylvanian Kansas-type cyclothems. Chem. Geol. 206 (3–4), 289–318.

Algeo, T.J., Schwark, L., Hower, J.C., 2004. High-resolution geochemistry and sequence stratigraphy of the Hushpuckney Shale (Swope Formation, eastern Kansas): implications for climato-environmental dynamics of the Late Pennsylvanian Midcontinent Seaway. Chem. Geol. 206 (3–4), 259–288.

Aller, J., Valín, M.L., García-López, S., Brime, C., Bastida, F., 2005. Superposition of tectono-thermal episodes in the southern Cantabrian Zone (foreland thrust and fold belt of the Iberian Variscides, NW Spain). Bull. Soc. Geol. Fr. 176 (6), 487–497.

Álvaro, J., Vennin, E., Muñoz, A., Sánchez-Valverde, B., Ojeda, J., 2000b. Interplay of orbital forcing and tectonic pulses in the Cambrian Iberian platform, NE Spain. Int. J. Earth Sci. 89 (2), 366–376.

Álvaro, J.J., Rouchy, J.M., Bechstädt, T., Boucot, A., Boyer, F., Debrenne, F., Moreno-Eiris, E., Perejón, A., Vennin, E., 2000a. Evaporitic constraints on the southwards drifting of the western Gondwana Margin during Early Cambrian times. Palaeogeogr. Palaeoclimatol. Palaeoecol. 160, 105–122.

Anderson, R.F., Winckler, G., 2005. Problems with paleoproductivity proxies. Paleoceanography 20 (3), PA3012.

Aramburu, C., Méndez-Bedia, I., Arbizu, M., 2002. The lower Palaeozoic in the Cantabrian zone (Cantabrian mountains, NW Spain). In: Garciá-López, S., Bastida, F. (Eds.), Palaeozoic Conodonts from Northern Spain, vol. 1. Insituto Geológico y Minero de Españia, Serie Cuadernos del Museo Geominero, Madrid, pp. 35–50.

Aramburu, C., Truyols, J., Arbizu, M., Méndez-Bedía, I., Zamarreño, I., García-Ramos, J.C., Suarez de Centi, C., Valenzuela, M., 1992. El Paleozoico Inferior de la Zona Cantábrica. In: Gutiérrez Marco, J.C., Saavedra, J., Rábano, I. (Eds.), Paleozoico Inferior de Ibero-America. Univ. of Extremadura, Badajoz, pp. 397–421.

Aramburu-Zabala Higuera, C., García Ramos, J., 1993. La sedimentación cambroordovícica en la Zona Cantábrica (NO de España). Trab. Geol. 19, 45–73.

Armstrong, H.A., Abbott, G.D., Turner, B.R., Makhlouf, I.M., Muhammad, A.B., Pedentchouk, N., Peters, H., 2009. Black shale deposition in an Upper Ordovician–Silurian permanently stratified, peri-glacial basin, southern Jordan. Palaeogeogr. Palaeoclimatol. Palaeoecol. 273 (3–4), 368–377.

Armstrong, H.A., Turner, B.R., Makhlouf, I.M., Weedon, G.P., Williams, M., Al Smadi, A., Abu Salah, A., 2005. Origin, sequence stratigraphy and depositional environment of an upper Ordovician (Hirnantian) deglacial black shale, Jordan. Palaeogeogr. Palaeoclimatol. Palaeoecol. 220 (3–4), 273–289.

Armstrong-Altrin, J.S., Verma, S.P., 2005. Critical evaluation of six tectonic setting discrimination diagrams using geochemical data of Neogene sediments from known tectonic settings. Sediment. Geol. 177, 115–129.

Arnaboldi, M., Meyers, P.A., 2006. 10. Data report: multiproxy geochemical characterization of OAE-related black shales at site 1276, Newfoundland basin. In: Tucholke, J., Sibuet, J., Klaus, A. (Eds.), Proceedings of the Ocean Drilling Program, Scientific Results, , 210 ed.vol. 210, pp. 1–16.

Arribas, J., Critelli, S., Johnnson, M.J., 2007. Sedimentary provenance and Petrogenesis: perspectives from petrography and geochemistry. In: Geological Society of America Special Paper 420. The Geological Society of America, Boulder, Colorado, p. 396.

Arthur, M.A., Sageman, B.B., 1994. Marine black shales: depositional mechanisms and environments of ancient deposits. Annu. Rev. Earth Planet. Sci. 22 (1), 499–551.

Arthur, M.A., 1979. 36. North Atlantic cretaceous black shales: the record at site 398 and a brief comparison with other occurrences. In: Sibuet, J.C., et al. (Eds.), Initial Reports of the Deep Sea Drilling Project, vol. 47 (2), pp. 719–751.

Barba, P., Fernández, L.P., 1991. Estratigrafía. Escala 1:50.000, No.103 (La Pola de Gordón). In: Memoria del mapa geologico de España, 2. Seria (MAGNA) ed. ITGE, Madrid.

Barbera, G., Mazzoleni, P., Critelli, S., Pappalardo, A., Giudice, A.L., Cirrincione, R., 2006. Provenance of shales and sedimentary history of the Monte Soro Unit, Sicily. Period. Mineral. 75 (2–3), 313–330.

Barrois, C.E., 1882. Recherches sur les terrains anciens des Asturies et de la Galice. Mémoires Société Géologique Nord 2 (1), 1–630.

Bassett, M.G. (Ed.), 2009. Early Palaeozoic Peri-gondwana Terranes: New Insights from Tectonics and Biogeography, 325. Geological Society, London, Special Publications, pp. 1–2.

Bernárdez, P., González-Álvarez, R., Francés, G., Prego, R., Bárcena, M., Romero, O., 2008. Palaeoproductivity changes and upwelling variability in the Galicia Mud Patch during the last 5000 years: geochemical and microfloral evidence. The Holocene 18 (8), 1207–1218.

Berry, W., Gutiérrez-Marco, J., Rábano, I., 1998. The Arabian sea: a modern analogue for North African–Southern European Silurian organic-rich graptolites-bearing shales, Proceedings of the 6th International Graptolite Conference (GWG-IPA) and 1998 field meeting. In: Gutiérrez-Marco, J.C., Rabano, I. (Eds.), IUGS Subcommission on Silurian Stratigraphy. Temas Geologico-Mineros ITGE, pp. 57–59.

Berry, W.B., Wilde, P., 1978. Progressive ventilation of the oceans; an explanation for the distribution of the lower Paleozoic black shales. Am. J. Sci. 278 (3), 257–275.

Bhatia, M.R., 1983. Plate tectonics and geochemical composition of sandstones. J. Geol. 91 (6), 611–627.

Blanchard, M., Alfredsson, M., Brodholt, J., Wright, K., Richard, C., Catlow, A., 2007. Arsenic incorporation into FeS2 pyrite and its influence on dissolution: a DFT study. Geochimica Cosmochimica Acta 71 (3), 624–630.

Bloch, J.D., Schroder-Adams, C., Leckie, D., 1999. Sedimentology, micropaleontology, geochemistry, and hydrocarbon potential of shale from the Cretaceous Lower Colorado Group in Western Canada. Bull. Canadan Geol. Surv. 531, 185.

Boggs, S., 2009. Petrology of Sedimentary Rocks. Cambridge University Press, Cambridge, p. 600.

Bohacs, K.M., Grabowski Jr., G.J., Carroll, A.R., Mankiewicz, P.J., Miskell-Gerhardt, K.J., Schwalbach, J.R., Wegner, M.B., Simo, J.T., 2005. Production, destruction, and dilution—the many paths to source-rock development. In: Harris, N.B. (Ed.), Deposition of Organic-carbon-rich Sediments: Models, Mechanisms, and Consequences, pp. 61–101.

Bonn, W.J., Gingele, F.X., Grobe, H., Mackensen, A., Fütterer, D.K., 1998. Palaeoproductivity at the Antarctic continental margin: opal and barium records for the last 400 ka. Palaeogeogr. Palaeoclimatol. Palaeoecol. 139 (3), 195–211.

Bosch, van den W.J., 1969. Geology of the Luna-Sil region, Cantabrian mountains (NW Spain). Leidse Geol. Meded. 44, 137–225.

Boström, K., 1973. The origin and fate of ferromanganoan active ridge sediments. Stock. Contributions Geol. 27, 149–243.

Brumsack, H., 1981. Geochemistry of Cretaceous black shales from the Atlantic Pcean (DSDP Legs 11, 14, 36 and 41). Chem. Geol. 31, 1–25.

Brumsack, H.J., 1986. The inorganic geochemistry of Cretaceous black shales (DSDP Leg 41) in comparison to modern upwelling sediments from the Gulf of California. Geol. Soc. Lond. Spec. Publ. 21 (1), 447–462.

Burgan, A.M., Ali, C.A., Tahir, S.H., 2008. Chemical composition of the Tertiary black shales of West Sabah, East Malaysia. Chin. J. Geochem. 27 (1), 28–35.

Butcher, A., 2009. Early Llandovery chitinozoans from Jordan. Palaeontology 52 (3), 593–629.

Calvert, S., Fontugne, M., 2001. On the late Pleistocene-Holocene sapropel record of climatic and oceanographic variability in the eastern Mediterranean. Paleoceanography 16 (1), 78–94.

Calvert, S.E., Pedersen, T.F., 1993. Geochemistry of Recent oxic and anoxic marine sediments: implications for the geological record. Mar. Geol. 113 (1–2), 67–88.

Calvert, S.E., Pedersen, T.F., 2007. Elemental proxies for palaeoclimatic and palaeoceanographic variability in marine sediments: interpretation and application. In: Hillaire-Marcel, C., De Vernal, A. (Eds.), Proxies in Late Cenozoic Paleoceanography. Elsevier Science, p. 518.

Campbell, F., Williams, G., 1965. Chemical composition of shales of Mannville group (lower Cretaceous) of Central Alberta, Canada. AAPG Bull. 49 (1), 81–87.

Campos Alvarez, N.O., Roser, B.P., 2007. Geochemistry of black shales from the Lower Cretaceous Paja Formation, Eastern Cordillera, Colombia: source weathering, provenance, and tectonic setting. J. S. Am. Earth Sci. 23 (4), 271–289.

Challands, T.J., Armstrong, H.A., Maloney, D.P., Davies, J.R., Wilson, D., Owen, A.W., 2009. Organic-carbon deposition and coastal upwelling at mid-latitude during the Upper Ordovician (Late Katian): a case study from the Welsh Basin, UK. Palaeogeogr. Palaeoclimatol. Palaeoecol. 273 (3–4), 395–410.

Chamley, H., 1989. Clay Sedimentology. Springer-Verlag, New York, p. 623.

Chang, C.W., Iizuka, Y., Tzeng, W.N., 2004. Migratory environmental history of the grey mullet *Mugil cephalus* as revealed by autolith Sr:Ca ratios. Mar. Ecol. Prog. Ser. 269, 277–288.

Charriau, A., Lesven, L., Gao, Y., Leermakers, M., Baeyens, W., Ouddane, B., Billon, G., 2011. Trace metal behaviour in riverine sediments: role of organic matter and Sulphides. Appl. Geochem. 26 (1), 80–90.
Chester, R., 1990. Marine Geochemistry. Wiley-Blackwell, London, p. 698.
Chun, C.O., Delaney, M.L., 2006. 8. Phosphorus, barium, manganese, and uranium concentrations and geochemistry, Nazca ridge site 1237 sediments. Proc. Ocean Drill. Program 202, 1–19.
Ciglenečki, I., Pichler, S., Prohić, E., Ćosović, B., 2006. Distribution of redox-sensitive elements in bottom waters, porewaters and sediments of Rogoznica Lake (Croatia) in both oxic and anoxic conditions. Water, Air Soil Pollut. 6, 173–181.
Cocks, L.R.M., 2007. Blowing hot and cold in the Palaeozoic. Proceeding Geologists' Assoc. 118, 225–237.
Cocks, L.R.M., Torsvik, T.H., 2002. Earth geography from 500–400 million years ago: a faunal and palaeomagnetic review. J. Geol. Soc. 159, 631–644.
Cocks, L.R.M., Torsvik, T.H., 2006. European geography in a global context from the Vendian to the end of the Palaeozoic. In: Gee, D.G., Stephenson, R.A. (Eds.), European Lithosphere Dynamics, 32. Memoirs, Geological Society, London, pp. 83–95.
Combaz, A., 1967. Leiosphaeridaceae Eisenack, 1954, et Protoleiosphaeridae Timofeev, 1959 — leurs affinités, leur rôle sédimentologique et géologique. Rev. Palaeobot. Palynol. 1 (1–4), 309–321.
Combaz, A., 1986. Les 'Zones Gamma' du Silurian des region Sahariennes; Contenu organique et conditions de depot. In: Breheret, J.G. (Ed.), Les couches riches en matierés organiques et leur conditions de depot. Documents du BRGM, 110, pp. 239–258.
Comte, P., 1937. La Serie cambrienne et silurienne du Leon, 202. Comptes Rendus de l'Academie des Sciences, Paris, pp. 337–341.
Comte, P., 1959. Recherches sur les terrains anciens de la Cordillére Cantabrique, 60. Memorias, IGME, pp. 1–440.
Cox, R., Lowe, D.R., 1995. A conceptual review of regional-scale controls on the composition of clastic sediment and the co-evolution of continental blocks and their sedimentary cover. J. Sediment. Res. 65 (1).
Cox, R., Lowe, D.R., Cullers, R.L., 1995. The influence of sediment recycling and basement composition on evolution of mudrock chemistry in the southwestern United States. Geochimica Cosmochimica Acta 59 (14), 2919–2940.
Craig, J., Sutcliffe, O.E., Lüning, S., Le Heron, D.P., Whittington, R.J., 2002. Ice sheets and hot rocks: Unravelling the glacial signature in the late Ordovician reservoirs of North Africa. Petrol. Explor. Soc. G.B. Newsl. 6–11. Dec.
Cramer, F., Diez, M., 1978. Iberian chitinozoans I. Introduction and summary of pre-Devonian data. Palinol. número Extraordin. 1, 149–201.
Cullen, J.T., Lane, T.W., Morel, F.M.M., Sherrell, R.M., 1999. Modulation of cadmium uptake in phytoplankton by sea-water CO_2 concentration. Nature 402, 165–167.
Cullers, R.L., Podkovyrov, V.N., 2000. Geochemistry of the Mesoproterozoic Lakhanda shales in southeastern Yakutia, Russia: implications for mineralogical and provenance control, and recycling. Precambrian Res. 104 (1–2), 77–93.
De Barry Barnett, E., Wilson, C.L., 1960. Inorganic Chemistry: A Text-book for Advanced Students. Wiley, London, p. 599.
De Vos, W., Tarvainen, T., Salminen, R., Reeder, S., Demetriadds, D.V., Pircs, A., Batista, M.J., Marsina, K., Ottesen, R.T., O'Conner, P.J., Bidovec, M., Lima, A., Siewers, U., Smith, B., Taylor, H., Shaw, R., Salpeteur, I., Gregorauskiene, V., Halamic, J., Slaninka, I., Lax, K., Gravesen, P., Birke, M., Breward, N., Ander, E.L., Jorden, G., Duris, M., Klein, P., Locutura, J., Bellan, A., Pasieczna, A., Lis, J., Mazreku, A., Gilucis, A., Heitzmann, P., Klaver, J., Petersell, V., 2006. Geochemical

atlas of Europe. Part II – interpretation of geochemical maps, additional tables, figures, maps, and related publications. In: Association of Geological Surveys of the European Union and IUGS/IAGC 'Global Geochemical Base Lines' Project, 2, p. 692 (Espoo, Geological Survey of Finland).

Dean, W., Arthur, M., Stow, D., 1984. Origin and geochemistry of Cretaceous deep-sea black shales and multicolored claystones, with emphasis on Deep Sea Drilling Project Site 530, southern Angola Basin. In: Initial Reports of the Deep Sea Drilling Project, 75, pp. 819–844.

Dean, W.E., Arthur, M.A., 1989. Iron-sulfur-carbon relationships in organic-carbon-rich sequences; I, Cretaceous western interior Seaway. Am. J. Sci. 289 (6), 708–743.

Deer, W.A., Howie, R.A., Zussman, J., 1966. An Introduction to the Rock-Forming Minerals. John Wiley and Sons, New York, p. 528.

Delaney, M., 1998. Phosphorus accumulation in marine sediments and the oceanic phosphorus cycle. Glob. Biogeochem. Cycles 12 (4), 563–572.

Dellwig, O., Hinrichs, J., Hild, A., Brumsack, H., 2000. Changing sedimentation in tidal flat sediments of the southern North Sea from the Holocene to the present: a geochemical approach. J. Sea Res. 44 (3–4), 195–208.

Demaison, G., 1993. Contributions of Geochemistry to Exploration Strategy. Applied Petroleum Geochemistry. Éditions Technip, Paris, pp. 489–503.

Demaison, G.J., Moore, G.T., 1980. Anoxic environments and oil source bed genesis. AAPG Bull. 64 (8), 1179–1209.

Dias, R., Ribeiro, A., 1995. The Ibero-Armorican Arc: a collision effect against an irregular continent? Tectonophysics 246 (1–3), 113–128.

Diester-Haass, L., Robert, C., Chamley, H., 1998. Paleoproductivity and climate variations during sapropel deposition in the eastern Mediterranean Sea. In: Proceedings of the Ocean Drilling Program. Scientific Results, 1998, Ocean Drilling Program, pp. 227–248.

Djarnia, M.R., Fekirine, B., 1998. Sedimentological and Diagenetic Controls on Cambro-ordovician Reservoir Quality in the Southern Hassi Messaoud Area (Saharan Platform, Algeria). Geol. Soc., Lond. Spec. Publ. 132 (1), 167–174.

Dodd, J.R., Crisp, E.L., 1982. Non-linear variation with salinity of Sr/Ca and Mg/Ca ratios in water and aragonitic bivalve shells and implications for paleosalinity studies. Palaeogeogr. Palaeoclimatol. Palaeoecol. 38 (1–2), 45–56.

Dymond, J., Suess, E., Lyle, M., 1992. Barium in deep-sea sediment: a geochemical proxy for paleoproductivity. Paleoceanography 7 (2), 163–181.

Dypvik, H., Harris, N.B., 2001. Geochemical facies analysis of fine-grained siliciclastics using Th/U, Zr/Rb and (Zr+Rb)/Sr ratios. Chem. Geol. 181 (1–4), 131–146.

Eickmann, B., Bach, W., Kiel, S., Reitner, J., Peckmann, J., 2009. Evidence for cryptoendolithic life in Devonian pillow basalts of Variscan orogens, Germany. Palaeogeogr. Palaeoclimatol. Palaeoecol. 283 (3–4), 120–125.

Einsele, G., 2000. Sedimentary Basins: Evolution, Facies, and Sediment Budget. Springer-Verlag, Berlin, Heidelberg, New York, p. 792.

Elbaz-Poulicheet, F., Nagy, A., Cserny, T., 1997. The distribution of redox sensitive elements (U, As, Sb, V and Mo) along a river-wetland-lake system (Balaton Region, Hungary). Aquat. Geochem. 3 (3), 267–282.

Elbaz-Poulichet, F., Seidel, J.L., Jézéquel, D., Metzger, E., Prévot, F., Simonucci, C., Sarazin, G., Viollier, E., Etcheber, H., Jouanneau, J., Weber, O., Radakovitch, O., 2005. Sedimentary record of redox-sensitive elements (U, Mn, Mo) in a transitory anoxic basin (the Thau lagoon, France). Mar. Chem. 95 (3–4), 271–281.

Eppley, R., Peterson, B.J., 1979. Particulate organic matter flux and planktonic new production in the deep ocean. Nature 282, 677–680.

Erbacher, J., Foreerich, O., Wilson, P.A., Birch, H., Mutterlose, J., 2005. Stable organic carbon isotope stratigraphy across Oceanic Anoxic Event 2 of Demerara Rise, western tropical Atlantic. Geochem. Geophys. Geosys. 6 (6), Q06010. http://dx.doi.org/10.1029/2004GC00850.

Erbacher, J., Mosher, D.C., Malone, M.J., Berti, D., Bice, K.L., Bostock, H., Brumsack, H., Danelian, T., Forster, A., Glatz, C., Heidersdorf, F., Henderiks, J., Janecek, T.R., Junium, C., Le Callonnec, L., MacLeod, K., Meyers, P.A., Mutterlose, H.J., Nishi, H., Norris, R.D., Ogg, J.G., O'Regan, A.M., Rea, B., Sexton, P., Sturt, H., Suganuma, Y., Thurow, J.W., Wilson, P.A., Wise, S.W., 2004. Demerara Rise: Equatorial Cretateous and Paleogene Paleoceanographic Transect, Western Atlantic. Proceedings of the Ocean Drilling Program. Initial Reports, College Station. Texas Vol. 207. http://dx.doi.org/10.2973/odp.proc.ir.207.2004.

Evers, H.J., 1967. Geology of the Leonides between the Bernesga and Porma rivers, Cantabrian mountains (NW Spain). Leidse Geol. Meded. 41, 83–151.

Fedo, C.M., Young, G.M., Nesbitt, H.W., Hanchar, J.M., 1997. Potassic and sodic metasomatism in the southern province of the Canadian shield: evidence from the paleoproterozoic serpent formation, Huronian supergroup, Canada. Precambrian Res. 84 (1–2), 17–36.

Fekirine, B., Abdallah, H., 1998. Palaeozoic lithofacies correlatives and sequence stratigraphy of the Saharan Platform, Algeria. Geol. Soc. Lond. Spec. Publ. 132 (1), 97–108.

Fello, N., Luning, S., Storch, P., Redfern, J., 2006. Identification of early Llandovery (Silurian) anoxic palaeo-depressions at the western margin of the Murzuq Basin,(southwest-Libya) based on gamma-ray spectrometry in surface exposures. Geoarabia-Manama 11 (3), 101.

Fernández-Suárez, J., Dunning, G.R., Jenner, G.A., Gutiérrez-Alonso, G., 2000. Variscan collisional magmatism and deformation in NW Iberia: constraints from U–Pb geochronology of granitoids. J. Geol. Soc. 157 (3), 565–576.

Fertl, W., Chilingar, G., 1988. Total organic carbon content determined from well logs. SPE Form. Eval. 3 (2), 407–419.

Fertl, W.H., 1979. Gamma ray spectral data assists in complex formation evaluation. The Log Analyst 20 (5), 3–37.

Fisher, Q.J., Wignall, P.B., 2001. Palaeoenvironmental controls on the uranium distribution in an upper Carboniferous black shale (Gastrioceras listeri Marine Band) and associated strata; England. Chem. Geol. 175 (3–4), 605–621.

Floyd, P., Winchester, J., Park, R., 1989. Geochemistry and tectonic setting of Lewisian clastic metasediments from the early Proterozoic Loch Maree Group of Gairloch, NW Scotland. Precambrian Res. 45 (1), 203–214.

Floyd, P.A., Winchester, J.A., 1978. Identification and discrimination of altered and metamorphosed volcanic rocks using immobile elements. Chem. Geol. 21 (3–4), 291–306.

Fortey, R.A., Cocks, L.R.M., 2003. Palaeontological evidence bearing on global Ordovician-Silurian continental reconstructions. Earth-Science Rev. 61, 245–307.

Francois, R., Honjo, S., Manganini, S.J., Ravizza, G.E., 1995. Biogenic barium fluxes to the deep sea: implications for paleoproductivity reconstruction. Glob. Biogeochem. Cycles 9 (2), 289–303.

Froelich, P., Klinkhammer, G., Bender, M.A.A., Luedtke, N., Heath, G.R., Cullen, D., Dauphin, P., Hammond, D., Hartman, B., Maynard, V., 1979. Early oxidation of organic matter in pelagic sediments of the eastern equatorial Atlantic: suboxic diagenesis. Geochimica Cosmochimica Acta 43 (7), 1075–1090.

Galindo, C., Mougin, L., Fakhi, S., Nourreddine, A., Lamghari, A., Hannache, H., 2007. Distribution of naturally occurring radionuclides (U, Th) in Timahdit black shale (Morocco). J. Environ. Radioact. 92 (1), 41–54.

Garciá-López, S., Bastida, F. (Eds.), 2002. Palaeozoic Conodonts from northern Spain, 1. Insituto Geológico y Minero de Españia, Serie Cuadernos del Museo Geominero, Madrid, p. 438.

García-López, S., Brime, C., Valín, M.L., Sanz-López, J., Bastida, F., Aller, J., Blanco-Ferrera, S., 2007. Tectonothermal evolution of a foreland fold and thrust belt: the Cantabrian Zone (Iberian Variscan belt, NW Spain). Terra Nova 19 (6), 469–475.

García-Ramos, J., Valenzuela, M., Suárez De Centi, C., 1989. Estructuras de deformación asociadas a tempestivas: ejemplos en series siliciclásticas y carbonatadas someras del Paleozoico y Jurásico de Asturias. Trab. Geol. 18 (18), 65–77.

Garnier, V., Giuliani, G., Ohnenstetter, D., Fallick, A.E., Dubessy, J., Banks, D., Vinh, H.Q., Lhomme, T., Maluski, H., Pêcher, A., Bakhsh, K.A., Long, P.V., Trinh, P.T., Schwarz, D., 2008. Marble-hosted ruby deposits from Central and Southeast Asia: towards a new genetic model. Ore Geol. Rev. 34 (1–2), 169–191.

Gladney, E.S., Roelandts, I., 1988. 1987 compilation of elemental concentration data for USGS BHVO-1, MAG-1, QLO-1, RGM-1, SCo-1, SDC-1, SGR-1 and STM-1. Geostand. Newsl. 12 (2), 253–362.

González, R.M.R., 1983. Palinología de las formaciones del Silúrico Superior-Devónico Inferior de la Cordillera Cantábrica. Institución Fray Bernardino de Sahagún de la Excelentísima Diputación Provincial de León y del Servicio de Publicaciones de la Universidad de León, p. 231 (Universidad de León).

Gordon, R.M., Coale, K.H., Johnson, K.S., 1997. Iron distributions in the equatorial Pacific: implications for new production. Limnol. Oceanogr. 42 (3), 419–431.

Govindaraju, K., 1989. 1989 compilation of working values and sample description for 272 geostandards. Geostand. Newsl. 13 (s1), 1–113.

Govindaraju, K., 1994. 1994 compilation of working values and sample description for 383 geostandards. Geostand. Newsl. 18 (S1), 1–158.

Gradstein, F.M., Ogg, J.G., Schmitz, M.D., Ogg, G.M. (Eds.), 2012. The Geological Time Scale 2012, vols. 1 & 2. Elsevier, Amsterdam, p. 1144.

Gromet, L.P., Haskin, L.A., Korotev, R.L., Dymek, R.F., 1984. The "North American shale composite": its compilation, major and trace element characteristics. Geochimica Cosmochimica Acta 48 (12), 2469–2482.

Gutiérrez-Alonso, G., Johnston, S., Weil, A., Pastor-Galán, D., Fernández-Suárez, J., Autin, W.J., Holbrook, J.M., 2012. Buckling an orogen: the Cantabrian orocline. GSA Today 22 (7), 4–9.

Gutiérrez-Marco, J.C., Rábano, I. (Eds.), 1998. Proceedings of the Sixth International Graptolite Conference of the GWG (IPA) and the SW Iberia Field Meeting 1998 of the International Subcommission on Silurian Stratigraphy (ICS-IUGS), vol. 23. Insituto Technological Geominero de Epspania, Temas Geológico-Mineros, Madrid, p. p. 337.

Gutiérrez-Marco, J.C., Ghienne, J., Bernárdez, E., Hacar, M.P., 2010. Did the Late Ordovician African ice sheet reach Europe? Geology 38 (3), 279–282.

Handoh, I.C., Bigg, G.R., Jones, E.J.W., 2003. Evolution of upwelling in the Atlantic Ocean basin. Palaeogeogr. Palaeoclimatol. Palaeoecol. 202 (1–2), 31–58.

Hans Wedepohl, K., 1995. The composition of the continental crust. Geochimica Cosmochimica Acta 59 (7), 1217–1232.

Harland, W.B., 1972. The Ordovician ice age. Geol. Mag. 109 (5), 451–456.

Harris, N.B., Freeman, K.H., Pancost, R.D., White, T.S., Mitchell, G.D., 2004. The character and origin of lacustrine source rocks in the Lower Cretaceous synrift section, Congo Basin, west Africa. AAPG Bull. 88 (8), 1163–1184.

Hatch, J., Leventhal, J., 1992. Relationship between inferred redox potential of the depositional environment and geochemistry of the Upper Pennsylvanian (Missourian) Stark Shale Member of the Dennis Limestone, Wabaunsee County, Kansas, USA. Chem. Geol. 99 (1), 65–82.

Hedges, J.I., Keil, R.G., 1995. Sedimentary organic matter preservation: an assessment and speculative synthesis. Mar. Chem. 49 (2–3), 81–115.
Heinrichs, H., Schulz-Dobrick, B., Wedepohl, K.H., 1980. Terrestrial geochemistry of Cd, Bi, Tl, Pb, Zn, and Rb. Geochim. Cosmochim. Acta 44, 1519–1533.
Herron, M.M., 1988. Geochemical classification of terrigenous sands and shales from core or log data. J. Sediment. Res. 58 (5), 820–829.
Herut, B., Sandler, A., 2006. Normalization methods for pollutants in marine sediments: review and recommendations for the Mediterranean. Israel Oceanographic & Limnological Research and Geological Survey of Israel IOLR Report H 1–18.
Hetzel, A., Brumsack, H., Schnetger, B., Böttcher, M.E., 2006. Chapter 8. Inorganic geochemical characterization of lithologic units recovered during ODP Leg 207 (Demerara Rise). In: Mosher, D.C., Erbacher, J., Malone, M.J. (Eds.), Proceeding of the Ocean Drilling Program, Scientific Results, 207, pp. 1–37.
Hetzel, A., März, C., Vogt, C., Brumsack, H., 2011. Geochemical environment of Cenomanian – Turonian black shale deposition at Wunstorf (northern Germany). Cretac. Res. 32 (4), 480–494.
Hild, E., Brumsack, H.J., 1998. Major and minor element geochemistry of lower Aptian sediments from the NW German basin (core Hohenegglesen KB 40). Cretac. Res. 19, 615–633.
Hillaire-Marcel, C., De Vernal, A., 2007. Methods in late Cenozoic Paleoceanography: introduction. In: Hillaire-Marcel, C., De Vernal, A. (Eds.), Proxies in Late Cenozoic Paleoceanography. Elsevier Science, pp. 1–15.
Hillebrand, W.F., 1907. The vanadium sulphide, patronite, and its mineral associates from Minasragra, Peru. J. Am. Chem. Soc. 29 (7), 1019–1029.
Jeppsson, L., 1990. An oceanic model for lithological and faunal changes tested on the Silurian record. J. Geol. Soc. 147 (4), 663–674.
Johnson, M.E., 2006. Relationship of Silurian sea-level fluctuations to oceanic episodes and events. GFF 128 (2), 115–121.
Johnson, M.E., 2010. Tracking Silurian eustasy: Alignment of empirical evidence or pursuit of deductive reasoning? Palaeogeogr. Palaeoclimatol. Palaeoecol. 296 (3), 276–284.
Johnsson, M.J., Basu, A., 1993. Processes controlling the composition of clastic sediments. In: Geological Society of America Special Paper 284, p. 352. The Geological Society of America, Boulder (Colorado).
Jones, B., Manning, D.A.C., 1994. Comparison of geochemical indices used for the interpretation of paleoredox conditions in ancient mudstones. Chem. Geol. 111, 111–129.
Jorissen, F.J., Fontanier, C., Thomas, E., 2007. Paleoceanographical proxies based on deep-sea benthic foraminiferal assemblage characteristics. In: Hillarie-Marcel, C., De Vernal, A. (Eds.), Proxies in Late Cenozoic Paleoceanography. Elsevier Science, pp. 263–325.
Julivert, M., Marcos, A., 1973. Superimposed folding under flexural conditions in the Cantabrian zone (Hercynian Cordillera, northwest Spain). Am. J. Sci. 273 (5), 353–375.
Julivert, M., 1971. Decollement tectonics in the Hercynian Cordillera of northwest Spain. Am. J. Sci. 270 (1), 1–29.
Julivert, M., 1978. Hercynian orogeny and Carboniferous paleogeography in Northwestern Spain: a model of deformation-sedimentation relationships. Z. Dtsch. Geol. Ges. 129, 565–592.
Julivert, M., Fontboté, J., Ribeiro, A., Conde, L., 1972. Mapa Tectonico de la Penınsula Iberica Y Baleares, Escala 1⁄1000000. Instituto Geológico y Minero de Espana, Madrid.
Kampunzu, A.B., Cailteux, J.L.H., Moine, B., Loris, H.N.B.T., 2005. Geochemical characterisation, provenance, source and depositional environment of 'Roches Argilo-Talqueuses' (RAT) and Mines Subgroups sedimentary rocks in the

Neoproterozoic Katangan Belt (Congo): lithostratigraphic implications. J. Afr. Earth Sci. 42 (1–5), 119–133.

Kato, Y., Nakao, K., Isozaki, Y., 2002. Geochemistry of Late Permian to Early Triassic pelagic cherts from southwest Japan: implications for an oceanic redox change. Chem. Geol. 182 (1), 15–34.

Kegel, W., 1929. Das Gotlandium in den kantabrischen Ketten Nordspaniens. Z. Dtsch. Geol. Ges. 81, 35–62.

Keller, M., Bahlburg, H., Reuther, C., 2008. The transition from passive to active margin sedimentation in the Cantabrian Mountains, Northern Spain: Devonian or Carboniferous? Tectonophysics 461 (1–4), 414–427.

Ketris, M.P., Yudovich, Y.E., 2009. Estimations of Clarkes for Carbonaceous biolithes: world averages for trace element contents in black shales and coals. Int. J. Coal Geol. 78 (2), 135–148.

Kiipli, E., Kiipli, T., Kallaste, T., 2004. Bioproductivity rise in the East Baltic epicontinental sea in the Aeronian (Early Silurian). Palaeogeogr. Palaeoclimatol. Palaeoecol. 205 (3–4), 255–272.

Kiipli, E., Kiipli, T., Kallaste, T., Siir, S., 2012. Al_2O_3/TiO_2 ratio of the clay fraction of Late Ordovician–Silurian carbonate rocks as an indicator of paleoclimate of the Fennoscandian Shield. Palaeogeogr. Palaeoclimatol. Palaeoecol. 365, 312–320.

Killops, S.D., Killops, V.J., 2009. An Introduction to Organic Geochemistry. Wiley-Blackwell.

Kimura, H., Watanabe, Y., 2001. Oceanic anoxia at the Precambrian-Cambrian boundary. Geology 29 (11), 995–998.

King, C., 2004. The Black Sea: A History. Oxford University Press, p. 296.

Konhauser, K.O., Urrutia, M.M., 1999. Bacterial clay authigenesis: a common biogeochemical process. Chem. Geol. 161 (4), 399–413.

Konhauser, K.O., 1998. Diversity of bacterial iron mineralization. Earth-Science Rev. 43 (3–4), 91–121.

Krauskopf, K.B., Burg, D.K., 1995. Introduction to Geochemistry, third ed. McGraw-Hill Inc, p. 647.

Kupers, M.M., Pancost, R.D., Nejenhuis, I.A., Damsté, J.S.S., 2002. Enhanced productivity led to increased organic carbon burial in the euxinic North Atlantic basin during the late Cenomanian oceanic anoxic event. Paleoceanography 17 (4), 3–13.

Kuroda, J., Ohkouchi, N., Ishii, T., Tokuyama, H., Taira, A., 2005. Lamina-scale analysis of sedimentary components in Cretaceous black shales by chemical compositional mapping: implications for paleoenvironmental changes during the Oceanic Anoxic Events. Geochimica Cosmochimica Acta 69 (6), 1479–1494.

Lane, T.W., Saito, M.A., George, G.N., Pickering, I.J., Prince, R.C., Morel, F.M.M., 2005. Biochemistry: a cadmium enzyme from a marine diatom. Nature 435, 42.

LaPorte, D.F., Holnden, C., Patterson, W.P., Loxton, J.D., Melchin, M.J., Mitchell, C.E., Finney, S.C., Sheets, H.D., 2009. Local and global perspectives on carbon and nitrogen cycling during the Hirnantian glaciation. Paleogeogr. Paleoclimatol. Paleoecol. 276, 182–195.

Le Heron, D., Sutcliffe, O., Bourgig, K., Craig, J., Visentin, C., Whittington, R., 2004. Sedimentary architecture of Upper Ordovician tunnel valleys, Gargaf Arch, Libya: implications for the genesis of a hydrocarbon reservoir. Geoarabia-Manama 9, 137–160.

Le Heron, D., Sutcliffe, O., Whittington, R., Craig, J., 2005. The origins of glacially related soft-sediment deformation structures in Upper Ordovician glaciogenic rocks: implication for ice-sheet dynamics. Palaeogeogr. Palaeoclimatol. Palaeoecol. 218 (1), 75–103.

Le Heron, D.P., Craig, J., 2008. First-order reconstructions of a Late Ordovician Saharan ice sheet. J. Geol. Soc. Lond. 165 (1), 19–29.

Le Heron, D.P., Craig, J., Etienne, J.L., 2009. Ancient glaciations and hydrocarbon accumulations in North Africa and the Middle East. Earth-Science Rev. 93 (3–4), 47–76.
Le Heron, D.P., Ghienne, J., El Houicha, M., Khoukhi, Y., Rubino, J., 2007. Maximum extent of ice sheets in Morocco during the Late Ordovician glaciation. Palaeogeogr. Palaeoclimatol. Palaeoecol. 245 (1–2), 200–226.
Le Heron, D.P., Meinhold, G., Page, A., Whitham, A., 2013. Did lingering ice sheets moderate anoxia in the Early Palaeozoic of Libya? J. Geol. Soc. 170 (2), 327–339.
Leton, T.M., Crouch, M., Johnson, M., Pires, N., Dolan, L., 2012. First plants cooled the Ordovician. Nat. Geosci. 5, 86–89.
Lev, S.M., Filer, J.K., 2004. Assessing the impact of black shale processes on REE and the U–Pb isotope system in the southern Appalachian Basin. Chem. Geol. 206 (3–4), 393–406.
Limburg, K.E., 1995. Otolith strontium traces environmental history of subyearling American shad *Alosa sapidissima*. Mar. Ecol. Prog. Ser. 119, 25–35.
Lipinski, M., Warning, B., Brumsack, H., 2003. Trace metal signatures of Jurassic/Cretaceous black shales from the Norwegian shelf and the Barents sea. Palaeogeogr. Palaeoclimatol. Palaeoecol. 190 (0), 459–475.
Lotze, F., 1945. Zur Gliederung der Varisziden der Iberischen Meseta. Geotekt. Forschungen 6, 78–92.
Loydell, D., Butcher, A., Frýda, J., Lüning, S., Fowler, M., 2009. Lower Silurian "hot shales" in Jordan: a new depositional model. J. Petrol. Geol. 32 (3), 261–270.
Loydell, D.K., 1998. Early Silurian sea-level changes. Geol. Mag. 135 (4), 447–471.
Lüning, S., Kolonic, S., 2003. Uranium spectral gamma-ray response as a proxy for organic richness in black shales: applicability and limitations. J. Petrol. Geol. 153–174.
Lüning, S., Kolonic, S., 2006. Uranium as a proxy for organic richness in black shales: applicability and limitations. J. Petrol. Geol. 26 (2), 153–174.
Lüning, S., Archer, R., Craig, J., Loydell, D., 2003a. The lower Silurian 'Hot Shales' and 'Double Hot Shales' in North Africa and Arabia. In: The geology of northwest Libya (Ghadamis, Jifarah, Tarabulus and Sabratah basins): Tripoli, Earth Science Society of Libya, 3, pp. 91–105.
Lüning, S., Craig, J., Loydell, D.K., Štorch, P., Fitches, B., 2000. Lower Silurian 'hot shales' in North Africa and Arabia: regional distribution and depositional model. Earth-Science Rev. 49 (1–4), 121–200.
Lüning, S., Kolonic, S., Loydell, D.K., Craig, J., 2003b. Reconstruction of the original organic richness in weathered Silurian shale outcrops (Murzuq and Kufra basins, southern Libya). GeoArabia 8 (3), 299–308.
Lüning, S., Loydell, D.K., Štorch, P., Shahin, Y., Craig, J., 2006. Origin, sequence stratigraphy and depositional environment of an Upper Ordovician (Hirnantian) deglacial black shale, Jordan—Discussion. Palaeogeogr. Palaeoclimatol. Palaeoecol. 230 (3–4), 352–355.
Lüning, S., Shahin, Y., Loydell, D.K., Al-Rabi, H.T., Masri, A., Tarawneh, B., Kolonic, S., 2005. Anatomy of a world-class source rock: distribution and depositional model of Silurian organic rich shales in Jordan and implications for hydrocarbon potential. AAPG Bull. 89 (10), 1397–1427.
Lyons, T.W., Werne, J.P., Hollander, D.J., Murray, R.W., 2003. Contrasting sulfur geochemistry and Fe/Al and Mo/Al ratios across the last oxic-to-anoxic transition in the Cariaco Basin, Venezuela. Chem. Geol. 195 (1–4), 131–157.
Mac Niocaill, C.M., Van der Pluijm, B.A., Van der Voo, R., 1997. Ordovician palegeography and the evolution of the Iapetus Ocean. Geology 25 (2), 159–162.
Małkowski, K., Racki, G., 2009. A global biogeochemical perturbation across the Silurian-Devonian boundary: ocean-continent-biosphere feedbacks. Paleogeogr. Paleoclimatol. Paleoecol. 276 (1–4), 244–254.

Marcos, A., Pulgar, J.A., 1982. An approach to the tectonostratigraphic evolution of the Cantabrian foreland thrust and fold belt, Hercynian Cordillera of NW Spain. Neues Jahrb. Geol. Paläont. Abh. 163, 256–260.

Marshall, J.E.A., 2000. Palynofacies and orbital cyclicity: an example from the Silurian of Saudi Arabia. In: Al-Hajri, S., Owens, B. (Eds.), Stratigraphic Palynology of the Palaeozoic of Saudi Arabia. Gulf PetroLink, Bahrain, pp. 216–231.

Martens, C.S., Haddad, R.I., Chanton, J.P., 1992. Organic matter accumulation, remineralization, and burial in an anoxic coastal sediment. In: Whelan, J., Farrington, J.W. (Eds.), Organic Matter – Productivity, Accumulation, and Preservation in Recent and Ancient Sediments. Columbia University Press, pp. 82–98.

Martínez Catalán, J.R., Aller, J., Alonso, J.L., Bastida, F., 2009a. The Iberian Variscan orogen. In: Cortes, A.G., Villar, J.A., Suarez-Valgrande, J.P., Gonzalez, C.L.S. (Eds.), Spanish Geological Frameworks and Geosites: An Approach to Spanish Geological Heritage of International Relevance. Instituto Geológico y Minero de España, Madrid, pp. 13–30.

Martínez Catalán, J.R., Arenas, R., Abati, J., Martínez, S.S., García, F.D., Suárez, J.F., Cuadra, P.G., Castiñeiras, P., Barreiro, J.G., Montes, A.D., Clavijo, E.G., Pascual, F.J.R., Andonaegui, P., Jeffries, T.E., Alcock, J.E., Fernández, R.D., Carmona, A.L., 2009b. A rootless suture and the loss of the roots of a mountain chain: the Variscan belt of NW Iberia. Comptes Rendus Geosci. 341 (2–3), 114–126.

März, C., Beckmann, B., Franke, C., Vogt, C., Wagner, T., Kasten, S., 2009. Geochemical environment of the Coniacian–Santonian western tropical Atlantic at Demerara rise. Palaeogeogr. Palaeoclimatol. Palaeoecol. 273 (3–4), 286–301.

März, C., Poulton, S.W., Beckmann, B., Küster, K., Wagner, T., Kasten, S., 2008. Redox sensitivity of P cycling during marine black shale formation: dynamics of sulphidic and anoxic, non-sulphidic bottom waters. Geochimica Cosmochimica Acta 72, 3703–3717.

McManus, J., Berelson, W.M., Klinkhammer, G.P., Johnson, K.S., Coale, K.H., Anderson, R.F., Kumar, N., Burdige, D.J., Hammond, D.E., Brumsack, H.J., McCorkle, D.C., Rushdi, A., 1998. Geochemistry of barium in marine sediments: implications for its use as a paleoproxy. Geochimica Cosmochimica Acta 62 (21–22), 3453–3473.

Meyers, P.A., 1997. Organic geochemical proxies of paleoceanographic, paleolimnologic, and paleoclimatic processes. Org. Geochem. 27 (5–6), 213–250.

Meyers, P.A., 2006. Paleoceanographic and paleoclimatic similarities between Mediterranean sapropels and Cretaceous black shales. Palaeogeogr. Palaeoclimatol. Palaeoecol. 235 (1–3), 305–320.

Meyers, S.R., Sageman, B.B., Lyons, T.W., 2005. Organic carbon burial rate and the molybdenum proxy: theoretical framework and application to Cenomanian-Turonian oceanic anoxic event 2. Paleoceanography 20 (2), PA2002.

Mongelli, G., Critelli, S., Perri, F., Sonnino, M., Perrone, V., 2006. Sedimentary recycling, provenance and paleoweathering from chemistry and mineralogy of Mesozoic continental redbed mudrocks, Peloritani mountains, southern Italy. Geochem. J. 40 (2), 197–209.

Moore, G.T., Hayashida, D.N., Ross, C.A., 1993. Late Early Silurian (Wenlockian) general circulation model-generated upwelling, graptolitic black shales, and organic-rich source rocks—An accident of plate tectonics? Geology 21 (1), 17–20.

Moosavirad, S.M., Janardhana, M.R., Sethumadhav, M.S., Moghadam, M.R., Shankara, M., 2011. Geochemistry of lower Jurassic shales of the Shemshak Formation, Kerman Province, Central Iran: provenance, source weathering and tectonic setting. Chem. Erde – Geochem. 71 (3), 279–288.

Müller, P.J., Suess, E., 1979. Productivity, sedimentation rate, and sedimentary organic matter in the oceans—I. Organic carbon preservation. Deep Sea Res. Oceanogr. Res. Pap. 26 (12), 1347–1362.

Murphy, A.E., Sageman, B.B., Hollander, D.J., Lyons, T.W., Brett, C.E., 2000. Black shale deposition and faunal overturn in the Devonian Appalachian basin: clastic starvation, seasonal water-column mixing, and efficient biolimiting nutrient recycling. Paleoceanography 15 (3), 280–291.

Murray, R.W., Leinen, M., 1996. Scavenged excess aluminum and its relationship to bulk titanium in biogenic sediment from the central equatorial Pacific Ocean. Geochimica Cosmochimica Acta 60 (20), 3869–3878.

Murray, R., Leinen, M., Isern, A., 1993. Biogenic flux of Al to sediment in the central equatorial Pacific Ocean: evidence for increased productivity during glacial periods. Paleoceanography 8 (5), 651–670.

Nagarajan, R., Madhavaraju, J., Nagendra, R., Armstrong-Altrin, J.S., Moutte, J., 2007. Geochemistry of Neoproterozoic shales of the Rabanpalli formation, Bhima basin, northern Karnataka, southern India: implications for provenance and paleoredox conditions. Rev. Mex. Ciencias Geol. 24 (2), 150–160.

Nance, R.D., Gutiérrez-Alonso, G., Keppie, J.D., Linnemann, U., Murphy, J.B., Quesada, C., Strachan, R.A., Woodcock, N.H., 2010. Evolution of the Rheic Ocean. Gondwana Res. 17 (2–3), 194–222.

Nielsen, A.T., 2004. Ordovician sea level changes: a Baltoscandian perspective. In: Webby, B.D., Paris, F., Droser, M.L., Percival, I.G. (Eds.), The Great Ordovician Biodiversification Event. Columbia University Press, New York, pp. 84–93.

Nijenhuis, I.A., Bosch, H.-, Sinninghe Damsté, J.S., Brumsack, H.-, De Lange, G.J., 1999. Organic matter and trace element rich sapropels and black shales: a geochemical comparison. Earth Planet. Sci. Lett. 169 (3–4), 277–290.

Nyakairu, G.W., Koeberl, C., 2001. Mineralogical and chemical composition and distribution of rare earth elements in clay-rich sediments from central Uganda. Geochem. J. 35 (1), 13–28.

Okunlola, O., Adeigbe, O., Oluwatoke, O., 2009. Compositional and Petrogenetic features of schistose rocks of Ibadan area, Southwestern Nigeria. Earth Sci. Res. J. 13 (2), 119–133.

Oschmann, W., 2000. Microbes and black shales. In: Riding, R., Awramik, S.M. (Eds.), Microbial Sediments. Springer, Berlin, pp. 137–148.

Page, A., Zalasiewicz, J.A., Williams, M., Popov, L.E., 2007. Were transgressive black shales a negative feedback modulating glacioeustasy in the Early Palaeozoic Icehouse? In: Williams, M., Haywood, A.M., Gregory, F.J., Schmidt, D.N. (Eds.), Deep-time Perspectives on Climate Change: Marrying the Signal from Computer Models and Biological Proxies. The Micropalaeontological Society, Special Publications, pp. 123–156.

Parés, J.M., Van Der Voo, R., 1992. Paleozoic paleomagnetism of Almaden, Spain – a cautionary note. J. Geophys. Res. 97, 9353–9356.

Paris, F., Robardet, M., 1990. Early Palaeozoic palaeobiogeography of the Variscan regions. Tectonophysics 177 (1–3), 193–213.

Paris, F., Elaouad-Debbaj, Z., Jaglin, J., Massa, D., Oulebsir, L., 1995. Chitinozoans and Late Ordovician glacial events on Gondwana. In: Cooper, J.C., Droser, M.L., Finney, S. (Eds.), Ordovician Oddessey: Short Papers for the 7th ISOS. Society for Sedimentary Geology, Las Vegas, pp. 171–176. Pacific Section. 77.

Paris, F., Thusu, B., Rasul, S., Meinhold, G., Strogen, D., Howard, J.P., Abutarruma, Y., Elgadry, M., Whitham, A.G., 2012. Palynological and palynofacies analysis of early Silurian shales from borehole CDEG-2a in Dor el Gussa, eastern Murzuq Basin, Libya. Rev. Palaeobot. Palynol. 174, 1–26.

Pastor-Galán, D., Gutiérrez-Alonso, G., Murphy, J.B., Fernández-Suárez, J., Hofmann, M., Linnemann, U., 2013. Provenance analysis of the Paleozoic sequences of the northern Gondwana margin in NW Iberia: passive margin to Variscan collision and orocline development. Gondwana Res. 23 (3), 1089–1103.

Paytan, A., Gornitz, V., 2008. Ocean Paleoproductivity. Encyclopedia of Paleoclimatology and Ancient Environments. Springer, New York, pp. 644–650.

Pearce, R., Clayton, T., Kemp, A.E., 1991. Illitization and organic maturity in Silurian sediments from the Southern Uplands of Scotland. Clay Miner. 26 (2), 199–210.

Pearce, T., McLean, D., Martin, J.H., Ratcliffe, K., Wray, D., 2010. A whole-rock geochemical approach to the recognition and correlation of "marine bands". In: Application of Modern Stratigraphic Techniques: Theory and Case Histories, vol. 94. SEPM Special Publication, pp. 221–238.

Pedersen, T.F., Calvert, S.E., 1990. Anoxia vs productivity; what controls the formation of organic-carbon-rich sediments and sedimentary rocks? AAPG Bull. 74 (4), 454–466.

Pérez-Estaún, A., Bastida, F., Alonso, J.L., Marquínez, J., Aller, J., Alvarez-Marrón, J., Marcos, A., Pulgar, J.A., 1988. A thin-skinned tectonics model for an arcuate fold and thrust belt: the Cantabrian Zone (Variscan Ibero-Armorican Arc). Tectonics 7 (3), 517–537.

Perroud, H., Bonhommet, N., 2007. A Devonian palaeomagnetic pole for Armorica. Geophys. J. R. Astron. Soc. 77 (3), 839–845.

Perroud, H., Calza, F., Khattach, D., 1991. Paleomagnetism of the Silurian volcanism at Almaden, southern Spain. J. Geophys. Res. 96 (B2), 1949–1962.

Peterson, B.J., 1979. Particulate organic matter flux and planktonic new production in the deep ocean. Nature 282, 677–680.

Pettijohn, F.J., 1975. Sedimentary Rocks, third ed. Harper and Row, New York, p. 628.

Pettijohn, F.J., Potter, P.E., Siever, R., 1987. Sand and Sandstone. Springer, New York, p. 553.

Pfeifer, K., Kasten, S., Hensen, C., Schulz, H.D., 2001. Reconstruction of primary productivity from the barium contents in surface sediments of the South Atlantic Ocean. Mar. Geol. 177 (1–2), 13–24.

Pi, D., Liu, C., Shields-Zhou, G.A., Jiang, S., 2013. Trace and rare earth element geochemistry of black shale and kerogen in the early Cambrian Niutitang Formation in Guizhou province, South China: constraints for redox environments and origin of metal enrichments. Precambrian Res. 225 (0), 218–229.

Piper, D.Z., Calvert, S.E., 2009. A marine biogeochemical perspective on black shale deposition. Earth-Science Rev. 95 (1–2), 63–96.

Prakash Babu, C., Brumsack, H., Schnetger, B., Böttcher, M.E., 2002. Barium as a productivity proxy in continental margin sediments: a study from the eastern Arabian Sea. Mar. Geol. 184 (3–4), 189–206.

Reitz, A., Pfeifer, K., De Lange, G., Klump, J., 2004. Biogenic barium and the detrital Ba/Al ratio: a comparison of their direct and indirect determination. Mar. Geol. 204 (3), 289–300.

Richardson, J., Rodriguez, R., Sutherland, S., 2001. Palynological zonation of Mid-Palaeozoic sequences from the Cantabrian Mountains, NW Spain: implications for inter-regional and interfacies correlation of the Ludford/Prídolí and Silurian/Devonian boundaries, and plant dispersal patterns. Bulletin-Natural Hist. Mus. Geol. Ser. 57 (2), 115–162.

Riding, R.E., Awramik, S.M., 2000. Microbial Sediments. Springer (Heidelberg), p. 331.

Rieke, H.H., Chilingarian, G.V., 1974. Compaction of Argillaceous Sediments. Elsevier, Amsterdam, p. 424.

Rimmer, S.M., 2004. Geochemical paleoredox indicators in Devonian–Mississippian black shales, Central Appalachian Basin (USA). Chem. Geol. 206 (3–4), 373–391.

Rinna, J., Warning, B., Meyers, P., Brumsack, H., Rullkötter, J., 2002. Combined organic and inorganic geochemical reconstruction of paleodepositional conditions of a Pliocene sapropel from the eastern Mediterranean Sea. Geochimica Cosmochimica Acta 66 (11), 1969–1986.

Robardet, M., Gutierrez-marco, J., 2002. Silurian. In: Gibbons, W., Moreno, T. (Eds.), The Geology of Spain. Geological Society, London, pp. 51–66.

Robardet, M., 2003. The Armorica 'microplate': fact or fiction? Critical review of the concept and contradictory palaeobiogeographical data. Palaeogeogr. Palaeoclimatol. Palaeoecol. 195 (1–2), 125–148.
Rollinson, H.R., 1993. Using Geochemical Data: Evaluation, Presentation, Interpretation. pp. 352. Pearson Prentice Hall (Harlow).
Romer, R.L., Hahne, K., 2010. Life of the Rheic Ocean: scrolling through the shale record. Gondwana Res. 17 (2–3), 236–253.
Roser, B., Korsch, R., 1988. Provenance signatures of sandstone-mudstone suites determined using discriminant function analysis of major-element data. Chem. Geol. 67 (1), 119–139.
Ross, D.J.K., Bustin, R.M., 2009. Investigating the use of sedimentary geochemical proxies for paleoenvironment interpretation of thermally mature organic-rich strata: examples from the Devonian–Mississippian shales, Western Canadian Sedimentary Basin. Chem. Geol. 260 (1–2), 1–19.
Rubino, J., Deynoux, M., Ghienne, J.-F., Moreau, J., Blanpied, C., Lafont, F., Andres-Calatrava, R., Galeazzi, S., Mynth, T., Sommer, F., 2003. Late Ordovician glaciation in Northern Gondwana, Reappraisal and petroleum implications. In: AAPG Search and Discovery Article #90016, AAPG Hedberg Conference, 2003. Algeria, Algiers.
Rule, J.H., 1986. Assessment of trace element geochemistry of Hampton Roads Harbor and lower Chesapeake Bay area sediments. Environ. Geol. 8 (4), 209–219.
Rupke, J., 1965. The Esla nappe, Cantabrian mountains (Spain). Leidse Geol. Meded. 32, 1–74.
Sachanski, V., Göncüoglu, M.C., Gedik, I., 2010. Late Telychian (early Silurian) graptolitic shales and the maximum Silurian highstand in the NW Anatolian Palaeozoic terranes. Palaeogeogr. Palaeoclimatol. Palaeoecol. 291 (3–4), 419–428.
Sageman, B.B., Murphy, A.E., Werne, J.P., Ver Straeten, C.A., Hollander, D.J., Lyons, T.W., 2003. A tale of shales: the relative roles of production, decomposition, and dilution in the accumulation of organic-rich strata, Middle–Upper Devonian, Appalachian basin. Chem. Geol. 195 (1–4), 229–273.
Sarmiento, G., Méndez-Bedia, I., Aramburu, C., Arbizu, M., Truyols, J., 1994. Early Silurian conodonts from the Cantabrian zone, NW Spain. Geobios 27 (4), 507–522.
Schieber, J., 2003. Black shales. In: Middleton, G.V., Church, M.J., Coniglio, M., Hardie, L.A., Longstaffe, F.J. (Eds.), Encyclopedia of Sediments and Sedimentary Rocks. Springer, Heidelberg, pp. 128–132.
Schlumberger, 2009. Log Interpretation Charts, 2009 Ed. Schlumberger, Texas.
Schmitz, B., Charisi, S.D., Thompson, E.I., Speijer, R.P., 1997. Barium, SiO_2 (excess), and P_2O_5 as proxies of biological productivity in the Middle East during the palaeocene and the latest palaeocene benthic extinction event. Terra Nova 9 (2), 95–99.
Schneider, J., de Wall, H., Kontny, A., Bechstädt, T., 2004. Magnetic susceptibility variations in carbonates of the La Vid Group (Cantabrian Zone, NW-Spain) related to burial diagenesis. Sediment. Geol. 166 (1–2), 73–88.
Schröder, S., Grotzinger, J., 2007. Evidence for anoxia at the Ediacaran–Cambrian boundary: the record of redox-sensitive trace elements and rare earth elements in Oman. J. Geol. Soc. 164 (1), 175–187.
Schultz, R.B., 2004. Geochemical relationships of Late Paleozoic carbon-rich shales of the Midcontinent, USA: a compendium of results advocating changeable geochemical conditions. Chem. Geol. 206 (3–4), 347–372.
Schweineberg, J., 1987. Silurische Chitinozoen aus der Provinz Palencia (Kantabrisches Gebirge, N-Spanien). Göttinger Arb. zur Geol. Paläontologie 33, 1–94.
Scopelliti, G., Bellanca, A., Neri, R., Baudin, F., Coccioni, R., 2006. Comparative high-resolution chemostratigraphy of the Bonarelli level from the reference Bottaccione

section (Umbria–Marche Apennines) and from an equivalent section in NW Sicily: consistent and contrasting responses to the OAE2. Chem. Geol. 228 (4), 266–285.
Scotese, C.R., Boucot, A.J., McKerrow, W.S., 1999. Gondwanan palaeogeography and paleoclimatology. J. Afr. Earth Sci. 28 (1), 99–114.
Serra, O., 1984. Fundamentals of Well-log Interpretation: The Acquisition of Logging Data. Developments in Petrolium Science. No.15A. Elsevier, Amsterdam, p. 423.
Servais, T., Sintubin, M., 2009. Avalonia, Armorica, Perunica: terranes, microcontinents, microplates or palaeobiogeographical provinces? In: Bassett, M.G. (Ed.), Early Palaeozoic Peri-gondwana Terranes: New Insights from Tectonics and Biogeography, 325. Special Publications, Geological Society, London, pp. 103–115.
Sethi, P.S., Schieber, J., 1998. Economic aspects of shales and clays: an overview. In: Schieber, J., Zimmerle, W., Sethi, P. (Eds.), Shales and Mudstones: Petrography, Petrophysics, Geochemistry and Economic Geology, pp. 237–253.
Shaw, D., Reilly, G., Muysson, J., Pattenden, G., Campbell, F., 1967. An estimate of the chemical composition of the Canadian Precambrian Shield. Can. J. Earth Sci. 4 (5), 829–853.
Shaw, D.M., Dostal, J., Keays, R.R., 1976. Additional estimates of continental surface Precambrian shield composition in Canada. Geochimica Cosmochimica Acta 40 (1), 73–83.
Sheehan, P.M., 2001. The late Ordovician mass extinction. Annu. Rev. Earth Planet. Sci. 29 (1), 331–364.
Stanley, D.J., Blanpied, C., 1980. Late quaternary water exchange between the eastern Mediterranean and the black sea. Nature 285, 537–541.
Stanley, S.M., 1988. Paleozoic mass extinctions: shared patterns suggest global cooling as a common cause. Am. J. Sci. 288, 334–352.
Stein, R., 1990. Organic carbon content/sedimentation rate relationship and its paleoenvironmental significance for marine sediments. Geo-Marine Lett. 10, 37–44.
Štorch, P., 1998. Biostratigraphy, palaeobiogeographical links and environmental interpretation of the Llandovery and Wenlock graptolite faunas of peri-Gondwanan Europe. In: Gutiérrez-Marco, J.C. (Ed.), Proceedings of the Sixth International Graptolite Conference (GWG-IPA) and 1998 Field Meeting, IUGS Subcommison on Silurian Stratigraphy. In: Rábano, I. (Ed.), 23. Insitituto Tecnológico Geominero de Españia, Temas Geológico-Mineros, pp. 126–129.
Stow, D.A.V., Huc, A.-, Bertrand, P., 2001. Depositional processes of black shales in deep water. Mar. Petrol. Geol. 18 (4), 491–498.
Suarez de Centi, C., 1988. Estratigrafia, Sedimentologia y Paleogeografia de la Fm. Furada/San Pedro. Zona Cantabrica (NW de Espana) (Ph.D. thesis). Departimento de Geología, Universidad de Oviedo, p. 530.
Suárez de Centi, C., García-Ramos, J., Valenzuela, M., 1989. Icnofósiles del Silúrico de la Zona Cantábrica (NO de España). Bol. Geol. Min. 100 (3), 339–394.
Sun, S.S., 1982. Chemical composition and origin of the Earth's primitive mantle. Geochimica Cosmochim. Acta 46 (2), 179–192.
Sutcliffe, O.E., Dowdeswell, J.A., Whittington, R.J., Theron, J.N., Craig, J., 2000. Calibrating the Late Ordovician glaciation and mass extinction by the eccentricity cycles of Earth's orbit. Geology 28 (11), 967–970.
Suttner, L.J., Dutta, P.K., 1986. Alluvial sandstone composition and paleoclimate, I. Framework mineralogy. J. Sediment. Res. 56 (3), 329–345.
Svendsen, J.B., Hartley, N.R., 2001. Comparison between outcrop-spectral gamma ray logging and whole rock geochemistry: implications for quantitative reservoir characterisation in continental sequences. Mar. Petrol. Geol. 18 (6), 657–670.
Swanson, V.E., 1961. Geology and geochemistry of uranium in marine black shales-a review. U.S. Geological Surv. Prof. Pap. 356 (C), 67–112.

Tait, J., Bachtadse, V., Franke, W., Soffel, H., 1997. Geodynamic evolution of the European Variscan fold belt: palaeomagnetic and geological constraints. Geol. Rundsch. 86 (3), 585–598.

Taylor, S.R., McLennan, S.M., 1985. The Continental Crust: Its Composition and Evolution. Blackwell Scientifc Publications, p. 328.

Taylor, S.R., McLennan, S.M., 1995. The geochemical evolution of the continental crust. Rev. Geophys. 33 (2), 241–265.

Taylor, S.R., McLennan, S.M., McCulloch, M.T., 1983. Geochemistry of loess continental crust composition and crustal model ages. Geochimica Cosmochimica Acta 47 (11), 1897–1905.

Thompson, R.N., 1982. Magmatism of the British Tertiary province. Scott. J. Geol. 18, 49–107.

Torsvik, T.H., Van der Voo, R., 2002. Refining Gondwana and Pangea palaeogeography: estimates of Phanerozoic non-depole (octupole) fields. Geophys. J. Int. 151, 771–794.

Tourtelot, H.A., 1979. Black shale; its deposition and diagenesis. Clays Clay Minerals 27 (5), 313–321.

Tribovillard, N., Algeo, T.J., Lyons, T., Riboulleau, A., 2006. Trace metals as paleoredox and paleoproductivity proxies: an update. Chem. Geol. 232 (1–2), 12–32.

Trotter, J., Barnes, W.H., 1958. The structure of vanadinite. Can. Mineralogist 6 (2), 161–173.

Truyols, J., Julivert, M., 1983. El Silúrico en el Macizo Ibérico. Libro Jubilar JM Ríos. Geol. España 1, 246–265.

Truyols, J., Arbizu, M., García-Alcalde, J.L., García-López, S., Méndez Bedia, I., Soto, F., Truyols-Massoni, M., 1990. Cantabrian and Palentian zones; stratigraphy: the Asturian-Leonese domain (Cantabrian zone). In: Dallmeyer, R.D., Martínez-Garcia, E. (Eds.), Pre-Mesozoic Geology of Iberia. Springer, Berlin, pp. 10–19.

Truyols, J., Philippot, A., Julivert, M., 1974. Les formations siluriennes de la Zone Cantabrique et leurs faunes. Bulletin de la Societé Géologique de France. 7ème sèrie 16, 25–35.

Turekian, K., 1955. Paleoecological significance of the Sr/Ca ratio in fossils and sediments. Bull. Geol. Soc. Am. 66, 155–158.

Turekian, K.K., Wedepohl, K.H., 1961. Distribution of the elements in some major units of the earth's crust. Geol. Soc. Am. Bull. 72 (2), 175–192.

Turekian, K.K., 1964. The marine geochemistry of strontium. Geochimica Cosmochimica Acta 28, 1479–1496.

Turgeon, S., Brumsack, H., 2006. Anoxic vs. dysoxic events reflected in sediment geochemistry during the Cenomanian–Turonian boundary event (Cretaceous) in the Umbria–Marche basin of central Italy. Chem. Geol. 234 (3–4), 321–339.

Tyson, R., 1987. The genesis and palynofacies characteristics of marine petroleum source rocks. Geol. Soc. Lond. Spec. Publ. 26 (1), 47–67.

Tyson, R.V., Pearson, T.H., 1991. The modern and ancient continental shelf anoxia: an overview. Geol. Soc. Spec. Publ. 58, 1–24.

Tyson, R.V., 1995. Sedimentary Organic Matter: Organic Facies and Palynofacies. Chapman & Hall, London, p. 615.

Valladares, M.I., Barba, P., Ugidos, J.M., 2002. Precambrian. In: Gibbons, W., Moreno, T. (Eds.), The Geology of Spain. Geological Society, London, pp. 7–16.

Van der Voo, R., 2005. Paleomagnetism of the Atlantic, Tethys and Iapetus Oceans. Cambridge University Press.

Van der Weijden, C.H., 2002. Pitfalls of normalization of marine geochemical data using a common divisor. Mar. Geol. 184 (3–4), 167–187.

Veizer, J., Demovič, R., 1974. Strontium as a tool in facies analysis. J. Sediment. Petrol. 44, 93–115.

Verniers, J., Nestor, V., Paris, F., Dufka, P., Sutherland, S., Van Grootel, G., 1995. A global Chitinozoa biozonation for the Silurian. Geol. Mag. 132 (06), 651–666.
Veselovsky, Z., 2004 (Ph.D. thesis). Integrated numerical modelling of a polyhistory basin, southern Cantabrian basin (Palaeozoic, NW-Spain), vol. 13. Gaea Heidelb., p. 225
Vine, J.D., Tourtelot, E.B., 1970. Geochemistry of black shale deposits; a summary report. Econ. Geol. 65 (3), 253–272.
Wahlberg, J.S., Baker, J.H., Vernon, R.W., Dewar, R.S., 1965. Exchange adsorption of strontium on clay minerals. Geol. Surv. Bull. 1140-C, 1–26.
Walliser, O.H., 1996. Global Events and Event Stratigraphy in the Phanerozoic. Springer (Berlin and Heidelberg), p. 341.
Walther, J.V., 2009. Essentials of Geochemistry. Jones & Bartlett Learning, Sudbury (Massachusetts).
Warning, B., Brumsack, H.J., 2000. Trace metal signatures of eastern Mediterranean sapropels. Palaeogeogr. Palaeoclimatol. Palaeoecol. 158, 293–309.
Weaver, C.E., 1989. Clays, muds, and shales. Dev. Sedimentol. 44, 819. Elsevier (Amsterdam).
Wedepohl, K., 1971. Environmental influences on the chemical composition of shales and clays. Phys. Chem. Earth 8, 305–333.
Wedepohl, K., 1991. The composition of the upper earth's crust and the natural cycles of selected metals. In: Merrian, E. (Ed.), Metals in Natural Raw Materials. Metals and Their Compounds in the Environment. Occurrence, Analysis, and Biological Relevance. Wiley-VCH, Weinheim, pp. 3–18.
Wei, G., Liu, Y., Li, X., Shao, L., Liang, X., 2003. Climatic impact on Al, K, Sc and Ti in marine sediments: evidence from ODP site 1144, South China Sea. Geochem. J.-Jap. 37 (5), 593–602.
Wei, H., Chen, D., Wang, J., Yu, H., Tucker, M.E., 2012. Organic accumulation in the lower Chihsia Formation (Middle Permian) of South China: constraints from pyrite morphology and multiple geochemical proxies. Palaeogeogr. Palaeoclimatol. Palaeoecol. 353, 73–86.
Weil, A., Gutiérrez-alonso, G., Conan, J., 2010. New time constraints on lithospheric-scale oroclinal bending of the Ibero-Armorican Arc: a palaeomagnetic study of earliest Permian rocks from Iberia. J. Geol. Soc. 167 (1), 127–143.
Weil, A.B., Van der Voo, R., van der Pluijm, B.A., 2001. Oroclinal bending and evidence against the Pangea megashear: the Cantabria-Asturias arc (northern Spain). Geology 29 (11), 991–994.
Weil, A.B., Van der Voo, R., van der Pluijm, B.A., Parés, J.M., 2000. The formation of an orocline by multiphase deformation: a paleomagnetic investigation of the Cantabria–Asturias Arc (northern Spain). J. Struct. Geol. 22 (6), 735–756.
Werne, J.P., Sageman, B.B., Lyons, T.W., Hollander, D.J., 2002. An integrated assessment of a "type euxinic" deposit: evidence for multiple controls on black shale deposition in the Middle Devonian Oatka Creek Formation. Am. J. Sci. 302 (2), 110–143.
Wignall, P.B., Hallam, A., 1991. Biofacies, stratigraphic distribution and depositional models of British onshore Jurassic black shales. Geol. Soc. Lond. Spec. Publ. 58 (1), 291–309.
Wignall, P.B., Maynard, J.R., 1993. The sequence stratigraphy of transgressive black shales. In: Source Rocks in a Sequence Stratigraphic Framework: American Association of Petroleum Geologists, Studies in Geology, 37, pp. 35–47.
Wignall, P.B., Myers, K.J., 1988. Interpreting benthic oxygen levels in mudrocks: a new approach. Geology 16 (5), 452–455.
Wignall, P.B., Newton, R., 2001. Black shales on the basin margin: a model based on examples from the Upper Jurassic of the Boulonnais, northern France. Sediment. Geol. 144 (3–4), 335–356.

Wignall, P.B., 1994. Black Shales. Clarendon Press Oxford, UK, p. 127.
Wilde, P., Berry, W., Quinby-Hunt, M., 1991. Silurian oceanic and atmospheric circulation and chemistry. Special Pap. Palaeontol. 44, 123–143.
Wilde, P., Lyons, T.W., Quinby-Hunt, M.S., 2004. Organic carbon proxies in black shales: molybdenum. Chem. Geol. 206 (3–4), 167–176.
Wright, J., Colling, A., 1995. Seawater: Its Composition, Properties, and Behaviour. Pergamon Press (in association with the Open University), p. 168.
Xie, X., Ellis, A., Wang, Y., Xie, Z., Duan, M., Su, C., 2009. Geochemistry of redox-sensitive elements and sulfur isotopes in the high arsenic groundwater system of Datong Basin, China. Sci. Total Environ. 407 (12), 3823–3835.
Xu, H., Liu, B., Wu, F., 2010. Spatial and temporal variations of Rb/Sr ratios of the bulk surface sediments in Lake Qinghai. Geochem. Trans. 11, 3.
Xu, Y., Morel, F.M.M., 2013. Chapter 16 Cadmium in marine phytoplankton. In: Sigel, A., Sigel, H., Sigel, R.K.O. (Eds.), Cadmium: From Toxicity to Essentiality, Metal Irons in Life Sciences, vol. 11, pp. 509–528.
Zhang, S., Barnes, C.R., Jowett, D.M.S., 2006. The paradox of the global standard Late Ordovician–Early Silurian sea level curve: evidence from conodont community analysis from both Canadian Arctic and Appalachian margins. Palaeogeogr. Palaeoclimatol. Palaeoecol. 236 (3–4), 246–271.

CHAPTER FOUR

Macroevolution and Biostratigraphy of Paleozoic Foraminifers

D. Vachard
Université de Lille, UMR CNRS 8198 Evolution, Ecologie et Paléontologie, Villeneuve d'Ascq cedex, France
E-mails: Daniel.Vachard@univ-lille1.fr; Daniel.Vachard@free.fr

Contents

Abstract

A new classification presented for the Paleozoic foraminifers encompasses six classes: Allogromiata, Astrorhizata, Fusulinata, Miliolata, Nodosariata, and Textulariata. The three main classes, Fusulinata, Miliolata, and Nodosariata, include 10 orders, 17 suborders, and 32 superfamilies. The biostratigraphy of these subdivisions and main genera is summarized. Some paleobiogeographical implications are given. Emended and new taxa are: Tuberitinoidea emend.; Calcisphaeroidea n. superfam.; Pseudoammodiscina emend.; Tournayellina n. subord.; Tournayelloidea emend.; Lasiodiscina emend.; Earlandiina n. subord.; Caligellina n. subord.; Eonodosariina n. subord.; Pseudopalmulida emend.; Pseudopalmuloidea emend.; Fusulinata emend.; Fusulinana emend.; Lituobellina n. subord.; Staffellina emend.; Calcivertelloidea n. superfam.

1. INTRODUCTION

With the Frasnian–Famennian crisis (FF), the first evolutive phase of the foraminifers was completed, and the second phase started and lasted until the Permian–Triassic boundary (PTB), where life on the Earth nearly disappeared completely. The initial phase of the foraminiferal history, from

Stratigraphy & Timescales, Volume 1
ISSN 2468-5178
http://dx.doi.org/10.1016/bs.sats.2016.10.005

Cambrian to early Devonian, was dominated by agglutinated tests. The Givetian (middle Devonian) revolution resulted in the replacement of these agglutinated tests by calcareous secreted tests (Vachard et al., 2010). The lower Paleozoic agglutinated foraminifers are considered as belonging either to the classes Textulariata (Loeblich and Tappan, 1964, 1987) and/or Astrorhizata (Vdovenko et al., 1993; Mikhalevich, 2003), but these so-called agglutinates could also have resulted from recrystallizations of secreted tests (Fig. 1).

Numerous discussions exist with respect to the oldest representatives of the foraminiferal phylum; especially about, (1) the bilocular, undivided, tubular forms; (2) the unilocular (or monothalamous) forms; and (3) the planispirally coiled, evolute, undivided tests. In this latter group, the first taxon that appears as early as the Cambrian was called *Ammodiscus* Reuss, an extant genus with a complicate taxonomy (Loeblich and Tappan, 1954). At the Permian–Triassic boundary, a taxon was successively called *Ammodiscus*; *Cornuspira* Schulze; *Rectocornuspira* Warthin; *Postcladella* Krainer and Vachard, and then again *Ammodiscus* (Nestell et al., 2015). Likewise, Paleozoic *Hyperammina* (*auctorum non* Brady), became *Earlandia* Plummer by the latter part of 20th century or renamed *Sansabaina* Loeblich and Tappan (according to the coarseness of their agglutinates), are currently reassigned to *Hyperammina* (Nestell et al., 2015).

2. PALEOZOIC FORAMINIFERAL CLASSES

The foraminifers have long been considered as an order or a class in the western European or North American literature (Sigal, 1952; Ciry, 1952; Loeblich and Tappan, 1964, 1987, 1992) and split into orders or superorders. In contrast in the Russian literature, it was recognized long ago that similar divisions were mostly subclasses and classes (Miklukho-Maklay et al., 1958, 1959; Rauzer-Chernousova and Fursenko, 1959; Rozovskaya, 1975; Mikhalevich, 1980, 2003; Vdovenko et al., 1993; Rauzer-Chernousova et al., 1996). Modern biological investigations have demonstrated that the foraminifers constitute a phylum or subphylum (Cavalier Smith, 2002, 2003). Personal observations of Paleozoic foraminifers during almost 50 years (Vachard, 1974a,b; Vachard et al., 2010, 2013, 2015, 2016, 2017) agree with this status of phylum and these subdivisions in classes, but are not consistent with some phylogenic hypotheses and attempts of dating by using molecular biology (Flakowski et al., 2005; Groussin et al., 2011; Pawlowski et al., 2013).

FORAMINIFERAL CLASSES	Microstructures of walls	Main arrangements of chambers	Architectures of tests
ALLOGROMIATA	Organic-walled	Spherical to polygonal	Unilocular
ASTRORHIZATA	Agglutinated	Spherical to polygonal	Unilocular to bilocular
FUSULINATA	Low Mg calcitic + rare aragonitic	Plurilocular: endothyroidally coiled; planispiral involute or rarely evolute; biserially coiled, rectilinear uniseriate or biseriate	Unilocular, bilocular, plurilocular
MILIOLATA	High Mg calcitic	Streptospiral, planispiral evolute, attached serpuliform	Bilocular
NODOSARIATA	Low Mg calcitic	Rectilinear uniseriate	Plurilocular
TEXTULARIATA	Agglutinated	Rectilinear biseriate	Plurilocular

Figure 1 Main characters of the six classes of Paleozoic foraminifers.

This chapter tries to synthetize these observations. The transitional and provisional character of this attempt is obvious, pending additional molecular biology studies, and the challenge of several foraminiferal paradigms such as the wall microstructures, morphologies of tests, types of coilings and apertures, morphogroups, or microhabitats (Rigaud et al., 2015; Rigaud and Martini, 2016; Vachard, 2016).

The aim of this chapter is not only to present a new classification, but also to advocate a more logical synthesis about Paleozoic foraminifers. The proposed classification (Fig. 2) would result in several geological implications, principally biostratigraphic, which complement the work of

PHYLUM	CLASSES	SUBCLASSES	ORDERS	SUBORDERS	SUPERFAMILIES
FORAMINIFERA	ALLOGROMIATA				
	ASTRORHIZATA		SACCAMMINIDA		
			PARATHURAMMINIDA		
	FUSULINATA	AFUSULINANA	ARCHAEDISCIDA	PSEUDOAMMODISCINA	PSEUDOAMMODISCOIDEA
					"TOLYPAMMINA" GROUP
				TOURNAYELLINA	TOURNAYELLOIDEA
				ARCHAEDISCINA	ARCHAEDISCOIDEA
				LASIODISCINA	LASIODISCOIDEA
			EARLANDIIDA	EARLANDIINA	EARLANDIOIDEA
				CALIGELLINA	CALIGELLOIDEA
				EONODOSARIINA	EONODOSARIOIDEA
			PSEUDOPALMULIDA		PSEUDOPALMULOIDEA
		FUSULINANA	ENDOTHYRIDA	LITUOTUBELLINA	LITUOTUBELLOIDEA
					SEPTABRUNSIINOIDEA
					MSTINIOIDEA
					QUASIENDOTHYROIDEA
				ENDOTHYRINA	ENDOTHYROIDEA
					BRADYINOIDEA
					LOEBLICHIOIDEA
				PALAEOTEXTULARIINA	PALAEOTEXTULARIOIDEA
					ENDOTEBOIDEA
					TETRATAXOIDEA
					GLOBIVALVULINOIDEA
			FUSULINIDA	STAFFELLINA	STAFFELLOIDEA
				FUSULININA	OZAWAINELLOIDEA
					SCHUBERTELLOIDEA
					FUSULINOIDEA
					SCHWAGERINOIDEA
					NEOSCHWAGERINOIDEA
	MILIOLATA		CORNUSPIRIDA	NUBECULARIINA	CALCIVERTELLOIDEA
					NUBECULARIOIDEA
				CORNUSPIRINA	CORNUSPIROIDEA
					NEODISCOIDEA
	NODOSARIATA		NODOSARIIDA	GEINITZININA	GEINITZINOIDEA
				NODOSARIINA	NODOSARIOIDEA
	TEXTULARIATA		TEXTULARIIDA		
			VERNEUILIDA		
			LITUOLIDA		

Figure 2 Proposed classification of the Paleozoic foraminifers.

Conil et al. (1991), Poty et al. (2006), Cózar et al. (2008a,b, 2011, 2014), Vachard et al. (2010), Hance et al. (2011), Davydov et al. (2012), and Henderson et al. (2012). Other implications concern the paleoecology, paleobiogeography, and geodynamics.

3. PALEOZOIC FORAMINIFERAL CLASSIFICATION

Subkingdom RHIZARIA Cavalier-Smith, 2002

Phylum FORAMINIFERA d'Orbigny 1826 emend. Cavalier-Smith, 2003

Class ALLOGROMIATA Fursenko, 1958

Description: Foraminifers generally free and unilocular, with an organic-walled test. Aperture terminal simple.

Remarks: The status of this class during the Paleozoic has remained unchanged to what it was in the classical treatises and handbooks (Sigal, 1952; Loeblich and Tappan, 1964, 1987; Vdovenko et al., 1993).

Occurrence: ?Late Proterozoic, Cambrian to Holocene; rare in fossil record but probably cosmopolitan.

Class ASTRORHIZATA Saidova, 1981 emend. Mikhalevich, 1995

Description: Foraminifers generally free, unilocular, or bilocular. Agglutinated wall with an inner organic lining. Aperture terminal simple.

Remarks: The existence of the Paleozoic Astrorhizata remains hypothetical and largely theoretical. It depends on the interpretation of the "agglutinates," the homeomorphy with extant taxa, and on how they are studied. For instance, when they are obtained from conodont residues or other extractions, these forms received a name of Astrorhizata (e.g., Gutschick, 1962; Conkin and Conkin, 1964, 1970; Herbig, 2006a; Holcová, 2004; Holcová and Slavík, 2013); when they are illustrated in thin sections, they received a name of Fusulinata (Lipina, 1955; Conil and Lys, 1964; Herbig, 2006b; Krainer and Vachard, 2011).

Astrorhizata from the Carboniferous and Permian have been described by Conkin and Conkin (1964, 1970), Langer (1969), Sosipatrova (1970), Ebner (1973), Eickhoff (1974), Ukharsaya (1975, 1978), Loeblich and Tappan (1987), Vdovenko et al. (1993), Pronina (1999), Nestell et al. (2009, 2011); despite that, the presence of orders Ammodiscida Fursenko, 1958 and Hyperamminida Vdovenko et al.,

1993 remains discussed during the Paleozoic (Vachard et al., 2010). They are either recrystallized Fusulinata (Pseudoammodiscida or Earlandiida) or representatives of a not yet described class. Many genera of Paleozoic Astrorhizida Lankester, 1885 nom. correct. Calkins, 1909, Saccamminida, and Parathuramminida likely belong to algae rather than foraminifers. Hence these two latter orders are hypothetically included in the description of the Paleozoic Astrorhizata of this chapter.

The inner lining of the foraminiferal wall is generally interpreted as a primitive character inherited from the Allogromiata. However, the first unquestionable inner organic linings of foraminifers are late Permian in age (Stancliffe, 1989; Groves et al., 2004; a dating that probably partly corresponds to the modern middle Permian), and often called "microforaminifers" within the palynologic preparations. Such organic layers are often described in the Cambrian to early Devonian literature (Winchester-Seeto and Bell, 1994; Scott et al., 2003), where they are generally questionable because not associated in situ with tests, and being plurilaminar and not monolaminar. Moreover, these organic layers are totally absent from the Carboniferous, where the palynologic studies, due to the coal and petroleum investigations, are very numerous. Appeared relatively tardive, the organic basal layer is therefore an evolved character rather than a primitive character.

Occurrence: Cambrian–Holocene; probably cosmopolitan.

Order SACCAMMINIDA Lankester, 1885 (*auctorum*)

Description: Unilocular Astrorhizata, spherical to polygonal, with a terminal aperture, generally with a neck.

Remarks: Many unilocular Devonian and Pre-Devonian foraminifers, more or less spherical and with a rounded aperture, are considered as foraminifers of the classes Textulariata or Astrorhizata, and even as representatives of extant genera, such as *Saccammina* Carpenter, *Lagenammina* Rhumbler, *Thurammina* Brady, *Hyperammina* Brady, and *Sorosphaera* Brady (Loeblich and Tappan, 1964, 1987; Poyarkov, 1969, 1977, 1979; Ross and Ross, 1991; Vdovenko et al., 1993). However, other assignments have been suggested: (1) volvocale algae, as for the calcisphaeraceans and radiosphaeraceans (Kazmierczak, 1976); (2) pseudoforaminifers (Vachard, 1994); (3) calcitarcha (i.e., calcareous acritarcha; Versteegh et al., 2009); (4) thaumatoporellacean incertae sedis algae (Schlagwinteit et al., 2013b); (5) "radiolarians, the tests of which were calcified after diagenesis" (Vishnevskaya and Sedaeva, 2002a,b; Nestell et al., 2011; Afanasieva and Amon, 2011), as a return to the lapsed hypotheses of Williamson (1881) and Pia (1937); (6) thecamoebian protozoans [E. Armynot du Châtelet (pers. comm., March 2015)]; (7) tintinnids, which are occasionally agglutinating (Henjes and Assmy, 2008), can also be mentioned.

Occurrence: Ordovician-Holocene; cosmopolitan.

Order PARATHURAMMINIDA Mikhalevich, 1980 (*auctorum*)

Description: Unilocular, free to temporarily attached tests, cysts or thalli. Large central chamber. Neither bilocular nor plurilocular, but clusters of chambers can exist (*Rauserina* Antropov, *Uralinella* Bykova, and various tuberitinids). Apertures are (1) typical at the extremity of hollow neck; (2) present as numerous minute pores through the wall; or (3) inconspicuous. Wall thin to moderately thick, dark microgranular; occasionally bilayered with an inner, hyaline pseudofibrous layer; rarely more differentiated (*Tubesphaera* Vachard and some parathuramminids) (Figs. 3(1–25, 27–36) and 4(1)).

Composition: Parathuramminoidea Rauzer-Chernousova and Fursenko, 1959; Irregularinoidea Gaillot and Vachard, 2007; Tuberitinoidea Gaillot and Vachard, 2007 emend.; ?Calcisphaeroidea n. superfam.

Remarks: Appeared in the Cambrian, common in the Givetian and Frasnian, the parathuramminids diversified from the Famennian to the Visean. The last abundant and typical parathuramminoid is probably the Serpukhovian genus *Hemithurammina* Mamet (see Perret and Vachard, 1977); however, rare and questionable parathuramminoids and irregularinoids were mentioned up to the middle Permian (Nguyen Duc Tien, 1966; Nestell and Nestell, 2006), and the tuberitinoids are present during the entire range with *Eotuberitina* Miklukho-Maklay, and even up to the earliest Triassic in South China (Song et al., 2007, 2011).

Occurrence: Late Cambrian–earliest Triassic; either cosmopolitan or endemic genera.

Superfamily PARATHURAMMINOIDEA Fursenko in Rauzer-Chernousova and Fursenko, 1959

Description: Unilocular, free tests. Broad central chamber, spherical to polygonal. Apertures inconspicuous (*Vicinesphaera*) or located at the extremity of radiate necks. Wall thin to moderately thick, dark microgranular, occasionally bilayered with an inner hyaline pseudofibrous layer, rarely more differentiated.

Composition: Eovolutinidae Loeblich and Tappan, 1986; Ivanovellidae Chuvashov and Yuferev in Zadorozhnyi and Yuferev, 1984;

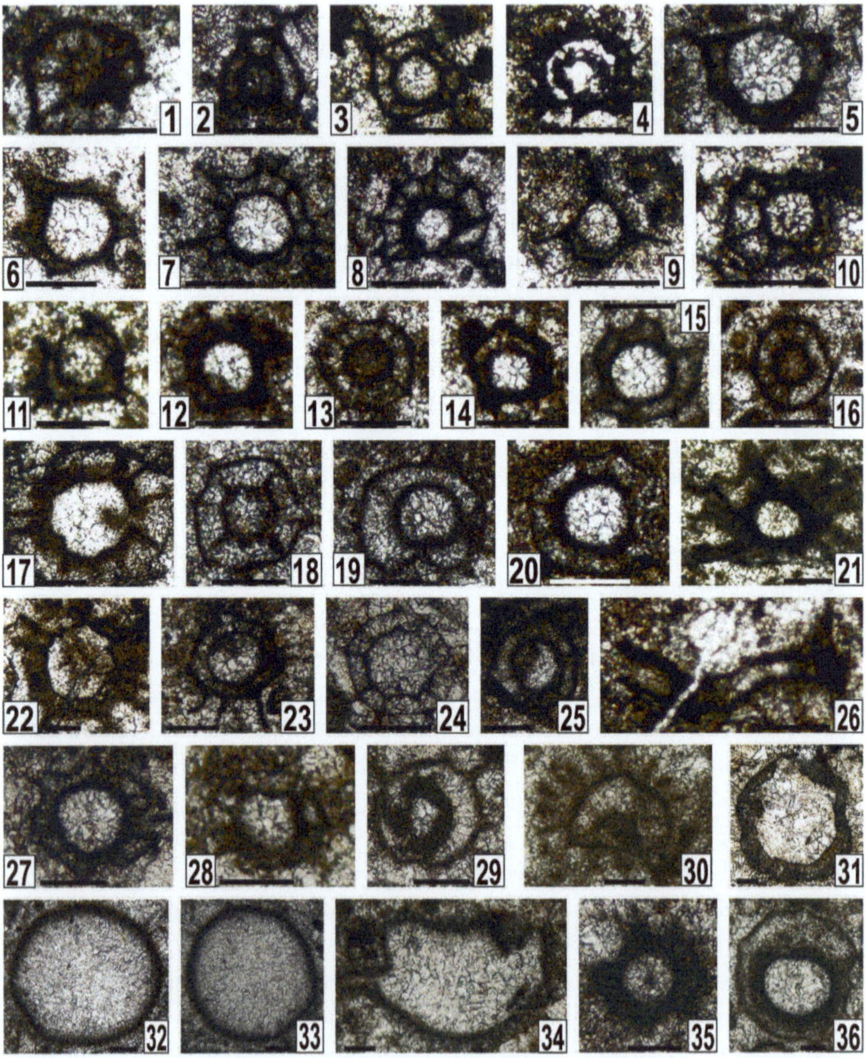

Figure 3 Parathuramminoidea and Irregularinoidea (Givetian; Carnic Alps, Austria). Scale bar = 0.100 mm. **1–3, 4?, 7–8, 10–11, 12–14, 16, 18, 24?, 25.** *Uralinella* spp. **5–6, 9, 15, 17–20, 21?, 22–23, 27–28, 35.** *Elenella* spp. **26.** *Paracaligella* sp. **29.** *Bithurammina*? sp. **30.** *Auroria* sp. **31, 36?** *Apertauroria*? sp. **32–33.** *Cribrosphaeroides* sp. **34.** *Uslonia* sp.

Parathuramminidae Bykova in Bykova and Polenova, 1955; Uralinellidae Chuvashov, Yuferev, and Zadorozhnyi in Zadorozhnyi and Yuferev, 1984.

Occurrence: *Vicinesphaera* Antropov appears in the late Cambrian of Kazakhstan and early Ordovician of Mexico (Vachard et al., 2017); the parathuramminoids have their acme in the late Silurian–Mississippian (their last typical genus, *Hemithurammina* Mamet, is Serpukhovian in age); they are rare to very rare in Pennsylvanian–Permian. They are probably cosmopolitan from the Silurian to the Tournaisian–early Visean.

Superfamily IRREGULARINOIDEA Gaillot and Vachard, 2007

Description: Tests free, spherical to polygonal. Wall simple, dark microgranular, rarely bilayered with a pseudofibrous inner layer (large Tournaisian *Bisphaera* Birina) or with more diversified but poorly described wall textures (*Auroria* Poyarkov). No necks. Apertures inconspicuous or perforated walls (*Uslonia* Antropov; *Cribrosphaeroides* Reitlinger); or with a supposed terminal aperture (*Apertauroria* Sabirov).

Composition: Irregularinidae Zadorozhnyi and Yuferev, 1984; Usloniidae Conil and Longerstaey in Conil et al., 1980; Cribrosphaeroididae Sabirov in Zadorozhnyi and Yuferev, 1984 nom. correct. Sabirov, 1987; Auroriidae Loeblich and Tappan, 1986; Bisphaeridae Sabirov, 1987.

Remarks: *Bisphaera* is an irregularinoid genus, which was recently synonymized with the algal or cyanobacterial genus *Thaumatoporella* Pia (Schlagintweit et al., 2013a,b) as well as the radiolaria *Trochodiscus* Haeckel (Afanavieva and Amon, 2011). The discussion is therefore hard.

Occurrence: Late Silurian–early Tournaisian. Middle Permian *Bisphaera*? *improvisa* Nestell and Nestell, 2006, as indicated by its authors, is questionable.

Superfamily TUBERITINOIDEA Gaillot and Vachard, 2007 emend. herein

Emended diagnosis: Tests unilocular, but often composed of a cluster of chambers. The cycle of life includes two generations (Conil and Lys in Conil et al., 1977), with a diplospherin free stage and a tuberitin attached stage. Both stages are probably present in all genera, but not always mentioned in the diagnoses. The attached tuberitin stages (*Eotuberitina*, *Tuberitina* Galloway and Harlton, *Mendipsia* Conil and

Longerstaey) display a concave face of fixation; the free diplospherin stages have one (*Diplosphaerina* Derville) or several polar spheres (*Polysphaerinella* Mamet). Wall dark microgranular (*Eotuberitina, Tuberitina*), bilayered with a thin dark microgranular outer tectum and a thicker, hyaline pseudofibrous inner layer (*Polysphaerinella*; Fig. 4(1)), or more differentiated (*Mendipsia*; *Tubeporina* Pronina; *Tubesphaera*; or *Atjusella* Petrova). Apertures absent; or numerous microperforations through the wall, but generally inconspicuous and/or diagenetically obliterated.

Composition: Tuberitinidae Miklukho-Maklay, 1958 and several unpublished other families based on distinct types of walls (Vachard, 2016): Tubeporininae Zadorozhnyi and Yuferev, 1984; *Tubesphaera* group; and Atjusellinae Zadorozhnyi and Yuferev, 1984.

Occurrence: Silurian–latest Permian (Vdovenko et al., 1993); rare in the earliest Triassic (Song et al., 2007, 2011; Okuyucu et al., 2014; Vachard et al., 2015; Vachard, 2016); cosmopolitan.

?Superfamily CALCISPHAEROIDEA n. superfam.

Diagnosis: Tests (or thalli) spherical, unilocular. Thick, dark microgranular walls (*Calcisphaera* Williamson = *Pachysphaerina* Conil and Lys), pseudofibrous (*Palaeocancellus* Derville), bilayered (*Radiosphaera* Reitlinger; *Polyderma* Derville), or complex (*Asterosphaera* Reitlinger). No apertures visible or minute radiate microperforations in *Calcisphaera*.

Composition: Calcisphaeridae Williamson, 1881 emend. Vachard and Téllez-Girón, 1986 is the unique family during the Paleozoic, but other families exist during the Mesozoic.

Remarks: *Radiosphaera* and *Calcisphaera* were two genera, which attained their hour of glory in the 1960s. They are currently often neglected. The genus *Pachythurammina* Vachard is morphologically transitional between *Pachysphaerina* and *Salpingothurammina* Poyarkov, whereas *Tubesphaera* is microstructurally transitional between tuberitinoids and calcisphaeroids (Vachard, 1994). Hence Calcisphaeroidea could belong to Parathuramminida, even if, because of their spherical shape, they were preferentially assigned to volvocale green algae or calcitarcha (Versteegh et al., 2009).

Occurrence: Givetian to middle Pennsylvanian (?late Pennsylvanian–Permian); cosmopolitan.

Class FUSULINATA Gaillot and Vachard, 2007 emend. herein

Emended diagnosis: Foraminifers with a homogeneous, microgranular and/or pseudofibrous, primary test wall of low-Mg calcite in which crystal units are optically unordered, more or less equidimensional, and only a few micrometers in size. Rare Fusulinata have possibly an aragonitic wall.

Composition: Afusulinana Vachard et al., 2010; Fusulinana Maslakova, 1990 emend. herein.

Remarks: The name Fusulinata introduced by Maslakova (1990) and very briefly defined by Maslakova et al. (1995) is in fact applied to a subclass (especially, because in both papers, the unique-described class is Foraminifera itself). In contrast, the name of subclass of Maslakova (1990) has formally priority upon Fusulinana Vachard et al., 2010, despite the very short definition and absence of discussion of Maslakova et al. (1995).

Occurrence: ?Cambrian—?early Silurian (depending on the assignment of some taxa to Astrorhizata or Fusulinata; see Vachard et al., 2010); late Silurian—latest Permian; ?Triassic—?Cretaceous (because of questionable earlandiids; see Vachard et al., 2010); either cosmopolitan or endemic.

Subclass AFUSULINANA Vachard et al., 2010

Description: Bilocular Fusulinata generally undivided, except in the three more evolved groups: Tournayellina, Eonodosariina, and Semitextulariida.

Composition: Archaediscida Poyarkov and Skvortsov, 1979; Earlandiida Loeblich and Tappan, 1982; Semitextulariida Poyarkov, 1979.

Occurrence: As for the class.

Order ARCHAEDISCIDA Poyarkov and Skvortsov, 1979

Synonym: Pseudoammodiscida Gaillot and Vachard, 2007.

Description: Test bilocular. Proloculus followed by an undivided tubular chamber, diversely coiled: streptospiral, oscillating, sigmoidal, eosigmoilinoid, planispiral, helical. Involute or evolute. Only a suborder, Tournayellina n. suborder, is pseudoseptate. Umbilical pillars exist in helical forms (Lasiodiscina). Wall often bilayered with hyaline pseudofibrous and dark microgranular layers, or unilayered with one of these two layers. Aperture terminal simple; sutural apertures with appendices can exist (Lasiodiscina).

Composition: Pseudoammodiscina Pronina, 1994 emend. herein; Tournayellina n. subord. (which not to be confused with the italicized genus name *Tournayellina* Lipina); Archaediscina Haynes, 1981; Lasiodiscina Mikhalevich, 1993 emend. herein.

Remarks: The order Moravamminida Termier et al., 1975 non Maslakova, 1990 previously attributed to the foraminifers is currently assigned to incertae sedis algae (Vachard and Cózar, 2010).

Occurrence: ?Cambrian—?Silurian (see the previously mentioned class). Middle Devonian—late Permian; shelves of the Tethys and Ural oceans, or rarely cosmopolitan.

Suborder PSEUDAMMODISCINA Pronina, 1994 emend. herein

Emended diagnosis: Free or attached tests, bilocular with an undivided second chamber, streptospirally to planispirally coiled, eventually uncoiled in the terminal stage. Wall dark microgranular, passing to brownish granular, and calcareously agglutinated. Aperture terminal simple.

Composition: Pseudoammodiscoidea Vdovenko et al., 1993; "*Tolypammina*" *auctorum* non Rhumblergroup.

Remarks: Except for several mysterious, extraordinarily large tubes, such as *Platysolenites* Eichwald and various "*Bathysiphon*" *auctorum* non Sars, the oldest pseudoammodiscid and the oldest typical foraminifer is "*Ammodiscus*" *diai* Culver, which appears in the late early Cambrian. Other "*Ammodiscus*"-like (i.e., evolute and undivided) tests were outlined throughout the Paleozoic, and occasionally received other generic names, especially when the last part of the tubular chamber was uncoiled: *Rectoammodiscus* Reitlinger or *Bifurcammina* Ireland during the Silurian, *Pseudocornuspira* Reitlinger during the Frasnian, *Eotournayella* Lipina and Pronina during the Famennian, *Pseudoammodiscus* Conil and Lys during the Visean, and *Postcladella* Krainer and Vachard (which has its acme in the early Triassic but first appears in the late Permian). For several authors, these "*Ammodiscus*" or "*Pseudoammodiscus*" are present up to the PTB (Nestell et al., 2015), but more likely, they disappear during the Serpukhovian or Bashkirian, and are replaced by homeomorphous cornuspirid miliolates. Moreover, *Pseudoammodiscus* is not the ancestor of the Miliolata because the first porcelaneous types are attached forms such as *Calcivertella* Cushman and Waters and *Ammovertella* Cushman (see Cózar et al., 2014). Then, *Hemigordiellina* Marie and *Cornuspira* really appear (Vachard et al., 2010; Cózar et al., 2014).

Similarly, many Paleozoic "*Tolypammina*" differ from the modern, homeomorphous Astrorhizata by their type of dark microgranular wall, which is often recrystallized in iron- and manganese oxides, with silica grains mimicking an agglutinate. They are considered

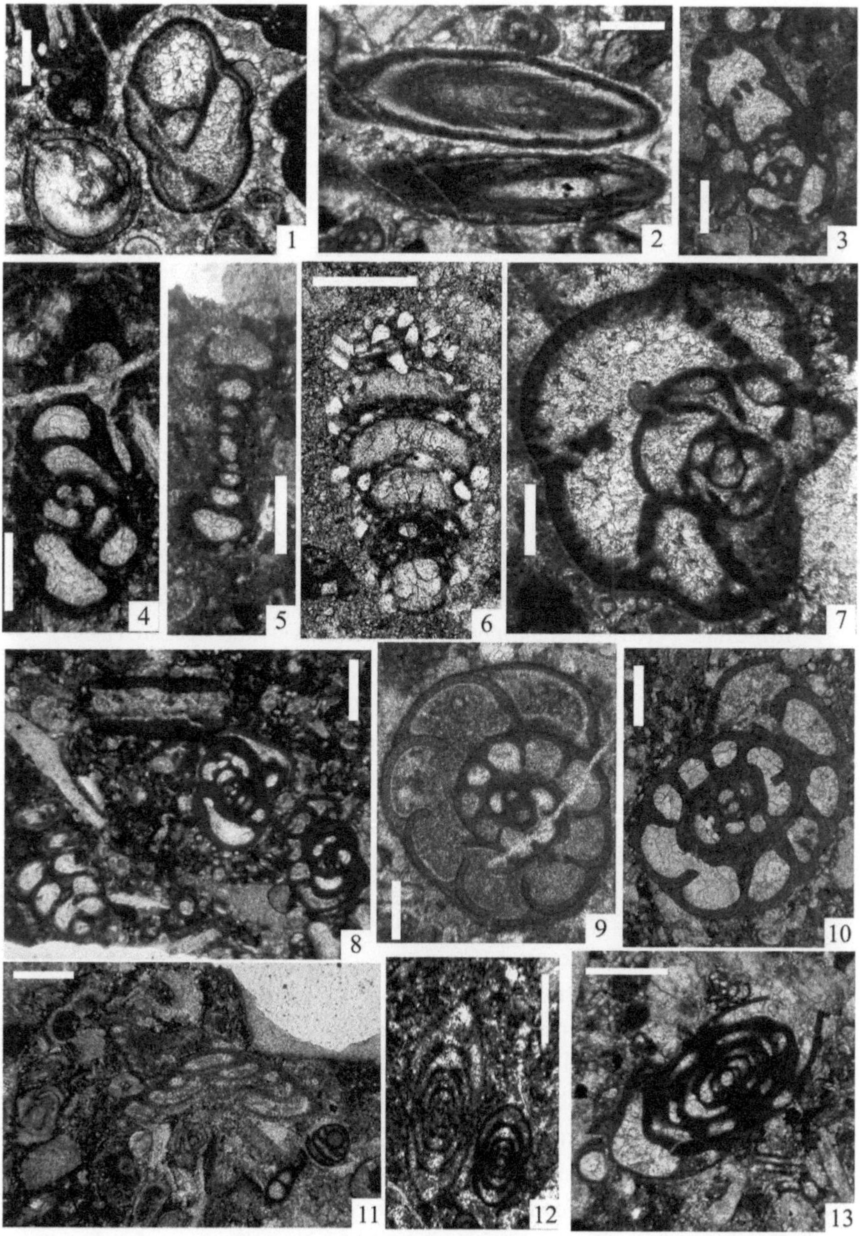

Figure 4 Tethyan Endothyrida and primitive Fusulinida (1–5, 7–11, 13: latest Visean of southern France; 6 Artinskian of the Carnic Alps; 12: late Visean of England). Scale bar = 0.250 mm. **1.** *Polysphaerinella* sp. (left) and *Ademassa* sp. (right). **2.** Two *Magnitella* sp. **3.** *Mstinia* sp. **4.** *Lituotubella* sp. **5.** *Forschia* sp. **6.** Palaeotextulariid with abundant siliceous agglutinate. **7.** *Bradyina* sp. **8.** Two *Omphalotis* sp. (right) with *Earlandia* sp. (top) and *Palaeotextularia* sp. (bottom, left). **9.** *Semiendothyra* sp. **10.** *Omphalotis* sp. **11.** *Tetrataxis* (top left and center) and *Howchinia* (right). **12.** *Pseudoendothyra* sp. (left) and *Eostaffella* sp. 1 (right). **13.** *Eostaffella* sp. 2.

herein as attached pseudoammodiscoids. However, other false Paleozoic "*Tolypammina*," such as those described by Ireland (1956) in the Pennsylvanian-Cisuralian of the USA could be calcivertelloid miliolates. Other homeomorphous miliolates could be "*Tolypammina*" *fortis* Reitlinger and related taxa in the Pennsylvanian of Russia, as well as these "*Tolypammina*" able to construct Triassic microreefs (Wendt, 1969).

Superfamily PSEUDOAMMODISCOIDEA Vdovenko et al., 1993

Description: Free Pseudoammodiscina. Wall microgranular. Spherical proloculus followed by a planispirally or streptospirally coiled, undivided, tubular chamber. Aperture terminal simple, at the extremity of the tubular chamber.

Composition: Pseudoammodiscidae Conil and Lys in Conil and Pirlet, 1970.

Occurrence: Silurian?—Serpukhovian, cosmopolitan.

Suborder TOURNAYELLINA n. subord.

Diagnosis: Test planispirally coiled evolute; rarely uncoiled and uniseriate (*Rectoseptatournayella* Brazhnikova and Rostovtseva) or uncoiled (*Forschiella* Mikhailov). Septa almost inconspicuous (*Eotournayella*) to well developed (*Septatournayella* Lipina, *Septaforschia* Conil and Lys). Supplementary deposits absent or present as nodes, crusts, and spines. Wall dark microgranular (*Tournayella* Dain) to granular with a calcareous agglutinate (*Forschia* Mikhailov) (Fig. 4(5)). Aperture terminal simple or rarely cribrate (*Forschia*, *Forschiella*).

Composition: Tournayelloidea Rauzer-Chernousova and Fursenko, 1959 emend. Hance et al., 2011.

Remarks: This suborder exhibits coiled tests with a pseudoseptation or sometimes a real septation. It derives from the pseudoammodiscoids *Pseudocornuspira* and *Eotournayella*. The representatives are especially homogeneous in shape, remaining planispirally evolute; but with two evolutionary trends: (1) to a relative gigantism and a change from thin dark microgranular walls to thick walls with a calcareous agglutinate, with *Eotournayella*-*Tournayella*-*Eoforschia* Mamet-*Forschia*-*Forschiella*-*Septaforschia*-*Viseina* Conil and Lys; (2) to a development of septa and terminal uncoiling within small forms: *Eotournayella*-*Eoseptatournayella* Lipina-*Septatournayella*-*Rectoseptatournayella*.

Occurrence: Late Famennian—late Serpukhovian; cosmopolitan and then restricted in Ural and Paleotethys oceans.

Superfamily TOURNAYELLOIDEA Rauzer-Chernousova and Fursenko, 1959 emend. Hance et al., 2011

Description: As for the suborder.

Composition: Tournayellidae Dain in Rauzer-Chernousova and Fursenko, 1959.

Remarks: Hance et al. (2011) restricted the concept of Tournayellida, but still included Globoendothyridae Hance et al. in this group. Tournayelloidea is reduced herein to the Tournayellidae. Globoendothyrids are not related with the tournayellids but with the septabrunsiinoids; i.e., with the Lituotubellina (see later).

Occurrence: Frasnian—late Serpukhovian, cosmopolitan in the Tournaisian; Tethyan during the Visean—Serpukhovian.

Suborder ARCHAEDISCINA Haynes, 1981

Description: Discoidal to inflated lenticular Archaediscata. Coiling planispiral, oscillating, sigmoidal, or eosigmoilinoid. Involute, rarely evolute. Lumen of tubular chambers at different evolutionary stages eventually filled with basal nodosities. Wall bilayered dark microgranular and hyaline pseudofibrous, or unilayered and pseudofibrous (Fig. 5(1—24)).

Composition: Archaediscoidea Piller, 1978.

Occurrence: From the second Visean foraminiferal biozone *sensu* Poty et al., 2006 (MFZ10) to earliest Moscovian (early Vereian).

Superfamily ARCHAEDISCOIDEA Piller, 1978

Description: As for the suborder.

Composition: Ammarchaediscidae Vachard in Vachard et al., 2004a, Archaediscidae Chernysheva, 1948, Eosigmoilinidae Vachard in Vachard et al., 2004a.

Remarks: The classifications of Pirlet and Conil (1974), Zaninetti and Altıner (1979), Conil et al. (1980), Brenckle et al. (1987), Vachard (1988), Cózar (2000), Hance et al. (2011), and Zandkarimi et al. (2016) are followed herein. The superfamily is characterized by (1) an evolution of the type of wall; (2) an evolution of the shape of the lumen of the tubular chamber; (3) planispiral or not planispiral coilings; (4) development of basal nodosities, which progressively occlude the lumen.

The evolution of the wall begins with an ancestral type, *Lapparentidiscus* Vachard, with a wall entirely dark microgranular. Then, with the first typical archaediscoid genus *Ammarchaediscus* Conil and Pirlet, a pseudofibrous plug appears in the umbilici. It

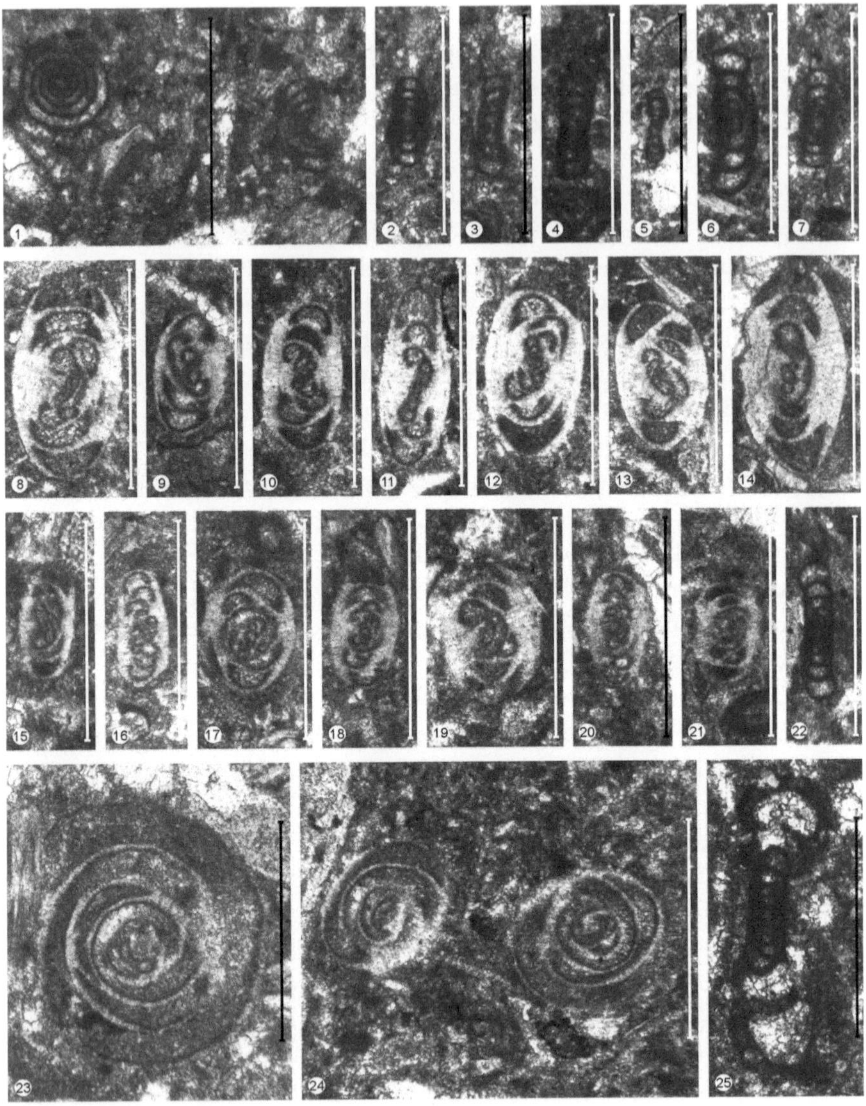

Figure 5 Archaediscoidea (Visean, Antalya Nappes, Turkey). Scale bar = 0.500 mm. **1–2, 4, 6–7, 22,** *Lapparentidiscus* sp. **3, 5.** *Viseidiscus* sp. **8–14, 17.** *Glomodiscus* sp. **15–16, 18?, 20?,** *Paraarchaediscus* sp. **19?, 21?, 23–24.** *Archaediscus* sp. **24, 25.** Gen. indet. (*Lapparentidiscus* passing to *Forschia*).

progressively covers all whorls, but the last whorl remains with a dark microgranular wall (*Viseidiscus* Mamet emend. Hance et al.; *Planoarchaediscus* Mikluko-Maklay); then, the test is entirely covered by the pseudofibrous layer (*Uralodiscus* Malakhova; *Glomodiscus*

Malakhova); inversely, the dark microgranular inner layer is progressively reduced (Archaediscidae) and finally disappears (Eosigmoilinidae). The tubular-coiled chamber remains always undivided (except for very rare teratogenic forms; especially *Tournarchaediscus* Conil and Pirlet). The coiling, rarely streptospiral or planispiral, is commonly oscillating or sigmoidal. The size of the tests remains small to moderate, except for some species (especially, almost the millimetric type species of *Archaediscus* Brady). The evolution of lumina includes various stages that are successively: ogival, involutus, concavus, concavo-angulatus, angulatus, evolutus, tenuis (Pirlet and Conil, 1974; Conil et al., 1980; Vachard, 1988; Hance et al., 2011; Zandkarimi et al., 2016). The ogival stage is represented by the genera *Uralodiscus* and *Glomodiscus* (= *Nudarchaediscus* Conil and Pirlet); the involutus stage corresponds to *Propermodiscus* Miklukho-Maklay, *Archaediscus* stage involutus, and *Conilidiscus* Vachard; the concavus stage corresponds to *Archaediscus* stage concavus, *Paraarchaediscus* Orlova, and *Pirletidiscus* Vachard; the concavo-angulatus stage to *Archaediscus sensu stricto*; the angulatus stage to *Archaediscus* stage angulatus and *Tubispirodiscus* Browne and Pohl; the evolutus stage to *Betpakodiscus* Marfenkova; the tenuis stage to *Browneidiscus* Brenckle, Ramsbottom, and Marchant and *Eosigmoilina* Ganelina.

The genera with nodosities are often considered as forming a subfamily, but they are interpreted herein as derived from every stages described earlier: concavus with nodosities (*Nodosarchaediscus* Conil and Pirlet; *Permodiscus* Dutkevich in Chernysheva *sensu* Pirlet and Conil); concavo-angulatus with nodosities (*Rugosoarchaediscus* Miklukho-Maklay, *Nodasperodiscus* Conil and Pirlet); angulatus with nodosities (*Neoarchaediscus* Miklukho-Maklay, *Asteroarchaediscus* Miklukho-Maklay, *Planospirodiscus* Sosipatrova); evolutus with nodosities (*Kasachstanodiscus* Marfenkova); tenuis with nodosities (*Brenckleina* Zaninetti and Altıner).

Occurrence: From the second Visean foraminiferal biozone (MFZ10; Poty et al., 2006) up to the earliest Moscovian (early Vereian).

Suborder LASIODISCINA Mikhalevich, 1993 emend. herein

Emended diagnosis: Test tubular, undivided, planar, plano-convex, or high conical. Wall dark microgranular and hyaline pseudofibrous. The hyaline layer is structureless or can show individualized pseudopillars. Aperture terminal simple, often with additional sutural apertures, which can be simple or protected by various appendices especially in the Permian genera.

Composition: Lasiodiscoidea Reitlinger in Vdovenko et al., 1993.

Occurrence: Late Visean foraminiferal biozone (MFZ13)—latest Permian (Changhsingian).

Superfamily LASIODISCOIDEA Reitlinger in Vdovenko et al., 1993

Description: This superfamily encompasses three families having in common a tubular chamber, a bilayered wall (microgranular and pseudofibrous), with umbilical pseudofibrous fillings and/or pseudopillars. The chamber remains undivided and the aperture is always terminal simple.

Composition: Howchiniidae Martini and Zaninetti, 1988; Lasiodiscidae Reitlinger, 1956 emend. Gaillot and Vachard, 2007; Pseudovidalinidae Altıner, 1988.

Phylogeny: (1) The ancestor derives from a single pseudoammodiscid becoming high conical in shape. (2) The deep umbilicus is then filled with a hyaline plug and/or more individualized pseudopillars [*Howchinia* Cushman (Fig. 4(11), right); *Planohowchinia* Cózar and Mamet]. (3) Lasiodiscoids, with secondary sutural apertures with appendices, appear in forms in which the conical test is progressively less high and finally almost plano-convex (*Hemidiscopsis* Cózar, *Monotaxinoides* Brazhnikova and Yartseva, *Eolasiodiscus* Reitlinger, *Hemidiscus* Schellwien emend. Vachard and Krainer; *Mesolasiodiscus* Rauzer-Chernousova and Chermnykh; *Lasiodiscus* Reichel; *Lasiotrochus* Reichel). (4) The family Pseudovidalinidae exhibits many convergences with the Archaediscoidea (Vachard, 2016). Vachard and Beckary (1991) speculated that the pseudofibrous layer, present in only one side of the Lasiodiscidae, becomes symmetrical and present on both sides of the Pseudovidalinidae. However, no transitional stages have been found, and the discovery of *Asselodiscus* Mamet and Pinard indicated that the family could be directly derived from another ancestor among the pseudoammodiscids.

The phylogeny of the Pseudovidalinidae is as follows: after *Asselodiscus*, *Raphconilia* Brenckle and Wahlman (= *Pseudovidalina auctorum*), ?*Falsodiscus* Davydov, *Pseudovidalina* Sosnina *sensu stricto*, *Xingshangdiscus* Zheng, and *Altineria* Özdikmen appear. As for the Archaediscoidea, these different genera correspond to different stages of the development of the lumina: involutus, concavus, angulatus, evolutus, and tenuis (Vachard, 2016).

Occurrence: As for the suborder.

Order EARLANDIIDA Sabirov in Vdovenko et al., 1993 emend. Vachard et al., 2010

Description: Test cylindrical to tapering, rectilinear to oscillating, undivided to divided, and uniseriate. Wall dark microgranular, rarely bilayered with a pseudofibrous layer. Aperture terminal simple.

Composition: Earlandiina n. subord.; Caligellina n. subord.; Eonodosariina n. subord.

Occurrence: Late Silurian—?early Cretaceous (see previously mentioned), cosmopolitan.

Suborder EARLANDIINA n. subord.

Diagnosis: Earlandiida cylindrical to tapering, undivided to pseudoseptate. Wall dark microgranular, rarely bilayered with an additional pseudofibrous layer. Aperture terminal simple.

Composition: Earlandioidea Loeblich and Tappan, 1982.

Occurrence: Late Silurian—?early Cretaceous, cosmopolitan.

Superfamily EARLANDIOIDEA Loeblich and Tappan, 1982

Description: As for the suborder.

Composition: Earlandiidae Cummings, 1955b; Paratikhinellidae Loeblich and Tappan, 1984.

Remarks: The wall is thin and dark microgranular to thick, brownish, granular, with an arrangement in paraboloids (Vachard et al., 2010; Fig. 4(8)). A bilayered wall exists in *Magnitella* Malakhova, which constitutes probably an aborted attempt to constitute a clear outer layer (Fig. 4(2)). That attempt was successful with the fibrous layer of *Syzrania* Reitlinger, which was at the origin of the Nodosariata. The pseudoseptate tubular forms are called *Paratikhinella* Reitlinger. Other tapering tests are *Lobatiquinella* Vachard and *Reitlingearlandia* Vachard. The Earlandioidea are homeomorphous of the agglutinated taxa called *Hyperammina*, *Sansabaina*, and *Kechenotiske* Loeblich and Tappan. Although these latter genera theoretically belong to another class, they are often difficult to differentiate from *Earlandia*.

Occurrence: Probably cosmopolitan during the Paleozoic; the superfamily seems to be present up to the Cretaceous? and to be restricted to some Tethyan areas.

Suborder CALIGELLINA n. subord.

Diagnosis: Irregular tests or permanent cysts of naked foraminifers surrounding their cytoplasm. Wall thin to thick, eventually with bricks composed of agglutinated or linked together tests of smaller foraminifers. Proloculus and apertures inconspicuous or questionable. Wall dark microgranular (Fig. 3(26)), often with a calcareous agglutinate and bilayered, dark microgranular and hyaline pseudofibrous (*Ademassa*; Fig. 4(1)).

Composition: Caligelloidea Gaillot and Vachard, 2007.

Remarks: It is possible to speculate that the *Earlandia*, as the true modern Hyperamminidae, belonged to the morphogroup B of foraminifers (Jones and Charnock, 1985; Murray, 1991; Murray et al., 2011) and have been able to belong to the morphogroup C of the same authors after an adaptation to the infaunal life. This phylogeny has been reconstructed by Vachard and Cózar (2004) and Özkan and Vachard (2015). At the beginning, the Earlandiida, appeared in the late Silurian, became infaunal during the early Devonian with *Eocaligella* Pronina. The genera *Paracaligella* Lipina and *Caligella* Antropov appeared in the Eifelian–Givetian and in the Frasnian, respectively. Then, the irregular *Baituganella* Lipina dominated, with their morphology very atypical for foraminifers. Finally, during the late Visean–Serpukhovian, a bilayered wall appeared in *Ademassa* Vachard (Fig. 4(1)).

The caligelloid tests are strongly deformed by the infaunal life; and some genera are so atypical that they were interpreted as syzygial cyst (Groves, 1987) or permanent cyst (Vachard and Cózar, 2004).

Even if, in some instances, this idea can be speculative, two lineages can be derived from Caligelloidea: (1) the lineage of *Halevikia* established by Özkan and Vachard (2015) can lead to the Tournayellinidae; (2) the family Insolentithecidae Gaillot and Vachard, 2007, which corresponds more probably to permanent cysts rather than real tests.

Occurrence: Late Silurian–lower Permian (Cisuralian); generally cosmopolitan (except the Tournayellinidae).

Superfamily CALIGELLOIDEA Gaillot and Vachard, 2007

Description: As for the suborder.

Composition: Caligellidae Reitlinger in Rauzer-Chernousova and Fursenko, 1959; Insolentithecidae Hance et al., 2011; ?Tournayellinidae Hance et al., 2011.

Occurrence: As for the suborder.

Suborder EONODOSARIINA n. subord.

Diagnosis: Test uniseriate, tapering. Chambers hemispherical to arcuate, undivided or with radiate septula. Wall bilayered, microgranular, and pseudofibrous, rarely only dark microgranular. Aperture simple, terminal, central.

Composition: Eonodosarioidea Rauzer-Chernousova in Vdovenko et al., 1993 emend. herein.

Remarks: The wall is first unilayered, dark microgranular as in the earlandioid ancestors; then, bilayered with an additional layer, which is pseudofibrous. This superfamily evolved very quickly during the Frasnian. Only *Tikhinella* was able to survive the FF crisis and to be found in the early Famennian. Furthermore, the last *Eonodosaria* are present in the Famennian of the Czech Republic (Kalvoda, 2001).

Occurrence: Early—late Frasnian; very rare in early Famennian; Paleotethys, Urals, and Siberia (including some terranes of Canadian Arctic).

Superfamily EONODOSARIOIDEA Rauzer-Chernousova in Vdovenko et al., 1993.

Emended diagnosis: As for the suborder.

Composition: *Tikhinella* Bykova group (*Tikhinella*, *Tikhinella*? Vachard in Clausen et al., unpublished data, and *Juferevella* Zadorozhnyi); *Multiseptida* group (*Frondilina* Bykova, *Multiseptida* Bykova); Eonodosariidae Rauzer-Chernousova in Vdovenko et al., 1993 (= Eogeinitzinidae Vachard, 1994; with *Eonodosaria* Lipina, and *Eogeinitzina* Lipina).

Occurrence: As for the suborder.

Order PSEUDOPALMULIDA Mikhalevich, 1993 emend. herein

Emended diagnosis: Tests planispirally coiled, biseriate, biseriate to uniseriate, or uniseriate. Chambers undivided or rarely with pillars (*Semitextularia* Miller and Carmer), tapering compressed to palmate. Aperture simple, basal (*Nanicella* Henbest), or terminal (*Paratextularia* Pokorny, *Pseudopalmula* Cushman and Stainbrook) or rarely multiple (*Semitextularia*).

Composition: Pseudopalmuloidea Vachard et al., 2010 emend. herein.

Occurrence: Devonian: Emsian (rare), Eifelian, Givetian, Frasnian, poorly known geographical distribution.

Superfamily PSEUDOPALMULOIDEA Vachard et al., 2010 emend. herein

Emended diagnosis: As for the order.

Composition: Semitextulariidae Pokorny, 1956; Pseudopalmulidae Bykova in Rauzer-Chernousova and Fursenko, 1959; Nanicellidae Fursenko, 1959.

Remarks: These forms are in my opinion the first true plurilocular foraminifers. The other forms interpreted as plurilocular [for example,

Reophax (*auctorum* non Montfort) and *Oxinoxis* Gutschick] are most probably connected moravamminacean or saccamminopsidacean incertae sedis algae, clusters of monolocular pelosinid foraminifers, and even just folded fragments of organic walls (Scott et al., 2003). The nanicellids adopt a planispiral coiling, whereas the Semitextulariidae and Pseudopalmulidae display some shapes that will only reappear in the Permian and the Mesozoic. Rare in Emsian and Eifelian, present in Givetian, and relatively common in Frasnian, the group disappears at the FF crisis. The pseudopalmuloid walls have a poorly known mineralogy, generally described as pseudofibrous; they may recrystallize into microsparite (Vachard, 1994) and consequently reveal a possible aragonitic and/or high-Mg calcitic composition.

Occurrence: As for the order.

Subclass FUSULINANA Maslakova, 1990 emend. herein

Emended diagnosis: Tests nautiloid, lenticular, discoid, subquadrate, inflated fusiform, elongate fusiform, subrhombic, subcylindrical. Proloculus spherical, small to large, and reniform or rectangular. Juvenaria often present (*Schubertella* Staff and Wedekind, *Dunbarula* Ciry). Endothyroidally or planispirally coiled tests, rarely uncoiled (*Mikhailovella* Ganelina, *Endothyranella* Galloway and Harlton, *Neoendothyranella* Nestell and Nestell, *Nipponitella* Hanzawa, *Reichelina* Erk, *Codonofusiella* Dunbar and Skinner). Chambers globular and not numerous (endothyrids), to quadratic to fusiform and numerous (fusulinids). Septa planar to moderately folded to strongly folded. Endoskeleton always developed (crusts, hooks, pseudochomata, chomata). Wall dark microgranular occasionally bilayered (*Omphalotis* Shlykova) to multilayered (globoendothyrids, ozawainelloids) often with a dark-microgranular thin tectum, and a differentiated inner layer (schubertelloids with primatheca; fusulinoids with diaphanotheca; schwagerinoids with keriotheca; neoschwagerinoids with "fine keriotheca"). Aperture terminal simple, basal, rarely cribrate, occasionally reduced to septal pores, cuniculi or foramina.

Composition: Endothyrida Fursenko, 1958 (including "Tournayellida" *auctorum*, and Palaeotextulariina Hohenegger and Piller, 1975 emend. herein); Fusulinida Fursenko, 1958.

Occurrence: Tournaisian–Changsinghian; complete phylogenetic lineages in the Tethys and Ural shelves, and strong North American and Siberian endemisms and/or migrations.

Order ENDOTHYRIDA Fursenko, 1958

Description: Tests nautiloid, lenticular, discoid. Proloculus small and spherical. Endothyroidally or planispirally coiled tests, rarely uncoiled (*Mikhailovella*, *Endothyranella*, *Neoendothyranella*). Chambers globular and not numerous. Septa planar. Endoskeleton generally developed (crusts, hooks, pseudochomata). Wall dark microgranular, occasionally bilayered (*Omphalotis*) to multilayered (globoendothyrids). Aperture terminal simple, basal, rarely cribrate (*Mikhailovella*) or central (*Endothyranella*).

Composition: Lituobellina n. subord.; Endothyrina Bogush, 1985; Palaeotextulariina Hohenegger and Piller, 1975 emend. herein.

Remarks: Hance et al. (2011) have concluded that the traditional subdivision Tournayellida (*sensu* Hohenegger and Piller, 1975; Rauzer-Chernousova et al., 1996) was not based on a phylogenetic reality, because: (1) the Tournayellida was only a group of evolute forms, which had evolved from *Eotournayella*; (2) some "families" among the two orders "Tournayellida" and Endothyrida share pseudoseptate and septate representatives (e.g., the septabrunsiinoids and the lituotubelloids), and this criterion is therefore not used for an order.

Occurrence: Late Famennian–Tournaisian; acme in the Visean–Gzhelian; rare up to the Cisuralian (lower Permian); cosmopolitan.

Suborder LITUOTUBELLINA n. subord.

Diagnosis: Tests free, streptospirally to endothyroidally coiled; rarely with a planispirally coiled terminal part (*Septabrunsiina* Lipina) or an uncoiled terminal part (*Lituotubella* Rauzer-Chernousova, *Rectoseptabrunsiina* Lipina, mstinioids) (Fig. 4(4)). Wall dark microgranular to granular with calcareous agglutinates (*Lituotubella*, *Mstinia* Dain). Aperture terminal, simple, and basal; rarely cribrate (*Lituotubella*, *Mstinia*).

Composition: Lituotubelloidea Hance et al., 2011; Septabrunsiinoidea Colpaert and Vachard in Colpaert et al., 2017; Mstinioidea Vachard et al., 2010 emend. Hance et al., 2011; Quasiendothyroidea Hance et al., 2011.

Occurrence: Givetian–Moscovian; Tethyan and Uralian oceans; only the Septabrunsiinoidea are cosmopolitan.

Superfamily LITUOTUBELLOIDEA Gaillot and Vachard, 2007 emend. Hance et al., 2011

Description: Free to attached (pseudolituotubids) Lituotubellina streptospirally coiled; rarely or with uncoiled terminal part (*Lituotubella*). Wall dark microgranular to granular with calcareous agglutinates (*Lituotubella*). Wall dark microgranular to granular with calcareous agglutinates. Aperture terminal, simple, and basal; rarely cribrate (*Lituotubella*).

Composition: Lituotubellidae; Laxoendothyridae Hance et al., 2011; Palaeospiroplectamminidae Loeblich and Tappan, 1984; Chernyshinellidae Lipina and Reitlinger in Rauzer-Chernousova et al., 1996; Pseudolituotubidae Conil and Longerstaey in Conil et al., 1980.

Occurrence: Middle Givetian—early Serpukhovian, shelves of the Ural and Tethys oceans; Tournaisian of North America.

Superfamily SEPTABRUNSIINOIDEA Colpaert and Vachard in Colpaert et al., 2017

Description: Test discoid to subnautiloid, biumbilicate or with prominent flanks. Pseudoseptate to septate from early to last whorls, or rarely completely septate. Supplementary deposits often absent, or represented by corner fillings, or rarely by curved spines in the last chambers. Wall dark microgranular forming tectum (*Laxoseptabrunsiina* Vachard), or differentiated with clearer particles (*Septabrunsiina*), or rarely multilayered (*Globoendothyra* Reitlinger).

Composition: Septabrunsiinidae Conil and Lys, 1977; Globoendothyridae; *Laxoseptabrunsiina* group; *Eblanaia* Conil and Marchant group; *Inflatoendothyra* Lipina group; *Urbanella* Malakhova group.

Remarks: All these forms have in common a juvenarium similar to the adult stage of *Septabrunsiina*. That feature is especially evident for the globoendothyrids (Hance et al., 2011).

Comparisons: The septabrunsiinoids differ from the lituotubelloids by the discoid, planispiral, and often evolute last whorls, and from the endothyroids by a less developed septation (except in *Globoendothyra*), less developed supplementary deposits generally limited to a spine in the last chamber, and more differentiated walls.

Occurrence: Famennian (DFZ3)—early Serpukhovian (MFZ16); cosmopolitan but with more or less delayed migrations.

Superfamily MSTINIOIDEA Vachard et al., 2010 emend. Hance et al., 2011

Synonym: Haplophragminoidea Vachard et al., 2014.

Description: Lituotubellina morphologically diversified (planispirally coiled, uniseriate, biseriate, biseriate coiled), having in common the thick walls, chernyshinellid chambers, and well-developed calcareous agglutinates (Fig. 4(3)).

Composition: Mstiniidae Hance et al., 2011; Haplophragminidae Reitlinger, 1950.

Occurrence: Latest Tournaisian–Moscovian; shelves of the Ural and Tethys oceans.

Superfamily QUASIENDOTHYROIDEA Hance et al., 2011

Description: Endothyroid to planispiral coiling, involute to seminvolute. Chambers numerous, quadratic, without sutures. Secondary deposits strong: crusts, pseudochomata, chomata. Aperture simple, terminal, basal.

Composition: Quasiendothyridae Rozovskaya, 1961 (*Baelenia* Conil and Lys, *Eoquasiendothyra* Durkina, *Eoendothyra* Miklukho-Maklay, *Quasiendothyra* Rauzer-Chernousova, *Klubovella* Lebedeva).

Occurrence: Late early Famennian (DFZ3 = uppermost *marginifera* Zone) to latest Famennian (DFZ8 = *praesulcata* Zone). Rarely mentioned in the lowermost *sulcata* Zone (= earliest Tournaisian); only Tethyan (but from Belgium to NW Australia) and Uralian (including some terranes of Canadian Arctic). Two other allochthonous localities are mentioned in Corsica (Krylatov and Mamet, 1966) and central Iran (Bagheri and Stampfli, 2008).

Suborder ENDOTHYRINA Bogush, 1985

Description: Endothyroid or planispiral coiling at least in the initial stage, rarely becoming uniseriate or biseriate. Supplementary deposits well developed (Endothyroidea, Loeblichioidea) or almost absent (Bradyinoidea). Wall dark microgranular to granular with calcareous agglutinates; tectum and inner layer in some genera (*Semiendothyra* Reitlinger, *Omphalotis*). Aperture simple basal, rarely central, rarely cribrate.

Composition: Endothyroidea Glaessner, 1945; Bradyinoidea Rauzer-Chernousova et al., 1996; Loeblichioidea Vachard et al., 2010 emend. Hance et al., 2011.

Occurrence: Late Tournaisian–late Permian; genera either cosmopolitan or endemic.

Superfamily ENDOTHYROIDEA Glaessner, 1945

Description: Endothyroid coiling at least in the initial stage, rarely becoming uniseriate or biseriate. Supplementary deposits well developed. Wall dark microgranular or with tectum and a differentiated inner layer (*Semiendothyra*, *Omphalotis*; Fig. 4(8–10)). Aperture simple basal, rarely central, rarely cribrate.

Composition: Endothyridae Rhumbler, 1895.

Remark: The superfamily name was introduced by Glaessner (1945) and not by Loeblich and Tappan (1961).

Occurrence: Tournaisian—late Permian; cosmopolitan.

Superfamily BRADYINOIDEA Rauzer-Chernousova et al., 1996

Description: See Hance et al. (2011).

Composition: Bradyinidae Reitlinger, 1958; Endothyranopsidae Rauzer-Chernousova et al., 1996; Eoendothyranopsidae Hance et al., 2011; Janischewskinidae Reitlinger in Rauzer-Chernousova et al., 1996.

Remarks: The evolution of the genus *Bradyina* Möller (Fig. 4(7)) is slow during the Carbonifeous and the Permian, but some species (including those of *Bradyinelloides* Mamet and Pinard) permit local zonations. The genera *Endothyranopsis* Cummings, *Janischewskina* Mikhailov, and *Postendothyra* Lin are good biostratigraphical markers.

Occurrence: Visean (middle part of MFZ14)—late Permian; first endemic, then, cosmopolitan, and finally, endemic.

Superfamily LOEBLICHIOIDEA Vachard et al., 2010 emend. Hance et al., 2011

Description: See Hance et al. (2011).

Composition: Dainellidae Hance et al., 2011; Loeblichiidae Cummings, 1955a.

Occurrence: Late Tournaisian—Visean; endemic in Tethys and Ural ocean shelves (including Northern Alaska terranes).

Suborder PALAEOTEXTULARIINA Hohenegger and Piller, 1975

Description: See Hance et al. (2011).

Composition: Palaeotextularioidea Habeeb, 1979; Endoteboidea Vachard et al., 2012; Tetrataxoidea Haynes, 1981; Globivalvulinoidea Hance et al., 2011.

Occurrence: Early Tournaisian—latest Permian; cosmopolitan.

Superfamily PALAEOTEXTULARIOIDEA Habeeb, 1979

Description: Entirely biseriate Palaeotextulariina.

Composition: Palaeotextulariidae Wedekind, 1937; Koskinotexulariidae Hance et al., 2011.

Remarks: The Palaeotextulariidae have typically a bilayered wall with a dark microgranular or granular outer layer with calcareous agglutinate, and an inner pseudofibrous layer. However, some siliceous agglutinates can be sporadically present (Fig. 4(6); see later for the Textulariida).

Occurrence: Middle Visean—late Changhsingian; cosmopolitan.

Superfamily ENDOTEBOIDEA Vachard et al., 2012

Description: Planispirally to endothyroidally coiled Palaeotextulariina with a biseriate terminal part or only endothyroidally coiled.

Composition: Endotebidae Vachard et al., 1994; Spireitlinidae Vachard et al., 2013.

Occurrence: The Spireitlinidae appear in the latest Visean–Serpukhovian of southern France (Vachard et al., 2016) and Tian Shan (Kulagina et al., 1992); they are relatively common and cosmopolitan from the Bashkirian to Wordian (Vachard et al., 2012; Fig. 5(5)), and then, rare in the Capitanian (Nestell and Nestell, 2006; Nestell et al., 2006) and the Wuchiapingian (Ueno et al., 2010). The Endotebidae first occur in the late Cisuralian (Vachard et al., 2002). They are abundant in the Capitanian, rare in the late Permian and earliest Triassic, and diversify again in the middle Triassic (Vachard et al., 1994).

Subfamily TETRATAXOIDEA Haynes, 1981

Diagnosis: Test conical, trochospiral, with a planar to slightly depressed face (Fig. 4(11)). Umbilical face depressed to planar or forming more or less developed umbilical pivots (*Globotetrataxis* Brazhnikova, some *Tetrataxis* Ehrenberg). Mode of life attached to vagile (Vachard et al., 2010), epiphytic on algae or corals. Four or five chambers in each whorl. Undivided chambers, or with secondary partitions (*Valvulinella* Schubert, *Abadehella* Okimura and Ishii). Aperture in the umbilicus, occasionally protected by a micro-aquarium (Vachard and Krainer, 2001a; Vachard et al., 2010; Schlagintweit, 2012). Wall thin dark microgranular, thick with the hyaline microgranular always as the outer layer and only located in the umbilicus, and not on the flanks.

Composition: Tetrataxidae Pokorny, 1958 (see Hance et al., 2011).

Occurrence: Late Tournaisian (MFZ6)–Changhsingian; generally cosmopolitan. Triassic forms need further studies; Jurassic forms as *Moehlerina* Bucur et al. appear closely related to Tetrataxidae (Schlagwinteit, 2012).

Superfamily GLOBIVALVULINOIDEA Hance et al., 2011

Description: Test free, biserially coiled to biseriate. Few whorls, few chambers increasing rapidly in haight. An oral valvula near the aperture. Supplementary deposits absent or very rare. Supplementary chamberlets occasionally present. Wall unilayerd dark microgranular or granular with calcareous agglutinate, or bilayered with an inner pseudofibrous layer. Aperture simple basal.

Composition: Globivalvulinidae Hance et al., 2011; ?Biseriamminidae Chernysheva, 1941 emend. Hance et al., 2011; ?Koktjubinidae Marfenkova, 1991 (see Hance et al., 2011; Ebrahim Nedjad et al., 2015; Vachard et al., 2014; Vachard, 2016).

Occurrence: Latest Tournaisian (MFZ8)—latest Permian (Changhsingian); rarely cosmopolitan, common in the Tethyan and Uralian oceans. Triassic forms have been mentioned in South China (Song et al., 2007).

Order FUSULINIDA Fursenko, 1958

Description: See Vachard et al. (2013; Figs. 4(12 and 13), 6(1—4), 7(1—11), 8(1—4), and 9(1—4)).

Composition: Staffellina (aragonitic wall); Fusulinina (calcitic wall).

Occurrence: Latest Tournaisian—latest Permian; generally endemic with various migrations.

Suborder STAFFELLINA Zhang et al., 1981 emend. herein

Emended diagnosis: Tests planispiral, involute, lenticular or spherical, very rarely subfusiform or uncoiled. The septa remain planar during all the history of the suborder (in contrast to the other suborder, Fusulinina). Chomata are often present but poorly developed. Some structures confused with parachomata are occasionally described. Wall calcareous secreted, probably initially dark aragonitic (probably fibrous and finely perforated; see Kochansky-Devidé, 1966), then neomicrosparitized from aragonite (Fig. 6(1 and 2)). Aperture simple terminal basal.

Composition: Staffelloidea Solovieva, 1978 emend. Vachard et al., 2013.

Occurrence: As for the order.

Superfamily STAFFELLOIDEA Solovieva, 1978 emend. Vachard et al., 2013

Description: As for the suborder.

Composition: Pseudoendothyridae Mamet in Mamet et al., 1970; Staffellidae Miklukho-Maklay, 1949; and Nankinellidae Miklukho-Maklay, 1963. Short fusiform genera such as *Leella* Dunbar and Skinner and *Jinzhangia* Ueno, eventually constitute another family.

Occurrence: The oldest ancestor *Eoparastaffellina* Vdovenko is known in latest Tournaisian. The diversification remains weak during the Mississippian with the lineage *Eoparastaffella* Vdovenko-*Pseudoendothyra* Mikhailov-*Reitlingerina* Rauzer-Chernousova. Almost concomitantly,

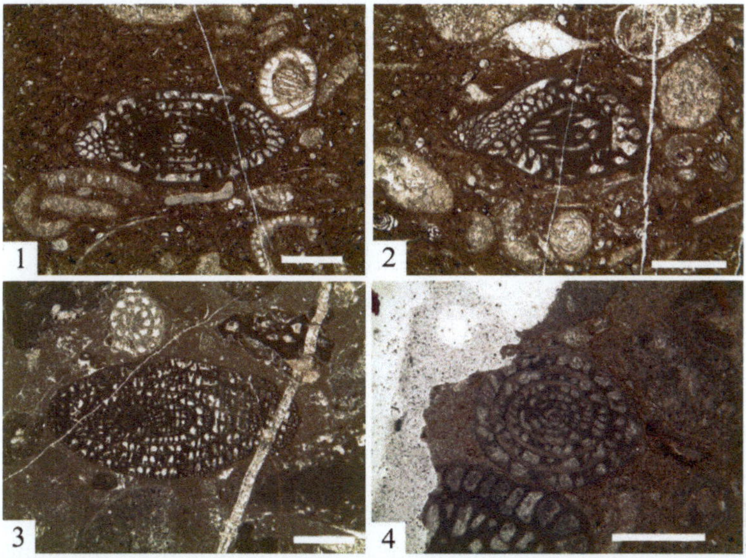

Figure 6 Staffellins and fusulinins (Cisuralian–Guadalupian) of central Afghanistan. Scale bar = 0.500 mm. **1.** *Skinnerella* sp. 1 (center), *Staffella* sp., and aragonitic algae showing the same recrystallization. **2.** *Skinnerella* sp. 2, *Staffella* sp., *Pseudovermiporella* sp., and *Agathammina*. **3.** *Staffella* sp. (top) compared with a neoschwagerinoid (center). **4.** *Cancellina* sp. with fine keriotheca (top) compared with a parafusulinid keriotheca (bottom).

Eoparastaffella gives rise to *Eostaffella* Rauzer-Chernousova, and therefore to the superfamily Ozawainelloidea of the order Fusulinida (Fig. 4(12 and 13)). The distribution of the typical staffellids is from the Bashkirian to the Changhsingian with cosmopolitan or endemic genera (Rozovskaya, 1975; Rauzer-Chernousova et al., 1996; Leven and Gorgij, 2011).

Suborder FUSULININA Wedekind, 1937 nomen correct. Loeblich and Tappan, 1961.

Description: The suborder is subdivided according to its type of wall: dark microgranular tectum, occasionally weakly differentiated (Ozawainelloidea Solovieva, 1978); bilayered with tectum and protheca (Schubertelloidea Vachard in Vachard et al., 1993); multilayered with tectum, diaphanotheca and tectoria (Fusulinoidea Ciry, 1952); coarsely keriothecal (Schwagerinoidea Solovieva, 1978); and finely keriothecal (Neoschwagerinoidea Solovieva, 1978). In this suborder, some other evolutive criteria exist; for example, from the late Mississippian to the middle Permian, the tests are successsively planispiral lenticular, spherical, inflated fusiform, elongate fusiform, or cigar-shaped; they are very rarely uncoiled (two genera of Schwagerinoidea and two genera of Schubertelloidea). The Ozawainelloidea and Schubertelloidea are medium sized; the Fusulinoidea are large; the Schwagerinoidea and Neoschwagerinoidea are large to giant. The gigantic taxa adopt two types of mechanical reinforcements, either very strong septal folding (advanced Fusulinoidea and Schwagerinoidea), or a developed endoskeleton with septula and parachomata, whereas the septa remain planar (Neoschwagerinoidea). The secondary deposits made up of pseudochomata, chomata, axial fillings, parachomata, and septula of first and second orders. Aperture terminal, simple, basal, rarely central, never cribrate; eventually reduced to secondary, very small openings: septal pores, foramina *sensu* Ciry (1952), or extremities of cuniculi (Fig. 9(1 and 2)).

Composition: Ozawainelloidea, Schubertelloidea, Fusulinoidea, Schwagerinoidea Solovieva, 1978, and Neoschwagerinoidea (= Verbeekinoidea Miklukho-Maklay et al., 1958).

Remarks: The two latter superfamilies might be considered as independent orders in the future, Schwagerinida Solovieva in Rauzer-Chernousova et al., 1996 and Neoschwagerinida Minato and Honjo, 1966. The first superfamily, the Schwagerinoidea, is monophyletic, as well as the Ozawainelloidea and the Fusulinoidea, but its lineages are so numerous that its status of superfamily seems to be insufficient. In contrast, the Verbeekinoidea are relatively homogeneous, but do not derive from the former lineage but from the Schubertelloidea, and a diphyletism exists. In contrast, the Verbeekinoidea are also considered deriving from the Staffelloidea (Rozovskaya, 1975; Rauzer-Chernousova et al., 1996).

Two groups conventionally classified with the Ozawainelloidea (Miklukho-Maklay et al., 1958, 1959; Rozovskaya, 1975) are puzzling because they flourish in the Guadalupian–Lopingian, whereas the Ozawainelloidea probably disappeared, with *Eostaffella* and/or *Pseudoacutella* Vachard et al., during the Sakmarian–Artinskian (Cisuralian) (Alipour et al., 2013; Yarahmazdari and Vachard, 2014). These Capitanian–Lopingian specialized forms, the reichelinins and kahlerinins, were microstructurally linked with the Schubertelloidea by Vachard et al. (1993), but no additional data have confirmed this hypothesis, especially for the kahlerinins, which are perhaps endothyroid or bradyinoid forms.

Occurrence: The distribution is Visean with *Eostaffella*; then, during the Pennsylvanian–Permian, the biodiversification is important with cosmopolitan or endemic genera (Rozovskaya, 1975; Rauzer-Chernousova et al., 1996; Leven and Gorgij, 2011; Vachard et al., 2013).

Superfamily OZAWAINELLOIDEA Solovieva, 1978

Description: Test free, planispirally coiled, nautiloid to subspherical, lenticular to rhombic to discoid, entirely involute or becoming evolute (*Millerella* Thompson), and vice versa (*Pseudonovella* Kireeva), subcarinate to carinate. Some deviations of coiling (*Plectostaffella* Reitlinger) present in rare tests. Proloculus, small, spherical. Juvenarium rarely developed. Pseudochomata developing to chomata, weak to strong (*Ozawainella* Thompson). Septa planar, perpendicular to the wall or falciform (*Millerella*). Chambers quadratic, relatively numerous. No sutures. Wall thin dark microgranular or layered with tectum and inner less dark layer (type species of *Eostaffella*; Fig. 4(13)). Regular tunnel up to a basal, simple, terminal aperture.

Composition: Ozawainellidae Miklukho-Maklay et al., 1958; Eostaffellidae Mamet in Mamet et al., 1970.

Occurrence: Lower Visean (upper MFZ9)–lower Sakmarian; complete lineages in Tethys and Urals oceans; episodic migrations to Northern America.

Superfamily SCHUBERTELLOIDEA Vachard in Vachard et al., 1993

Description: Test of medium size, short fusiform, inflated fusiform to elongated fusiform. Proloculus spherical. Juvenarium generally lenticular deviated at 90 degrees with the adult whorls. Wall unilayered in the ancestral genus *Eoschubertella* Thompson (= *Schubertina* Marshall *sensu* Davydov, 2011), or more frequently, bilayered with an outer, dark microgranular tectum and an inner, thicker, yellowish, microgranular layer, called the protheca; very finely porous wall rarely present (*Biwaella* Morikawa and Isomi). Septa faintly to moderately folded at the poles, planar in the central parts of the chambers. Tunnel obvious. Chomata small to moderate. Rare genera display cuniculi.

Composition: Schubertellidae Skinner, 1931; Boultoniidae Skinner and Wilde, 1954; Yangchieniidae Leven, 1987; Dunbarulinidae Vachard in Vachard et al., 1993; Palaeofusulinidae Leven, 1987; Reichelinidae Vachard in Vachard et al., 1993 (partim); *Biwaella* group.

Occurrence: Middle Bashkirian to late Changhsinghian; endemic or cosmopolitan genera.

Superfamily FUSULINOIDEA Ciry, 1952

Description: Globular, subquadratic, and fusiform Fusulinida with a wall including a diaphanotheca; the primitive forms (*Pseudostaffella* Thompson, *Profusulinella* Rauzer-Chernousova and Belyaev) having still a dark microgranular of their ozawainelloid ancestors. Very finely porous walls rarely present (Vachard et al., 2004b). Septa planar to partly folded and finally, entirely folded. Chomata always present, but diversely developed.

Composition: ?Pseudostaffellidae Putrya, 1956; Profusulinellidae Solovieva in Rauzer-Chernousova et al., 1996; Aljutovellidae

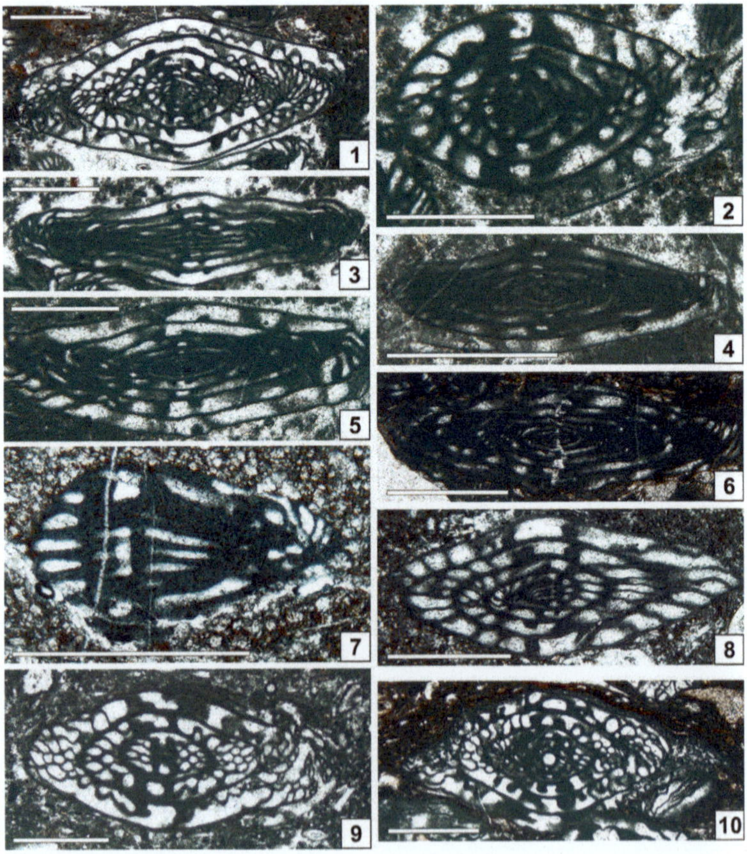

Figure 7 Late Moscovian (Desmoinesian) fusulinids (Newwell Peak, New Mexico). Scale bar = 1 mm. **1–2, 9–10.** *Beedeina* spp. **3–6.** *Wedekindellina* sp. **7–8.** *Zellerella* sp.

Solovieva in Rauzer-Chernousova et al., 1996; Fusulinellidae Staff and Wedekind, 1910; Fusulinidae Möller, 1878.

Occurrence: Latest Bashkirian. Acme during the Moscovian; rare in the late Pennsylvanian and the early Permian (up to the Artinskian); either cosmopolitan or endemic genera.

Superfamily SCHWAGERINOIDEA Solovieva, 1978

Description: Elongate fusiform to inflated fusiform fusulinids with keriothecal wall. Chomata developed or absent. Axial fillings developed or absent. Septal folding present in the entire chambers or only at the pole. Aperture median simple with tunnel or absent. Supplementary small apertures as septal pores or cuniculi (Figs. 8(1 and 4) and 9(1 and 2)).

Composition: Triticitidae Davydov in Rauzer-Chernousova et al., 1996; Schwagerinidae Dunbar and Henbest, 1930; Pseudofusulinidae Bensh, 1987; *Pseudoschwagerina* Dunbar and Skinner group; *Rugosofusulina* Rauzer-Chernousova group; *Chusenella* Hsu group; Polydiexodinidae Miklukho-Maklay, 1953 (Figs. 6(1 and 2), 8(1–4), and 9(1–4)).

Remarks: This superfamily is so diversified that it is possibly an independent order Schwagerinida (Solovieva in Epshteyn et al., 1985; Rauzer-Chernousova et al., 1996), but it remains a monophyletic group (Vachard et al., 2013). Even if the keriothecal wall appeared sporadically in several superfamilies of Fusulinida (Vachard et al., 2004a, 2013), all the Schwagerinoidea were derived from *Protriticites* Putrya. A biphyletism with a parallel lineage including

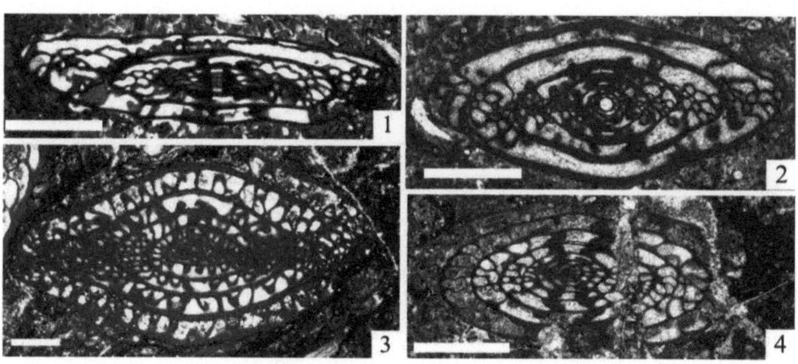

Figure 8 Triticitidae from New Mexico (8.1–8.3) and Karawanken (Austria; 8.4). Scale bars = 0.500 mm. **1.** Missourian *Triticites* sp. 1. **2.** Virgilian *Triticites* sp. 2. **3.** Newellian *Triticites* sp. 3 transitional to *Thompsonites* sp. **4.** Artinskian *Darvasites* sp.

Dutkevichites Davydov, *Biwaella*, and *Sphaeroschwagerina* Miklukho-Maklay (Davydov, 1984, 2011) is very unlikely. *Sphaeroschwagerina* evolved more probably from *Pseudoschwagerina* (Vachard et al., 2013) in a center of radiation located in the Canadian Arctic Archipelago (Rui Lin and Nassichuk, 1994). A Japanese radiation is also possible from *Carbonoschwagerina* Ozawa et al. (Vachard, unpublished data). A monophyly is also consistent with the phylogeny passing from *Occidentoschwagerina* Miklukho-Maklay to *Sphaeroschwagerina* (Vilesov, 1998).

The parafusulinids and polydiexodinids are the oldest foraminifers, which exhibit undisputable micro- and macrospheric generations (Vachard, 1980; Vachard and Gaillot, 2006; Fig. 9(1 and 2)).

Occurrence: Latest Moscovian–latest Capitanian; either cosmopolitan or endemic genera.

Superfamily NEOSCHWAGERINOIDEA Solovieva, 1978

Description: Test large-inflated fusiform to spherical (*Brevaxina* Schenck and Thompson, *Verbeekina* Staff), planispirally coiled and involute. Supplementary deposits as parachomata often present. Numerous genera display chambers divided into chamberlets by transverse septula of first order (which connect with the parachomata in neoschwagerinids: *Neoschwagerina* Yabe, *Yabeina* Deprat, and *Lepidolina* Lee) and shorter transverse septula of second order, as well as axial septula (Figs. 6(3 and 4) and 9(4)). Wall microgranular and finely keriothecal or re-becoming apparently dark microgranular (sumatrinins). Poorly developed apertures as small foramina between the parachomata and the chamberlets.

Composition: Misellinidae Leven, 1982; Verbeekinidae Miklukho-Maklay, 1957; Neoschwagerinidae Dunbar and Condra, 1927.

Occurrence: Middle Kungurian–late Capitanian; Tethyan and North American terranes; very rare in the North American craton.

Class MILIOLATA Saidova, 1981

Description: See Loeblich and Tappan (1987, p. 309).

Composition: All Paleozoic taxa belong to only one order, Cornuspirida.

Occurrence: Serpukhovian–Holocene; either cosmopolitan or endemic.

Order CORNUSPIRIDA Mikhalevich, 1980

Description: All late Paleozoic miliolates exhibit an undivided, tubular chamber; hence they constitute a unique order

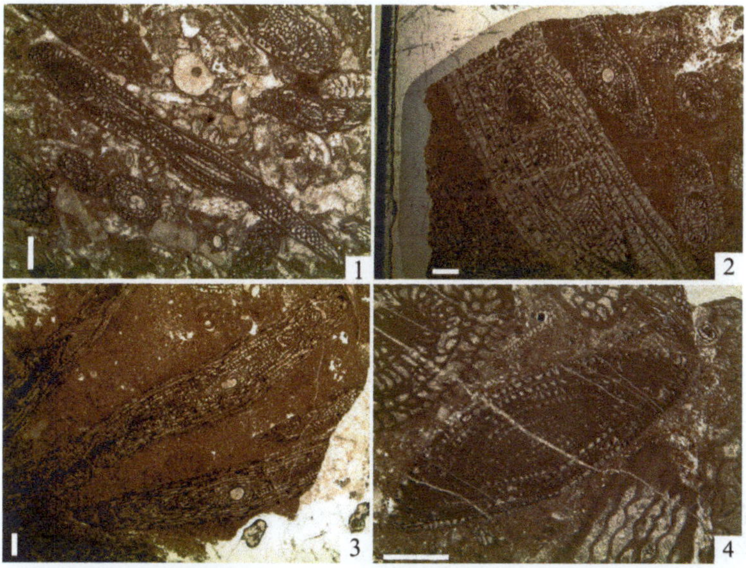

Figure 9 Giant schwagerinoids and neoschwagerinoids (Capitanian, northern Afghanistan). Scale bar = 0.500 mm. **1.** *Parafusulina* sp. (two generations), *Dunbarula* sp., and *Climacammina* with siliceous agglutinate. **2.** *Eopolydiexodina* sp. 1 (two generations). **3.** Macrospheric generation of *Eopolydiexodina* sp. 2. **4.** *Sumatrina* sp. associated with *Parafusulina* sp., *Chusenella* sp., and *Codonofusiella* sp.

Cornuspirida. Two suborders are distinguished herein, Nubeculariina for the attached families and Cornuspirina for the free families. *Composition*: Nubeculariina Saidova, 1981; Cornuspirina Jirovec, 1953.

Remarks: According to Pronina (1994), a porcelaneous wall is "black in thin section." It is difficult in this case to see the differences with Fusulinata tests as well as many tests of Astrorhizata and Textulariata where the agglutinate is poorly developed. Indeed, many of these tests appear dark in thin sections and even microgranular when they are studied with a SEM (Vachard et al., 2015).

Gaillot and Vachard (2007) proposed another definition of porcelaneous wall, and a new classification of the Permian Miliolata based on the increasing size of the taxa and the differentiation or not of the wall. Such a classification leads to separate genera visibly phylogenetically associated, such as *Agathammina* Neumayr and *Septagathammina* Lin or *Septigordius* Gaillot and Vachard, and *Baisalina* Reitlinger. Consequently, a more traditional classification is used herein, which re-includes the families Agathamminidae and Baisalinidae. Paleozoic miliolates derived from the Lituotubelloidea *Scalebrina* Conil and Longerstaey, at the end of the Visean (Gaillot and Vachard, 2007; Vachard, 2016). In contrast, the class of the involutinids can derive from a planispiral evolute ancestor, either *Pseudoammodiscus* or *Postcladella* (see discussions in Gargouri and Vachard, 1988; Vachard et al., 1993; Groves and Altıner, 2005; Gaillot and Vachard, 2007; Krainer and Vachard, 2011).

Occurrence: Earliest Serpukhovian–Holocene. Paleozoic Nubeculariina are common in the Serpukhovian–Bashkirian. Primitive Cornuspirina, *Cornuspira* and *Hemigordiellina*, are only known during and/or after the Serpukhovian. More complex coilings appear with *Brunsiella* Reitlinger and then *Hemigordius* Schubert in the Moscovian; the diversity is already important in the early Permian, but increases gradually during the middle Permian where appear giant forms such as *Hemigordiopsis* Reichel, *Shanita* Brönnimann, Whittaker, and Zaninetti, *Neodiscus* Miklukho-Maklay, *Neohemigordius* Wang and Sun, and *Kamurana* Altıner and Zaninetti. The Guadalupian–Lopingian is an episode of maximal diversity for the group.

Suborder NUBECULARIINA Saidova, 1981

Description: Attached Cornuspirida.

Composition: Calcivertelloidea n. superfam. The family Nubeculariidae Jones in Griffith and Henfrey, 1875 is only known from the Triassic.

Occurrence: As for the order.

Superfamily CALCIVERTELLOIDEA n. superfam.

Diagnosis: Attached, undivided tubes, with initial coiling and terminal uncoiling more or less zigzagging. The foraminiferal shape can be very much transformed in the family Tubiphytidae, and in the groups of *Pseudovermiporella* Elliott and *Ellesmerella* Mamet and Roux, often interpreted as algal thalli. Wall porcelaneous, smooth or, rarely, with pits (Vachard and Krainer, 2001a; Vachard, 2016).

Composition: Calcivertellidae Reitlinger in Vdovenko et al., 1993 emend. Gaillot and Vachard, 2007; Tubiphytidae Vachard et al., 2012; *Pseudovermiporella* group; *Ellesmerella* group.

Remarks: Even if the Calcivertellidae are typical foraminifers, some authors speculate that their descendents, tubiphytids, pseudovermiporellids, and ellesmerellids transform entirely their structure and come to look like green algal or cyanobacterial morphologies (Vachard, 2016 with references therein). However, no modern foraminifers exhibit similar convergences; hence these groups have been commonly interpreted as microproblematica, algae, or cyanobacteria. Nevertheless, there are not unique atypical forms among the porcelaneous foraminifers, since strange taxa are known in Triassic times (Senowbari-Daryan and Zaninetti, 1986).

Occurrence: Early Serpukhovian—latest Permian, cosmopolitan.

Suborder CORNUSPIRINA Jirovec, 1953

Synonyms: Hemigordiopsina Pronina, 1994; Hemigordiopsida Mikhalevich, 1988 (partim); Pseudoammodiscina Pronina, 1994 (partim).

Remarks: The Cornuspirina are classically divided into two superfamilies: Cornuspiroidea (discussed in the following) and Neodiscoidea (with Neodiscidae: homeomorphs of the previous ones, but repeating their evolutive trends with a wall with buttresses; and Hemigordiopsidae: the most complex, with frequent flosculinisations or other modifications of the wall).

Occurrence: Serpukhovian—Holocene; cosmopolitan or endemic.

Superfamily CORNUSPIROIDEA Bogdanovich in Subbotina et al., 1981

Synonyms: Hemigordiopsina Mikhalevich, 1988; Hemigordiopsidea Nikitina, 1969 (partim); Shanitoidea Loeblich and Tappan, 1986; Pseudoammodiscacea Vdovenko et al., 1993 (partim).

Description: Tests free with porcelaneous wall composed of a single tube undivided coiled in one or various planes.

Composition: Cornuspiridae Schultze, 1854; Hemigordiidae Reitlinger in Vdovenko et al., 1993; Neodiscidae Gaillot and Vachard, 2007; Hemigordiopsidae Nikitina, 1969.

Occurrence: Early Serpukhovian—Holocene, cosmopolitan.

Class NODOSARIATA Mikhalevich, 1993

Subclass NODOSARIANA Mikhalevich, 1993

Order NODOSARIIDA Calkins, 1926

Description: Foraminifers with a monolamellar wall of radiate calcite, with crystal *c*-axis perpendicular to the surface and without secondary lamination (Loeblich and Tappan, 1987).

Remarks: The Paleozoic Nodosariida are divided herein into two superfamilies: Nodosarioidea Norvang, 1957 and Geinitzinoidea Loeblich and Tappan, 1984 (= Robuloidoidea *sensu* Gaillot and

Vachard, 2007). The two superfamilies are distinguished by the presence of a primitive round or oval or slitlike aperture (Geinitzinoidea) or a typical stellate aperture (Nodosarioidea). Numerous homeomorphs differing only by these types of an aperture exist in both superfamilies and even within the same genus. The exact status of the ancient family Nodosinellidae, and its possible priority, depends on the revision of its generotype *Nodosinella* Cummings, the descriptions of which by Cummings (1955b), Foster et al. (1985), and Pinard and Mamet (1998) remain incomplete.

Occurrence: Late early Moscovian–Holocene, either cosmopolitan or endemic genera.

Superfamily GEINITZINOIDEA Loeblich and Tappan, 1984

Synonym: Robuloidoidea *sensu* Gaillot and Vachard, 2007.

Description: Test subcylindrical, uniseriate, rarely coiled, with a primitive round, oval, or slitlike terminal aperture. Wall fibrous unilayered, sometimes bilayered (microgranular and hyaline fibrous), rarely unilayered dark fibrous (Frondinidae).

Composition: Syzraniidae Vachard in Vachard and Montenat, 1981; Protonodosariidae Mamet and Pinard, 1992; Geinitzinidae Bozorgnia, 1973; Robuloididae Loeblich and Tappan, 1984; Partisaniidae Loeblich and Tappan, 1984; Frondinidae Gaillot and Vachard, 2007; Colaniellidae Fursenko in Rauzer-Chernousova and Fursenko, 1959.

Occurrence: The oldest fibrous forms (Syzraniidae) appear in the late early Moscovian; the complete septation appears in the late Pennsylvanian; the acme of the group is Permian but it subsists in the Triassic and even in the Jurassic.

Superfamily NODOSARIOIDEA Norvang, 1957

Description: Test uniseriate, thick walled, with stellate aperture. Frontal and sagittal axial sections are very different in aspect. Sagittal axial sections show enveloping curved septa, whereas frontal axial sections show the thickened walls.

Composition: Nodosariidae Ehrenberg, 1839; Pachyphloiidae Loeblich and Tappan, 1984; Ichthyolariidae Loeblich and Tappan, 1986.

Occurrence: Artinskian–Holocene; cosmopolitan.

Class TEXTULARIATA Mikhalevich, 1980

Description: See Mikhalevich (1980) and the superfamily Textulariacea in Loeblich and Tappan (1987).

Remarks: The Paleozoic taxa described with an agglutinated test, and called as *Ammodiscus*, *Glomospira* Rhezak, *Tolypammina*, *Hyperammina*, *Reophax*, and *Ammobaculites* Cushman (Ukharskaya, 1970; Sosipatrova, 1970; Pronina, 1998) can be also interpreted as fusulinates and/or miliolates (Vachard et al., 2010; Hance et al., 2011; Vachard, 2016). The same taxa have subsequently been named as *Pseudoammodiscus*, *Pseudoglomospira* Bykova, or *Earlandia*; admitting that they possessed not an agglutinated but calcitic, dark-migrogranular wall. Recently, the pendulum swung back and the names *Ammodiscus* and *Hyperammina* were again proposed (Nestell et al., 2015).

As indicated by Vachard (2016), three hypotheses can explain these striking wall similarities: (1) Textulariida and Fusulinana have experienced a convergent evolution; (2) the palaeotextulariids and endotebids belong to the Textulariida and have been erroneously included in the Fusulinana; and (3) all Fusulinana have agglutinated microparticles, but the nature and size of the grains can vary to microgranular to obviously agglutinated. In this latter case, the Fusulinana would have survived the Permian—Triassic and the Triassic—Jurassic major extinction events and still have representatives among some lineages of Textulariata. In consequence, there would be no major extinct groups of foraminifers, and molecular phylogeny, well combined with fossil data, should permit to retrace the foraminiferal evolution to an advanced level.

Due to (1) the monophyletism of the Rotaliata (Flakowski et al., 2005), (2) the possible connection between Fusulinata and Rotaliata via the Nodosariata (Gaillot and Vachard, 2007), and (3) the connection between Fusulinata and Textulariata admitted herein, it seems difficult to include the Textulariida within the Rotaliata as proposed by Mikhalevich (2003).

Occurrence: Rare in the Permian. Triassic—Holocene.

Order TEXTULARIIDA Lankester, 1885

Description: See Textulariacea in Loeblich and Tappan (1987).

Remarks: Vachard et al. (2010) believed that the siliceous agglutinate should be reserved to the Textulariata from Triassic onwards, whereas the rare agglutinates of the Fusulinata were exclusively calcareous. Nevertheless, there are several palaeotextularioids, endoteboids, and endothyroids that unquestionably possess a siliceous agglutinated wall; especially, *Palaeobigenerina* Galloway in the Carnic Alps (Vachard and Krainer, 2001b; Fig. 4(6)) and northern Afghanistan (Vachard, 2016), *Deckerella* sp. 1 in Texas (Nestell et al., 2006), many Kazanian *Textularia* Defrance of Russia (Ukharskaya, 1970; Pronina, 1998), and *Labiodagmarita* Gaillot and Vachard, which is probably a palaeotextulariid (Altıner and Özkan-Altıner, 2010).

The principal difference between Palaeotextulariina and Textulariida is the absence of pseudofibrous inner layer in Textulariida.

Occcurrence: As for the class.

Order VERNEUILINIDA Mikhalevich and Kaminski in Mikhalevich, 2003

Description: See Mikhalevich (2003) and the superfamily Verneuilinacea in Loeblich and Tappan (1987).

Remarks: The first Verneuilinida appear in late Kungurian—early Roadian of Tajikistan (Angiolini et al., 2015), Capitanian of Oman (Vachard et al., 2002), Kazanian of Russia (Ukharskaya, 1970; Pronina, 1998), Permian of Australia (Crespin, 1958), middle Permian of Croatia (Flügel et al., 1984), and middle Permian of Sicily (Flügel et al., 1991).

Occcurrence: As for the class.

Order LITUOLIDA Lankester, 1885

Description: See the superfamily Lituolacea in Loeblich and Tappan (1987).

Remark: *Vachardella* Nestell and Nestell might be transitional between endoteboids and lituolids or be the first representative of this order.

Occcurrence: Questionable in the Capitanian (late middle Permian). Triassic—Holocene.

4. DISCUSSION

The phylum Foraminifera was known for five evolutionary periods: (1) lower Paleozoic—lower Devonian: "agglutinated" foraminifers are rare, and secreted foraminifers nearly absent; all the typical forms are uni- or bilocular; (2) Givetian—Frasnian: plurilocular, secreted foraminifers became abundant in shallow-water carbonates (concomitantly with the development of reefs); many modern architectures of a test are represented: planispiral involute, uniseriate and biseriate; (3) latest Famennian—latest Permian (this paper), after a recovery period in the Famennian, as early as latest Famennian, microgranular forms are very numerous in marine shallow waters; from latest Famennian to early Bashkirian, the tests remain small and endothyroid in shape (nautiloid with permanent deviation of the axis of coiling); from the late Bashkirian to middle/late Permian, the coiling becomes planispiral, and the shape fusiform; the increasing in size continues and is progressive up to the gigantism of the middle Permian and the first group of larger foraminifers, with the giant schwagerinoids (Colpaert et al., 2015); (4) the Triassic is a peculiar period dominated by involutinids and miliolates; (5) from the Jurassic to Holocene, the trends are repetitive, with a diversification of the Textulariata, Miliolata, and Rotaliata Mikhalevich, 1980, with the true agglutinated, porcelaneous, and hyaline fibrous foraminiferal tests.

Because of such a history, the principal evolutionary character seems to be the wall microstructure; even if this idea (omnipresent in the work of Loeblich and Tappan, 1964, 1987, 1992), it is currently challenged in both ways: (1) in classical morphological studies (Mikhalevich, 2003; Nestell and Nestell, 2006) and (2) in molecular studies (Pawlowski et al., 2013). Some papers of Rigaud (Rigaud and Martini, 2016; with references therein) try to find a synthesis.

The paleoecology of the oldest foraminifers is poorly known (Vachard et al., 2017). It appears that the Givetian revolution is largely correlative of the development of reefs and of terrestrial plants, as sources of organic carbon. The oldest large foraminifers are middle and late Pennsylvanian in age. No cosmogenic cause can be correlated with that. Similarly, and despite many attempts (Groves and Wang, 2009; Davydov, 2013), no correlation has been proved between the Late Paleozoic Ice Age, and the evolutionary trends of the foraminifers, which were, in contrast, very responsive to the mass extinction crises (Vachard et al., 2010).

The principal question remains that of the agglutinated tests. The polyphyletism of the Textulariata has been known for a long time (see Rigaud et al., 2015); but it seems that in some Paleozoic foraminiferal genera, the wall can be dark microgranular as well as agglutinated. The dark microgranular forms probably inhabit the shallow seas, whereas the agglutinated forms inhabit deeper water; examples can be encountered in *Pseudoammodiscus*, *Earlandia*, *Tikhinella*, *Eotournayella* (= *Pseudocornuspira*), *Eoseptatournayella*, *Septabrunsiina*, and various parathuramminids. Generally, this fact is difficult to establish when the tests are unilocular (parathuramminids) or bilocular (earlandiids), but it becomes undisputable for some plurilocular genera, especially when no homeomorphs exist in the entire history of the foraminifers; for example, the Famennian *Rectoseptatournayella*, the morphology of which (planispirally coiled and evolute, becoming uncoiled uniseriate) is so particular, that it cannot be confused with the agglutinated *Ammobaculites* [see in particular, *Rectoseptatournayella chappelensis* (Gutschick et al.)].

It is admitted that the Allogromiata evolved during the late Proterozoic. All the agglutinated foraminifers from the latest Proterozoic–early Cambrian are generally disputable (Scott et al., 2003; Vachard et al., 2017). The oldest unquestionable foraminiferal assemblage remains that discovered 25 years ago by Culver (1991). The oldest typical foraminifer *Rectoammodiscus diai* is already a paragon of this group, with a proloculus followed by a planispirally coiled chamber, a terminal aperture, and an agglutinated wall (Fig. 10). As early as the Ordovician–Silurian, a dozen of foraminiferal genera are known, and the calcareous secreted tests have appeared (Vachard et al., 2017), even if some works are disputable (Sabirov and Gushchin, 2006; Vachard et al., 2010, 2017; Nestell et al., 2011). These first secreted tests increase in number from the late Silurian to early middle Devonian, Ludlovian–Eifelian (Fig. 10). The oldest planispiral and plurilocular foraminifer *Nanicella* is present as early as the Pragian (Vachard and Massa, 1989). It occurs during the Givetian–Frasnian foraminiferal phase and probably gives rise to all the plurilocular tests of this period, the Semitextularioidea. The genus "*Reophax*" mentioned in the Paleozoic literature (Gutschick, 1986) is to be reinterpreted; in particular, some Devonian forms can be microproblematica similar to *Baculella* Conil and Dreesen or *Saccamminopsis* Sollas, whereas Tournaisian forms are mstinioid foraminifers. Consequently, *Reophax* probably did not exist during the Paleozoic, and only appeared during the Triassic or Jurassic.

- Famennian. In the Famennian, five microgranular-walled, bilocular, undivided to pseudoseptated groups evolved: (1) *Pseudocornuspira/Eotournayella*, ancestors of the suborder Tournayellina; (2) *Pseudoglomospira* gives rise to *Glomospiranella* Lipina and all the Lituotubellina, Endothyrina, and Palaeotextulariina, and then to all the Fusulinida; (3) *Pseudoglomospira* and *Glomospiranella* generate also the Quasiendothyroidea, the evolution of which is meteoric in the late Famennian (Fig. 10). Finally, the Earlandiida survived but did not present important evolutionary events prior to the Nodosariata derivation during the Moscovian.

  The history from Famennian to Permian of the possibly agglutinated test is poorly known. Theoretically, the groups of agglutinated

CHRONOSTRATIGRAPHY	BIOSTRATIGRAPHY	FORAMINIFERAL MARKERS
FAMENNIAN	DFZ8	*Tournayellina pseudobeata*, *Laxoendothyra* spp., *Septabrunsiina kingirica*, last *Uralinella* spp. last *Eoendothyra communis*
	DFZ7	*Quasiendothyra kobeitusana*, *Q. konensis*, *Klubovella* spp., *Laxoendothyra parakosvensis*, *Paracaligelloides* spp., *Eoendothyra communis*
	DFZ6	*Eoendothyra radiata*, *Septaglomospiranella bouckaerti*
	DFZ5	*Eoendothyra regularis*, *Avesnella* spp., *Rectoavesnella* spp., *Rectoglomospiranella* spp., first *Laxoendothra* (= *Endoglomospiranella*) spp.
	DFZ4	First *Eoendothyra communis*, first *Tournayella* spp.
	DFZ3	*Eoquasiendothyra bella*, *Septabrunsiina baeleni*, *Baelenia gosseleti*
	DFZ2	*Eoseptatournayella* spp., *Septatournayella* spp., *Septaglomospiranella* spp., *Eotournayella* spp., last *Tikhinella* sp.
FRASNIAN	DFZ1	*Eonodosaria* spp., *Eogeinitzina* spp., *Nanicella* spp. *Multiseptida* spp., *Nanicella* spp., first *Eotournayella* (= *Pseudocornuspira*) spp. Nanicellidae and Semitextulariidae
GIVETIAN		*Auroria* spp., *Apertauroria* spp., *Uslonia* spp., rare *Nanicella* spp., *Parathurammina* spp., *Bykovaella* spp., *Uralinella* spp., *Bisphaera* spp. (first acme of the moravamminale algae)
EIFELIAN		*Ivdelina* spp., *Cribrosphaeroides* spp., first *Parathurammina* spp., first *Bykovaella* spp., first Semitextulariidae
LOWER DEVONIAN		*Ivanovella* spp., *Cribrosphaeroides* spp. , *Elenella* spp., *Neoarchaesphaera* spp., *Parathuramminites* spp., first *Nanicella* spp.
UPPER SILURIAN		*Polygonella* spp., *Ivanovella* spp., first *Cribrosphaeroides* spp., first Caligellidae
LOWER SILURIAN		First *Earlandia* spp.
MIDDLE-UPPER ORDOVICIAN		*Amphitremoidea* spp., *Rauserina* spp.
LOWER ORDOVICIAN		*Vicinesphaera* sp. in North America
CAMBRIAN		*Rectoammodiscus diai*, "saccamminids", *Vicinephaera* sp. in the Urals (Russia)

Figure 10 Foraminiferal biostratigraphy during the early and middle Paleozoic.

foraminifers did not evolve up from the Paleozoic to modern representatives, with sometimes the same genera (Kaminski et al., 2010). It is noteworthy also that the "agglutinated foraminifera" described by Thorez and Dreesen (1986) were first named "silicified foraminifers" by Dreesen et al. (1985). In this case, due to a recent restudy, it is possible to confirm that the foraminifers are indeed secondarily silicified and not agglutinated (Vachard et al., unpublished data).

- Tournaisian. Many new trends appear, but Lituotubellina remain streptospiral, or streptospiral and becoming planispiral. The terminal uncoiling appears in several families (*Rectoseptaglomospiranella* Lipina, *Rectoseptatournayella*, *Rectoseptabrunsiina*). Such forms probably give rise to the ancestor of the Palaeotextulariina with *Palaeospiroplectammina* Lipina, in the middle Tournaisian. The first epiphytic foraminifers appear in the late Tournaisian with the Tetrataxoidea because of their conical, patelliform shape (Vachard et al., 2010). They probably generated the Globivalvulinoidea in the latest Tournaisian (with *Parabiseriella* Cózar and Somerville), which nearly do not evolve up to the middle–late Permian. Except for the Tetrataxoidea, all these foraminifers are probably infaunal (Vachard et al., 2010). After various attempts in different lineages, the endothyroid coiling is definitively realized during the late Tournaisian, coming from the middle Tournaisian chernyshinellids (Fig. 11).

- Visean. During the Visean, the first groups of Fusulinida (eostaffellids and pseudoendothyrids) appear; archaediscoids and howchiniids are biostratigraphically important, as well as still poorly known loeblichioids (endothyrids with numerous chambers and very strong deviations of coiling axis) (Fig. 11). Palaeotextulariins and mstinioids are secondary markers. The Palaeotextulariina need further studies because they can include the oldest Textulariata, and consequently, perhaps differ from the Endothyrida. Similarly, the Tetrataxoidea, which are very particular and apparently survived during the Triassic, can constitute another distinct suborder or order. Finally, the Globivalvulinoidea could be also separated as a suborder, especially because their coiled biseriate tests constitute a very rare type of architecture among the foraminifers (Vachard et al., 2010). Loeblichioids are relatively characteristic of the Visean. They are probably the true ancestors of the fusulinids (Vachard et al., 2010, 2013), unlike Rozovskaya (1975) who thought that the quasiendothyrids were these ancestors. The true ancestor genera are *Eoparastaffellina* and *Eoparastaffella* (characteristic of the latest Tournaisian and earliest Visean biozones MFZ8 and MFZ9), with the double potentially, via primitive *Eostaffella* and *Pseudoendothyra*, to give rise to both suborders Staffellina and Fusulinina. It is possible that *Eoparastaffellina* and *Eoparastaffella* that have a dark microgranular wall with clearer particles can have a double composition, with a dark calcitic wall and clear particles of aragonite; the wall will remain calcitic in the Fusulinina, whereas it will become progressively aragonitic in the Staffellina; the transitional Pseudoendothyridae show only a clear, aragonitic? luminotheca within the wall.
- Serpukhovian–Bashkirian. These stages constitute a transitional period for the foraminifers, in which many Visean genera survive, but where the Fusulinida begin their evolution (Figs. 11 and 12).
- Moscovian. The first large fusulinids appear and with them the first larger foraminifers. Except for some mstinioids, all Visean forms disappear, especially the archaediscoids (Fig. 12).
- Late Pennsylvanian–Cisuralian. That is the age of the fusulinids, with the succession Ozawainelloidea-Schubertelloidea-Fusulinoidea-Schwagerinoidea (Figs. 12 and 13). The complexification of fusulinids, especially schwagerinoids, is huge. Unfortunately, modern models for

STAGES	SUBSTAGES	BIOSTRATIGRAPHY	FORAMINIFERAL MARKERS
SERPUKHOVIAN	**Zapaltyubinian**	MFZ17	*Monotaxinoides* ex gr. *transtorius*, first *Varvariella* spp., last *Loeblichia minima*, first *Plectostaffella* spp.
	Protvian	?Upper MFZ16	*Eosigmoilina* spp., *Brenckleina* spp., first *Insolentitheca* spp., *Bradyina cribrostomata*, *Eostaffellina paraprotvae*, *Globivalvulina moderata*, last *Janischewskina* spp.
	Steshevian	?Lower MFZ16	First Nubeculariina: *Calcivertella* spp., *Ammovertella* sp., ?last *Forschia* spp., *Asteroarchaediscus* ex gr. *parvus*, *Endothyranopsis plana*, *Reitlingeropsis sphaerica*, *Eostaffella parastruvei*, *Eostaffellina paraprotvae*
	Tarusian	?	*Zellerinella tortula*, *Biseriella parva*, *Rectoendothyra* spp., *Koktjubina* spp., *Asteroarchaediscus* ex gr. *postrugosus*, *A.* ex gr. *baschkiricus*
VISEAN	**Venevian**	MFZ15	*Hemidiscopsis* spp., *Magnitella* spp., *Spinothyra* spp., *Semiendothyra* spp., first *Janischewskina* spp.
	Mikhailovian		*Asteroarchaediscus* ex gr. *rugosus*, *Climacammina* spp., transition *Cribrospira* sp./*Janischewskina* sp.
	Aleksinian	MFZ14 (A and B)	*Cribrospira* spp., *Bradyina rotula*, *Nodasperodiscus* spp., *Rugosoarchaediscus* spp., [first algae *Ungdarella uralica*]
			Howchinia bradyana, *Neoarchaediscus* sensu stricto (= *Asperodiscus*) spp., *Cribrostomum* spp., *Endothyranopsis crassa*, *Archaediscus karreri* (= *A. gigas*).
		MFZ13 (A and B)	First *Pseudoendothyra* spp.
			Vissariotaxis spp., *Archaediscus moelleri*, *Endothyranopsis compressa*, first *Rugosoarchaediscus* spp.
	Tulian (= Livian)	MFZ12	*Pojarkovella* spp., *Koskinotextularia* spp., *Paraarchaediscus* spp., *Consobrinellopsis* spp., *Nodosarchaediscus* spp., *Endothyranopsis compressa*
	Moliniacian	MFZ11 (A, B and C)	*Conilidiscus* spp., *Paraarchaediscus* spp. (= concavus stage), *Archaediscus krestovnikovi* (= concavo-angulatus stage); first *Consobrinellopsis* spp.
			Uralodiscus spp., *Conilidiscus* spp., *Propermodiscus* spp (= involutus stage)
			Uralodiscus spp., *Glomodiscus* spp., last *Eotextularia* spp.
		MFZ10	*Ammarchaediscus transitus*, first *Planoarchaedicus* (= *Brunsiarchaediscus*) spp., first *Glomodiscus* (= *Nudarchaediscus*) spp., *Viseidiscus primaevus*
		MFZ9 (A and B)	First unilayered *Eostaffella* spp.
			Eoparastaffella simplex, *Eoendothyranopsis donica robusta*
TOURNAISIAN	**Ivorian**	MFZ8	*Parabiseriella bristolensis*, *Eoparastaffellina* spp., *Brenckleites* spp., first *Mediocris* spp., *Lysella* spp.
		MFZ7	*Darjella monilis*, *Dainella* spp., *Bessiella* spp., *Elevenella* spp., *Florennella* spp., *Omphalotis* spp.
		MFZ6	First *Eotextularia diversa*, *Tetrataxis* spp., *Brunsia* spp., *Endospiroplectammina conili*, first primitive *Valvulinella* spp.
		MFZ5	*Eoforschia moelleri*, *Granuliferella* spp., *Latiendothyra* spp., *Paraendothyra nalivkini*
	Hastarian	MFZ4	*Granuliferella nalivkini*, *Spinochernella* spp., *Tuberendothyra tuberculata*
		MFZ3	*Chernyshinella glomiformis*, *Crassiseptella* spp., *Palaeospiroplectammina tschernyshinensis*
		MFZ2	*Tournayellina beata*, *Septaglomospiranella rudis*, *Inflatoendothyra* spp., *Septabrunsiina minuta*
		MFZ1	*Bisphaera* spp., *Earlandia* spp.

Figure 11 Foraminiferal biostratigraphy during the Mississippian.

such large foraminifers are lacking, and their paleobiologic knowledge remains limited. Among the smaller foraminifers, many late Pennsylvanian survivals are present up to the Artinskian. Several species of *Nodosinelloides* Mamet and Pinard and *Geinitzina* Spandel are biomarkers. *Pseudovermiporella* appears at the Sakmarian/Artinskian boundary, and becomes rapidly abundant. The Artinskian sees the disappearance of *Bradyina* and the possible appearance of true *Pachyphloia* Lange (Fig. 13).

- Kungurian (Fig. 13). The main bioevents are the disappearance of several lasiodiscids and pseudovidalinids. Various species of *Globivalvulina* are noticeable, which herald the relative gigantism of this genus in the Capitanian–Lopingian. Species of true *Nodosaria* Ehrenberg, with the oldest stellate apertures in nodosariates, appear (Fig. 13).
- Roadian. Smaller foraminifers are poorly known during this period (Fig. 13). New species of *Nodosaria* and the oldest typical species of *Septaglobivalvulina* appear in many areas of western Paleotethys, and probably all around the world. The gigantism of the Miliolata and Nodosariata could begin during this period.
- Wordian. Wordian smaller foraminifers are also poorly known and often confused with the Capitanian taxa (partly because the two stages were before linked under the name Murgabian or Murghabian). Giant Nodosariata diversify with *Langella* Sellier de Civrieux and Dessauvagie and large *Nodosaria*. The appearances of *Dagmarita* Reitlinger and *Hemigordiopsis* seem characteristic (Fig. 13).
- Capitanian. The gigantism is huge among the schwagerinoid and neoschwagerinoid fusulinids, and some genera reach the most important size for unicellular microfossils (Vachard and Bouyx, 2002), especially at the end of the middle Permian (Colpaert et al., 2015). Numerous appearances of smaller foraminifer genera occur during this time, and especially during the Capitanian (Fig. 13). There is also the acme of *Hemigordiopsis* and *Endoteba* Vachard and Razgallah, especially in Tebaga (Tunisia). The oldest colaniellids and youngest pseudovidalinids *Altineria* appear near the end of the late Capitanian.
- Middle/late Permian boundary interval (Fig. 13). A possible significant time to distinguish in the Permian history is the Capitanian/Wuchiapingian boundary interval, in areas such as Guadalupe Mountains (USA), Japan, Turkey, Armenia, Azerbaijan, Iran (Abadeh), and Tunisia (Tebaga) (see Gaillot and Vachard, 2007; Ghazzay et al., 2015; Ghazzay-Souli et al., 2015, with references therein).

STAGES	SUBSTAGES	SMALLER FORAMINIFERS	FUSULINIDS TETHYS-URALS	FUSULINIDS AMERICAS
GZHELIAN	uppermost	*Protonodosaria* spp., *Nodosinelloides* spp., first *Geinitzina primitiva*	*Ultradaixina* spp.	*Trititicites cellamagnus*, *Leptotriticites* spp., *Thompsonites* spp., *Pseudofusulina robleda*, *Schubertella kingi*
			Daixina sokensis	
	upper	First *Protonodosaria* spp.	*Daixina* spp., *Jigulites* spp.	*Schubertella ciscoensis*, *Dunbarinella* spp., acme of *Triticites secalicus*, *T. beedei*
	lower	First *Raphconilia* spp.	*Daixina fragilis*, *Shagonella* spp., *Rauserites* spp., *Ferganites* spp., *Jigulites* spp.	First *Triticites secalicus T. jacksboroensis*, *Waeringella spivyei*
KASIMOVIAN	upper	*Lateenoglobivalvulina pergrata*	*Rauserites quasiarcticus*, *R. rossicus*, *Triticites acutus*	*Kansanella* spp., *Triticites hobblensis*, *T. provoensis*
	middle	First *Spireitlina baschkirica*	*Montiparus* spp.	*Triticites ohioensis*, *T. celebroides*
	lower	First *Asselodiscus primitivus* and first *Lateenoglobivalvulina spiralis*	*Protriticites* spp., *Obsoletes* spp., *Quasifusulinoides* spp.	*Eowaeringella* spp., *Oketaella* spp.
MOSCOVIAN	"N3"	First *Latitubiphytes rauserae*	First *Quasifusulinoides* spp., *Fusulinella? kumpani*, *Praeobsoletes burkemensis*	*Fusulina* spp., *Zellerella* spp.
	Myachkovian	*Globivalvulina syzranica*, *G. mosquensis*	*Fusulina cylindrica*, *Fusulinella bocki*, *Neostaffella* spp., *Protriticites ovatus*	*Fusulina* spp., *Bartramella* spp., *Zellerella* spp.
	Podolskian	Last *Haplophragmina* spp., *Globivalvulina pulchra*	*Fusulina chernovi*, *Fusulinella colaniae*, *Beedeina ulitinensis*, *Putrella brazhnikovae*	*Beedeina? insolita*, *Beedeina pumila*, *B. leei*, *B. euryteines*, *Wedekindellina* spp.
	Kashirian	*Globivalvulina kamensis*, *G. granulosa*, first *Syzrania* spp.	*Hemifusulina vozhgalica*, *Moellerites praecolaniae*, *Pulchrella subpulchra*, *Beedeina pseudoelegans*, *Priscoidella priscoidea*	*Fusulinella acuminata*, *F. devexa*, *F.* spp. [acme of *Komia* algae and chaetetid ischyrosponges]
	Vereian	Last archaediscoids	*Ovatella arta*, *Aljutovella aljutovica*, *Depratina* spp., *Eofusulina* spp., *Citronites* spp.	Last *Profusulinella* spp., first *Fusulinella* spp.
BASHKIRIAN	Melekessian (= Asatauskian)	Last *Biseriella* ex gr. *parva*	*Verella* spp., *Tikhonovella tikhonovichi*	*Profusulinella apodacensis*
			Profusulinella rhombiformis	*Profusulinella copiosa*
	Tashastinkian	First *Endothyranella* spp.	*Ozawainella pararhomboidalis*, *Profusulinella parva*, *Neostaffella* spp., *Pseudostaffella gorskyi*	*Millerella marblensis*
	Askynbashian	Last *Endotaxis* ex gr. *brazhnikovae*	*Ozawainella umbonata*, *Pseudostaffella grandis*, *P. praegorskyi*, first *Neostaffella* spp., *Staffellaeformes* spp.	*Millerella* spp.
	Akavassian	Last *Bradyina cribrostomata*	*Pseudostaffella antiqua*, *Varistaffella* spp.	*Millerella* spp., *Eostaffella* spp.
	Voznessenian (= Syuranian)	First *Globivalvulina* ex gr. *bulloides*	*Semistaffella variabilis*, *S. minuscularia*, *Plectostaffella jakhensis*, *P. bogdanovkensis*, *Varvariella varvariensis*	*Eostaffella* spp.

Figure 12 Foraminiferal biostratigraphy during the Pennsylvanian.

This time interval corresponds to the Khachik Formation in Armenia (Leven, 1998), *Sphaerulina* beds of Abadeh in Iran (Iranian-Japanese Research Group, 1981; Kobayashi and Ishii, 2003; Davydov and Arefifard, 2013), the "barren interval" of Isozaki et al. (2007) and *Neoendothyra permica* Zone in Japan (Kobayashi, 2012). Some smaller foraminifers are also characteristic of this interval, especially *Shanita*, *Altineria*, and a possible transitional taxon between *Syzrania* and *Rectostipulina* Jenny-Deshusses, as well as the oldest *Rectostipulina* (Fig. 13).

- Wuchiapingian and early Changhsingian. After the end–mid Permian crisis, during which disappeared the giant fusulinids, smaller fusulinids evolved within the superfamily Schubertelloidea (appearance of true *Codonofusiella* and true *Reichelina*), and disappearance of *Chusenella*, whereas the Staffellina remain abundant but do not evolve. Characteristic Tethyan smaller foraminifers are some globivalvulinoids (in particular the oldest *Paradagmarita* Lys), neodiscids, and colaniellids (Gaillot and Vachard, 2007; Fig. 13).
- Late Changhsingian. This substage is dominated by the appearance, followed by the rapid disappearance of colaniellids, paradagmaritins, and giant globivalvulinids (Fig. 13). The recovery of the smaller foraminifers leads to relative gigantic specimens of Miliolata and Globivalvulinoidea. Conversely, the late Permian Palaeotextularioidea and Tetrataxoidea that have a recrudescence are perfectly homeomorphous of lower–middle Pennsylvanian species (Gaillot and Vachard, 2007; Vachard et al., 2010).

The biozonation based on foraminifers of the Devonian, Tournaisian, and Visean is well established (Conil et al., 1991; Kulagina et al. (2003); Poty et al., 2006; Hance et al., 2011; Cózar et al., 2008a,b, 2011, 2014; Vachard et al., 2016; this work: Figs. 10–13). The Serpukhovian is known thanks to the work of Aizenverg et al. (1983); Kulagina et al. (2003); Cózar et al. (2011, 2014, 2015); Cózar and Somerville (2014, 2016). The Bashkirian was not recently studied, it was known by the work of Groves (1988) and Kulagina et al. (1992). The Moscovian is also classical and well known in Russia, Spain, Libya, Iran, and Tunisia (Leven and Gorgij, 2011; Ghazzay-Souli et al., 2015). The late Pennsylvanian is the interval with the maximal foraminiferal endemism; the fusulinids of the Kasimovian and Gzhelian of Russia (Leven and Davydov, 2001; Davydov and Leven, 2003), northern Spain (van Ginkel and Villa, 1999), Turkey (Kobayashi and Altıner, 2008; Okuyucu, 2009; Vachard and Moix, 2011), the Carnic Alps (Vachard and Krainer, 2001a; Davydov and

STAGES	SUB-STAGES	SMALLER FORAMINIFERS	FUSULINIDS TETHYS-URALS	FUSULINIDS AMERICAS
CHANGHSINGIAN	upper	*Paradagmarita monodi, Colaniella parva, Paraglobivalvulinoides septulifer*	*Palaeofusulina sinensis*	unknown
	lower	First primitive *Paradagmarita* spp.	*Nanlingella meridionalis, Shindella* spp., *Codonofusiella kwangsiana, Paradunbarula* spp., *Baudiella* spp., *Pseudodunbarula* spp.	unknown
WUCHIAPINGIAN		First *Louisettita* spp., *Colaniella minima, Glomomidiellopsis uenoi, Charliella* spp.		
Capitanian/ Wuchiapingian boundary interval		*Shanita* spp., first *Rectostipulina* spp., first *Paraglobivalvulina* spp., *Altineria alpinotaurica*, last *Hemigordiopsis* spp.	Last *Chusenella* spp., last *Lepidiolina* spp. first *Codonofusiella* spp., acme of *Dunbarula* spp.	*Paraboultonia splendens, Reichelina lamarensis, Yabeina? texana*
CAPITANIAN	upper	First *Robuloides* spp., first *Altineria* spp., acme of *Endoteba controversa*	*Lepidolina* spp., *Yabeina opima, Y. syrtalis, Dunbarula matthieui*, first *Lantschichites* spp.	*Polydiexodina* spp., *Rauserella erratica, Codonofusiella paradoxica*
	lower	First *Multidiscus* (sensu stricto) spp., acme of *Hemigordiopsis renzi* [?umbellaceans *Sphaerionia* spp.]	*Yabeina archaica*, first *Dunbarula* spp., first *Kahlerina* spp., first *Reichelina* spp., first *Sumatrina* spp., last *Afghanella* spp., last *Eopolydiexodina* spp.	First *Polydiexodina* spp., last *Parafusulina* spp.
WORDIAN	upper	First *Hemigordiopsis* spp., first large *Pachyphloia* spp.	*Colania douvillei, Neoschwagerina craticulifera, N. margaritae, Afghanella schencki, Praedunbarula simplex*	*Parafusulina deliciasensis*
	lower	First *Dagmarita* spp., first *Calvezina* spp., first *Neodiscus* spp.	*Praesumatrina* spp., first *Eopolydiexodina* spp., *Neoschwagerina* ex gr. *craticulifera*	*Parafusulina rothi, P. lineata*
ROADIAN	upper	First *Septaglobivalvulina* spp.	*Neoschwagerina simplex, Cancellina* spp., first primitive *Eopolydiexodina* (= *Bidiexodina*) spp., *Parafusulina kaerimizensis*	*Parafusulina boesei*
	lower	First *Langella* (sensu stricto) spp., *Rectoglandulina* spp.	*Maklaya* spp., *Skinnerella* spp., *Misellina claudiae, M. ovalis, Armenina* spp.	*Skinnerina* spp., *Skinnerella* spp.
KUNGURIAN	upper	First *Rectoglandulina* spp., *Olgaorlovella* spp., large *Geinitzina* spp.	*Brevaxina* spp., *Leeina* spp., first *Neofusulinella* spp.	First *Skinnerella* spp., *Parafusulina? deltoides, P.? leonardensis, Schubertella melonica*
	lower	First *Agathammina* spp., first *Nodosaria* (sensu stricto) spp.	*Pamirina* spp., *Levenella* spp., *Depratella* spp.	
ARTINSKIAN		First primitive *Neoendothyra* spp., *Raphconilia?* spp, first *Praeneodiscus* spp., first *Pachyphloia* spp., last *Bradyina* spp.	*Chalaroschwageria insolita, C. vulgaris, Darvasites* spp., *Praeskinnerella* spp., *Robustoschwagerina schellwieni, Pseudoreichelina* spp.	*Pseudoreichelina* spp., *Praeskinerella crassitectoria, Robustoschwagerina stanislavi*
SAKMARIAN	upper	First *Pseudovermiporella* spp., acme of *Geinitzina* spp.	*Tastubella* spp., *Verneuilites verneuili, Mesoschubertella* spp.	
	lower	*Nodosinelloides pinardae, Globivalvulina kantharensis*	*Sakmarella moelleri, Paraschwagerina* spp., first *Robustoschwagerina* spp.	*Monodiexodina linearis;* last *Pseudoschwagerina* spp.
ASSELIAN	upper	*Cribrogenerina? sphaerica, Deckerella elegans,* first *Nodosinelloides netschajewi*	*Sphaeroschwagerina gigas, Schwagerina nux, Pseudoschwagerina saibulakensis*	*Pseudoschwagerina beedei, Paraschwagerina* spp., *Pseudofusulina huecoensis*
	lower	*Raphconilia* spp., *Nodosinelloides talimuensis, Geinitzina postcarbonica*	*Sphaeroschwagerina fusiformis, S. vulgaris, Likharevites* spp.	*Thompsonites* spp., last *Triticites* spp., *Pseudoschagerina* ex gr. *convexa*

Figure 13 Foraminiferal biostratigraphy during the Permian.

Krainer, 1999; Forke, 2002), and Tunisia (Ghazzay-Souli et al., 2015) totally differ from those of the Missourian and Virgilian of New Mexico (Wilde, 1990, 2006; Vachard et al., 2013; Lucas et al., 2015). The Permian biozonations have recently summarized (Henderson et al., 2012; Lucas et al., 2015; Vachard et al., 2015; Vachard, 2016), and no need of an extended description herein.

Many paleobiogeographical reconstructions have been proposed for the Guadalupian and Wuchiapingian, especially based on the distributions of brachiopods (Angiolini et al., 2013 with references therein), fusulinids (Kobayashi, 1999; Nestell, 1999; Kobayashi and Ishii, 2003; Ueno, 2003; Colpaert et al., 2015), and smaller foraminifers (Şengör et al., 1988; Nestell and Pronina, 1997; Altıner et al., 2000; Huang et al., 2007; Gaillot and Vachard, 2007; Vachard, 2014). *Hemigordiopsis* is intertropical, from the Primorye to Oman, and from Tunisia to Japan. Both borders of Paleotethys and Neotethys are generally difficult to be paleobiogeographically discriminated because they are located in the tropical realm, and because both Tethyan oceanic branches lead to a cul-de-sac and close westward (Ghazzay et al., 2015). All microfaunal and microfloral mixtures are therefore possible. Paleobiogeographic data about Panthalassa are rare, except for Japan (from the latest Visean) and exotic Rocky Mountain terranes. In contrast, Cimmerian assemblages are generally more accurately characterized (Gaetani et al., 2009; Angiolini et al., 2015).

Since the pioneer work of Şengör et al., 1988, *Shanita* has always been confirmed as an important paleobiogeographical marker (Sheng and He, 1983; Nestell and Pronina, 1997; Jin and Yang, 2004; Gaillot and Vachard, 2007; Huang et al., 2007; Ebrahim Nejad et al., 2015). It is considered as characteristic of the Cimmerian terranes (Nestell and Pronina, 1997; Ueno, 2003), and/or Perigondwanan terranes during the opening of the Neotethys (Vachard, 2016).

Vachard (2014) indicated also how *Colaniella* Likharev, by its evolution and migration, confirms this paleogeography with: (1) the Salt Range (Pakistan) as a radiation center; (2) a first migration along the Perigondwanan border with the primitive species of *Colaniella*; (3) a second migration to the Perilaurentian northern boundary of the Paleotethys and the shelves bordering the eastern Tethys and western Panthalassa.

The source of *Altineria* is the Taurus Mountains in southern Turkey; *Altineria* is also known in Armenia (Pronina and Gubenko, 1990; Pronina, 1996), Tunisia (Ghazzay et al., 2015), northwestern part of Chios Island (Greece), Karaburun Peninsula (Turkey), NW and central Iran (Abadeh area). Its possible presence in Hubei (South China) is easier to explain with the paleomap of Gaillot and Vachard (2007), where the South China and Iran terranes are closely located. In constrast, many exotic terranes of Crimea and northern Turkey, Far East, Japan, and the Rocky Mountains might possess a remote origin. Vachard and Ferrière (1991) indicated a northern origin at more than 3000 km of the New Zealand exotic blocks.

During the entire Permian, the southern part of the North American Craton (i.e., the modern states of Sonora, New Mexico, and Texas) remained relatively isolated from the Laurentian and Gondwanan continents. The endemism of the fusulinids and smaller foraminifers was important during the late Cisuralian (= Wolfcampian) (Lucas et al., 2015; Vachard et al., 2015), where several Tethyan groups are apparently lacking and probably endemic taxa exist among the globivalvulinoids, miliolates, and gymnocodiacean algae (Vachard et al., 2015). In contrast, several studies (Vachard et al., 1992; Nestell and Nestell, 2006; Nestell et al., 2006) have demonstrated that the endemism of the smaller foraminifers is lesser than for the large fusulinids, during the Cisuralian and Guadalupian. However, giant paraglobivalvulinins, baisalinids, hemigordiopsids, and langellins remain unknown in the Americas, as well as Lopingian foraminifers.

5. CONCLUSIONS

1. Six classes of organic-walled, secreted, and agglutinated foraminifers are present from Famennian to Permian: Allogromiata, Astrorhizata, Fusulinata, Miliolata, Nodosariata, and Textulariata.
2. Biostratigraphically, the most interesting groups are the Archaediscoidea, Lasiodiscoidea, Bradyinoidea, Globivalvulinoidea, and the entire order of the Fusulinida, among the Fusulinata; the Cornuspirida among the Miliolata; and the entire class of the Nodosariata. The appearance of the order Verneuilinida at the end of the Permian is probably another important datum.
3. The Archaediscoidea evolve as follows: from the ancestral type *Lapparentidiscus* having a wall entirely dark microgranular, a hyaline pseudofibrous plug appears in the umbilici of *Ammarchaediscus* (Visean biozone MFZ10). It covers progressively all the whorls, but the last whorl remains with a dark microgranular wall (*Viseidiscus*;

Planoarchaediscus); then the clear layer covers entirely the test (*Uralodiscus*; *Glomodiscus*; biozone MFZ11), after that it is progressively reduced (Archaediscidae; biozones MFZ12–MFZ15), and finally, disappears (Eosigmoilinidae; Serpukhovian). Forms with nodosities complete the biostratigraphical value of this group.

4. The Lasiodiscina are still poorly known during the Carboniferous–Permian. *Pseudovidalina* gives rise to two evolute forms, *Xingshandiscus* during the Kungurian, and *Altineria* at the Capitanian/Wuchiapingian boundary interval. *Lasiodiscus* and *Lasiotrochus* diversify during the Capitanian–Lopingian.
5. The Bradyinoidea evolve relatively rapid during the late Mississippian–early Pennsylvanian, and slowly during the middle–late Pennsylvanian and Permian.
6. The phylogeny of the globivalvulinoids, from late Tournaisian to Kungurian, was only based on the *Globivalvulina* lineages. During the Guadalupian–Lopingian, they diversify with new subfamilies, well known in Turkey and Iran (Vachard, 2016).
7. The Fusulinida evolution, well known for many years, is especially precise in the case of the Neoschwagerinoidea; however, their accuracy was perhaps overestimated, since regional stages correspond only to one neoschwagerinoid biozone [*Brevaxina* for Bolorian (= late Kungurian); *Yabeina* for Midian (= Capitanian)].
8. The Miliolata are poorly undertood during the Cisuralian. *Praeneodiscus* Vachard et al. and *Olgaorlovella* Vachard et al., recently defined, might have a biostratigraphic importance. The evolution of tubiphytids and ellesmerellids is yet to be discussed because these taxa are generally assigned to cyanobacteria and/or algae. The middle–late Permian is well characterized by giant genera of smaller foraminifers: *Neodiscus*, *Neohemigordius*, *Baisalina*, *Glomomidiellopsis* Gaillot and Vachard, *Hemigordiopsis*, and *Shanita*.
9. The biostratigraphically most interesting genera of Cisuralian Nodosariata are *Tezaquina* Vachard, *Nodosinelloides*, *Geinitzina*, *Polarisella*, and *Vervilleina* Groves, although they present many problems of nomenclature. The middle Permian is a period of generic diversification of nodosariates. The late Permian is characterized by evolved colaniellids and frondinids, whereas at the base of the Triassic several primitive Cisuralian genera reappeared, especially *Nodosinelloides* and *Polarisella*, which crossed through the Permian–Triassic boundary.

10. Paleogeographically, the most interesting markers are *Quasiendothyra*, *Chernyshinella* Reitlinger, *Uralodiscus* Malakhova, *Hemidiscopsis*, *Timanella* Reitlinger, *Sphaeroschwagerina*, *Robustoschwagerina* Miklukho-Maklay, *Cancellina* Hayden, *Lepidolina*, *Neoendothyranella*, *Hemigordius*, *Shanita*, *Colaniella*, and *Altineria* (Vachard, 2016).
11. *Hemigordiopsis* is intertropical, from the Primorye to Oman, and from Tunisia to Japan. *Shanita* is characteristic of Perigondwanan terranes, which moved to Cimmeria and Sibumasu microcontinents after the opening of the Neotethys. *Altineria* confirms such a paleogeographic and geodynamic evolution. *Colaniella* migrates from its radiation center in the Salt Range (Pakistan), Perigondwanan border, and then, to the shelves of the eastern Tethys and western Panthalassa.
12. The southern part of the North American Craton (Sonora, New Mexico, and Texas) remained relatively isolated from the Laurentian and Gondwanan continents. The endemism of the fusulinids was strong during the early Bashkirian, late Moscovian, late Pennsylvanian, and late Cisuralian.

ACKNOWLEDGMENTS

I warmly thank the reviewers G. Nestell, M. Nestell, I. Somerville, and P. Cózar, who improved the manuscript by their pertinent criticisms. Thanks to M. Montenari and H. Kabes for editorial advice, and to K. Zandkarimi (Tehran), S. Clausen and E. Locatelli (Villeneuve d'Ascq), K. Krainer (Innsbruck), and S. Lucas (Albuquerque), for their help.

REFERENCES

Afanasieva, M.S., Amon, E.O., 2011. Devonian radiolarians of Russia. Paleontol. J. 45, 1313–1352.

Aizenverg, D.E., Astakhova, T.V., Berchenko, O.I., Brazhnikova, N.E., Vdovenko, M.V., Dunaeva, N.N., Zernetskaya, N.V., Poletaev, V.I., Sergeeva, M.T., 1983. Late Serpukhovian Substage in the Donets Basin. Akademiya Nauk Ukrainskoi SSR, Institut Geologicheskikh Nauk, pp. 1–164 (in Russian).

Alipour, Z., Hosseini-Nezhad, S.M., Vachard, D., Rashidi, K., 2013. Biostratigraphy and description of smaller foraminifers of Dorud Group (uppermost Gzhelian-lower Sakmarian) in Tang-e Shamshirbor section Eastern Alborz, Iran. Geol. J. 48, 385–402.

Altıner, D., 1988. Pseudovidalinidae n. fam. and *Angelina* n. gen. from the upper permian of south and Southeast Turkey. Rev. Paléobiol. Spec. Vol. 2, 25–36. Benthos'86 Genève.

Altıner, D., Özkan-Altıner, S., 2010. *Danielita gailloti* n. gen., n. sp., within the evolutionary framework of middle-late Permian dagmaritins. Turk. J. Earth Sci. 19, 497–512.

Altıner, D., Özkan-Altıner, S., Koçyigit, A., 2000. Late Permian foraminiferal biofacies belts in Turkey: palaeogeographic and tectonic implications. In: Bozkurt, E., Winchester, J.A., Piper, J.D.A. (Eds.), Tectonics and Magmatism in Turkey and the Surrounding Area, vol. 173. Geological Society London, Special Publications, pp. 83–96.

Angiolini, L., Crippa, G., Muttoni, G., Pignatti, J., 2013. Guadalupian (middle Permian) paleobiogeography of the Neotethys ocean. Gondwana Res. 24, 173–184.
Angiolini, L., Zanchi, A., Zanchetta, S., Nicora, A., Vuolo, I., Berra, F., Henderson, C., Malaspina, N., Rettori, R., Vachard, D., Vezzoli, G., 2015. From rift to drift in South Pamir (Tajikistan): Permian evolution of a Cimmerian terrane. J. Asian Earth Sci. 102, 146–169.
Bagheri, S., Stampfli, G.M., 2008. The Anarak, Jandaq and Posht-e-Badam metamorphic complexes in central Iran: new geological data, relationships and tectonic implications. Tectonophysics 451, 123–155.
Bensh, F.R., 1987. Revised systematics of pseudofusulinid genera, *Pseudofusulina* Dunbar et Skinner and related genera. Vopr. Mikropaleont. 29, 20–53 (in Russian).
Bogush, O.I., 1985. Foraminifers and stratigraphy of the early Carboniferous from the western Siberia plain. In: Biostratigraphy of Paleozoic from Western Siberia, vol. 619. Akademiya Nauk SSSR, Sibirskoye Otdelenie, Trudy Instituta Geologii i Geofiziki, pp. 49–68 (in Russian).
Bozorgnia, F., 1973. Paleozoic Foraminiferal Biostratigraphy of Central and East Alborz Mountains, Iran, vol. 4. National Iranian Oil Company, Geological Laboratories Publication, pp. 1–185.
Brenckle, P.L., Ramsbottom, W.H.C., Marchant, T.R., 1987. Taxonomy and classification of Carboniferous archaediscacean foraminifers. Cour. Forschungsinst. Senckenberg 98, 11–24.
Bykova, E.V., Polenova, E.N., 1955. Foraminifers and radiolaria from the Volga-Ural area and the central Devonian field and their stratigraphical importance. Tr. VNIGRI 87, 1–141 (in Russian).
Calkins, G.N., 1909. Protozoology. Lea and Febiger, New York and Philadelphia.
Calkins, G.N., 1926. The Biology of the Protozoa. Lea and Febiger, Philadelphia.
Cavalier-Smith, T., 2002. The phagotrophic origin of eukaryotes and phylogenetic classification of Protozoa. Int. J. Syst. Evol. Microbiol. 52, 297–354.
Cavalier-Smith, T., 2003. Protists phylogeny and the high-level classification of protozoa. Eur. J. Protistol. 39, 338–348.
Chernysheva, N.E., 1941. New genus of foraminifera from the Tournaisian deposits of the Urals. Dokl. Akad. Nauk. SSSR 32, 69–70 (in Russian).
Chernysheva, N.E., 1948. On *Archaediscus* and Related Forms From the Early Carboniferous of the USSR, vol. 19. Akademiya Nauk SSSR, Trudy Instituta Geologicheskikh Nauk, 62, geologicheskaya seriya, pp. 150–158 (in Russian).
Ciry, R., 1952. Super-famille 4. Fusulinoidea. In: Piveteau, J. (Ed.), Traité de Paléontologie, vol. 1. Masson et Cie, Paris, pp. 179–191.
Colpaert, C., Monnet, C., Vachard, D., 2015. *Eopolydiexodina* (middle Permian giant fusulinids of Afghanistan): biometry, morphometry, palaeobiogeography and end-Guadalupian events. J. Asian Earth Sci. 102, 127–145.
Colpaert, C., Vachard, D., Monnet, C., Clausen, S., Timokhina, I., Obut, O., Nadezhda, I., 2017. Tournaisian (Early Carboniferous) Foraminifers of the Kuznetsk Basin (South-West Siberia, Russia). Bull. Soc. Géol. France (in press).
Conil, R., Lys, M., 1964. Matériaux pour l'étude micropaléontologique du Dinantien de la Belgique et de la France (Avesnois). Pt. 1, Algues et foraminifères; Pt. 2, Foraminifères (suite). Mém. Inst. Géol. Univ. Louvain 23, 1–372.
Conil, R., Lys, M., 1977. Les transgressions dinantiennes et leur influence sur la dispersion et l'évolution des foraminifères. Mém. Inst. Géol. Univ. Louvain 29, 9–55.
Conil, R., Pirlet, H., 1970. Le calcaire carbonifère du synclinorium de Dinant et le sommet du Famennien. Colloque sur la stratigraphie du Carbonifère, vol. 55. Congrès et Colloques Université Liège, pp. 47–63.

Conil, R., Groessens, E., Lys, M., 1977. Etude micropaléontologique de la tranchée d'Yves-Gomezée (Tn 3c-V 1-V 2, Belgique). Bull. Soc. Belge. Géol. 82 (1973), 201–239.

Conil, R., Longerstaey, P.J., Ramsbottom, W.H.C., 1980. Matériaux pour l'étude micropaléontologique du Dinantien de Grande-Bretagne. Mém. Inst. Géol. Univ. Louvain 30 (1979), 1–115.

Conil, R., Groessens, E., Laloux, M., Poty, E., Tourneur, F., 1991. Carboniferous guide foraminifera, corals and conodonts in the Franco-Belgian and Campine basins; their potential for widespread correlations. Cour. Forschungsinst. Senckenberg 130, 15–30.

Conkin, J.E., Conkin, B.M., 1964. Devonian foraminifera: Part I. The Louisiana limestones of Missouri and Illinois. Bull. Am. Paleontol. 47 (213), 49–105.

Conkin, J.E., Conkin, B.M., 1970. North American Kinderhookian (lower Mississippian) arenaceous foraminifers. Sixième Congrès International Stratigraphie Géologie Carbonifère, Sheffield 1967. C. R. 2, 575–583.

Cózar, P., 2000. Archaediscidae y Lasiodiscidae (Foraminiferida) del Mississippiense (Carbonífero) del Area del Guadiato (Suroeste de España). Rev. Esp. Micropaleontol. 32, 193–212.

Cózar, P., Somerville, I.D., 2014. Foraminiferal early Namurian/Serpukhovian (Carboniferous) correlations: implications for biostratigraphic and glacioeustatic correlations. Newsl. Stratigr. 47, 355–367.

Cózar, P., Somerville, I.D., 2016. Problems correlating the late Brigantian-Arnsbergian western European substages within northern England. Geol. J. 5, 817–840. http://dx.doi.org/10.1002/gj.2700.

Cózar, P., Vachard, D., Medina-Varea, P., Said, I., Somerville, I.D., Rodríguez, S., Berkhli, M., 2008a. Late Viséan-early Serpukhovian foraminiferans and calcareous algae from the Adarouch region (central Morocco). Geol. J. 43, 463–485.

Cózar, P., Somerville, I.D., Burgess, I., 2008b. New foraminifers in the Visean/Serpukhovian boundary interval of the lower limestone formation, Midland valley. Scotl. J. Paleontol. 82, 906–923.

Cózar, P., Said, I., Somerville, I.D., Vachard, D., Medina-Varea, P., Rodríguez, S., Berkhli, M., 2011. Potential foraminiferal markers for the Visean-Serpukhovian and Serpukhovian-Bashkirian boundaries—a case-study from Central Morocco. J. Paleontol. 85, 1105–1127.

Cózar, P., Medina-Varea, P., Somerville, I.D., Vachard, D., Rodriguez, S., Said, I., 2014. Foraminifers and conodonts from the late Viséan to early Bashkirian succession in the Saharan Tindouf Basin (southern Morocco): biostratigraphic refinements and implications for correlations in the western Palaeotethys. Geol. J. 49, 271–302.

Cózar, P., Sanz-López, J., Blanco-Ferrera, S., 2015. Lasiodiscid foraminifers during the late Viséan to Serpukhovian in Vegas de Sotres section (Cantabrian Mountains, NW Spain). Geobios 48, 213–238.

Crespin, I., 1958. Permian foraminifera of Australia. Commonwealth of Australia, Department National Development, Bureau Mineral Resources. Geol. Geophys. 48, 1–207.

Culver, S.J., 1991. Early Cambrian foraminifera from west Africa. Science 254, 689–691.

Cummings, R.H., 1955a. New genera of foraminifera from the British lower Carboniferous. J. Wash. Acad. Sci. 45, 1–8.

Cummings, R.H., 1955b. Nodosinella Brady, 1876 and associated upper Paleozoic genera. Micropaleontology 1, 221–238.

Davydov, V.I., 1984. K voprosy o proickhozhdenii shvagerin (on the question of the origin of schwagerins). Paleontol. Zh. 4, 3–16 (in Russian).

Davydov, V.I., 2011. Taxonomy, nomenclature and evolution of the early schubertellid fusulinids. Acta Paleontol. Pol. 56, 181–194.

Davydov, V.I., 2013. Climate fluctuations within the western Pangean tropical shelves – the Pennsylvanian/Permian record from benthic foraminifera. In: Lucas, S.G., et al. (Eds.), The Carboniferous/Permian Transition, vol. 60. New Mexico Museum of Natural History of Science, Bulletin, pp. 73–78.

Davydov, V.I., Arefifard, S., 2013. Middle Permian (Guadalupian) fusulinid taxonomy and biostratigraphy of the mid-latitude Dalan Basin, Zagros, Iran and their applications in paleoclimate dynamics and paleogeography. GeoArabia 18, 17–62.

Davydov, V.I., Krainer, 1999. Fusulinid assemblages and facies of the Bombaso formation and basal Meledis formation (Moscovian, Kasimovian) in the Carnic Alps (Austria/Italy). Facies 40, 157–196.

Davydov, V.I., Leven, E.Y., 2003. Correlation of upper Carboniferous (Pennsylvanian) and lower Permian (Cisuralian) marine deposits of the Peri-Tethys. Palaeogeogr. Palaeoclimatol. Palaeoecol. 196, 39–57.

Davydov, V.I., Korn, D., Schmitz, M.D., 2012. The Carboniferous period. In: Gradstein, F.M., Ogg, J.G., Schmitz, M.D., Ogg, G. (Eds.), The Geologic Time Scale 2012, vol. 2. Elsevier, Amsterdam, pp. 603–651.

Dreesen, R., Bless, M.J.M., Conil, R., Flajs, G., Laschet, Ch., 1985. Depositional environment, paleocology and diagenetic history of the "Marbre rouge à crinoïdes de Baelen" (Late Upper Devonian, Verviers Synclinorium, Eastern Belgium). Ann. Soc. Géol. Belg. 108, 311–359.

Dunbar, C.O., Condra, G.E., 1927. The Fusulinidae of the Pennsylvanian system in Nebraska. Neb. Geol. Surv. 2nd Ser. 2, 1–135.

Dunbar, C.O., Henbest, L.G., 1930. The fusulinid genera *Fusulina, Fusulinella* and *Wedekindelella.* Am. J. Sci. 5th Ser. 20, 357–364.

Duc Tien, N., 1966. Foraminifera and algae from the permian of Kampuchea. In: Fontaine, H., et al. (Eds.), The Permian of Southeast Asia, Appendix 2, vol. 18. CCOP Technical Bulletin, pp. 116–137.

Ebner, F., 1973. Foraminiferen aus dem Paläozoikum der Karnischen Alpen. Mitteilungen Abteilung Geologie. Paläontol. Bergbau Landesmus. Joanneum Graz 34, 5–36.

Ebrahim Nejad, E., Vachard, D., Siabeghodsy, A., Abbasi, S., 2015. Middle-Late Permian (Murgabian-Djulfian) foraminifers of the northern Maku area (western Azerbaijan, Iran). Palaeontol. Electron. 18.1.19A.

Ehrenberg, C.G., 1839. Über dem blossen Auge unsichtbare Kalkthierchen und Kieselthierchen als Hauptbestandtheile der Kreidegebirge. Ber. über Bekanntm. geeigneten Verh. königlichen Preuss. Akad. Wiss. Berl. 3 (1838), 192–200.

Eickhoff, G., 1974. Stratigraphic significance of silicified foraminifera in the upper Devonian of central Europe. In: Bouckaert, J., Streel, M. (Eds.), International Symposium on Namur, 1974; Belgian Micropaleontological Limits from Emsian to Visean, vol. 11. Publication, pp. 1–14.

Epshteyn, O.G., Tereshkova, T.P., Solovieva, M.N., 1985. Paleozoic from Koryak Nagorya (foraminiferal fauna, biostratigraphy). Vopr. Mikropaleont. 27, 47–77 (in Russian).

Flakowski, J., Bolivar, I., Fahrni, J., Pawlowski, J., 2005. Actin phylogeny in foraminifera. J. Foraminifer. Res. 35 (2), 93–102.

Flügel, E., Kochansky-Devidé, V., Ramovs, A., 1984. A middle Permian calcisponge/algal/cement reef: Straza near Bled, Slovenia. Facies 10, 179–256.

Flügel, E., Di Stefano, P., Senowbari-Daryan, B., 1991. Microfacies and depositional structure of allochtonous carbonate base-of-slope deposits. The late Permian Pietra di Salomone megablock, Sosio Valley (western Sicily). Facies 25, 147–186.

Forke, H.C., 2002. Biostratigraphic subdivision and correlation of uppermost Carboniferous/lower permian sediments in the southern Alps: fusulinoidean and conodont faunas from the Carnic Alps (Austria/Italy), Karavanke mountains (Slovenia), and southern Urals (Russia). Facies 47, 201–276.

Foster, C.B., Palmieri, V., Fleming, P.J.G., 1985. Plant microfossils, Foraminiferida, and Ostracoda, from the fossil Cliff formation (early permian, Sakmarian), Perth basin, western Australia. Spec. Publ. S. Aust. Dep. Mines Energy 5, 61–105.

Fursenko, A.V., 1958. Main Stages of Development of Foraminiferal Fauna in the Geological Past, vol. 1. Akademiya Nauk Beloruskoi SSR, Trudy Instituta Geologicheskikh Nauk, pp. 10–29 (in Russian).

Gaetani, M., Angiolini, L., Ueno, K., Nicora, A., Stephenson, M., Sciunnach, D., Rettori, R., Price, G., Sabouri, J., 2009. Pennsylvanian-Early Triassic Stratigraphy in the Alborz Mountains (Iran), vol. 312. Geological Society of London, Special Publications, pp. 79–128.

Gaillot, J., Vachard, D., 2007. The Khuff Formation (Middle East) and time-equivalents in Turkey and South China: biostratigraphy from Capitanian to Changhsingian times (Permian), new foraminiferal taxa and palaeogeographical implications. Coloq. Paleontol. 57, 37–223.

Gargouri, S., Vachard, D., 1988. Sur *Hemigordiopsis* et d'autres Foraminifères porcelanés du Murghabien du Tebaga (Permien supérieur de Tunisie). Rev. Paléobiol. Vol. Spéc. 2 1, 57–68. Benthos'86.

Ghazzay, W., Vachard, D., Razgallah, S., 2015. Revised fusulinid biostratigraphy of the middle-late Permian of Jebel Tebaga (Tunisia). Rev. Micropaléontol. 58, 57–83.

Ghazzay-Souli, W., Vachard, D., Razgallah, S., 2015. Carboniferous and Permian biostratigraphy by foraminifers and calcareous algae of Bir Mastoura (BMT-1) and related boreholes of southern Tunisia. Rev. Micropaléontol. 58, 239–265.

van Ginkel, A.C., Villa, E., 1999. Late fusulinellid and early schwagerinid foraminifera: relationship and occurrences in the Las Llacerias section (Moscovian/Kasimovian), Cantabrian Mountains, Spain. J. Foraminifer. Res. 29 (3), 263–290.

Glaessner, M.F., 1945. Principles of Micropaleontology. Melbourne University Press, pp. 1–269.

Griffith, J.W., Henfrey, A., 1875. The Micrographic Dictionary, third ed., vol. 1. van Voorst, London.

Groussin, M., Pawlowski, J., Yang, Z.-H., 2011. Bayesian relaxed clock estimation of divergence times in foraminifera. Mol. Phylogenet. Evol. 61, 157–166.

Groves, J.R., 1987. Evidence for syzygy in late Paleozoic calcareous foraminifers: early origin of an "advanced" reproductive strategy. Cour. Forschungsinst. Senckenberg 98, 5–10.

Groves, J.R., 1988. Calcareous foraminifers from the Bashkirian stratotype (Middle Carboniferous, South Urals) and their significance for intercontinental correlations and the evolution of the Fusulinidae. J. Paleontol. 62, 368–399.

Groves, J.R., Altıner, D., 2005. Survival and recovery of calcareous foraminifera pursuant to the end-Permian mass extinction. C. R. Palevol 4, 487–500.

Groves, J.R., Wang, Y., 2009. Timing and size selectivity of the Guadalupian (middle Permian) fusulinoidean extinction. J. Paleontol. 87, 183–196.

Groves, J.R., Rettori, R., Altıner, D., 2004. Wall structures in selected Paleozoic Lagenide foraminifera. J. Paleontol. 78, 245–256.

Gutschick, R.C., 1962. Arenaceous foraminifera from oncolites in the Mississippian Sappington formation. J. Paleontol. 36, 1291–1304.

Gutschick, R.C., 1986. Middle Ordovician agglutinated foraminifera including *Reophax* from the Mifflin formation, Platteville group of Illinois. J. Palaeontol. 60, 233–248.

Habeeb, K.H., May 19–26, 1979. SEM Study of Foraminiferal Wall Structure in the Lower Carboniferous Limestone. Abstracts of Papers, Ninth International Congress of Carboniferous Stratigraphy and Geology. University of Illinois at Urbana-Champaign, pp. 82–83.

Hance, L., Hou, H.F., Vachard, D., 2011. Upper Famennian to Visean Foraminifers and Some Carbonate Microproblematica from South China, Hunan, Guangxi and Guizhou. Geological Science Press, Beijing.

Haynes, J.R., 1981. Foraminifera. John Wiley and Sons, New York.

Henderson, C.M., Davydov, V.I., Wardlaw, B.R., Gradstein, F.M., Hammer, O., 2012. The permian period. In: Gradstein, F.M., Ogg, J.G., Schmitz, M.D., Ogg, G.M. (Eds.), The Geologic Time Scale 2012. Elsevier, Amsterdam, pp. 653–679.

Henjes, J., Assmy, P., 2008. Particle availability controls agglutination in Pelagic tintinnids in the southern ocean. Protist 159, 239–250.

Herbig, H.G., 2006a. 5.13 Sandschalig-agglutinierte Foraminiferen. Schriftenreihe der Deutschen Gesellschaft für Geowissenschaften (SDDG), Heft 41, Stratigraphie von Deutschland VI, Unterkarbon (Mississippium), pp. 243–249.

Herbig, H.G., 2006b. Kalkschalige Kleinforaminiferen. In: Amler, M.R.W., Stoppel, D. (Eds.), Stratigraphie von Deutschland VI. Unterkarbon (Mississippium), Schriftenreihe der Deutschen Gesellschaft für Geowissenschaften, vol. 41, pp. 250–270.

Hohenegger, J., Piller, W., 1975. Wandstrukturen und Grossgliederung der Foraminiferen. Sitzungsberichte der österreichischen Akademie der Wissenschaften, Mathematisch-naturwissenschaftliche Klasse. Abteilung I 184 (1–5), 67–96.

Holcová, K., 2004. Foraminifers from the lower/middle Devonian boundary beds of the Barrandian area, Czech Republic, and their paleoecology. J. Foraminifer. Res. 34, 214–231.

Holcová, K., Slavík, L., 2013. The morphogroups of small agglutinated foraminifera from the Devonian carbonate complex of the Prague Synform (Barrandian area, Czech Republic). Palaeogeogr. Palaeoclimatol. Palaeoecol. 386, 210–224.

Huang, H., Yang, X., Jin, X., 2007. The *Shanita* fauna (Permian foraminifera) from Baoshan area, western Yunnan province, China. Front. Biol. China 2, 114–124 (English translation from Acta Palaeontol. Sin. 44 (4), 544–555).

Iranian-Japanese Research Group, 1981. The Permian and the lower Triassic Systems in Abadeh region, Central Iran. Mem. Fac. Sci. Kyoto Univ. Ser. Geol. Mineral. 47, 61–133.

Ireland, H.A., 1956. Upper Pennsylvanian arenaceous foraminifers of Kansas. J. Paleontol. 30, 831–864.

Isozaki, Y., Kawahata, H., Minoshima, K., 2007. The Capitanian (Permian) Kamura cooling event: the beginning of the Paleozoic-Mesozoic transition. Palaeoworld 16, 16–30.

Jin, X., Yang, X., 2004. Paleogeographic implications of the *Shanita-Hemigordius* fauna (Permian foraminifer) in the reconstruction of Permian Tethys. Episodes 27, 273–278.

Jirovec, O., 1953. Protozoologie. Nakladatelstvi Ceskoslovenské Akademie Ved, Prague.

Jones, R.W., Charnock, M.A., 1985. Morphogroups of agglutinated foraminifera, their life positions and feeding and potential applicability in (paleo)ecological studies. Rev. Paléobiol. 4, 311–320.

Kalvoda, J., 2001. Upper Devonian-lower Carboniferous foraminiferal paleobiogeography and Perigondwana terranes at the Baltica-Gonwana interface. Geol. Carpathica 52, 205–215.

Kaminski, M.A., Setoyama, E., Cetean, C.G., 2010. The phanerozoic diversity of agglurtinated foraminifera: origination and extinction rates. Acta Palaeontol. Pol. 55, 529–539.

Kazmierczak, J., 1976. Volvocacean nature of Paleozoic non-radiosphaerid calcispheres and parathuramminid "foraminifera". Acta Palaeontol. Pol. 21, 245–258.

Kobayashi, F., 1999. Tethyan uppermost Permian (Dzhulfian and Dorashamian) foraminiferal faunas and their paleogeographic and tectonics implications. Palaeogeogr. Palaeoclimatol. Palaeoecol. 150, 279–307.

Kobayashi, F., 2012. Middle and late Permian foraminifers from the Chichibu Belt, Takachiho area, Kyushu, Japan: implications for faunal events. J. Paleontol. 86, 669–687.

Kobayashi, F., Altıner, 2008. Fusulinoidean faunas from the upper Carboniferous and lower Permian platform limestone in the Hadim area, central Taurides. Riv. Ital. Paleontol. Stratigr. 114, 191–232.

Kobayashi, F., Ishii, K., 2003. Permian fusulinaceans of the Surmaq formation in the Abadeh region, central Iran. Riv. Ital. Paleontol. Stratigr. 109, 307–337.

Kochansky-Devidé, V., 1966. Staffellids and other foraminifers with parachomata. Vopr. Mikropaleont. 10, 89–98 (in Russian).

Krainer, K., Vachard, D., 2011. The lower Triassic Werfen formation of the Karawanken Mountains (southern Austria) and its disaster survivor microfossils, with emphasis on *Postcladella* n. gen. (Foraminifera, Miliolata, Cornuspirida). Rev. Micropaléontol. 54, 59–85.

Krylatov, S., Mamet, B., 1966. Données nouvelles sur les terrains paléozoïques de l'Argentella-Tour Margine (Corse); attribution à la limite dévono-carbonifère du calcaire de Capitello. Bull. Soc. Géol. Fr. 7 (8), 73–79.

Kulagina, E.I., Gibshman, N.B., Pazukhin, V.N., 2003. Foraminiferal zonal standard for the lower Carboniferous of Russia zand its correlation with the conodont zonation. Riv. Ital. Paleontol. Stratigr. 109, 173–185.

Kulagina, E.I., Rumyantseva, Z.C., Pazukhin, V.N., Kochetova, N.N., 1992. Boundary Early-Middle Carboniferous in the Southern Urals and Central Tian-Shan. Rossiiskaya Akademiya Nauk, Uralskoe Otdelenie, Bashkirskii Nauchnyi Tsentr, Institut Geologii, Moskva, Nauka, pp. 1–112 (in Russian).

Langer, W., 1969. Foraminifera aus dem Alt-Paläozoikum der Karnischen Alpen. Carinthia II 79 (159), 34–60.

Lankester, E.R., 1885. Protozoa. In: Encyclopedia Britannica, ninth ed., vol. 19 London.

Leven, E.Y., 1982. Distribution and systematics of the fusulinid genus *Cancellina* Hayden. Vopr. Mikropaleont. 25, 40–51 (in Russian).

Leven, E.Y., 1987. Systematics and phylogeny of family Schubertellidae Skinner, 1931. Paleontol. Zh. 2, 30–40 (in Russian).

Leven, E.Y., 1998. Permian fusulinids assemblages and stratigraphy of Transcaucasia. Riv. Ital. Paleontol. Stratigr. 104, 299–328.

Leven, E.Ya, Davydov, V.I., 2001. Stratigraphy and fusulinids of the Kasimovian and upper Gzhelian (upper Carboniferous) in the Southwestern Darvaz (Pamir). Riv. Ital. Stratigr. Paleontol. 107, 3–46.

Leven, E.Ya, Gorgij, M.N., 2011. Fusulinids and stratigraphy of the Carboniferous and Permian in Iran. Stratigr. Geol. Correl. 19, 687–776.

Lipina, O.A., 1955. Upper Devonian and Tournaisian Foraminifers From the Volga-Ural Area and From the Western Slope of the Central Urals, vol. 70. Akademiya Nauk SSSR, Trudy Instituta Geologicheskikh Nauk, 163, geologichevskaya seriya, pp. 1–96 (in Russian).

Loeblich, A.R., Tappan, H., 1954. Emendation of the foraminiferal genera *Ammodiscus* Reuss, 1862 and *Involutina* Terquem, 1862. J. Wash. Acad. Sci. 44, 306–310.

Loeblich, A.R., Tappan, H., 1961. Suprageneric classification of the Rhizopodea. J. Paleontol. 35, 245–330.

Loeblich, A.R., Tappan, H., 1964. Sarcodina, chiefly "Thecamoebians" and Foraminiferida. In: Moore, R.C. (Ed.), Treatise of Invertebrate Paleontology, Part C, Protista 2, vol. 2. Geological Society of America and University of Kansas Press, pp. C1–C900.

Loeblich, A.R., Tappan, H., 1982. Classification of the Foraminiferida. In: Broadhead, T.W. (Ed.), Foraminifera, Notes for a Short Course Organized by M.A.

Buzas and B.K. Sen Gupta, vol. 6. University of Tennessee, Department of Geological Sciences, Studies in Geology, pp. 22–36.
Loeblich, A.R., Tappan, H., 1984. Suprageneric classification of the Foraminiferida (Protozoa). Micropaleontology 30, 1–70.
Loeblich, A.R., Tappan, H., 1986. Some new and redefined genera and families of Textulariina, Fusulinina, Involutinina and Miliolina (Foraminiferida). J. Foraminifer. Res. 16, 334–346.
Loeblich, A.R., Tappan, H., 1987. Foraminiferal Genera and Their Classification. Van Nostrand Reinhold Company Publisher, New York.
Loeblich, A.R., Tappan, H., 1992. Present Status of Foraminiferal Classification. Studies in Benthic Foraminifera, Benthos' 90, Sendai 1990. Tokai University Press, pp. 93–102.
Lucas, S.G., Krainer, K., Vachard, D., 2015. The lower permian Hueco group, Robledo Mountains, New Mexico (U.S.A.). In: Lucas, S.G., DiMichele, W.A. (Eds.), Carboniferous-permian Transition in the Robledo Mountains, vol. 65. New Mexico Museum of Natural History and Science Bulletin, Southern New Mexico, pp. 43–95.
Lin, R., Nassichuk, W.W., 1994. Fusulinaceans at the middle/upper Pennsylvanian (Desmoinesian/Missourian) boundary in the Canadian Arctic Archipelago. In: Embry, A.F., Beauchamp, B., Glass, D.J. (Eds.), Pangea: Global Environments and Resources, vol. 17. Canadian Society Petroleum Geologists, Memoir, pp. 907–925.
Mamet, B., Pinard, S., 1992. Note sur la taxonomie des petits foraminifères du Paléozoïque supérieur. Bull. Soc. Belge. Géol. 99 (1990), 373–398.
Mamet, B., Mikhailoff, N., Mortelmans, G., 1970. La stratigraphie du Tournaisien et du Viséen inférieur de Landelies. Comparaison avec les coupes du Tournaisis et du bord nord du synclinal de Namur. Mém. Soc. Belge. Géol. Paléontol. Hydrol. 8 (9), 1–81.
Marfenkova, M.M., 1991. Marine Carboniferous from Kazakhstan. Akademiya Nauk Kazakhskoi SSR, Institut Geologicheskikh Nauk, Alma-Ata. Zylym, 1–197 + 1–150 (in Russian).
Martini, R., Zaninetti, L., 1988. Structure et paléobiologie du foraminifère *Lasiodiscus* Reichel, 1946: étude d'après le matériel du Permien supérieur de Grèce. Rev. Paléobiol. 7, 289–300.
Maslakova, N.I., 1990. Criteria for establishing higher taxa in foraminifers. In: Menner, V.V. (Ed.), Systematics and Phylogeny of Invertebrates: The Criteria for Establishing Higher Taxa. Izdateltsvo Nauka, Moscow, pp. 22–27 (in Russian).
Maslakova, N.I., Gorbachik, T.N., Alekseev, A.S., Barskov, I.S., Golubev, S.N., Nazarov, B.B., Petruchevskaya, M.G., 1995. Micropaleontology. Izdatelstvo Moskovskogo Universita, Moscow (in Russian).
Mikhalevich, V.I., 1980. Systematics and Evolution of Foraminifera in the Light of New Data on Their Cytology and Ultrastructure, vol. 94. Akademiya Nauk SSSR, Trudy Zoologicheskogo Instituta, pp. 42–61 (in Russian).
Mikhalevich, V.I., 1988. System of the Subclass Miliolata (Foraminifera), vol. 184. Akademiya Nauk SSSR, Trudy Zoologicheskogo Instituta, pp. 77–110 (in Russian).
Mikhalevich, V.I., 1993. New higher taxa of the subclass Nodosariata (Foraminifera). Zoosyst. Ross. 2, 5–8.
Mikhalevich, V.I., 1995. A new classification of the class Astrorhizata (Foraminifera). Zoosyst. Ross. 3, 161–174.
Mikhalevich, V.I., 2003. System of the Four Foraminiferal Subclasses with the Agglutinated Shell Wall (Ammodiscana, Miliamminana, Hormosiniana, Textulariana) (Foraminifera), vol. 7. Rossiiskaya Akademiya Nauk, Izvestiya Zoologicheskogo Instituta, pp. 5–50.
Miklukho-Maklay, A.D., 1949. Late Paleozoic Fusulinids of Central Asia, Fergana, Darvaz and Pamir, vol. 3. Leningradskii Gosudarstvennyi Ordena Lenina, Universitet Imeni A.A. Zhdanova, Institut Zemnoi Kory, pp. 1–111 (in Russian).

Miklukho-Maklay, A.D., 1953. On the Systematics of the Family Fusulinidae Möller, vol. 159. Uchenye Zapiski Leningradskogo Ordena Lenina Gosudarstvennogo Universiteta, seriya geologicheskikh nauk, pp. 12–24 (in Russian).
Miklukho-Maklay, A.D., 1957. Some Permian Fusulinids of Crimea, vol. 225. Vestnik Leningradskogo Universiteta, seriya geologicheskikh nauk, pp. 93–159 (in Russian).
Miklukho-Maklay, A.D., 1958. A new foraminiferal family, Tuberitinidae M.Maclay fam. nov. Vopr. Mikropaleont. 2, 130–135 (in Russian).
Miklukho-Maklay, A.D., 1963. Upper Paleozoic of Central Asia. Izdateltsvo Leningradskogo Universiteta, pp. 1–328 (in Russian).
Miklukho-Maklay, A.D., Rauzer-Chernousova, D.M., Rozovskaya, S.E., 1958. Systematics and phylogeny of fusulinids. Vopr. Mikropaleont. 2, 5–21 (in Russian).
Miklukho-Maklay, A.D., Rauzer-Chernousova, D.M., Rozovskaya, S.E., 1959. Order Fusulinida. Bases Paleontol. 1, 201–225 (in Russian).
Minato, M., Honjo, S., 1966. Filogeniya podsemeystva Neoschwagerininae. Vopr. Mikropaleont. 10, 109–124 (in Russian).
Möller, V.von, 1878. Die spiral-gewundenen Foraminiferen des russischen Kohlenkalks. Mém. Acad. Imp. Sci. St. Pétersbg. 7th Ser. 25 (9), 1–147.
Murray, J.W., 1991. Ecology and Palaeoecology of Benthic Foraminifera. Longman Publisher, Hulow.
Murray, J.W., Alve, E., Jones, B.W., 2011. A new look at modern agglutinated benthic foraminiferal morphogroups: their value in palaeoecological interpretation. Palaeogeogr. Palaeoclimatol. Palaeoecol. 309, 229–241.
Nestell, G.P., Nestell, M.K., 2006. Middle Permian (late Guadalupian) foraminifers from Dark Canyon, Guadalupe mountains, New Mexico. Micropaleontology 52, 1–50.
Nestell, G.P., Mestre, A., Heredia, S., 2009. First Ordovician foraminifera from South America: a Darriwilian (middle Ordovician) fauna from the San Juan formation, Argentina. Micropaleontology 55, 329–344.
Nestell, G., Heredia, S., Mestre, A., Beresi, M., González, M., 2011. The oldest Ordovician foraminifers (*Oepikodus evae* conodont Zone, Floian) from South America. Geobios 44, 601–608.
Nestell, G.P., Nestell, M.K., Ellwood, B.B., Wardlaw, B.R., Basu, A.R., Ghosh, N., Lan, L.T.P., Rowe, H.D., Hunt, A., Tomkin, J.H., Ratcliffe, K.T., 2015. High influx of carbon in walls of agglutinated foraminifers during the Permian-Triassic transition in global oceans. Int. Geol. Rev. 57 (4), 411–427. http://dx.doi.org/10.1080/00206814.2015.1010610.
Nestell, M.K., 1999. Correlation of permian Tethyan Neoschwagerinida (fusulinaceans) from the United States. In: Proceedings of International Symposium "Upper Permian stratotypes of the Volga Region". GEOS, Moscow, pp. 161–171.
Nestell, M.K., Pronina, G.P., 1997. The distribution and age of the genus *Hemigordiopsis*. In: Ross, C.A., Ross, J.R.P., Brenckle, P.L. (Eds.), Late Paleozoic Foraminifera; Their Biostratigraphy, Evolution, and Paleoecology; and the Mid-carboniferous Boundary, vol. 36. Cushman Foundation for Foraminiferal Research, Special Publication, pp. 105–110.
Nestell, M.K., Nestell, G.P., Wardlaw, B.R., Sweatt, M.J., 2006. Integrated biostratigraphy of foraminifers, radiolarians and conodonts in shallow and deep water Middle Permian (Capitanian) deposits of the "Rader slide", Guadalupe Mountains, West Texas. Stratigraphy 3, 161–194.
Nikitina, A.P., 1969. Genus *Hemigordiopsis* (foraminifera) in the late permian of Primorye. Paleontol. Zh. 3, 63–69 (in Russian).
Norvang, A., 1957. The foraminifera of the Lias series in Jutland, Denmark. Medd. Dan. Geol. Foren. 13, 1–135.

Okuyucu, C., 2009. Systematics and biostratigraphic notes of the upper Moscovian-upper Gzhelian fusulinid foraminifers from the Anatolian platform in the southern Turkey. Geol. Balc. 38, 35–51.

Okuyucu, C., Ivanova, D., Bedi, Y., Ergen, A., 2014. Discovery of an earliest Triassic, post-extinction foraminiferal assemblage above the Permian-Triassic boundary, Strandzha nappes, north-west Turkey. Geol. Q. 58, 1–8.

d' Orbigny, A., 1826. Tableau méthodique de la classe des Céphalopodes. Ann. Sci. Nat. 7, 245–314.

Özkan, R., Vachard, D., 2015. Little-known and new foraminifers of the lower Frasnian (Upper Devonian) of eastern Taurus (southern Turkey). Rev. Micropaléontol. 58, 267–282.

Pawlowski, J., Holzmann, M., Tyszka, J., 2013. New supraordinal classification of Foraminifera: molecules meet morphology. Mar. Micropaleontol. 100, 1–10.

Perret, M.F., Vachard, D., 1977. Algues et pseudo-algues des calcaires serpoukhoviens d'Ardengost (Hautes-Pyrénées). Ann. Paléontol. Invertébr. 63, 85–156.

Pia, J.V., 1937. Die wichtigsten Kalkalgen des Jungpaläozoikums und ihre geologische Bedeutung. Compte Rendu 2e Congrès Avancement Etudes Stratigraphie Carbonifère. Heerlen 1935 (2), 765–856.

Piller, W., 1978. Involutinacea (Foraminifera) der Trias und des Lias. Beitr. Paläontol. Öst. 5, 1–164.

Pinard, S., Mamet, B., 1998. Taxonomie des petits foraminifères du Carbonifère supérieur-Permien inférieur du basin de Sverdrup, Arctique canadien. Palaeontogr. Can. 15, 1–253.

Pirlet, H., Conil, R., 1974. L'évolution des Archaediscidae viséens. Bull. Soc. Belge Géol. 1973 82 (2), 241–299 (published 1977, preprint 1974).

Pokorny, V., 1956. Semitextulariidae, a new family of foraminifera. Univ. Carol. Geol. 2, 279–286.

Pokorny, V., 1958. Grundzüge der zoologischen Mikropaläontologie. VEB Dtsch. Verl. Wiss. 1, 1–582.

Poty, E., Devuyst, F.X., Hance, L., 2006. Upper Devonian and Mississippian foraminiferal and rugose coral zonations of Belgium and Northern France, a tool for Eurasian correlations. Geol. Mag. 143, 829–857.

Poyarkov, B.V., 1969. Stratigraphy and Foraminifers From the Devonian Deposits of Tian-Shan. Akademiya Nauk Kirgizskoi SSR, Institut Geologii, Upravlennie Geologii Kirgizskoi SSR, Izdatelstvo "Ilim", Frunze, pp. 1–186 (in Russian).

Poyarkov, B.V., 1977. Biogeography of the Devonian foraminifera. In: Dubatolov, V.N., Yuferev, O.V. (Eds.), Divisions, Paleobiogeographic Subdivisions and Biostratigraphy, vol. 347. Akademiya Nauk SSSR, Sibirskoe Otdelenie, Trudy Instituta Geologii Geofiziki, "Nauka", pp. 8–28 (in Russian).

Poyarkov, B.V., 1979. Development and Diffusion of the Devonian Foraminifers. Akademiya Nauk SSSR, Dalnevostochnyi Nauchnyi Tsentr, Dalnevostochnyi Geologicheskii Insitut, Izdateltsvo "Nauka", pp. 1–172 (in Russian).

Poyarkov, B.V., Skvortsov, V.P., 1979. On methodological segregation of local epiboles and local biozones (exemplified by the Early Carboniferous of Tian Shan). In: Palaeontology and Stratigraphy of Far East. Akademiya Nauk SSR, Dalvnevostochnyi Nauchnyi Tsentr, Dalvnevostochnyi Geologicheskii Institut, pp. 5–27 (in Russian).

Pronina, G.P., 1994. Systematics and phylogeny of the order Hemigordiopsida (foraminifers). Paleontol. Zh. 3, 13–24 (in Russian).

Pronina, G.P., 1996. Genus *Sphairionia* and its stratigraphic significance. Reports of shallow Tethys 4, International Symposium, Albrechtsberg (Austria) 8-11 September 1994, supplement Annali Musei Civici Rovereto, Sezione Archeologia. Stor. Sci. Nat. 11 (1995), 105–118.

Pronina, G.P., 1998. Foraminifers. In: Esaulova, N.K., Lozovsky, V.R., Rozanov, A.Yu (Eds.), Stratotypes and Reference Sections of the Upper Permian in the Regions of the Volga and Kama Rivers. International Symposium Upper Permian Stratotypes of the Volga Region. GEOS, Moscow, pp. 163–169 (in Russian).
Pronina, G.P., 1999. Correlation of late Permian deposits of boreal areas by means of smaller foraminifers. In: Proceedings of International Symposium: Late Permian Stratotypes of Volga Region. Kazanskii Gosudarstvennyi Universitet, Tatarstan, Rossiya, Geos, pp. 182–191 (in Russian).
Pronina, G.P., Gubenko, T.A., 1990. Upper Permian planoarchaediscids from Transcaucasia. Paleontol. Zh. 2, 119–123.
Putrya, F.S., 1956. Middle Carboniferous stratigraphy and foraminifers of the eastern Donbass. Trudy VNIGRI, 98. Mikrofauna SSSR 8, 333–485 (in Russian).
Rauzer-Chernousova, D.M., Fursenko, A.B., 1959. Protozoa. In: Bases of Paleontology. Izdatelstvo Akademii Nauk SSSR, Moscow (in Russian).
Rauzer-Chernousova, D.M., Bensh, F.R., Vdovenko, M.V., Gibshman, N.B., Leven, E.Ya, Lipina, O.A., Reitlinger, E.A., Solovieva, M.N., Chediya, I.O., 1996. Reference-Book on the Systematics of Paleozoic Foraminifers; Endothyroida and Fusulinoida. Rossiiskaya Akademiya Nauk, Geologicheskii Institut, Moskva "Nauka", pp. 1–207 (in Russian).
Reitlinger, E.A., 1950. Middle Carboniferous Foraminifers of the Central Part of Russian Platform (Fusulinidae Excepted), vol. 47. Trudy Instituta Geologicheskikh Nauk 126, geologichevskaya seriya, pp. 1–126 (in Russian).
Reitlinger, E.A., 1956. New family Lasiodiscidae. Vopr. Mikropaleont. 1, 69–78 (in Russian).
Reitlinger, E.A., 1958. On the systematics and phylogeny of the superfamily Endothyridea. Vopr. Mikropaleont. 2, 53–73 (in Russian).
Rhumbler, L., 1895. Entwurf eines natürlichen Systems der Thalamophoren. Nachrichten Gesellschaft Wissenschaften Götttingen. Math Phys 1895 (1), 51–98.
Rigaud, S., Martini, R., 2016. Agglutinated or porcelaneous tests: where to draw the line? J. Foraminifer. Res. 46 (3), 334–345.
Rigaud, S., Vachard, D., Martini, R., 2015. Agglutinated versus microgranular foraminifers: end of a paradigm? J. Syst. Paleontol. 13, 75–95.
Ross, C.A., Ross, J.R.P., 1991. Paleozoic foraminifera. BioSystems 25, 39–51.
Rozovskaya, S.E., 1961. On the systematics of the families Endothyridae and Ozawainellidae. Paleontol. Zh. 3, 19–21 (in Russian).
Rozovskaya, S.E., 1975. Composition, Phylogeny and Systematics of Order Fusulinida, vol. 149. Akademiya Nauk SSSR, Trudy Paleontologicheskogo Instituta, pp. 1–267 (in Russian).
Sabirov, A.A., 1987. New foraminiferal species of the late Silurian and Devonian of Tajikistan. Paleontol. Zh. 2, 23–29 (in Russian).
Sabirov, A.A., Gushchin, S.B., 2006. New early Ordovician calcareous foraminifers of the middle tien shan. Paleontol. J. 40, 11–19.
Saidova, KhM., 1981. On an Up-to-date System of Supraspecific Taxonomy of Cenozoic Benthic Foraminifers. Institut Okeanologii P.P. Shirshova, Akademiya Nauk SSSR, Moscow (in Russian).
Schlagintweit, F., 2012. *Moehlerina basiliensis* (Mohler, 1938): a middle Jurassic-early Cretaceous facultative (?) epilithic foraminifer. Facies 58, 637–647.
Schlagintweit, F., Hladil, J., Nose, M., 2013a. New observations and interpretations of the enigmatic poorly known late Palaeozoic *Irregularina* Bykova, 1955 (incertae sedis)? Acta Paleontol. Rom. 9, 155–182.
Schlagintweit, F., Hladil, J., Nose, M., Salerno, C., 2013b. Palaeozoic record of *Thaumatoporella* Pia, 1927 (incertae sedis)? Geol. Croat. 66, 155–182.

Schultze, M.S., 1854. Über den Organismus der Polythalamien (Foraminiferen) nebst Bemerkungen über die Rhizopoden im allgemeinen. Wielhem Engelmann, Leipzig.
Scott, D.B., Medioli, F., Braund, R., 2003. Foraminifera from the Cambrian of Nova Scotia: the oldest multichambered foraminifera. Micropaleontology 49, 109–126.
Şengör, A.M.C., Altıner, D., Clin, A., Ustaösmer, T., Hsü, K.J., 1988. Origin and assembly of the Tethyside orogenic collage at the expense of Gondwana Land. In: Audley-Charles, M.G., Hallam, A. (Eds.), Gondwana and Tethys, vol. 37. Geological Society London Special Publication, pp. 119–181.
Senowbari-Daryan, B., Zaninetti, L., 1986. Taxonomic note on reefal Miliolacea (Protista: foraminifera) from the upper Triassic Tethys. Arch. Sci. Genève 39, 79–86.
Sheng, J.Z., He, Y., 1983. Permian *Shanita-Hemigordius (Hemigordiopsis)* (foraminifera) fauna in western Yunnan, China. Acta Palaeontol. Sin. 22, 55–59 (in Chinese).
Sigal, J., 1952. Ordre des Foraminifera. In: Piveteau, J. (Ed.), Traité de Paléontologie. Masson et Cie Editeurs Paris, pp. 132–301.
Skinner, J.W., 1931. Primitive fusulinids of the mid-continent region. J. Paleontol. 5, 253–259.
Skinner, J.W., Wilde, G.L., 1954. New early Pennsylvanian fusulinids from Texas. J. Paleontol. 28, 796–803.
Solovieva, M.N., 1978. On the systematics of the foraminifers (interpretation of the biological significance of structural feartures and spatial organization of the foraminifers). Byulletin Moskovskogo Obshchestva Ispytateley Prirody. Otd. Geol. 53, 159–160 (in Russian).
Song, H.J., Tong, J.N., Zhang, K.X., Wang, Q.X., Chen, Z.Q., 2007. Foraminiferal survivors from the Permian-Triassic mass extinction in the Meishan section, South China. Palaeoworld 16, 105–119.
Song, H., Tong, J., Chen, Z.Q., 2011. Evolutionary dynamics of the Permian-Triassic foraminifer size: evidence for Lilliput effect in the end-Permian mass extinction and its aftermath. Palaeogeogr. Palaeoclimatol. Palaeoecol. 308, 98–110.
Sosipatrova, G.P., 1970. Foraminifers. In: Menner, V.V., Sarytcheva, T.G., Chernyak, G.E. (Eds.), Stratigraphy of Carboniferous and Permian Deposits of the Northern Verkhoyansk Region, vol. 154. Trudy Nauchno-Issledovatelskogo Instituta Geologii Arktiki, Ministerstvo Geologii SSSR, Vsesoyuznyi Aerogeologischeskii Trest, Izdatelstvo "Nedra", Leninsgradskoe Otdelenie, Leningrad, pp. 56–70 (in Russian).
Staff, H., von Wedekind, R., 1910. Der oberkarbonische Foraminiferen-Sapropelit Spitzbergens, vol. 10. Geological Institute Bulletin, Upsala University, pp. 81–123.
Stancliffe, R.P.W., 1989. Microforaminiferal linings: their classification, biostratigraphy and paleoecology, with special reference to specimens from British Oxfordian sediments. Micropaleontology 35, 337–352.
Subbotina, N.N., Voloshinova, N.A., Azbel, A.Y., 1981. Introduction to the study of foraminifera (classification of calcareous foraminifera of the Meso-Cenozoic. VNIGRI Leningr. (in Russian).
Termier, H., Termier, G., Vachard, D., 1975. Recherches micropaléontologiques dans le Paléozoïque supérieur du Maroc central. Cah. Micropaléontol. 4, 1–99.
Thorez, J., Dreesen, R., 1986. A model of a regressive depositional system around the old Red continent as exemplified by a field trip in the upper Famennian "Psammites du Condroz" in Belgium. Ann. Soc. Géol. Belg. 109, 285–323.
Ueno, K., 2003. The Permian fusulinoidean faunas of the Sibumasu and Baoshan blocks: their implications for the paleogeographic and paleoclimatologic reconstruction of the Cimmerian Continent. Palaeogeogr. Palaeoclimatol. Palaeoecol. 193, 1–24.
Ueno, K., Miyahigashi, A., Chaoentitirat, T., 2010. The Lopingian (Late Permian) of mid-oceanic carbonates in the Eastern Palaeotethys: stratigraphical outline and foraminiferal faunal succession. Geol. J. 45, 285–307.

Ukharskaya, L.B., 1970. New arenaceous foraminifera from the Kazanian (permian) of the Russian platform. Paleontol. Zh. 4, 468–476 (in Russian).
Ukharsaya, L.B., 1975. On peculiarities of distribution of Kazanian and Zechstein agglutinating foraminifers. Vopr. Mikropaleont. 18, 167–170 (in Russian).
Ukharsaya, L.B., 1978. Discovery of Kazanian Agglutinated Foraminifers in Russian Platform, vol. 289. Trudy Vsesoyuznogo Ordena Lenina Nauchno-Issledovatel'skogo Geologicheskogo Instituta Imeni A.P. Karpinskogo, novaya seriya, pp. 26–38 (in Russian).
Vachard, D., 1974a. Remarques sur les Foraminifères des calcaires griottes *sensu lato* (Frasnien inférieur-Tournaisien inférieur) du versant méridional de la Montagne Noire (Aude-Hérault). C. R. Somm. Soc. géol. Fr. 1973 (4), 116–118.
Vachard, D., 1974b. Contribution à l'étude stratigraphique et micropaléontologique (algues et foraminifères) du Dévonien-Carbonifère inférieur de la partie orientale du versant méridional de la Montagne Noire (Hérault, France) (Ph.D. Paris).
Vachard, D., 1980. Téthys et Gondwana au Paléozoïque supérieur – les données afghanes-biostratigraphie, micropaléontologie, paléogéographie. Doc. Trav. IGAL 2, 1–463.
Vachard, D., 1988. Pour une classification raisonnée et raisonnable des Archaediscidae (Foraminifera, Carbonifère inférieur-moyen). Rev. Paléobiol. Vol. Spéc. 2. Benthos' 86, 103–123.
Vachard, D., 1994. Foraminifères et moravamminides du Givétien et du Frasnien du domaine Ligérien (Massif Armoricain, France). Palaeontogr. A 231 (1–3), 1–92.
Vachard, D., 2014. *Colaniella*, wrongly named, well-distributed Late Permian nodosariate foraminifers. Permophiles 60, 16–24.
Vachard, D., 2016. Permian smaller foraminifers; taxonomy, biostratigraphy and biogeography. In: Lucas, S.G., Shen, S.Z. (Eds.), The Permian Timescale, vol. 450. Geological Society London, Special Publications. http://dx.doi.org/10.1144/SP450.1 (in press).
Vachard, D., Beckary, S., 1991. Algues et foraminifères bachkiriens des coal balls de la Mine Rosario (Truebano, Léon, Espagne). Rev. Paléobiol. 10, 315–357.
Vachard, D., Bouyx, E., 2002. Les *Eopolydiexodina* géantes (Foraminiferida, Fusulinina) du Permien moyen d'Afghanistan, remarques préliminaires. Ann. Soc. Géol. Nord 9 (2), 163–189.
Vachard, D., Cózar, P., 2004. *Insolentitheca* emend. *Protoinsolentitheca* n. gen., and Caligellidae emend., permanent cysts of Paleozoic foraminifera? Riv. Ital. Paleontol. Stratigr. 110, 591–603.
Vachard, D., Cózar, P., 2010. An attempt of classification of the Palaeozoic incertae sedis Algospongia. Rev. Esp. Micropaleontol. 42, 129–241.
Vachard, D., Ferrière, J., 1991. Une association à *Yabeina* (foraminifère fusulinoïde) dans le Midien (Permien supérieur) de la région de Whangaroa (Baie d'Orua, Nouvelle-Zélande). Rev. Micropaléontol. 34, 201–230.
Vachard, D., Gaillot, J., 2006. Embryonic apparati and reproduction patterns in *Eopolydiexodina* (fusulinida, Schwagerinoidea, Guadalupian, middle permian). J. Foraminifer. Res. 36, 77–89.
Vachard, D., Krainer, K., 2001a. Smaller foraminifers of the upper Carboniferous Auernig group, Carnic Alps (Austria/Italy). Riv. Ital. Paleontol. Stratigr. 107, 147–168.
Vachard, D., Krainer, K., 2001b. Smaller foraminifers, characteristic algae and pseudo-algae of the latest Carboniferous/early permian Rattendorf group, Carnic Alps (Austria/Italy). Riv. Ital. Paleontol. Stratigr. 107, 169–195.
Vachard, D., Massa, D., 1989. Apparition précoce du genre *Nanicella* (Foraminifère) dans le Dévonien inférieur du Sud-Tunisien. Bull. Soc. Belge. Géol. 98, 287–293.
Vachard, D., Moix, P., 2011. Late Pennsylvanian to Middle Permian revised algal and foraminiferan biostratigraphy and palaeobiogeography of the Lycian Nappes (SW Turkey): palaeogeographic implications. Rev. Micropaléontol. 54, 141–174.

Vachard, D., Montenat, C., 1981. Biostratigraphie, micropaléontologie et paléogéographie du Permien de la région de Tezak (Montagnes Centrales d'Afghanistan). Palaeontogr. B 178, 1–88.

Vachard, D., Téllez-Girón, C., 1986. El género *Polyderma* y nuevas soluciones al problema de las Calcisferas (microproblematicas paleozoicas). Rev. Inst. Mex. Pet. 18, 6–44.

Vachard, D., Oviedo, A., Flores de Dios, A., Malpica, R., Brunner, P., Guerrero, M., Buitrón, B.E., 1992. Barranca d'Olinalá: une coupe de référence pour le Permien du Mexique central; étude préliminaire. Ann. Soc. Géol. Nord 2 (2ème série) 153–160.

Vachard, D., Clift, P., Decrouez, D., 1993. Une association à *Pseudodunbarula* (Fusulinoïde) du Permien supérieur (Djoulfien) remaniée dans le Jurassique d'Argolide (Grèce). Rev. Paléobiol. 12, 217–242.

Vachard, D., Martini, R., Rettori, R., Zaninetti, L., 1994. Nouvelle classification des Foraminifères Endothyroïdes du Trias. Geobios 27, 543–557.

Vachard, D., Hauser, M., Martini, R., Zaninetti, L., Matter, A., Peters, T., 2002. Middle permian (Midian/Capitanian) fusulinid assemblages from the Aseelah Unit (Batain group) in the Batain plain, East-Oman: their significance to Neotethys paleogeography. J. Foraminifer. Res. 32, 155–172.

Vachard, D., Laveine, J.P., Zhang, S., Huang, H., Zhan, M., Liu, L., Lemoigne, Y., 2004a. A rich assemblage with *Eoparastaffella* (foraminifera) from the lower Visean Cf4β (Mississippian) of Tianshui area (Guangxi Zhuang Autonomous region, South China). Rev. Paléobiol. 22, 753–759.

Vachard, D., Munnecke, A., Servais, T., 2004b. New SEM observations of keriothecal walls: implications for the evolution of fusulinida. J. Foraminifer. Res. 34, 232–242.

Vachard, D., Pille, L., Gaillot, J., 2010. Palaeozoic Foraminifera: systematics, palaeoecology and responses to the global changes. Rev. Micropaléontol. 53, 209–254.

Vachard, D., Krainer, K., Lucas, S., 2012. Pennsylvanian (late Carboniferous) calcareous microfossils from Cedro Peak (New Mexico, USA); Part 1: algae and microproblematica. Ann. Paléontol. 98, 225–252.

Vachard, D., Krainer, K., Lucas, S., 2013. Pennsylvanian (late Carboniferous) calcareous microfossils from Cedro Peak (New Mexico, USA); Part 2: smaller foraminifers and fusulinids. Ann. Paléontol. 99, 1–42.

Vachard, D., Haig, D.W., Mory, A.J., 2014. Lower Carboniferous (middle Visean) foraminifers and algae from an interior sea, southern Carnarvon basin, Australia. Geobios 47, 57–74.

Vachard, D., Krainer, K., Lucas, S., 2015. Late early permian (late Leonardian; Kungurian) algae, microproblematica, and smaller foraminifers from the Yeso group and san Andres formation (New Mexico; USA). Palaeontol. Electron. 18.1.21A.

Vachard, D., Cózar, P., Aretz, M., Izart, A., 2016. Late Visean-early Serpukhovian foraminifers in the Montagne Noire (France). Part 1, new species. Geobios. http://dx.doi.org/10.1016/j.geobios.2016.09.002.

Vachard, D., Clausen, S., Palafox, J.J., Buitrón, B.E., Devaere, L., Hayart, V., Régnier, S., 2017. Early Ordovician microfacies and microfossils from Cerro San Pedro (San Pedro de la Cueva, Sonora, Mexico; paleogeographical implications). Facies (in press).

Vdovenko, M.V., Rauzer-Chernousova, D.M., Reitlinger, E.A., Sabirov, A.A., 1993. Reference-Book on the Systematics of Paleozoic Smaller Foraminifers. Rossiiskaya Akademiya Nauk, Komissiya Mikropaleontologii, Moscow "Nauka", pp. 1–128 (in Russian).

Versteegh, G.J.M., Servais, T., Streng, M., Munnecke, A., Vachard, D., 2009. A discussion and proposal concerning the use of the term calcispheres. Palaeontology 52, 343–348.

Vilesov, A.P., 1998. The origin and taxonomic position of the genus *Sphaeroschwagerina* Miklukho-Maklay, 1956 (Foraminifera, Schwagerinida). Paleontol. J. 32, 437–446.
Vishnevskaya, V.S., Sedaeva, K.M., 2002a. A revision of some foraminiferal taxa of the Order Parathuramminida and discussion of foraminiferal and radiolarian evolution. Paleontol. J. 36, 581–590.
Vishnevskaya, V.S., Sedaeva, K.M., 2002b. On the zonal divisions of the deposits of the Devonian-Carboniferous boundary based on parathuramminids (foraminiferal or foraminifera-radiolarian scale? In: Chuvashov, B.I., Amon, E.O. (Eds.), Stratigraphy and Paleogeography of Carboniferous of Eurasia. Institute of Geology and Geochemistry. Akademiya Nauk, Mezhvedomstvennyi Stratigrafichevskii Komitet Rossii, Institut Geologii Geokhimii, UrO RAN, Ekaterinburg, pp. 53–60.
Wedekind, P.R., 1937. Einführung in die Grundlagen der historischen Geologie. Band II. Mikrobiostratigraphie die Korallen und Foraminiferenzeit. Ferdinand Enke, Stuttgart.
Wendt, J., 1969. Foraminiferen-"Riffe" im karnischen Hallstätter Kalk des Feuerkogels (Steiermark, Österreich). Paläontol. Z. 43, 177–193.
Wilde, G.L., 1990. Practical Fusulinid zonation: the species concept, with Permian Basin emphasis. West Tex. Geol. Soc. Bull. 29 (5–15), 28–34.
Wilde, G.L., 2006. Pennsylvanian-Permian Fusulinaceans of the Big Hatchet Mountains, NewMexico, vol. 38. New Mexico Museum Natural History and Science Bulletin, pp. 1–311.
Williamson, W.C., 1881. On the Organisation of the Fossil Plants of the Coal Measures, Pt X, Including an Examination of the Supposed Radiolarians of the Carboniferous Rocks. Philos. Trans. R. Soc. (1880) 171 (pt II), 493–539.
Winchester-Seeto, T., Bell, K.N., 1994. Microforaminiferal linings from the early Devonian of eastern Australia, and their generic placement. J. Paleontol. 68 (2), 200–207.
Yarahmadzahi, H., Vachard, D., 2014. Paleobiogeographic significance of a new ozawainelloid fusulinid *Pseudoacutella partoazari* n. sp., from the Asselian (lowermost Permian) of Gaduk (Central Alborz, Iran). Rev. Micropaléontol. 57, 117–124.
Zadorozhnyi, V.M., Yuferev, O.V., 1984. Phylum Protozoa, Class Sarcodina, Subclass Foraminifera, vol. 568. Akademiya Nauk SSR, Sibirskoe Otdelenie, Trudy Instituta Geologii Geofiziki, pp. 70–113 (in Russian).
Zandkarimi, K., Najafian, B., Vachard, D., Bahrammanesh, M., Vaziri, S.H., 2016. Latest Tournaisian-late Viséan foraminiferal biozonation (MFZ8-MFZ14) of the Valiabad area, northwestern Alborz (Iran): geological implications. Geol. J. 51, 125–142.
Zaninetti, L., Altıner, D., 1979. La famille des Archaediscidae (Foraminifères) : analyse taxonomique et propositions pour une nouvelle subdivision. Archives Sciences. Genève 32, 163–175.
Zhang, L.X., Wang, Y.J., Wang, J.H., 1981. Classification of Fusulinida. In: Selected Papers on the 1st Convention of Micropaleontological Society of China. Science Press, Peking, pp. 30–36.

CHAPTER FIVE

Ultra-High-Resolution Palynostratigraphy of the Early Bajocian Sauzei and Humphriesianum Zones (Middle Jurassic) from Outcrop Sections in the Upper Rhine Area, Southwest Germany

S. Feist-Burkhardt*,§,1 and A.E. Götz¶

*SFB Geological Consulting & Services, Ober-Ramstadt, Germany
§University of Geneva, Geneva, Switzerland
¶Keele University, Keele, United Kingdom
[1]Corresponding author: E-mail: feistburkhardt@gmail.com

Contents

Stratigraphy & Timescales, Volume 1
ISSN 2468-5178
http://dx.doi.org/10.1016/bs.sats.2016.10.001

Abstract

Marine and terrestrial palynomorphs of the Middle Jurassic, Early Bajocian are documented from three outcrop sections in the Upper Rhine area, southwest Germany. The studied part of the sections corresponds to the Early Bajocian Sauzei and Humphriesianum zones and is independently age dated down to ammonite subzone level. Palynomorph assemblages are quantitatively studied and detailed data are provided. The palynomorph assemblages are discussed and a succession of dinoflagellate cyst stratigraphical events is proposed. The dinoflagellate cyst assemblages are characterized by many first occurrences of known taxa and many new taxa left in open nomenclature. Most of them belong to the family of the Gonyaulacaceae that typically shows very fast evolution at this time. The findings are compared to coeval assemblages from Europe and Australasia. The study provides detailed data on the dinoflagellate cyst succession in a poorly studied time interval at a much finer scale than previously known. It will add valuable information to improve Middle Jurassic palynostratigraphical schemes.

1. INTRODUCTION

Dinoflagellate cysts are known to be excellent biostratigraphical markers in the Jurassic. After a slow beginning in the Lower Jurassic, they show a tremendous radiation in the Middle Jurassic, when dinoflagellate cyst families increase in diversity. One of the most interesting time intervals, when this radiation received a boost, is the Bajocian. Thanks to the rapid evolution of dinoflagellate cysts, detailed, high-resolution palynostratigraphical schemes have been developed that are continually improved. Palynostratigraphy is a recognized, valuable tool in hydrocarbon exploration and is also applied in many other fields of geological problem solving providing high-resolution age dating of strata.

The present contribution deals with the detailed palynostratigraphical analysis of three outcrop sections in the Upper Rhine area. The sections are well dated by ammonites and partly with foraminifers and ostracods. In this area the stratigraphical interval of the Sauzei and Humphriesianum zones is represented by a much thicker sedimentary succession than in other areas of southwest Germany. They thus provide a unique opportunity to study their palynomorph content at a much finer scale than previously known. This case study provides detailed data on the dinoflagellate cyst succession in a poorly studied time interval and will serve as a reference for future palynostratigraphical studies in the early Middle Jurassic.

2. LOCALITIES

The studied outcrop sections are located in southwest Germany, in the Upper Rhine area, about half way between Freiburg im Breisgau and Basel, Switzerland (Fig. 1). Middle Jurassic sediments crop out on the eastern flank of the Rhine Graben, between the Rhine Graben main fault in the west and the Black Forest to the east (Ohmert, 1994, Fig. 1). The three sections, Badenweiler, Egerten, and Hohlebach, together provide an almost complete succession from the upper part of the Wedelsandstein Formation and the Gosheim Formation up to the base of the Hauptrogenstein Formation.

For the present palynological study, 32 samples were collected from all accessible beds of the Sauzei and Humphriesianum zones. Sampling was carried out in 1995 together with W. Ohmert from the Geological Survey of Baden-Württemberg in Freiburg (Landesamt für Geologie, Rohstoffe und Bergbau). Outcrop conditions were good to satisfactory at the time of sampling, but may well be worse today.

2.1 Badenweiler

The section is located along the road called Schwärzestrasse connecting the communities Badenweiler and Britzingen, close to the Römerberg Clinic (German map coordinates r.: 34 00 900, h.: 52 98 020, 400–410m a.s.l.;

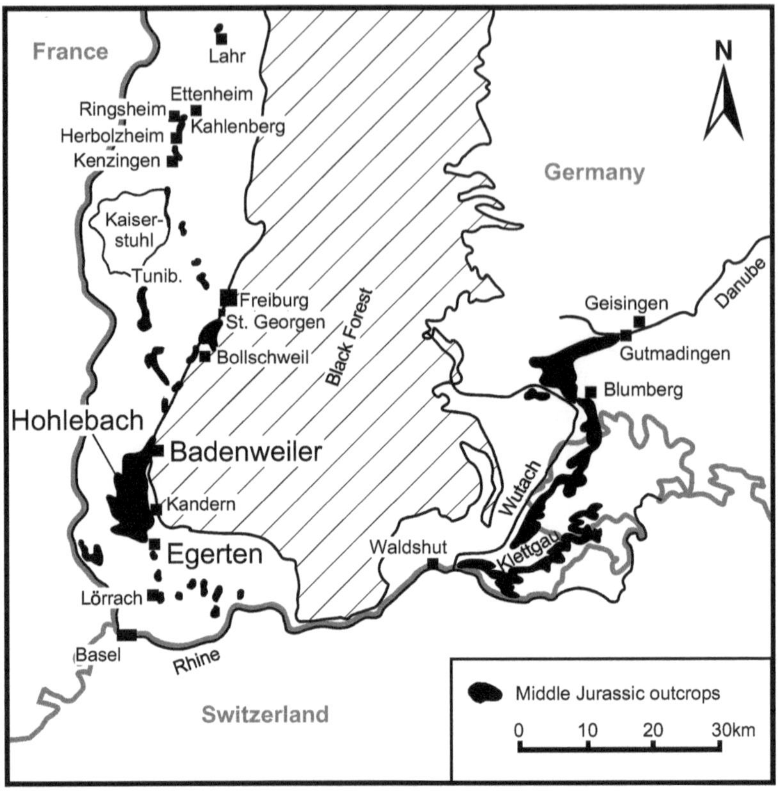

Figure 1 Location map and Middle Jurassic outcrops around the southern Black Forest. *Modified after Dietze, V., Bosch, K., Franz, M., Kutz, M., Schweigert, G., Wannenmacher, N., Studer, S., 2013. Die Humphriesianum-Zone (Unter-Bajocium, Mitteljura) am Kahlenberg bei Ringsheim (Oberrheingraben, SW Deutschland). Palaeodiversity 6, 29–61.*

WGS-84 coordinates: 47.812766°N, 7.675783°E). The section was described in detail by Ohmert (1988) including lithology, fossil contents, lithostratigraphy, and ammonite biostratigraphy down to subzone levels. For the present palynological study 12 samples were taken from the Demissusbänke Member of the Wedelsandstein Formation to the Gosheim Formation. Biostratigraphically the sampled interval corresponds to the Early Bajocian Sauzei Zone and Pinguis to Blagdeni subzones of the Humphriesianum Zone (Fig. 2).

2.2 Egerten

The section is located about 2 km southeast of Kandern and 1.5 km N of Egerten village, in the valley of the little river Wollbach (German map

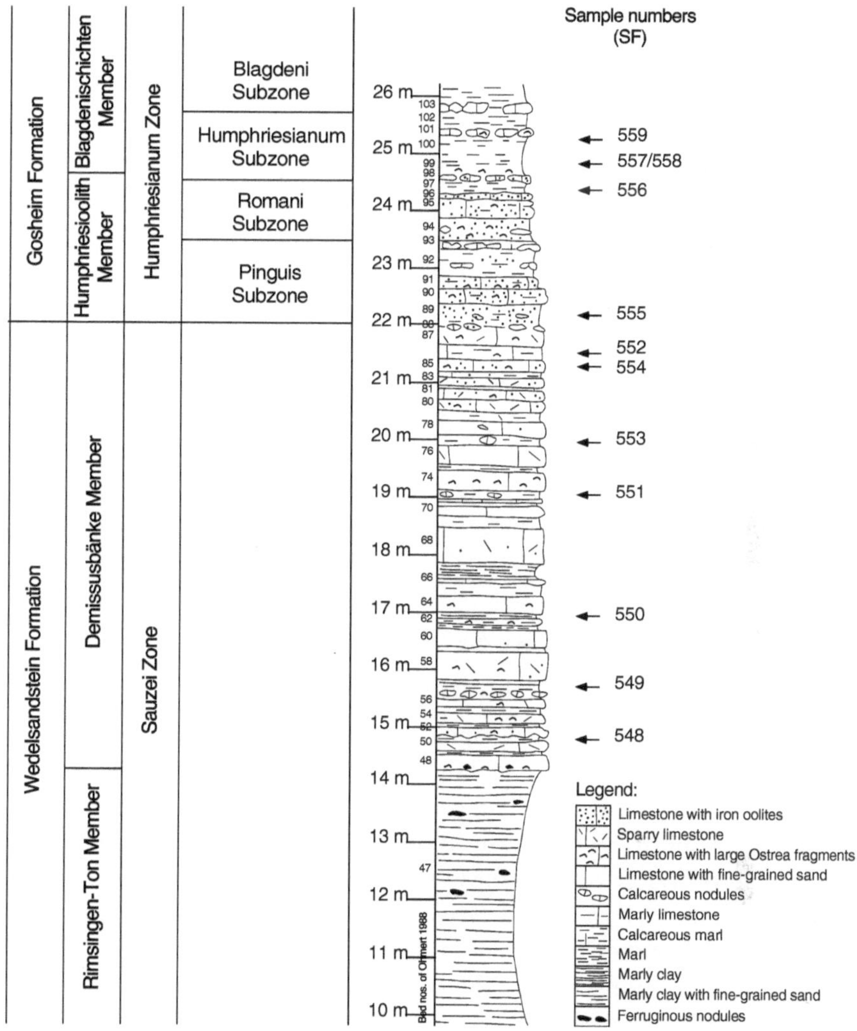

Figure 2 Detailed log and sample position of the Badenweiler section. *Modified after Ohmert, W., 1988. Das Unter-Bajocium von Badenweiler (Oberrhein), verglichen mit Nachbargebieten. Jh. Geol. Landesamt Baden-Württemberg 30, 315–347.*

coordinates r.: 34 01 449, h: 52 85 324, 445m a.s.l; WGS-84 coordinates: 47.6987°N, 7.686°E). The section was described in detail by Gassmann and Ohmert (1990) including lithology, fossil contents, lithostratigraphy, and ammonite biostratigraphy. Eighteen palynology samples were taken from the Demissusbänke Member of the Wedelsandstein Formation up to the Blagdenischichten Member of the Gosheim Formation, little below the boundary to the Hauptrogenstein Formation above. Biostratigraphically the sampled interval corresponds to the Early Bajocian Sauzei Zone to the Blagdeni Subzone of the Humphriesianum Zone (Figs. 3 and 4).

2.3 Hohlebach

The section is located about 5 km northwest of Kandern, in the Hohlebach River valley, half way between the villages Schlingen-Liel and Schlingen-Niedereggenen, about 200 m ENE of Kutzmühle (German map coordinates r.: 33 96 310, h.: 52 91 300, ca. 300m ü. NN; WGS-84 coordinates: 47.751621°N, 7.616119°E). In a short section along the river, the boundary between the Blagdenischichten Member of the Gosheim Formation and the Hauptrogenstein Formation crops out. Two palynological samples were taken, one from below and one from above the boundary between the two formations. Biostratigraphically the samples correspond to the Early Bajocian Humphriesianum Zone, Blagdeni Subzone.

3. STRATIGRAPHICAL SETTING

3.1 Litho- and Biostratigraphy

In 2002 the German Stratigraphic Commission introduced formal lithostratigraphical formations and published the first Stratigraphic Table of Germany (STD 2002; Deutsche Stratigraphische Kommission, 2002); this was followed by a short description by Bloos et al. (2005). These publications only give an overview of the general situation in Germany and, since their introduction, some formal units have been emended and renamed. An emended version of the Stratigraphic Table of Germany is expected for 2016.

In the present study, we use the current formal stratigraphical units for the Upper Rhine area that are found in the "Symbolschlüssel Geologie Baden-Württemberg" published by the Geological Survey of Baden-Württemberg in Freiburg (LGRB, Landesamt für Geologie, Rohstoffe und Bergbau) in its newest edition (LGRB, 2016). This list is regularly updated and is available for free download from the LGRB website.

Detailed lithological descriptions of the studied outcrops are found in Ohmert (1988) and Gassmann and Ohmert (1990). The studied sections correspond to the Early Bajocian Sauzei and Humphriesianum ammonite zones. Subdivision is down to subzone level and is based on direct ammonite evidence and in some cases complemented by micropaleontological data on foraminifera and ostracods (Gassmann and Ohmert, 1990; Ohmert, 1988).

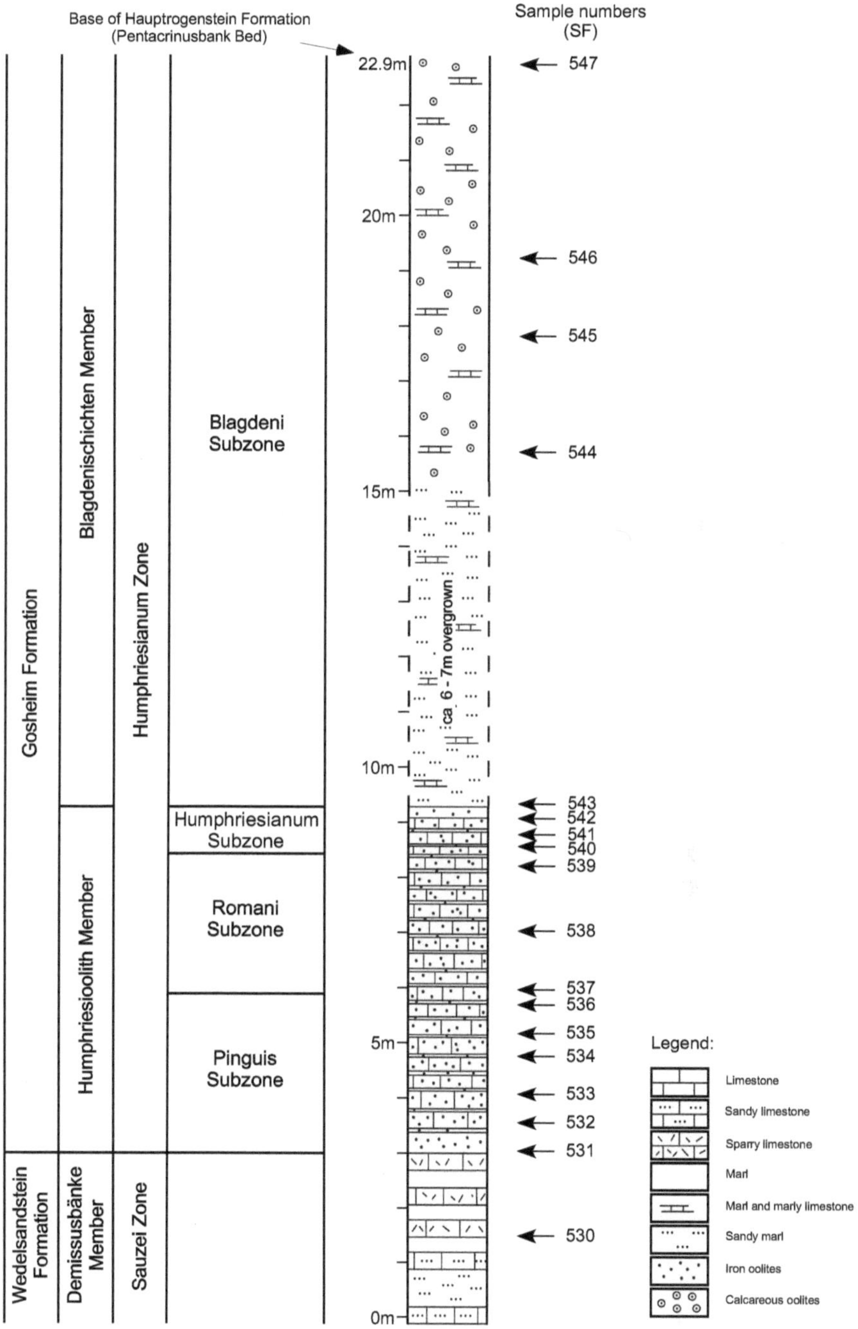

Figure 3 General log and sample position of the Egerten section. *Modified after Gassmann, G., Ohmert, W., 1990. Der Humphriesi-Oolith von Egerten im Wollbachtal (Oberrheingebiet N Lörrach). Jh. Geol. Landesamt Baden-Württemberg 32, 159–170. For details of the lower part of the section see Fig. 4.*

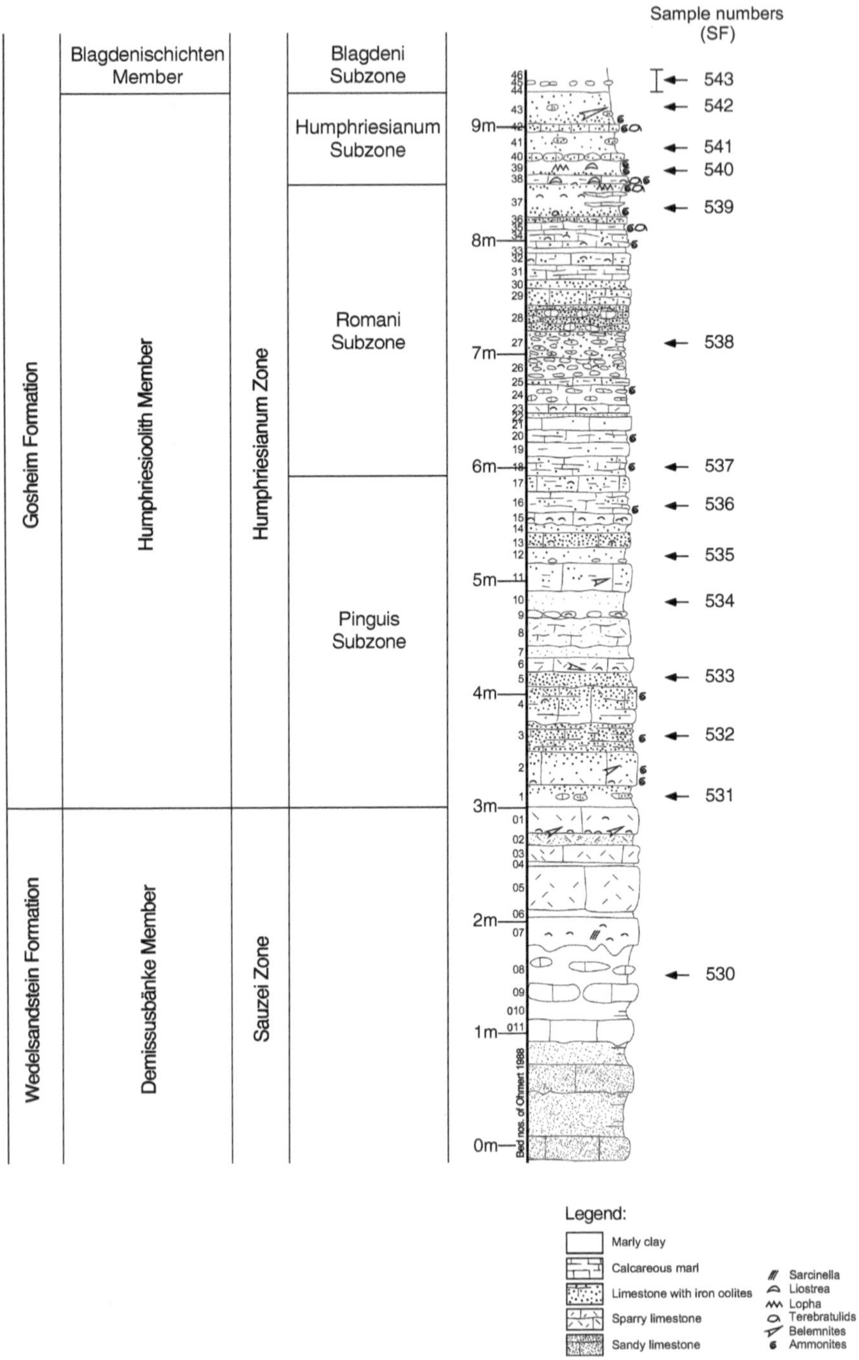

Figure 4 Detailed log and sample position of the Egerten section, lower part. *Modified after Gassmann, G., Ohmert, W., 1990. Der Humphriesi-Oolith von Egerten im Wollbachtal (Oberrheingebiet N Lörrach). Jh. Geol. Landesamt Baden-Württemberg 32, 159–170.*

3.1.1 *Wedelsandstein Formation, Demissusbänke Member*

We studied only the upper part of the Wedelsandstein Formation, the Demissusbänke Member. The underlying Rimsingen-Ton Member was not sampled. The Demissusbänke Member consists of alternating limestones and marls, with a maximum thickness in the Badenweiler area of about 7.5 m. The Demissusbänke Member corresponds to the later part of the Sauzei Zone.

3.1.2 *Gosheim Formation*

The Gosheim Formation was recently introduced by Dietze et al. (2015) and replaced the Humphriesioolith Formation of STD (2002) and Bloos et al. (2005) in the Upper Rhine area and the southwestern part of Swabia. The Gosheim Formation comprises the Humphriesioolith Member below and the Blagdenischichten Member above. The Humphriesioolith Member consists of fossiliferous limestones and calcareous marls, both with varying contents of iron oolites. The iron oolitic series varies from about 2.5 to 6.5 m thick. The Blagdenischichten Member has its base where the first beds without iron oolites begin. The lower part consists of sandy marls without oolites and some intercalated marly limestones. The upper part is formed of marls and marly limestones containing calcareous oolites. Thickness of the member in the Egerten area is about 13–14 m. The Gosheim Formation corresponds to most of the Humphriesianum Zone. The latest part of the Blagdeni Subzone reaches into the overlying Hauptrogenstein Formation (Ohmert, 1988).

3.1.3 *Hauptrogenstein Formation*

The Hauptrogenstein Formation consists of a thick succession of calcareous oolitic limestones. The base of the formation is formed of the "Untere Pentacrinusbank", a sparry limestone bed containing crinoids. The Hauptrogenstein Formation was deposited during the latest Early Bajocian to the Early Bathonian. The base of the Hauptrogenstein Formation in the studied area is near the top of the Humphriesianum Zone, Blagdeni Subzone (Gonzalez, 1996; Gonzalez and Wetzel, 1996; Ohmert, 1988).

3.2 Chronostratigraphy

The current chrono-/biostratigraphical subdivision of the Bajocian in southwest Germany has recently been summarized by Dietze et al. (2011). These authors list and describe the current standard zones and subzones of the Bajocian in southwest Germany as well as the ammonite faunal horizons for each unit that have been recorded in the area.

For international comparisons it is very important to be aware of the fact that in southwest Germany the Pinguis Subzone is dealt with as the basal subzone of the Humphriesianum Zone, whereas its equivalent, the Hebridica Subzone in many international listings (e.g., TimeScale Creator) is assigned to the Sauzei Zone (or its equivalent, the Propinquans Zone). In southwest Germany the Humphriesianum Zone therefore starts earlier and spans a longer time interval. Correspondingly, the Sauzei Zone (equivalent to the Propinquans Zone) finishes earlier and spans a shorter time interval. A juxtaposition of the different schemes and their assigned age according to *The Geologic Time Scale* (GTS 2012; Gradstein et al., 2012) is given in Fig. 5. TimeScale Creator v. 6.4 uses the ammonite zonation provided by the Groupe Français d'Etude du Jurassique 1997 (Rioult et al., 1997). The same authors explain and discuss the historical background of the problem with the Pinguis Zone and its placement in competing zonations (Rioult et al., 1997, pp. 47–48).

According to TimeScale Creator v. 6.4 using the latest Geologic Time Scale GTS 2012, the Sauzei and Humphriesianum zones together span an interval from 169.87 to 169.45 Ma, thus equal to a temporal duration of 420,000 years. This is drastically different to GTS 2004 (Gradstein et al., 2004). Using GTS 2004, the two zones span from 170.62 to 169.64 Ma, equivalent to a duration of 980,000 years, more than double the GTS 2012 calculated duration. This is, nonetheless, a very short time interval for biostratigraphical purposes in the Middle Jurassic.

Stage	Southwest Germany (Dietze et al. 2011) *(used herein)*		TimeScale Creator (GTS 2012)	
	Zone	Subzone	Zone (Ma)	Subzone (Ma)
Bajocian (pars)	Humphriesianum	Blagdeni	Humphriesianum (169.45–169.70)	Blagdeni (169.45–169.53)
	Humphriesianum	Humphriesianum	Humphriesianum	Humphriesianum (169.53–169.62)
	Humphriesianum	Romani	Humphriesianum	Romani (169.62–169.70)
	Humphriesianum	Pinguis	Propinquans (169.70–169.87)	Hebridica (169.70–169.79)
	Sauzei	*no subzones defined*	Propinquans	Sauzei (169.79–169.87)

Figure 5 Bio-/Chronostratigraphical standard zones and subzones in southwest Germany (Dietze et al., 2011) and those used in Time Scale Creator v. 6.4. Absolute ages are from The Geologic Time Scale (GTS 2012; Gradstein et al., 2012).

4. METHODS

A total of 32 samples were processed using standard palynological techniques (e.g., Wood et al., 1996). In a quantitative microscopical analysis two consecutive counts are carried out. In a first count 200 grains of all palynomorphs are counted and the number of dinoflagellate cysts is noted (column "DA = dinocyst abundance" in the distribution charts, Figs. 6–8). In a second count only dinoflagellate cysts are counted until a total of 100 dinoflagellate cysts are reached. The reminder of the slide is checked for additional species out of count. Taxa recorded out of count are marked with a "+" in the distribution charts (Figs. 6–8). The applied split count technique facilitates a quick assessment of the percentage of a taxon in relation to the dinoflagellate cyst assemblage or the total palynomorph assemblage from the counts in the distribution charts. The abundance of fungal remains is semiquantitatively assessed (R = rare, O = occasional, C = common, A = abundant, SA = superabundant). Results are illustrated using the software package StrataBugs v. 2.1 (Figs. 6–8). A selection of palynomorphs is illustrated on Plates (1–17).

5. SYSTEMATIC PALYNOLOGY

In the present study many dinoflagellate cysts have been recorded that differ from published taxa. These taxa are left in open nomenclature and are only briefly described in the following paragraphs pending further taxonomical studies (Feist-Burkhardt, in progress). The other taxa can be found in the online index of fossil dinoflagellate cysts DINOFLAJ2 (Fensome et al., 2008). For the descriptive terminology of dinoflagellate cysts we follow the modifications of Fensome et al. (2009). Main differences in the modified descriptive terminology are the use of the term archeopyle formula where it is appropriate, and not to use the sometimes cumbersome "para" terminology of Evitt et al. (1977), since we know that we are dealing with cysts and not motile stages of dinoflagellates.

Genus *Atopodinium* Drugg 1978 emend. Masure 1991
Atopodinium sp. 1 (*Atopodinium polygonale* precursor)
(Plate 11(5))

This species has a morphology which is intermediate between *Kallosphaeridium hypornatum* Prauss 1989 and *Atopodinium polygonale* (Beju 1983)

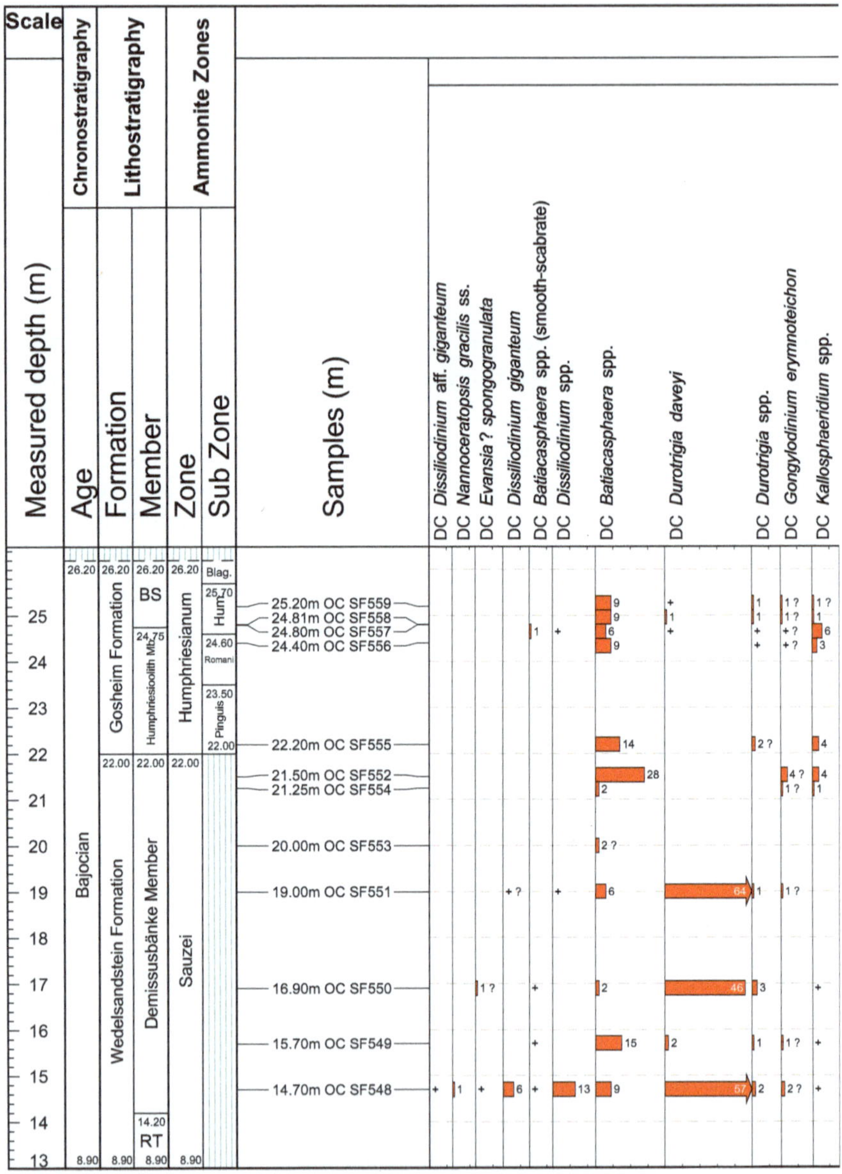

Figure 6 Quantitative distribution of palynomorphs in the Badenweiler section, important events, dinoflagellate cyst abundance (*DA*: 200 counts equals 100% of the total palynomorph assemblage) and percentages of all palynomorph categories (*AC*, Acritarchs; *ALBO*, Algae, Botryococcus; *ALPR*, Algae, Prasinophytes; *ALZY*, Algae, Zygnematophyceae; *DA*, Dinocyst abundance; *FT*, Foraminiferal test linings; *FU*, Fungi; *SP*, Spores and pollen). Semiquantitative assessment of fungal remains: *A*, abundant; *C*, common; *O*, occasional; *R*, rare; *SA*, superabundant. Abbreviations of lithostratigraphical units: *BS*, Blagdenischichten Member; *RT*, Rimsingen-Ton Member. Abbreviation of ammonite subzonal units: *Hum.*, Humphriesianum Subzone.

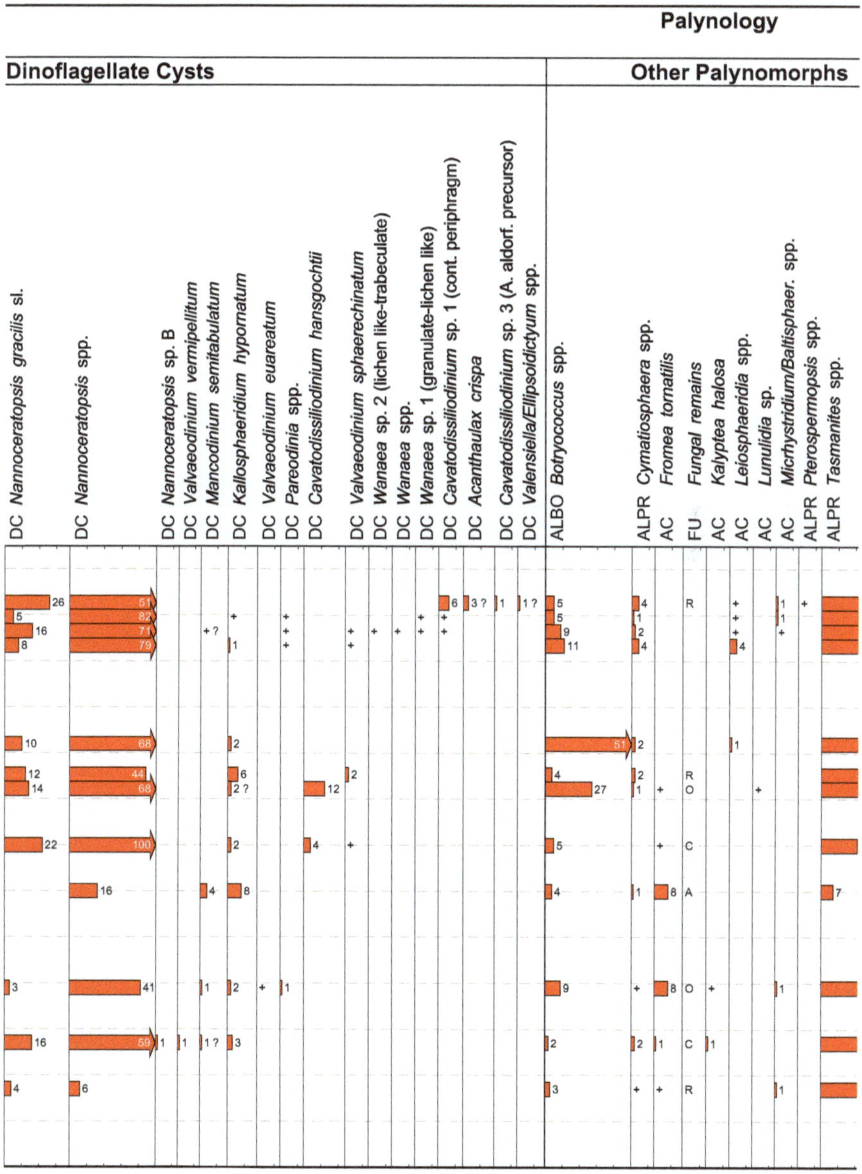

Figure 6 (continued).

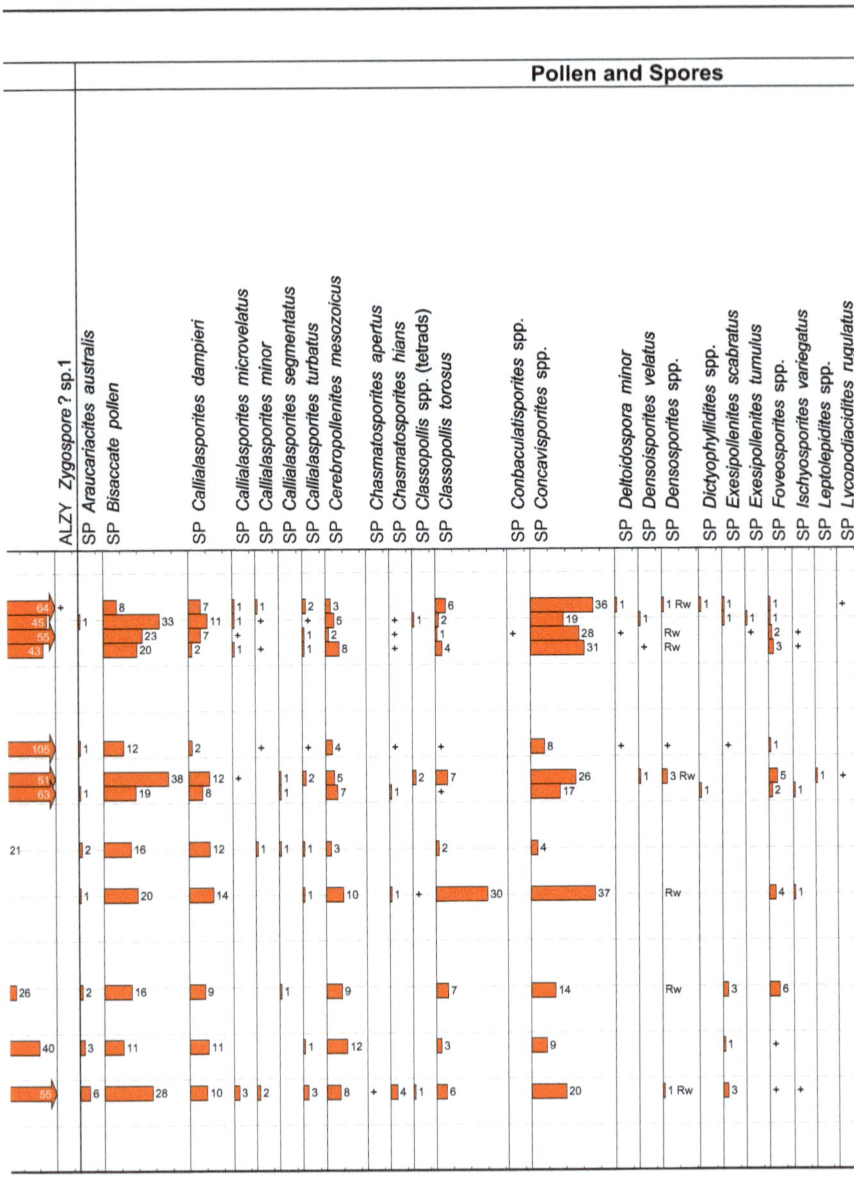

Figure 6 (*continued*).

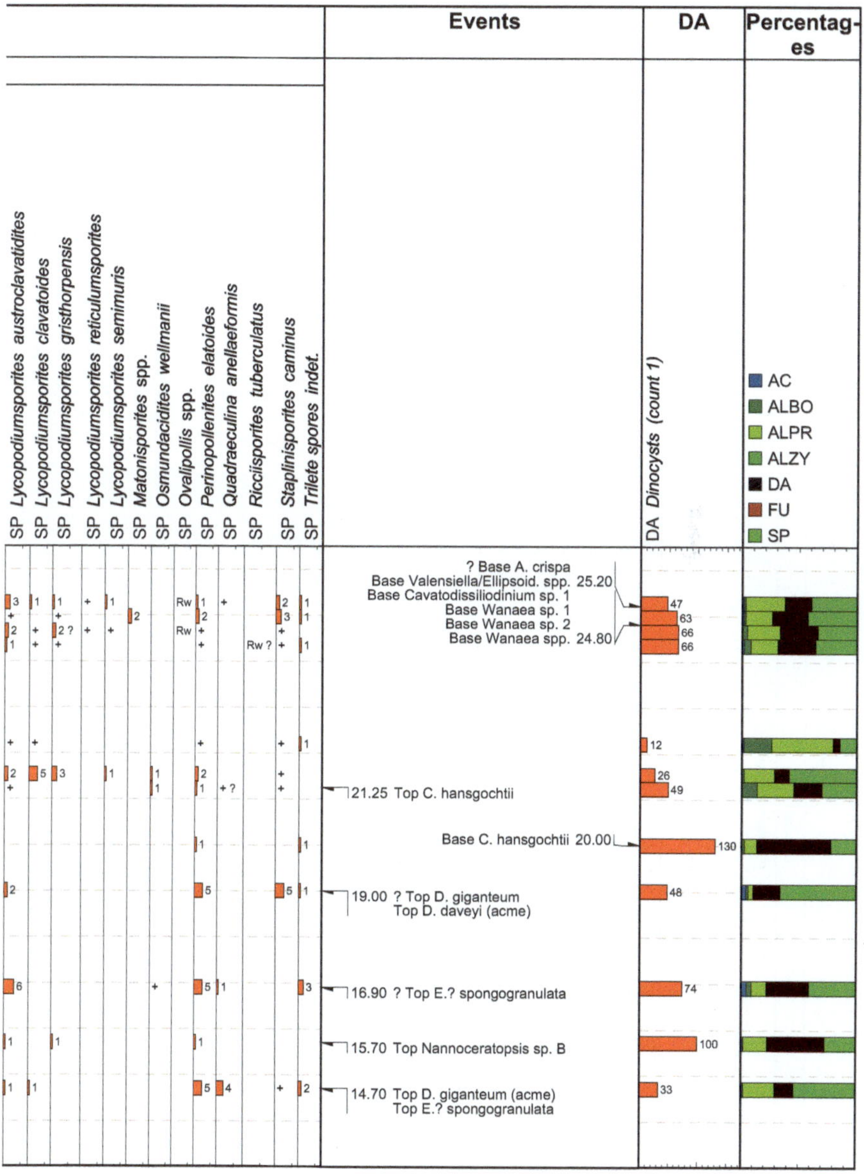

Figure 6 (*continued*).

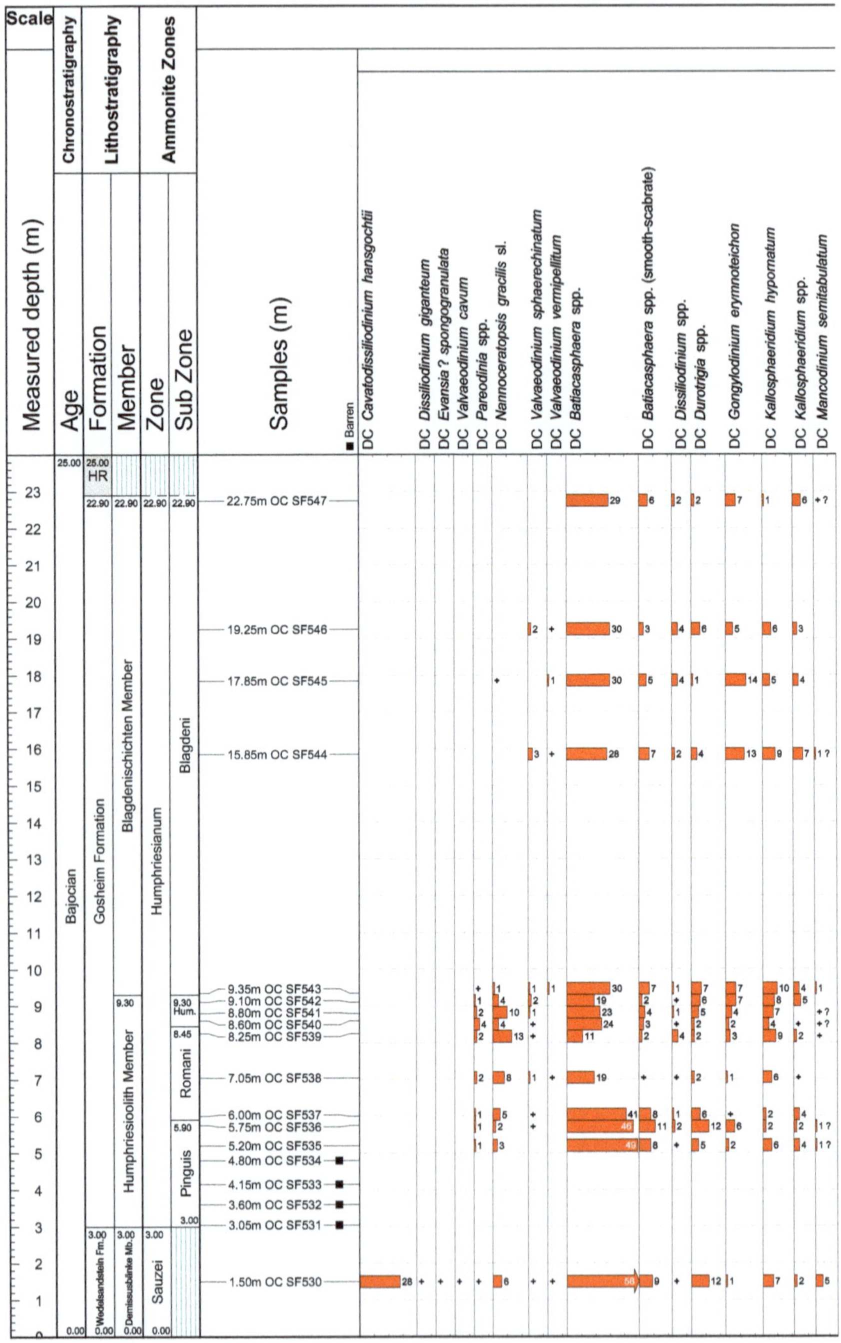

Figure 7 Quantitative distribution of palynomorphs in the Egerten section, important events, dinoflagellate cyst abundance (*DA*: 200 counts equals 100% of the total palynomorph assemblage) and percentages of all palynomorph categories (*AC*, Acritarchs; *ALBO*, Algae, Botryococcus; *ALPR*, Algae, Prasinophytes; *ALZY*, Algae, Zygnematophyceae; *DA*, Dinocyst abundance; *FT*, foraminiferal test linings; *FU*, fungi; *SP*, spores and pollen).

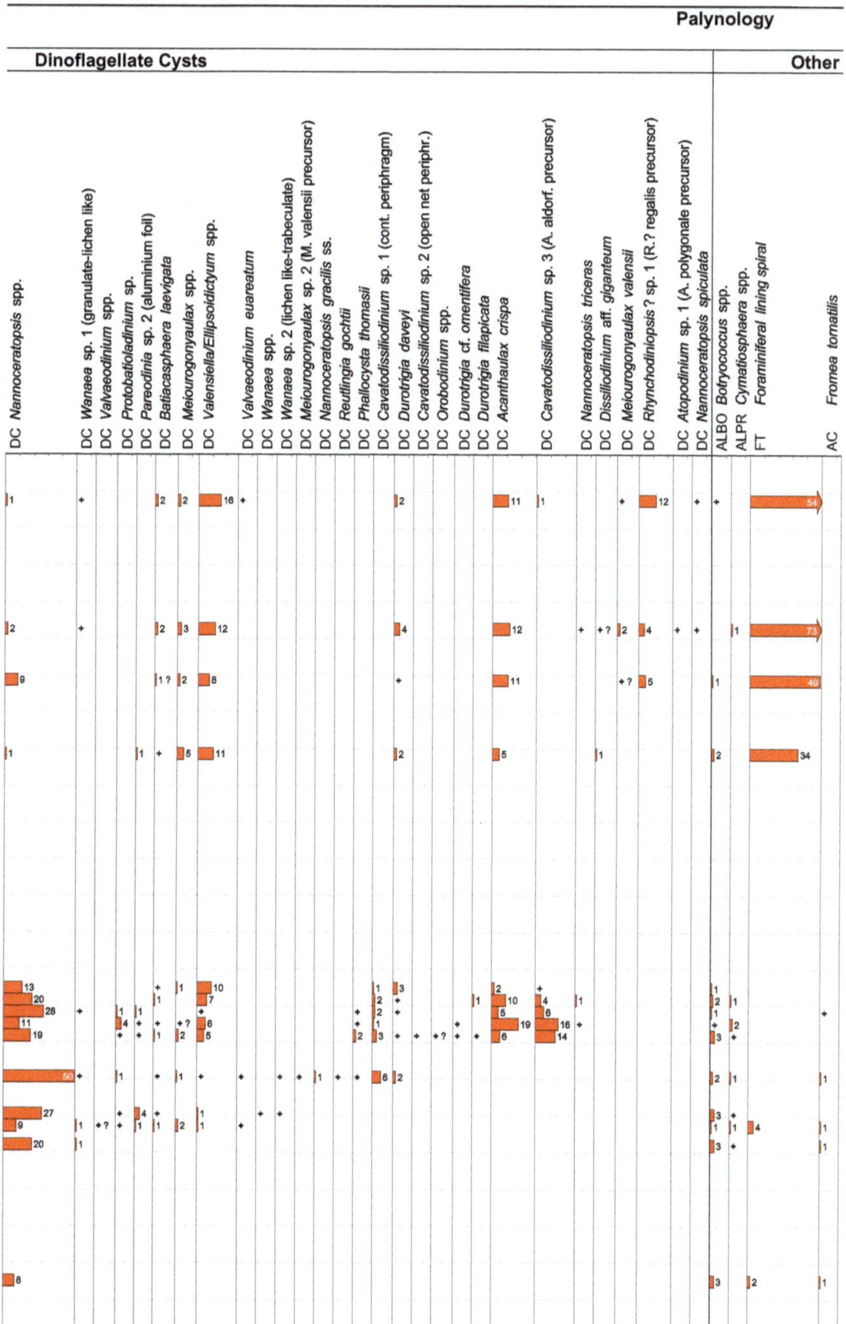

Figure 7 (*continued*) Semiquantitative assessment of fungal remains: *A*, abundant; *C*, common; *O*, occasional; *R*, rare; *SA*, superabundant. Abbreviation of lithostratigraphical units: *HR*, Hauptrogenstein Formation. Abbreviation of ammonite subzonal units: *Hum.*, Humphriesianum Subzone.

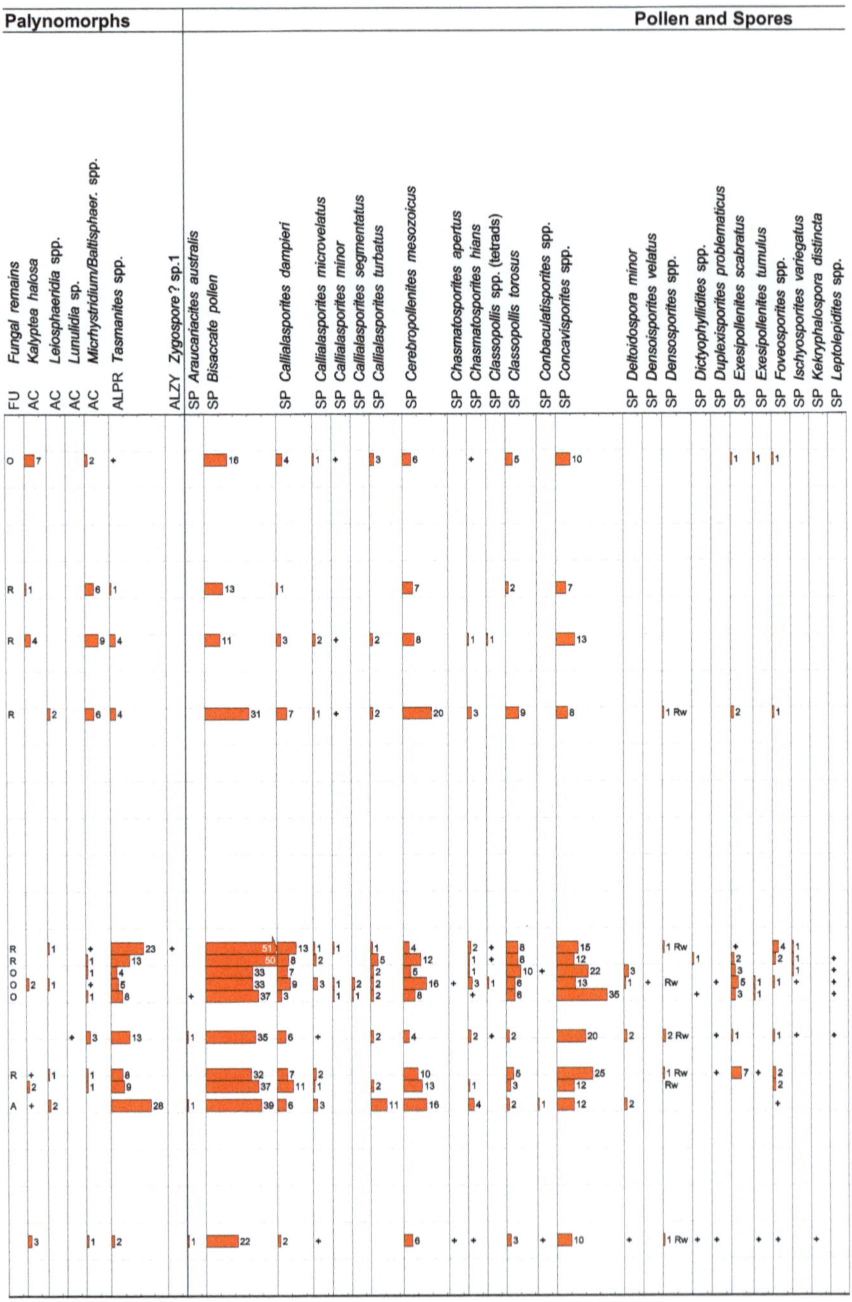

Figure 7 (continued).

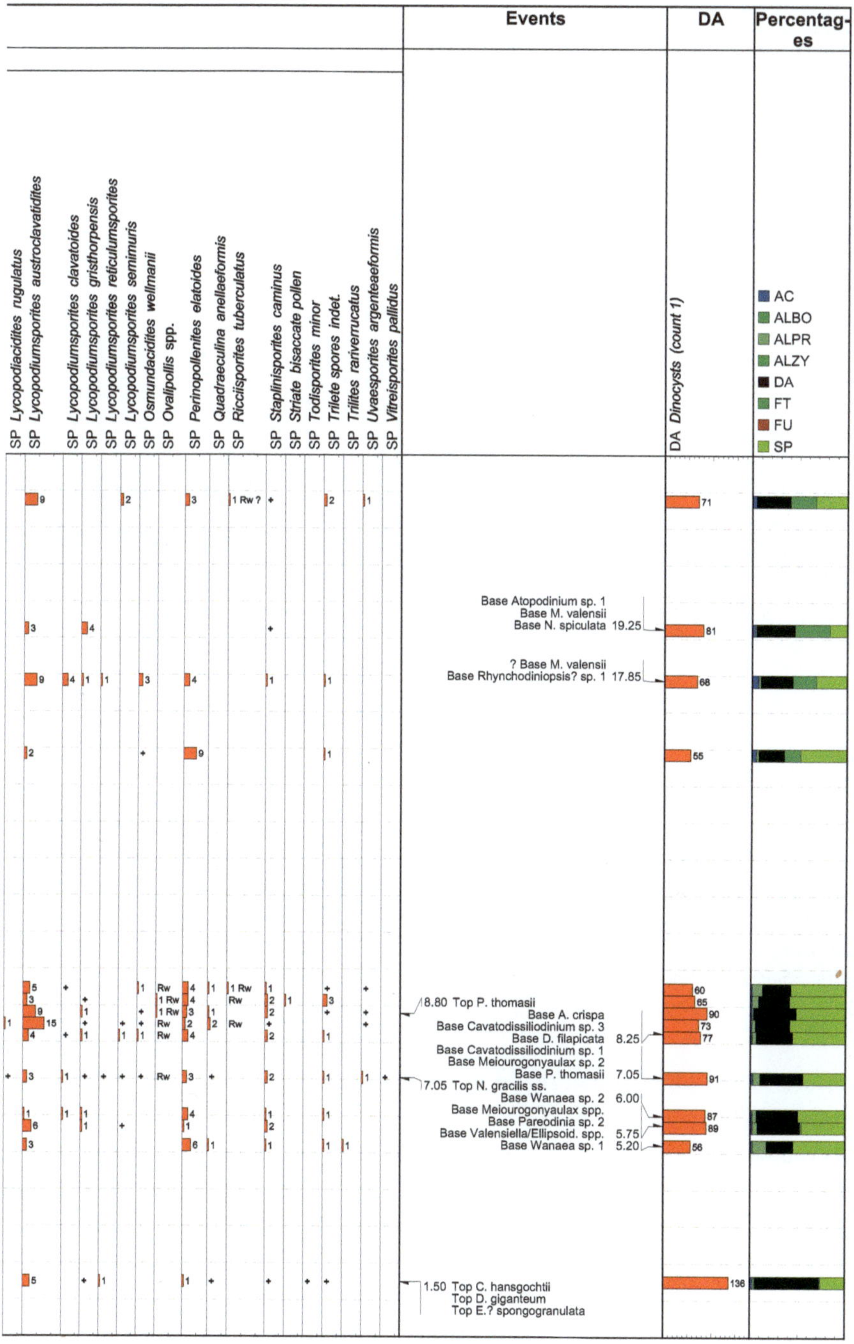

Figure 7 (*continued*).

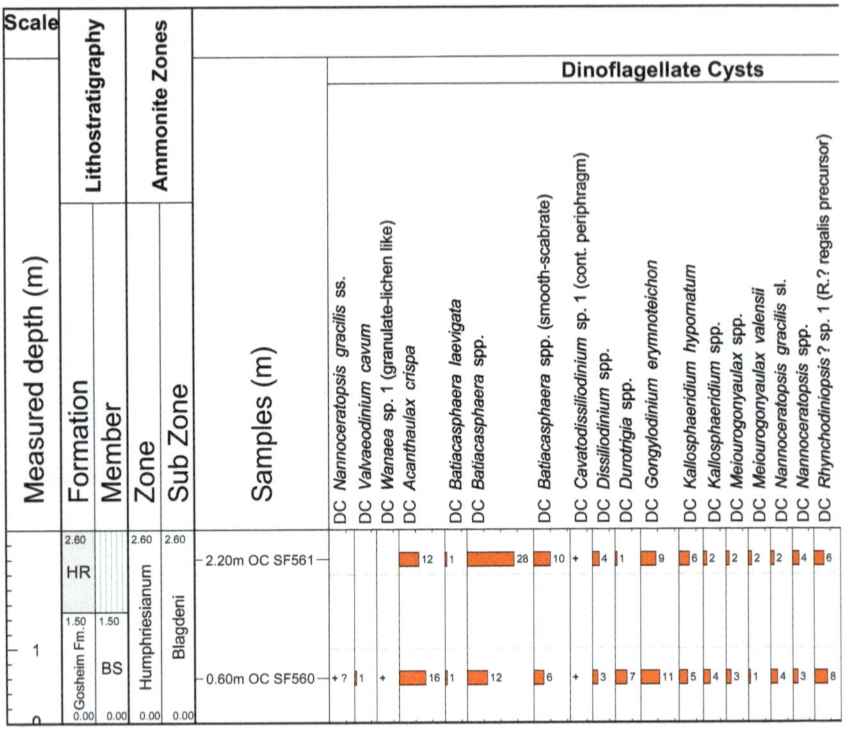

Figure 8 Quantitative distribution of palynomorphs in the Hohlebach section, important events, dinoflagellate cyst abundance (*DA*: 200 counts equals 100% of the total palynomorph assemblage) and percentages of all palynomorph categories (*AC*, Acritarchs; *ALBO*, Algae, Botryococcus; *ALPR*, Algae, Prasinophytes; *ALZY*, Algae, Zygnematophyceae; *DA*, Dinocyst abundance; *FT*, foraminiferal test linings; *FU*, fungi; *SP*, spores and pollen). Semiquantitative assessment of fungal remains: *A*, abundant; *C*, common; *O*, occasional; *R*, rare; *SA*, superabundant. Abbreviations of lithostratigraphical units: *BS*, Blagdenischichten Member; *HR*, Hauptrogenstein Formation.

Masure 1991. The general shape of the cyst and the archeopyle type is quite similar to *Kallosphaeridium*. The cyst wall is mainly thin and smooth, except for the antapical region where it is coarsely ornamented. Whereas *K. hypornatum* has a rounded antapex (e.g., Plate 11(1 and 2)), in *Atopodinium* sp. 1 the coarse ornamentation at the antapex starts becoming a rigid, angular shield, affecting the shape of the hypocyst, as it is known from *A. polygonale*. *Atopodinium* sp. 1 further differs from *A. polygonale* in lacking the typical elongate, polygonal hypocyst.

Atopodinium sp. 1 is considered to be a precursor of *A. polygonale*. It occurs in the Upper Rhine area samples in the Humphriesianum Zone,

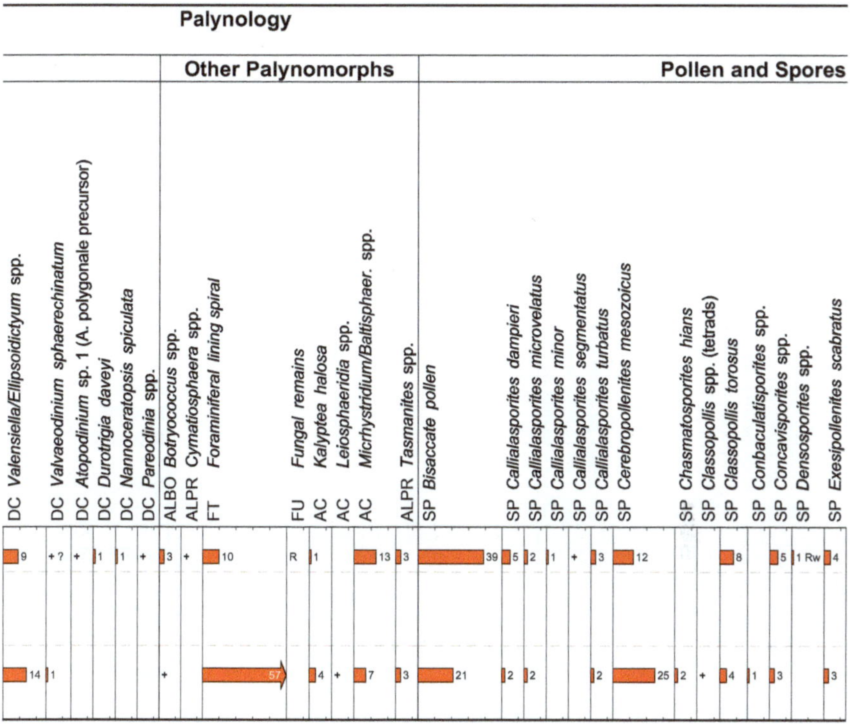

Figure 8 (*continued*).

Blagdeni Subzone. *A. polygonale* has its first appearance in southwest Germany slightly later in the Niortense Zone (Feist-Burkhardt and Wille, 1992; as *Bejuia polygonalis*).

Genus *Batiacasphaera* Drugg 1970
Batiacasphaera spp.
(Plates 8(5–8), 9(1–9), 10 (1 and 2))

Species of *Batiacasphaera* may exhibit a wide variety in both structure and ornamentation of the cyst wall, which makes it extremely difficult to speciate with confidence. There are species that are smooth to scabrate, strongly ornamented, granulate, verrucate, lichen-like ornamented, trabeculate, or pustulate. The cyst wall may be thin to thick, or spongy.

Species of *Batiacasphaera* may exhibit an ortho- or meta-type apical arrangement. Feist-Burkhardt and Monteil (1997, p. 40, 41) discussed the ortho- and meta-type apical arrangements in the genus *Batiacasphaera*. In the ortho-type, plate 2′ is in contact with 1″, causing a camerate 1″ and a

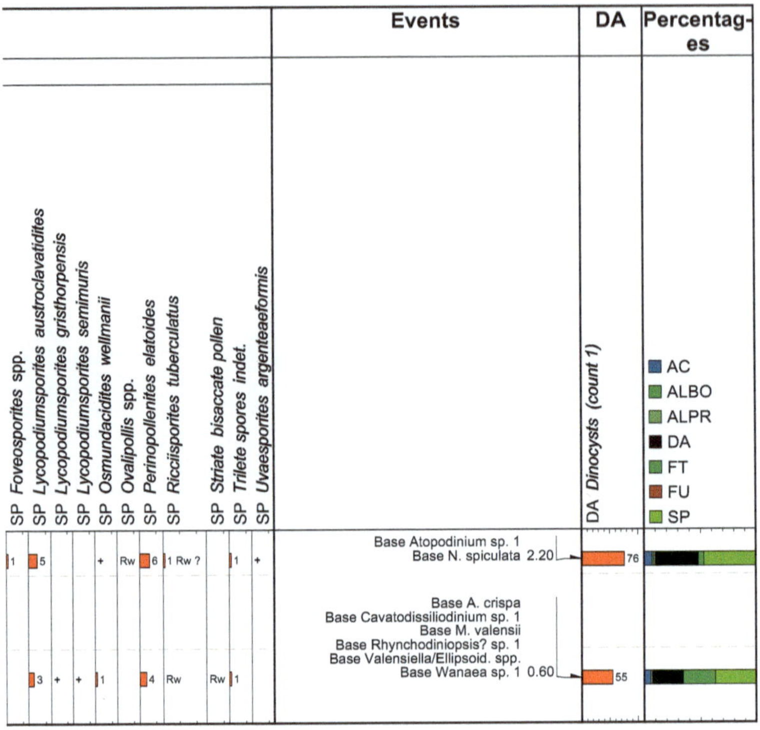

Figure 8 (*continued*).

planate 2″. In the meta-type, plate 1′ is in contact with 2″, causing a planate 1″ and a camerate 2″.

Ortho- and meta-type apical arrangement is a tabulation character to which not enough attention has been paid until now, but it may help to differentiate between species with few other morphological characters such as in *Batiacasphaera*.

Genus *Cavatodissiliodinium* Feist-Burkhardt & Monteil 2001
Cavatodissiliodinium hansgochtii Feist-Burkhardt & Monteil 2001
(Plate 2(1–6))

1990 *Scriniodinium* sp.—Feist-Burkhardt, Plate 4(4).

1992 *Endoscrinium* sp.—Feist-Burkhardt and Wille, Fig. 2.

2001 *Cavatodissiliodinium hansgochtii* n.g. et n.sp.—Feist-Burkhardt and Monteil, p. 52, 54, 56, Figs. 2/1-7, 18/1-8, 19/2.

This species and genus was originally described by Feist-Burkhardt and Monteil (2001) from the present material from Badenweiler and Egerten.

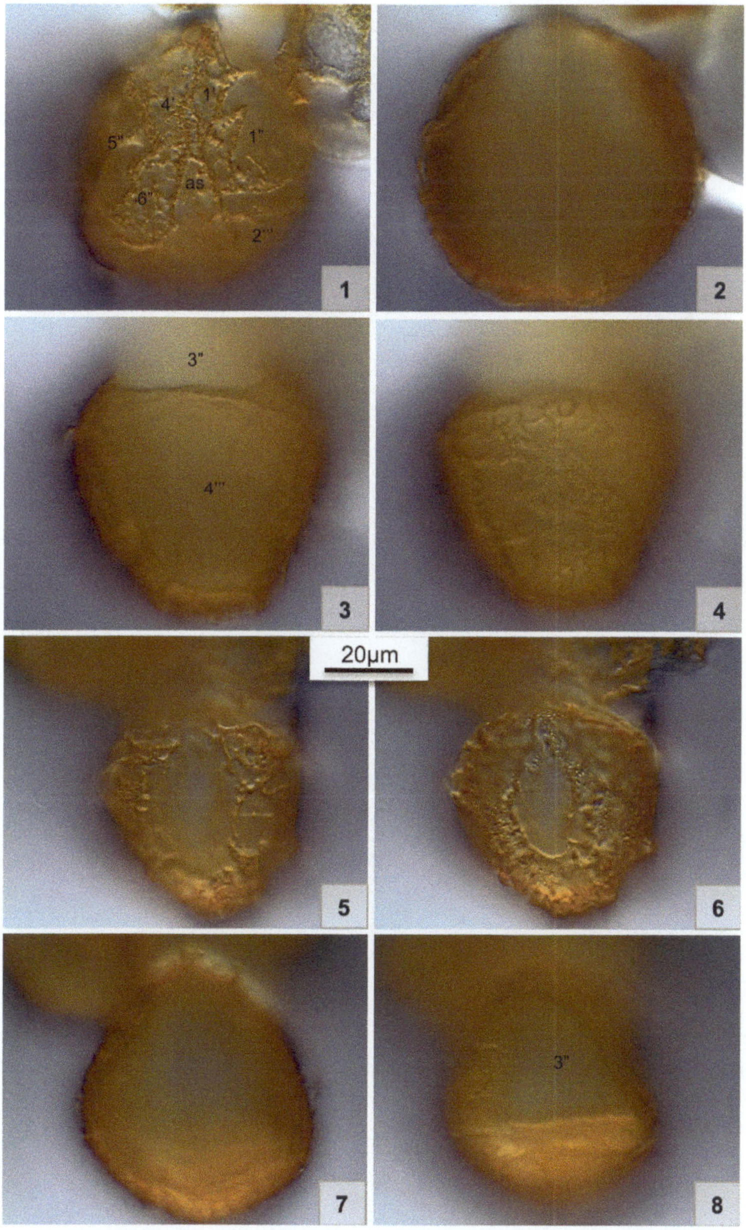

Plate 1. Dinoflagellate cysts from the Early Bajocian of the Upper Rhine area. All images were taken using differential interference contrast. The sample and slide number is followed by England Finder coordinates. All specimens are at the same magnification, scale bar = 20 μm.

(1–8) *Acanthaulax crispa* (Wetzel 1967) Woollam & Riding 1983.

(1–4) Specimen in ventral view. Descending focal series. Egerten SF 546-o-1, L37/2.

(5–8) Specimen in ventral view. Descending focal series. Egerten SF 546-o-1, G31.

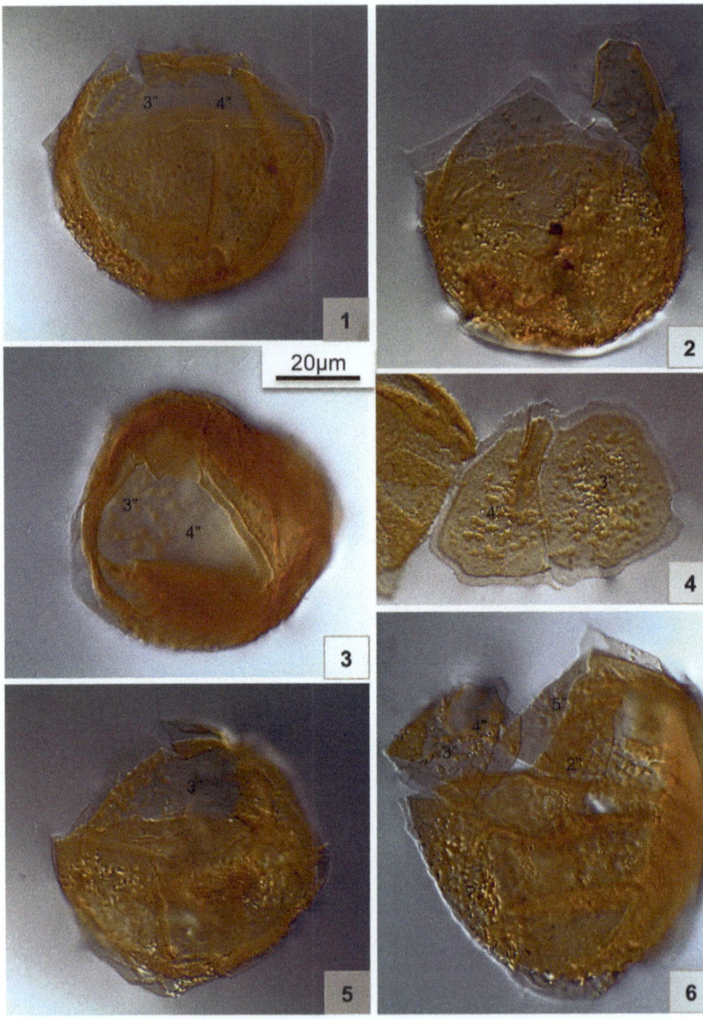

Plate 2. Dinoflagellate cysts from the Early Bajocian of the Upper Rhine area. All images were taken using differential interference contrast. The sample and slide number is followed by England Finder coordinates. All specimens are at the same magnification, scale bar = 20 µm.

(1–6) *Cavatodissiliodinium hansgochtii* Feist-Burkhardt & Monteil 2001.

(1) Dorsal face in high focus. Egerten SF 530-o-1, K17/1.

(2) Paratype 4. Specimen in right lateral view. Egerten SF 530-o-1, U40/1.

(3) Paratype 2. Dorsal face in high focus. Egerten SF 530-o-2, Q27/2.

(4) Paratype 3. Isolated opercular pieces 3″ and 4″ in internal view. Egerten SF 530-o-1, S25/1.

(5) Paratype1. Dorsal face in high focus. Egerten SF 530-o-1, Q17/2.

(6) Holotype. Specimen in right lateral view. Egerten SF 530-o-1, L23.

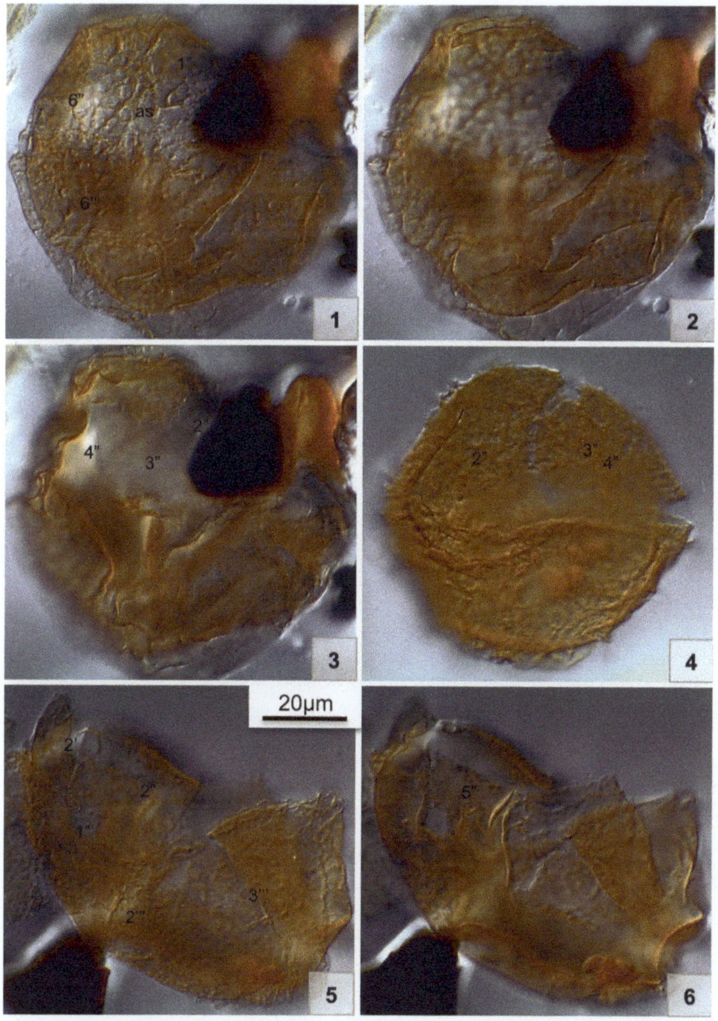

Plate 3. Dinoflagellate cysts from the Early Bajocian of the Upper Rhine area. All images were taken using differential interference contrast. The sample and slide number is followed by England Finder coordinates. All specimens are at the same magnification, scale bar = 20 μm.

(1–6) *Cavatodissiliodinium* sp. 1. A species with thin continuous periphragm. Note the varying extent of pericoel development in the different specimens.

(1–3) Specimen in ventral view, descending focal series. Opercular pieces 2″, 3″, and 4″ lost. Egerten SF 538-o-1, M35.

(4) Specimen in left lateral view. Opercular pieces 3″ and 4″ partially detached. Egerten SF 542-o-1, M37/1.

(5–6) Specimen in left lateral view, descending focal series. Opercular pieces 3″ and 4″ lost, 2″ partially detached. Egerten SF 538-o-1, O41/3.

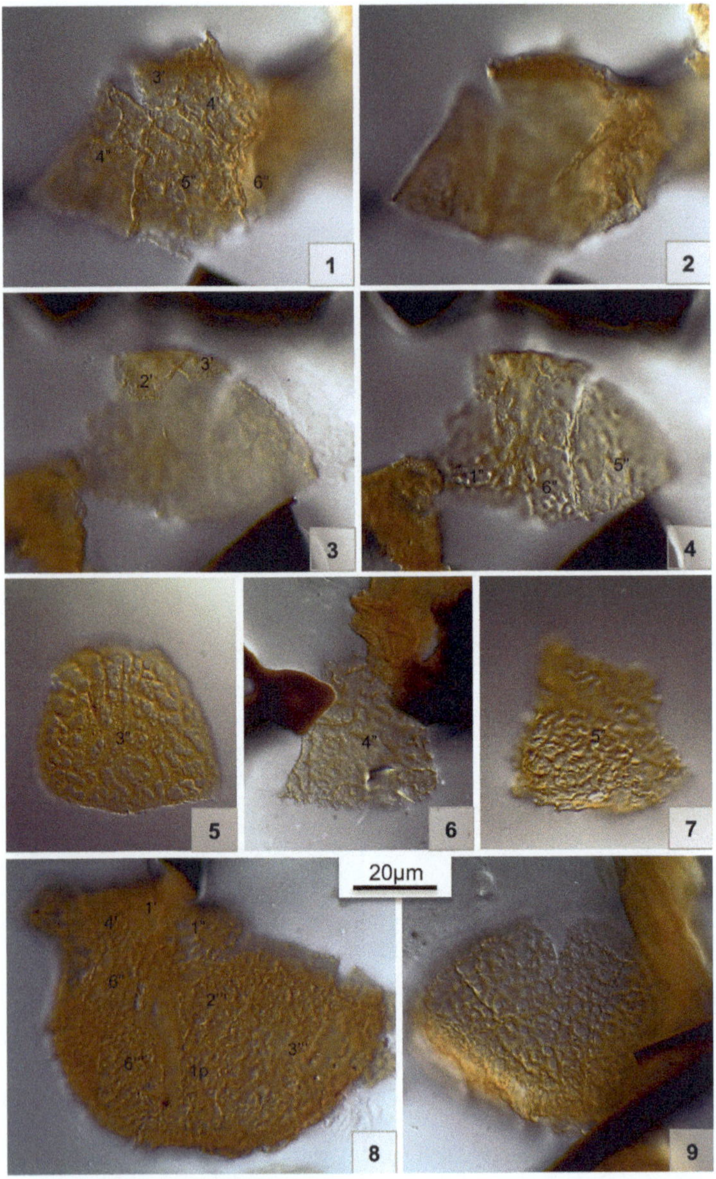

Plate 4. Dinoflagellate cysts from the Early Bajocian of the Upper Rhine area. All images were taken using differential interference contrast. The sample and slide number is followed by England Finder coordinates. All specimens are at the same magnification, scale bar = 20 μm.

(1–9) *Cavatodissiliodinium* sp. 1. A species with thin continuous periphragm. Note the varying extent of pericoel development in the different specimens.

The holotype (Plate 2(6)) and paratypes 1—4 (Plate 2(2—5)) are reillustrated herein.

Cavatodissiliodinium sp. 1 (with continuous periphragm)
(Plates 3(1—6), 4(1—9))

This is a species of *Cavatodissiliodinium* with a thin continuous periphragm. The cyst wall is ornamented with coalescing elements forming irregular ridges that may fuse to form circular structures. The extent of the pericoel varies largely. In some specimens the pericoel may be wide (Plate 3(1—3)), in other specimens the periphragm is more appressed to the endophragm forming only a narrow pericoel close to the plate boundaries. The species also has affinities to *Durotrigia*, as it shows tabulation and has a multiplate precingular archeopyle, but *Durotrigia* differs in its acavate nature.

Cavatodissiliodinium sp. 2 (open net periphragm)
(Plate 5(1—4))

This is a species similar to *Cavatodissiliodinium* sp. 1, but the periphragm is perforated. A network of open meshes forms the periphragm. The extent of the pericoel varies largely. This species also shares the same affinities with *Durotrigia*, but again lacks the acavate structure of that genus.

Cavatodissiliodinium sp. 3 (*Aldorfia aldorfensis* precursor)
(Plate 6(1—6))

This is a species of *Cavatodissiliodinium* that has a thick spongy cyst wall (differentiated autophragm) especially well developed in the apical and

(1 and 2) Isolated fragment of the epicyst, opercular pieces. Specimen in right lateral view, descending focal series. Egerten SF 539-o-2, M26.
(3 and 4). Isolated fragment of the epicyst, opercular pieces. Specimen in dorsal view, descending focal series. Egerten SF 538-o-1, L40/2.
(5) Isolated opercular piece, precingular plate 3″ in external view. Egerten SF 539-o-2, O34/2-4.
(6) Isolated opercular piece, precingular plate 4″ in external view. Egerten SF 538-o-1, M18.
(7) Isolated opercular piece, precingular plate 5″ in external view. Egerten SF 538-o-1, W35.
(8) Specimen in ventral view. Egerten SF 538-o-1, O13/2.
(9) Isolated fragment of the hypocyst. Specimen in dorsal view. Egerten SF 539-o-2, M26/2.

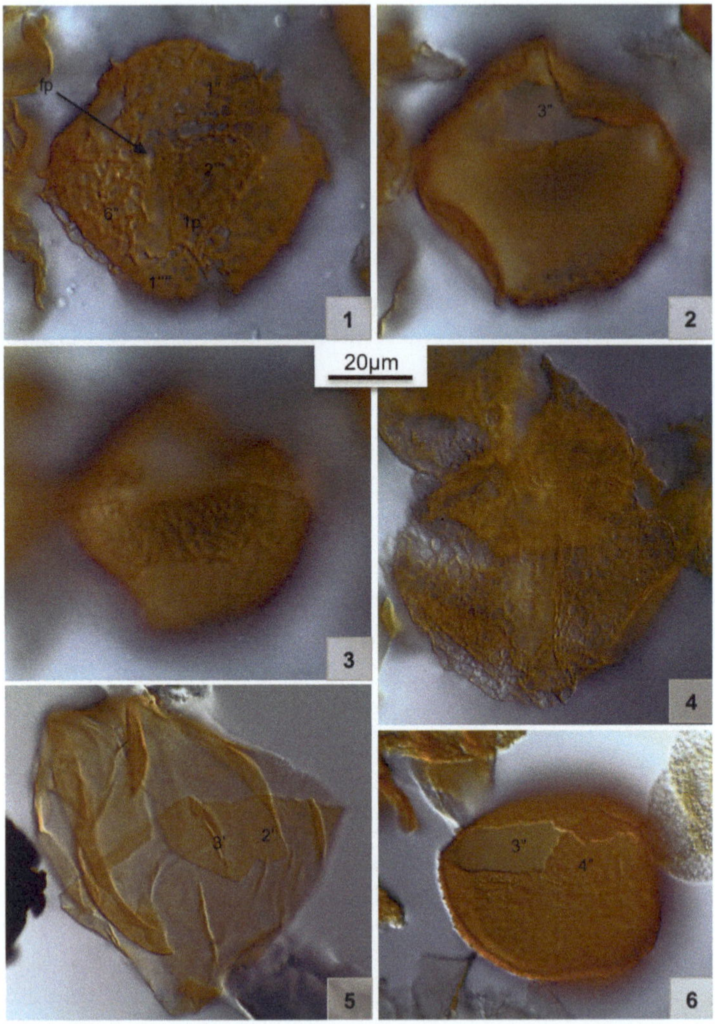

Plate 5. Dinoflagellate cysts from the Early Bajocian of the Upper Rhine area. All images were taken using differential interference contrast. The sample and slide number is followed by England Finder coordinates. All specimens are at the same magnification, scale bar = 20 μm.

(1–4) *Cavatodissiliodinium* sp. 2. A species with thin periphragm that is not continuous, but formed of a network of open meshes. Note the varying extent of pericoel development in the different specimens.

(1–3) Specimen in ventral view, descending focal series. Specimen with reduced pericoel. Note the open meshes of the periphragm. Arrow indicates flagellar pore (fp). Egerten SF 539-o-2, J46.

(4) Specimen with large pericoel. Note the open meshes of the periphragm. Egerten SF 539-o-2, M37/4.

(5) *Dissiliodinium giganteum* Feist-Burkhardt 1990. Apical cap in internal view. Badenweiler SF 548-o-2, O30.

(6) *Gongylodinium erymnoteichon* Fenton et al. 1980. Dorsal face in high focus. Egerten SF 540-o-1, L40.

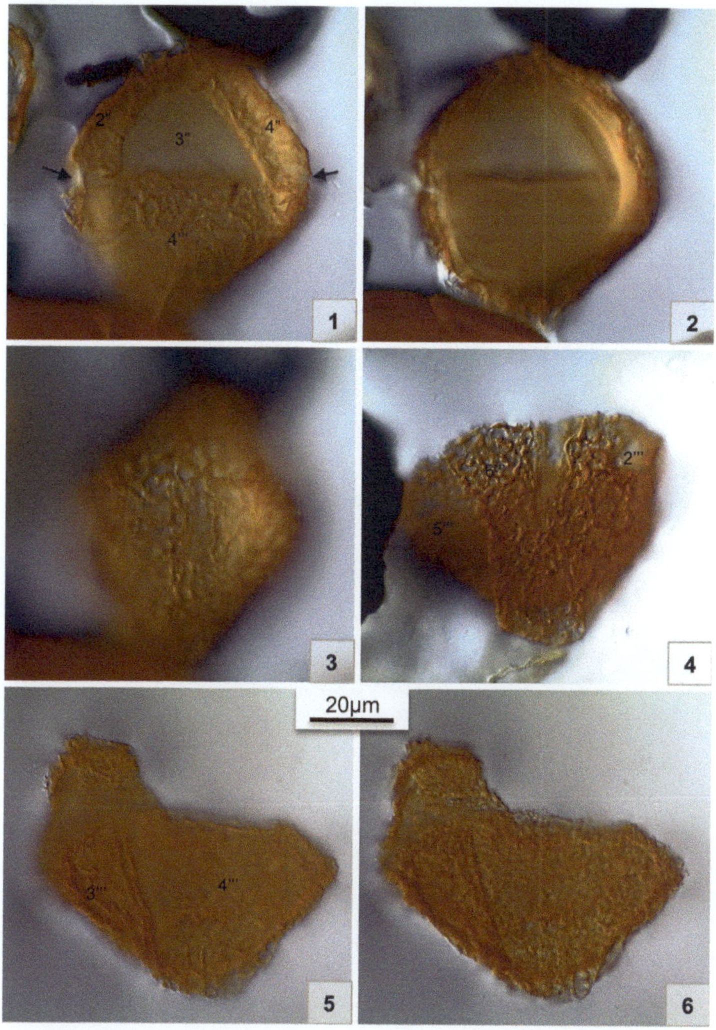

Plate 6. Dinoflagellate cysts from the Early Bajocian of the Upper Rhine area. All images were taken using differential interference contrast. The sample and slide number is followed by England Finder coordinates. All specimens are at the same magnification, scale bar = 20 μm.

(1–6) *Cavatodissiliodinium* sp. 3, a precursor of *Aldorfia aldorfensis* (Gocht 1970) Stover & Evitt 1978. Note the thick, "spongy" cyst wall (differentiated autophragm) especially well developed in the apical and antapical regions.

(1–3) Specimen in dorsal view, descending focal series. Note the split between the opercular pieces 2″ and 4″ and the anterior margin of the cingulum (arrows). Egerten SF 539-o-2, O31/1-2.

(4) Fragment, isolated hypocyst, in ventral view. Note the thickened and raised cyst wall in the antapical region. Egerten SF 539-o-2, W42/2.

(5 and 6) Specimen in oblique dorsal view, descending focal series. Opercular pieces 2″ to 5″ lost. Egerten SF 543-o-1, O27/1-3.

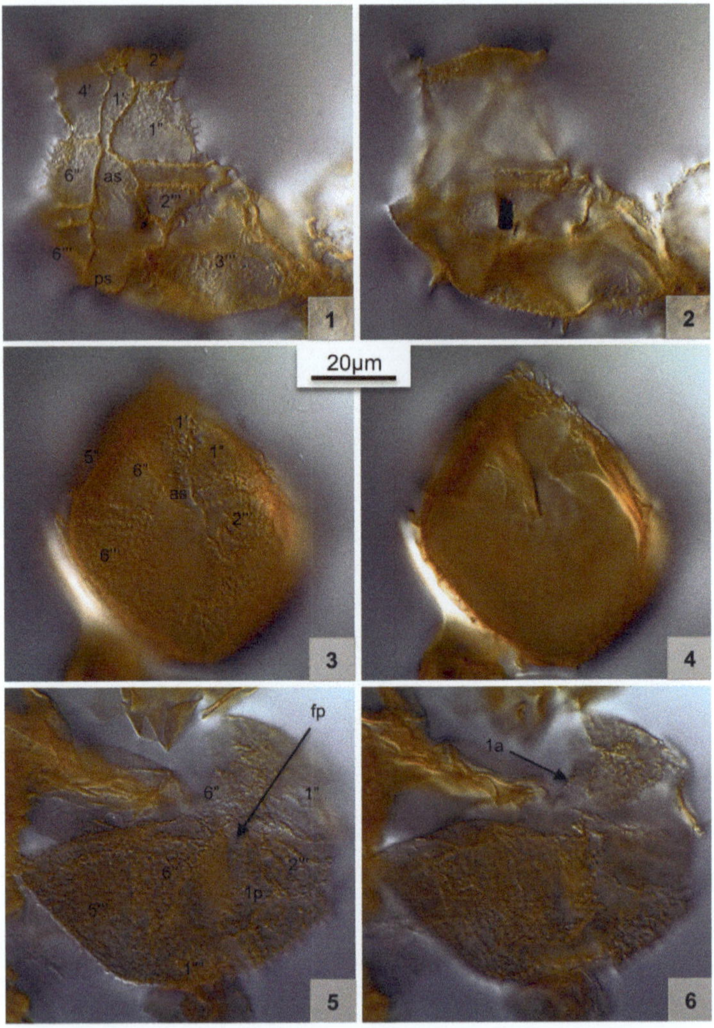

Plate 7. Dinoflagellate cysts from the Early Bajocian of the Upper Rhine area. All images were taken using differential interference contrast. The sample and slide number is followed by England Finder coordinates. All specimens are at the same magnification, scale bar = 20 μm.

(1 and 2) *Durotrigia daveyi* Bailey 1987. Specimen in ventral view, descending focal series. Opercular pieces 2″ to 5″ lost. Egerten SF 546-o-1, H41.

(3 and 4) *Durotrigia filapicata* (Gocht 1970) Riding & Bailey 1991. Specimen in ventral view, descending focal series. Note the trabeculate sutural crests and the fibrous anastomosing elements forming the apical horn (4). Egerten SF 539-o-2, R41/1.

(5 and 6) *Durotrigia* cf. *omentifera* Feist-Burkhardt & Monteil 2001. Specimen in oblique ventral view, descending focal series. Opercular pieces 2″ to 5″ lost. Egerten SF 539-o-2, K36/3.

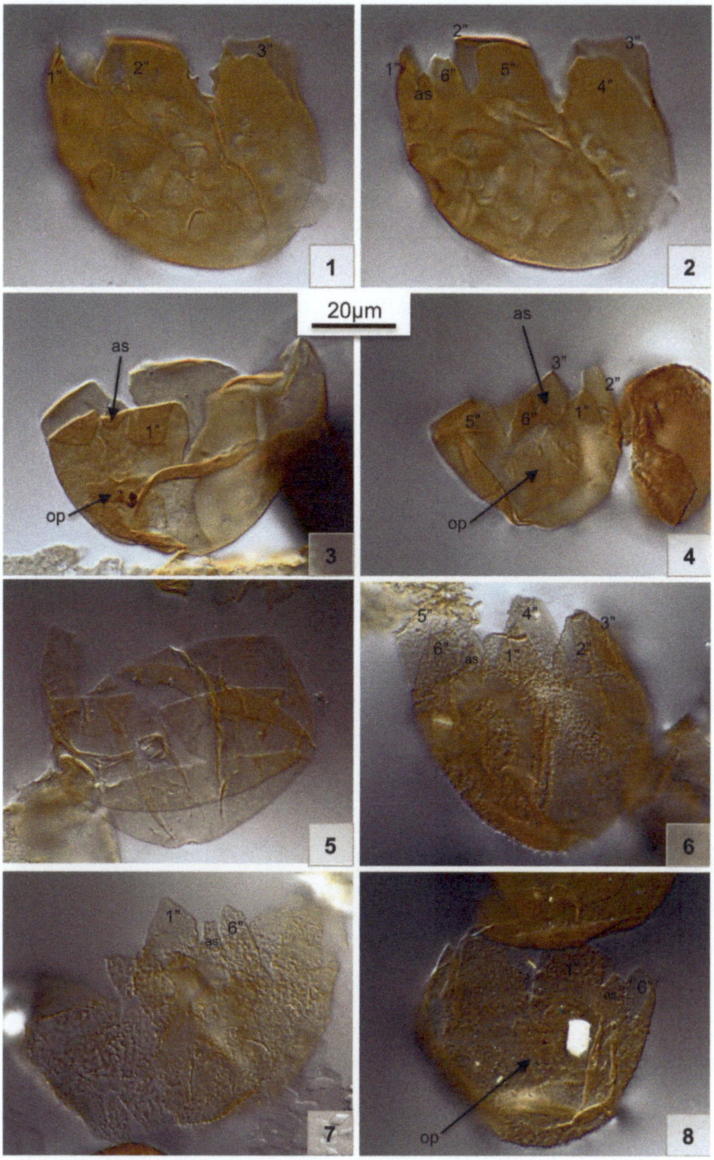

Plate 8. Dinoflagellate cysts from the Early Bajocian of the Upper Rhine area. All images were taken using differential interference contrast. The sample and slide number is followed by England Finder coordinates. All specimens are at the same magnification, scale bar = 20 μm.

(1–4) *Batiacasphaera laevigata* (Smelror 1988) Feist-Burkhardt and Monteil 1997. Note the camerate precingular plate 1″ that indicates an ortho-type apical arrangement.

(1 and 2) Specimen in left lateral view, descending focal series. Egerten SF 539-o-2, L20.

(3) Specimen in ventral view. Note the operculum inside the cyst. Egerten SF 536-o-2, N26/1-2.

antapical regions. Its cyst wall resembles that of *Aldorfia aldorfensis* (Gocht 1970) Stover & Evitt 1978, but the species differs in having a multiplate precingular archeopyle as typical in *Cavatodissiliodinium*, instead of a simple 1P precingular archeopyle as in *Aldorfia*.

Cavatodissiliodinium sp. 3 is considered to be a precursor of *A. aldorfensis* (Gocht 1970) Stover & Evitt 1978. It has its first occurrence in the Upper Rhine area samples in the Humphriesianum Zone, Romani Subzone. *A. aldorfensis* has its first appearance in southwest Germany slightly later, in the latest part of the Humphriesianum Zone (Feist-Burkhardt and Wille, 1992).

Genus *Kallosphaeridium* De Coninck 1969 emend. Jan du Chêne, Stover & De Coninck 1985
Kallosphaeridium spp.
(Plate 11(6–9))

Species of *Kallosphaeridium* show a wide variety in cyst wall structure and in ornamentation from very fine to coarse, which makes it extremely difficult to speciate with confidence. In this and other respects they are similar to the species in *Batiacasphaera*. As in *Batiacasphaera*, *Kallosphaeridium* species may show an ortho- or meta-type apical arrangement (see remarks under *Batiacasphaera* spp.). The apical cap is formed of the apical and intercalary plates; the apical cap is adnate.

Genus *Meiourogonyaulax* Sarjeant 1966
Meiourogonyaulax spp.
(Plate 10(3 and 4))

(4) Specimen in ventral view, with operculum inside the cyst. Egerten SF 539-o-2, K45.
(5–8) *Batiacasphaera* spp. Note the wide variety in cyst wall structure and in ornamentation from very fine to coarse.
(5) Specimen with a smooth to scabrate cyst wall. Egerten SF 530-o-1, Q32.
(6) Specimen in ventral view, with a granulate to lichen-like ornamentation. Egerten SF 530-o-1, S25/1.
(7) Specimen with ventral face in low focus. Specimen with lichen-like ornamentation. Note the possibly planate precingular plate 1″, that would indicate a meta-type apical arrangement. Egerten SF 538-o-1, J21/3-4.
(8) Specimen with ventral face in low focus, cyst wall with granulate to lichen-like ornamentation. Note the operculum inside the cyst. Egerten SF 530-o-1, N29.

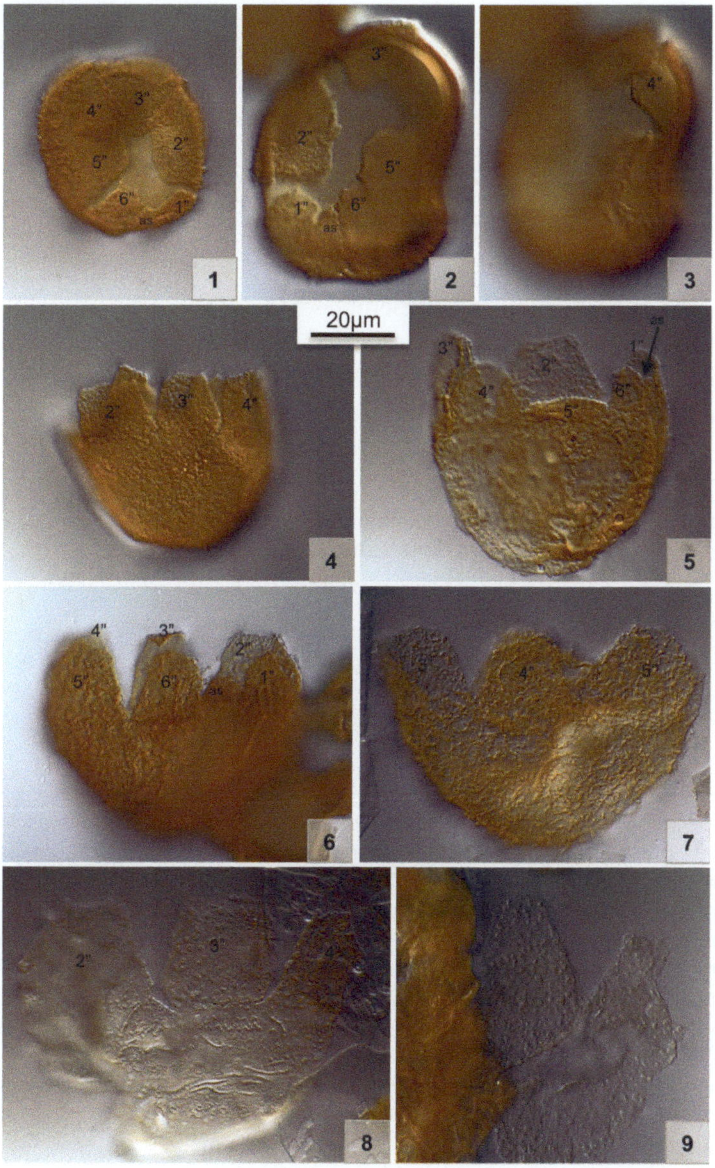

Plate 9. Dinoflagellate cysts from the Early Bajocian of the Upper Rhine area. All images were taken using differential interference contrast. The sample and slide number is followed by England Finder coordinates. All specimens are at the same magnification, scale bar = 20 μm.

(1–9) *Batiacasphaera* spp. Note the wide variety in cyst wall structure and in ornamentation from very fine to coarse.

The first dinoflagellate cysts with apical archeopyles and tabulation indicated by sutural crests probably developed from *Batiacasphaera* type precursors. Occurrence base of the first representatives of *Meiourogonyaulax* in the present material is at the top of the Pinguis Subzone.

Meiourogonyaulax sp. 2 (*Meiourogonyaulax valensii* precursor)
(Plate 10(5))

This species of *Meiourogonyaulax* resembles *Meiourogonyaulax valensii* Sarjeant 1966 in its ornamentation and the sutural crests, but differs in lacking the typical angular polygonal cyst shape. *Meiourogonyaulax* sp. 2 is considered to be a precursor of *M. valensii*.

Genus *Pareodinia* Deflandre 1947 emend. Below 1990
Pareodinia sp. 2 of Feist-Burkhardt and Monteil 1997
(Plate 15(8 and 9))

1992 *Pareodinia* sp.—Feist-Burkhardt and Wille, Plate 2(11).

(1) Specimen in apical view, with thick and coarsely ornamented cyst wall. Note the planate precingular plate 1″ and the camerate 2″ that indicate a meta-type apical arrangement. Egerten SF 535-o-1, M41.
(2 and 3) Specimen in antapical view, descending focal series. Apical face in low focus. Specimen with relatively thick and coarsely ornamented cyst wall. Note the planate precingular plate 1″ and the camerate 2″ that indicate a meta-type apical arrangement. Egerten SF 530-o-1, J23.
(4) Specimen in dorsal view, with thick and coarsely ornamented cyst wall. Egerten SF 530-o-1, Q17.
(5) Specimen in right lateral view, with granulate ornamentation of the cyst wall. Egerten SF 530-o-1, R18/2.
(6) Specimen in right lateral view, with coarsely ornamented cyst wall. Note the possibly planate precingular plate 1″, that would indicate a meta-type apical arrangement. Egerten SF 536-o-1, Q19/2.
(7) Specimen in oblique right lateral view, with coarsely lichen-like ornamentation of the cyst wall. Egerten SF 538-o-1, Q27/4.
(8) Fragment in dorsal view. Cyst wall ornamented with small pustules. Egerten SF 538-o-1, J41.
(9) Fragment of a specimen with a cyst wall ornamented with small pustules. Egerten SF 538-o-1, H26.

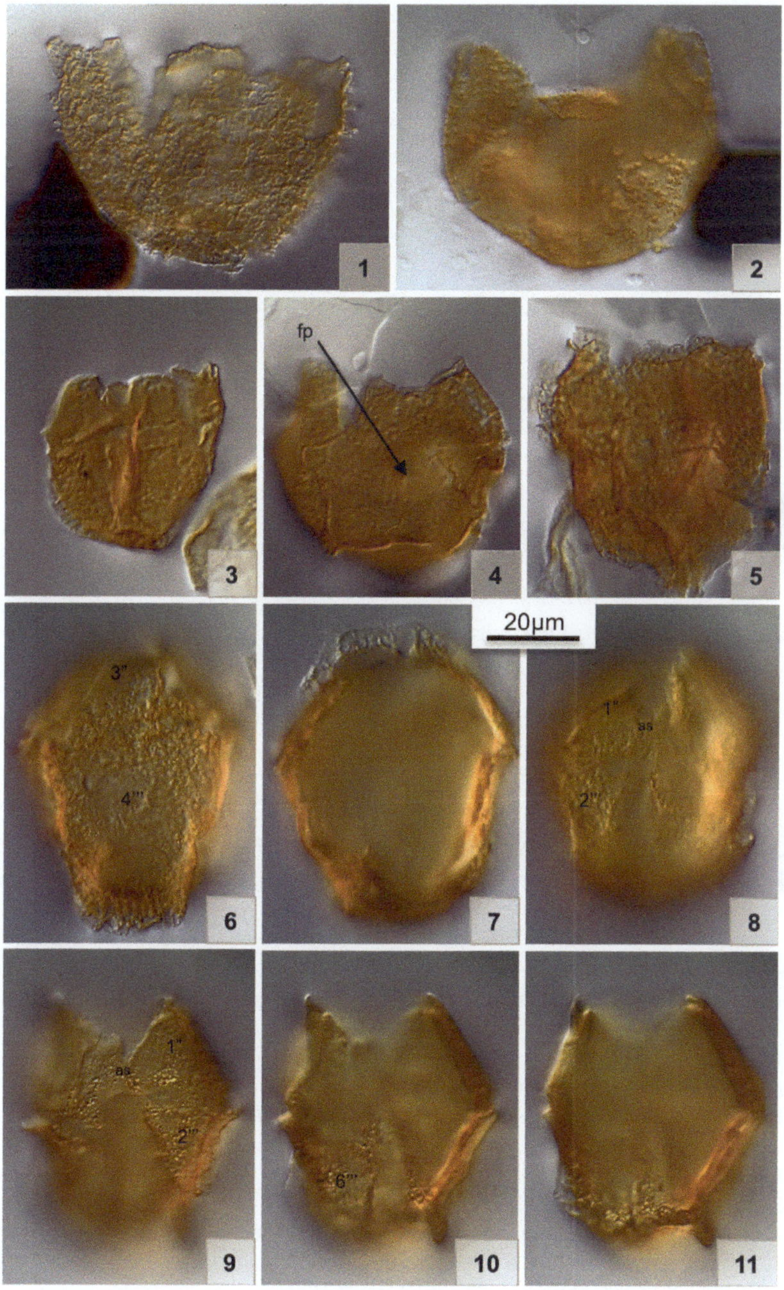

Plate 10. Dinoflagellate cysts from the Early Bajocian of the Upper Rhine area. All images were taken using differential interference contrast. The sample and slide number

1997 *Pareodinia* sp. 2—Feist-Burkhardt and Monteil, p. 40, Plate 4 (11–13).

This species has a cyst wall with irregular reticulation and a strongly reduced apical horn. Cyst wall ornamentation is reminiscent of crumpled aluminum foil.

Genus *Protobatioladinium* Nøhr-Hansen 1986
Protobatioladinium sp.
(Plate 15(11–13))

Protobatioladinium differs from *Pareodinia* in having a combined apical and intercalary archeopyle with the archeopyle formula $A_{(2'+3')} + I_{1a} + I_{2a}$. There are three opercular pieces, two intercalary plates 1a and 2a and the apical cap composed of two apical homologs (*2′+*3′) (see Feist-Burkhardt and Pross, 1999 for further details).

Occurrence in the present study is in the Pinguis Subzone to Humphriesianum Subzone of the Humphriesianum Zone. This is the oldest record so far for the genus.

Genus *Rhynchodiniopsis* Deflandre 1935 emend. Jan du Chêne, Fauconnier and Fenton 1985
Rhynchodiniopsis? sp. 1 (*Rhynchodiniopsis*? *regalis* precursor)
(Plate 12(1–4))

This species is characterized by the ornamentation of the cyst wall with many rod-like elements covering the cyst and forming the sutural crests. The species resembles *Rhynchodiniopsis*? *regalis* (Gocht 1970) Jan du Chêne et al. 1985 but differs in having a much denser cover of ornamental

is followed by England Finder coordinates. All specimens are at the same magnification, scale bar = 20 μm.

(1 and 2) *Batiacasphaera* spp. Note the wide variety in cyst wall structure and in ornamentation from very fine to coarse.
(1) Specimen with coarsely ornamented cyst wall. Egerten SF 538-o-1, W19/2.
(2) Specimen with coarsely ornamented cyst wall. Egerten SF 538-o-1, P34/4.
(3 and 4) *Meiourogonyaulax* spp.
(3) Egerten SF 538-o-1, M33/2.
(4) Specimen in ventral view. Egerten SF 538-o-1, K40.
(5) *Meiourogonyaulax* sp. 2, a precursor of *Meiourogonyaulax valensii* Sarjeant 1966. Egerten SF 538-o-1, O39/2.
(6–11) *Meiourogonyaulax valensii* Sarjeant 1966.
(6–8) Specimen in dorsal view, descending focal series. Egerten SF 546-o-1, M25/4.
(9–11) Specimen in ventral view, descending focal series. Egerten SF 546-o-1, H28/3.

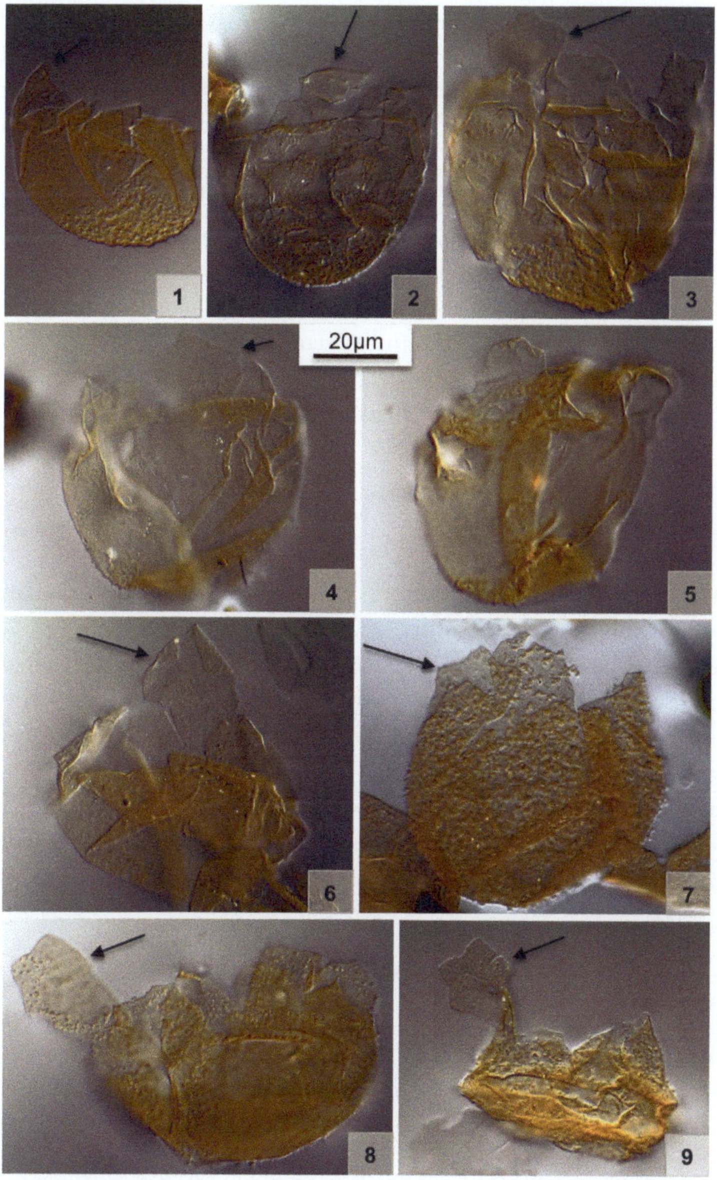

Plate 11. Dinoflagellate cysts from the Early Bajocian of the Upper Rhine area. All images were taken using differential interference contrast. The sample and slide number is followed by England Finder coordinates. All specimens are at the same magnification, scale bar = 20 μm.

(1–4) *Kallosphaeridium hypornatum* Prauss 1989. Arrow points to the adnate apical cap. Note the coarse ornamentation in the antapical region.

(1) Badenweiler SF 549-o-1, N43/1.

elements, and in the texture of the sutural crests. In *Rhynchodiniopsis*? sp. 1 the sutural crests are formed by the alignment of rods that are distally connected to form trabeculae. In contrast, *Rhynchodiniopsis*? *regalis* possesses sutural crests that are formed by the alignment of processes that are connected distally to form pointed arches. The crests in *Rhynchodiniopsis*? sp. 1 are distally trabeculate, whereas in *Rhynchodiniopsis*? *regalis* the crests are serrated.

Rhynchodiniopsis? sp. 1 is considered to be a precursor of *Rhynchodiniopsis*? *regalis*. It has its first occurrence in the Upper Rhine area samples in the later part of the Blagdeni Subzone, Humphriesianum Zone. *Rhychodiniopsis*? *regalis* has its first appearance in southwest Germany also in the latest part of the Humphriesianum Zone (Feist-Burkhardt and Wille, 1992).

Genus *Wanaea* Cookson & Eisenack 1958 emend. Riding and Helby 2001
Wanaea sp. 1 (granulate to lichen-like)
(Plate 13(1–4))

This species of *Wanaea* is characterized by granulate to lichen-like ornamentation of the cyst wall. Density of the ornamental elements varies. They may be rounded to angular, or irregular in shape and size, and may coalesce, which is reminiscent to the morphology of some lichen. The lichen-like ornamentation is similar to that in *Dissiliodinium lichenoides* Feist-Burkhardt and Monteil 2001.

Wanaea sp. 1 resembles *Wanaea indotata* Drugg 1978 but differs in having a coarser ornamentation. The more heavily ornamented specimens of *Wanaea* sp. 1 are very similar to the Australasian *Wanaea verrucosa* Riding

(2) Egerten SF 530-o-1, S22/2.
(3) Egerten SF 537-o-2, N19.
(4) Hohlebach SF 560-o-1, G19.
(5) *Atopodinium* sp. 1, a precursor of *Atopodinium polygonale* (Beju 1983) Masure 1991. Note the coarse ornamentation in the antapical region that forms a rigid shield. Hohlebach SF 561-o-1, W30/2-4.
(6–9) *Kallosphaeridium* spp. Note the wide variety in cyst wall structure and in ornamentation from very fine to coarse.
(6) Specimen with scabrate ornamentation. Egerten SF 536-o-1, V17/3.
(7) Specimen with coarse lichen-like ornamentation. Egerten SF 536-o-1, N40/2.
(8) Specimen in left lateral view, with granulate to lichen-like ornamentation. Egerten SF 539-o-2, K48/1.
(9) Specimen in left lateral view, with granulate ornamentation. Badenweiler SF 556-o-1, S40.

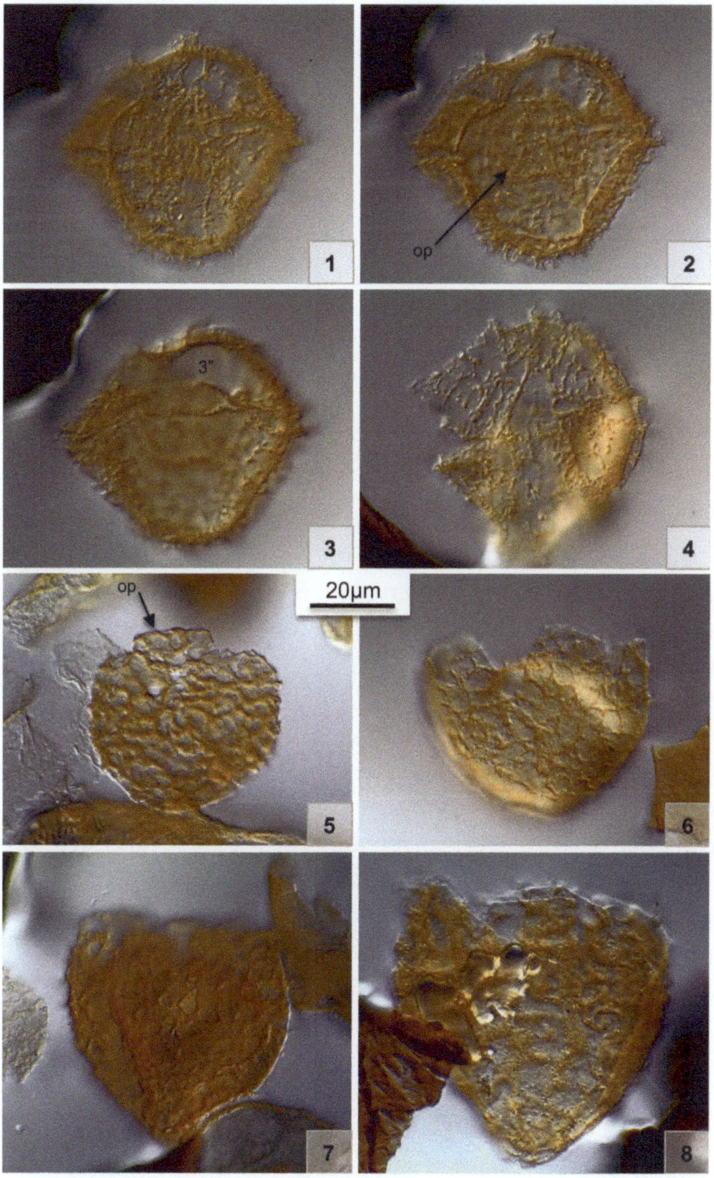

Plate 12. Dinoflagellate cysts from the Early Bajocian of the Upper Rhine area. All images were taken using differential interference contrast. The sample and slide number is followed by England Finder coordinates. All specimens are at the same magnification, scale bar = 20 μm.

(1–4) *Rhynchodiniopsis*? sp. 1, a precursor of *Rhynchodiniopsis? regalis* (Gocht 1970) Jan du Chêne et al. 1985.

and Helby 2001, and may even be conspecific. It is important to note that early forms of *Wanaea* in both hemispheres are strongly ornamented.

Wanaea sp. 2 (lichen-like to trabeculate)
(Plate 13(5 and 6))

This species of *Wanaea* is characterized by lichen-like to trabeculate ornamentation of the cyst wall. The ornamental elements in this species are longer and thinner than in *Wanaea* sp. 1 and tend more to interconnect forming trabeculate structures.

6. THE PALYNOMORPH ASSEMBLAGES

6.1 General Aspect of the Palynomorph Assemblages

The studied samples yielded mostly good palynological residues with rich and diverse palynofloras. Preservation is mostly good to moderate. Four samples from the base of the Humphriesioolith Member were barren of palynomorphs.

The palynomorph assemblages are mostly composed of dinoflagellate cysts and pollen and spores. In some samples there is a high proportion of prasinophytes and/or foraminiferal test linings comprising up to a third of the assemblage (Figs. 6—8). Minor components are acritarchs and green algae (e.g., *Botryococcus*). Fungal remains are recorded only semiquantitatively. A total of 46 dinoflagellate cyst taxa, 12 other aquatic palynomorphs, and 44 pollen and spore taxa are recorded. Some reworked sporomorphs (*Densosporites* spp., *Ovalipollis* spp., *Ricciisporites tuberculatus*, striate bisaccate pollen) indicate erosion of Triassic sediments in the source area.

(1—3) Specimen in ventral view, descending focal series. Note the operculum (precingular plate 3″) in the cyst (2). Hohlebach SF 561-o-1, Q17/1-2.
(4) Specimen in ventral view. Egerten SF 546-o-1, N29/2.
(5—8) *Valensiella/Ellipsoidictyum* spp. Note the variety in size, shape and ornamentation of the cyst wall.
(5) Arrow points to the apical operculum. Egerten SF 540-o-1, Q24/2.
(6) Egerten SF 546-o-1, N24/2-4.
(7) Egerten SF 539-o-2, Q35/2.
(8) Hohlebach SF 560-o-1, L21.

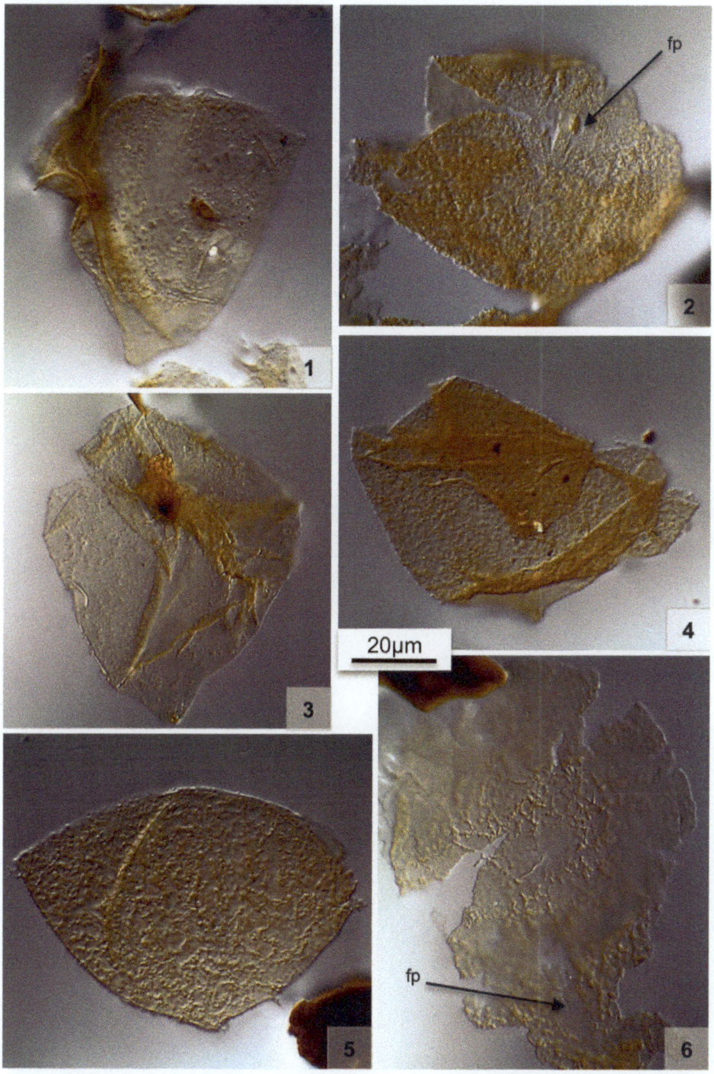

Plate 13. Dinoflagellate cysts from the Early Bajocian of the Upper Rhine area. All images were taken using Differential Interference Contrast. The sample and slide number is followed by England Finder coordinates. All specimens are at the same magnification, scale bar = 20 μm.

(1–4) *Wanaea* sp. 1. A species with granulate to lichen-like ornamentation.

(1) Specimen in left lateral view. Egerten SF 547-o-1, N29.

(2) Specimen in ventral view. Arrow indicates flagellar pore (fp). Egerten SF 536-o-2, M25.

(3) Hohlebach SF 560-o-1, X36.

(4) Egerten SF 535-p-1, H29/1-3.

(5 and 6) *Wanaea* sp. 2. A species with lichen-like to trabeculate ornamentation.

(5) Badenweiler SF 557-o-2, S29.

(6) Epicyst (adnate operculum) with parts of the sulcal area, in inner view. Arrow indicates flagellar pore (fp). Egerten SF 538-o-1, X36/2-4.

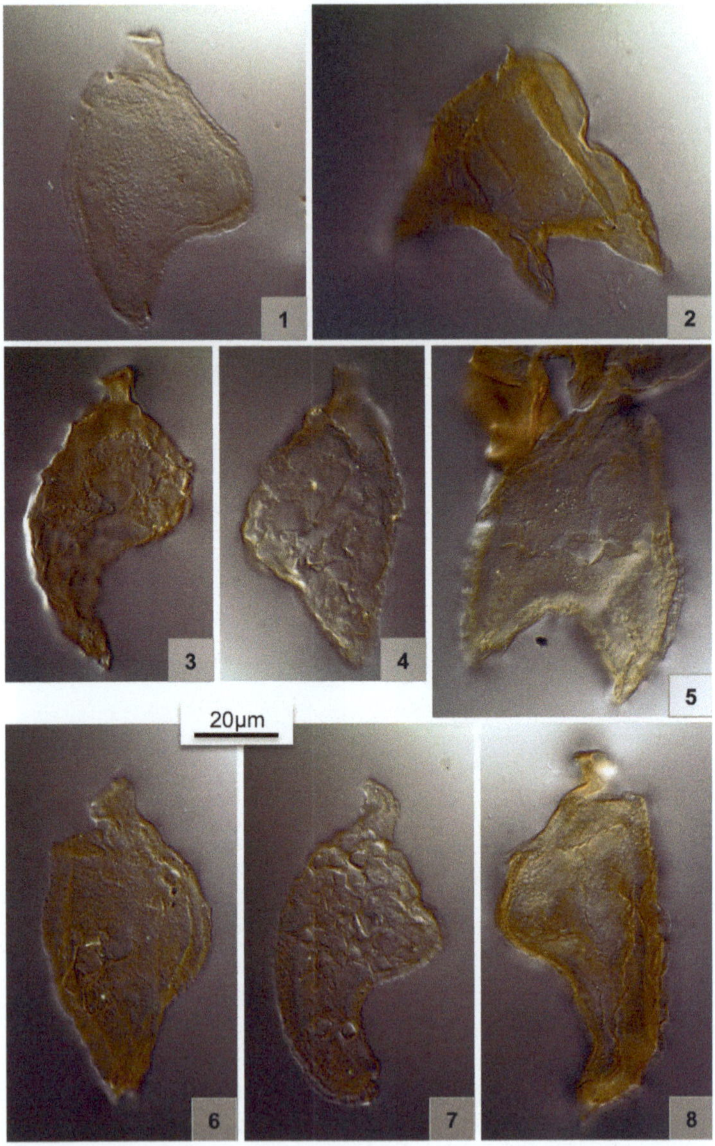

Plate 14. Dinoflagellate cysts from the Early Bajocian of the Upper Rhine area. All images were taken using differential interference contrast. The sample and slide number is followed by England Finder coordinates. All specimens are at the same magnification, scale bar = 20 μm.

(1) *Nannoceratopsis gracilis* Alberti 1961 sensu lato. Specimen in right lateral view. Badenweiler SF 549-o-1, Y36/2.

(2) *Nannoceratopsis triceras* Drugg 1978. Egerten SF 542-o-1, M23/4.

(3), (4), (6–8) *Nannoceratopsis* spp.

The pollen and spore assemblages are dominated by coniferous gymnosperm pollen. The most abundant are bisaccate pollen and diverse *Callialasporites* spp., followed by *Cerebropollenites mesozoicus*. The cheirolepidiacean, xerophytic *Classopollis torosus* is omnipresent but not particularly abundant. Taxodiacean *Exesipollenites* spp. and *Perinopollenites elatoides* are generally present but rare. Pteridophyte spores are less abundant and include the typical Jurassic smooth and ornamented trilete spore taxa *Concavisporites* spp., *Conbaculatisporites* spp., *Deltoidospora* spp., *Densoisporites velatus*, *Dictyophyllidites* spp., *Duplexisporites problematicus*, *Foveosporites* spp., *Ischyosporites variegatus*, *Kekryphalospora distincta*, *Leptolepidites* spp., *Lycopodiacidites* spp., *Lycopodiumsporites* spp., *Matonisporites* spp., *Osmundacidites wellmanii*, *Staplinisporites caminus*, *Todisporites* spp., *Trilites rariverrucatus*, and *Uvaesporites agenteaeformis*. Most of these are fern spores, while some belong to the lycopods.

6.2 Description of the Palynomorph Successions

The palynological count data are illustrated in distribution charts on Figs. 6–8. The main focus of this study is on the dinoflagellate cysts. This group allows excellent stratigraphical resolution in the Jurassic. They show rapid evolution during the Middle Jurassic and especially in the Bajocian, leading to many species having their first appearances in this interval. Some of the first and last occurrences observed herein have the potential to be used to refine or emend existing palynostratigraphical schemes.

6.2.1 Wedelsandstein Formation, Demissusbänke Member (Sauzei Zone)

The upper part of the Wedelsandstein Formation is the Demissusbänke Member. The sampled interval corresponds to the later part of the Early Bajocian Sauzei Zone. There are seven samples from Badenweiler (SF 548 to SF 552) and one sample from Egerten (SF 530) which, together, give a good picture of the late Sauzei Zone.

(3) Specimen in right lateral view. Badenweiler SF 550-o-1, V36/1.
(4) Specimen in left lateral view. Badenweiler SF 550-o-1, T34/3.
(6) Specimen in right lateral view. Badenweiler SF 550-o-1, T21/2.
(7) Specimen in right lateral view. Badenweiler SF 549-o-1, M23/1.
(8) Specimen in left lateral view. Badenweiler SF 556-o-1, L33/1.
(5) *Nannoceratopsis spiculata* Stover 1966. Specimen in left lateral view. Egerten SF 546-o-1, L36/1-2.

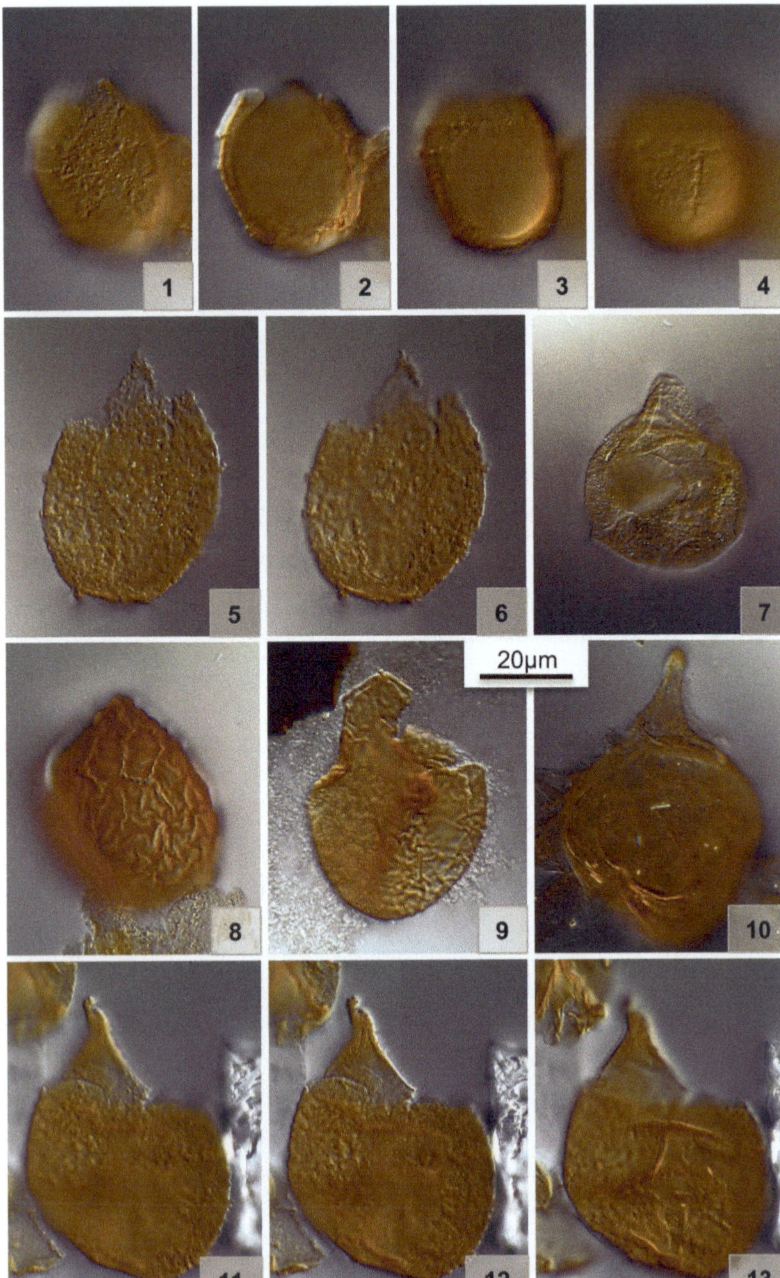

Plate 15. Dinoflagellate cysts from the Early Bajocian of the Upper Rhine area. All images were taken using differential interference contrast. The sample and slide number

Composition of the palynomorph assemblages varies from sample to sample across this interval. The proportion of dinoflagellate cysts varies from 13% to 68% of the total palynomorph assemblage, while the pollen and spores range from 22% to 66% of the assemblages. Prasinophytes are generally abundant to very abundant with *Tasmanites* spp. reaching a maximum proportion of 32% in sample SF 554. *Botryococcus* is typically scarce, but present in all samples and has a peak abundance of 14% in sample SF 554. Foraminiferal test linings are recorded only in Egerten with a few specimens in sample SF 530. Prasinophycean algae are often found in environments with poorly oxygenated bottom waters. Abundance of *Botryococcus* together with abundant prasinophycean algae indicates poor circulation and probable rafting of fresh water.

The most abundant dinoflagellate cyst genera in the Demissusbänke Member (Sauzei Zone) are *Batiacasphaera*, *Dissiliodinium*, *Durotrigia*, *Kallosphaeridium*, and *Nannoceratopsis*. All of them are difficult to speciate as the specimens show a wide variety of small morphological differences and morphologies grading from one end member of a morphological plexus to another. The dinoflagellate cyst assemblages are in fact much more diverse than a simple look on the distribution chart suggests.

In *Batiacasphaera* and *Kallosphaeridium* we find morphotypes from almost smooth to heavily ornamented, with intergrading stages of scabrate, granulate, lichen-like, pustulate, spongy, verrucate, trabeculate, and mixtures thereof (Plates 8(1–8), 9(1–9), 10 (1, 2), 11 (1–9)). Even the discrimination between the genera *Batiacasphaera* and *Kallosphaeridium* is often difficult, depending on preservation of the material and the orientation of the individual specimens. In *Batiacasphaera* the apical operculum is free, and in

◀———

is followed by England Finder coordinates. All specimens are at the same magnification, scale bar = 20 μm.

(1–6) Evansia? spongogranulata Below 1990.

(1–4) Specimen in ventral view, descending focal series. Egerten SF 530-o-1, K17/2.

(5 and 6) Specimen in ventral view, descending focal series. Egerten SF 530-o-1, O28/3.

(7) *Phallocysta thomasii* Smelror 1991. Egerten SF 530-o-1, O32/2-4.

(8 and 9) *Pareodinia* sp. 2 of Feist-Burkhardt and Monteil 1997. Cyst wall ornamentation is irregularly reticulate and reminiscent of crumpled aluminum foil.

(8) Egerten SF 536-o-1, V32/1.

(9) Egerten SF 537-o-1, O21.

(10) *Pareodinia* sp. Egerten SF 530-o-1, K30/3.

(11–13) *Protobatioladinium* sp. Specimen in ventral view, descending focal series. Egerten SF 538-o-1, S25.

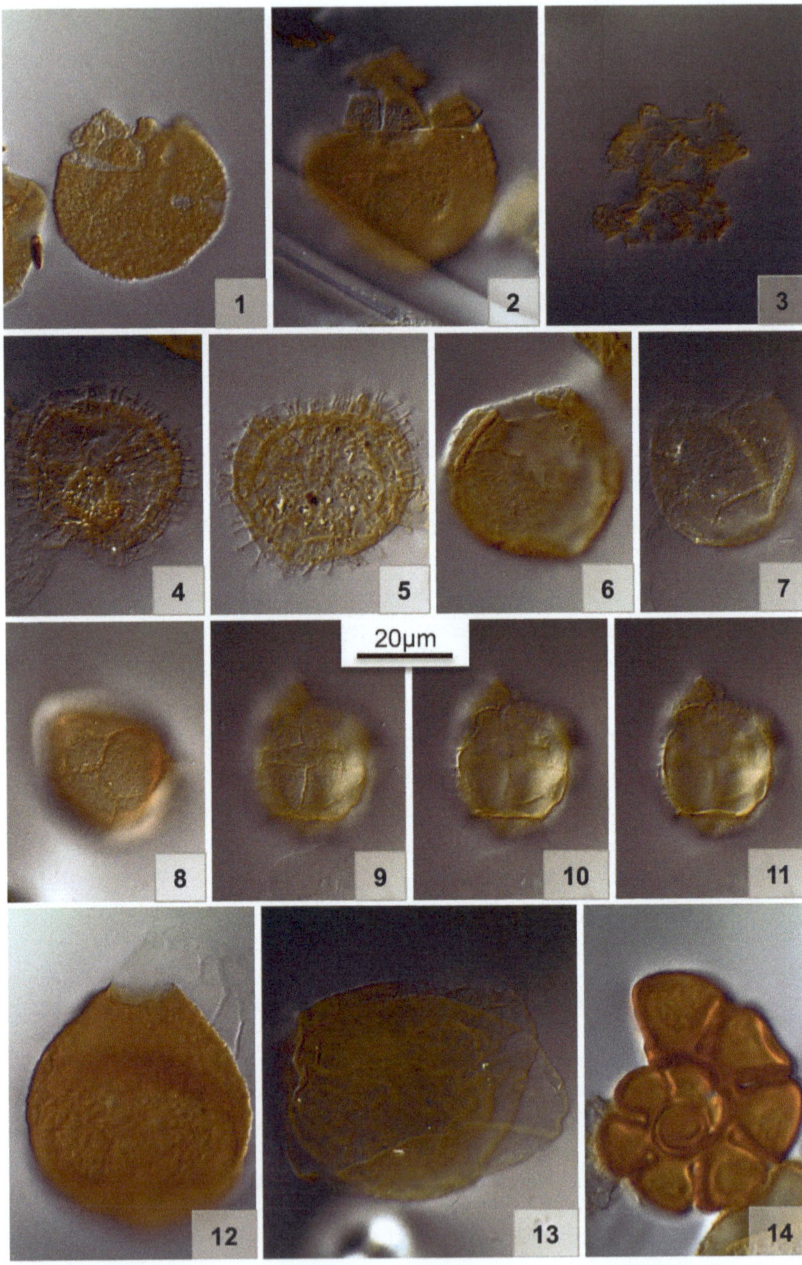

Plate 16. Palynomorphs from the Early Bajocian of the Upper Rhine area. All images were taken using differential interference contrast. The sample and slide number is followed by England Finder coordinates. All specimens are at the same magnification, scale bar = 20 μm.

Kallosphaeridium the apical operculum is adnate. End members in this plexus of forms that are identified are *Batiacasphaera* spp. (smooth to scabrate), *Batiacasphaera laevigata* (appearing later, in the Humphriesianum Zone) and *K. hypornatum*.

K. hypornatum occurs regularly throughout the succession and forms a typical component of the Sauzei and Humphriesianum zonal assemblages in the Upper Rhine area samples. In Badenweiler its first occurrence is within the Sauzei Zone, close to the base of the studied interval in sample SF 549. Its range base has been observed somewhat earlier in neighboring Switzerland in the lower part of the Wedelsandstein Formation, in the Laeviuscula Zone (Feist-Burkhardt, pers. obs.).

A similarly difficult generic plexus is that of *Dissiliodinium*, *Durotrigia*, and *Cavatodissiliodinium*. The principal difference between the genera is that in *Dissiliodinium* tabulation is not marked by sutural crests, in *Durotrigia* tabulation is marked by sutural crests, and *Cavatodissiliodinium* is cavate, that is it has more than one wall layer. But these seemingly clear-cut distinguishing features weaken when looking at the actual specimens present in the samples. In *Dissiliodinium* we find completely smooth morphotypes, and such that are ornamented with smaller or larger granules, verrucae or lichen-like interconnected ornamental elements. In smooth and weakly ornamented morphotypes the sutures may be indicated by more or less continuous or discontinuous grooves ("dotted lines") as in some *Dissiliodinium giganteum*

(1 and 2) *Mancodinium semitabulatum* Morgenroth 1970.
(1) Badenweiler SF 551-o-1, L24/4.
(2) Specimen in dorsal view. Note isolated pustules and groups of pustules in the center of the plates. Egerten SF 530-o-1, R41.
(3) *Reutlingia gochtii* Drugg 1978. Egerten SF 538-o-1, R18/3.
(4) *Valvaeodinium cavum* (Davies 1982) Below 1987. Egerten SF 530-o-1, N37.
(5) *Valvaeodinium sphaerechinatum* Below 1987. Egerten SF 544-o-1, K38.
(6 and 7) *Valvaeodinium vermipellitum* Below 1987.
(6) Egerten SF 530-o-1, T43.
(7) Egerten SF 543-o-1, K24.
(8–11) *Valvaeodinium euareatum* Prauss 1989.
(8) Specimen in high focus. Egerten SF 536-o-1, J44.
(9–11) Specimen in lateral view, descending focal series. Egerten SF 538-o-1, V20/2.
(12) *Fromea tornatilis* (Drugg 1978) Lentin & Williams 1981. Acritarch. Egerten SF 538-o-1, T35.
(13) Zygospore? sp. 1. Egerten SF 543-o-1, K26.
(14) Foraminiferal test lining, spiral. Egerten SF 546-o-1, M29/4.

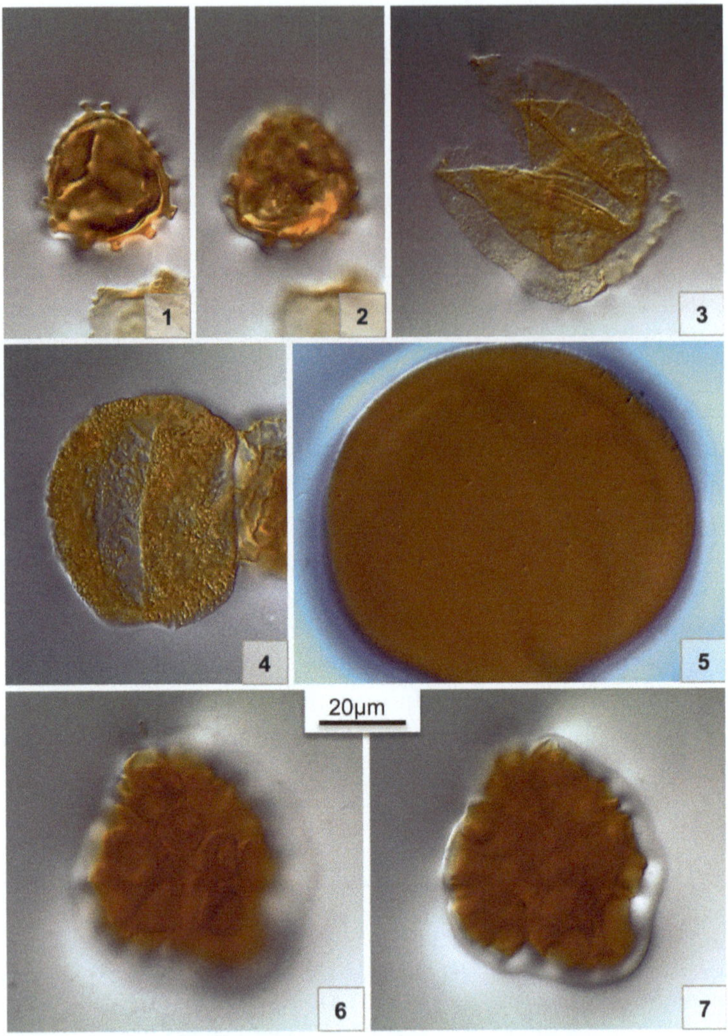

Plate 17. Palynomorphs from the Early Bajocian of the Upper Rhine area. All images were taken using differential interference contrast. The sample and slide number is followed by England Finder coordinates. All specimens are at the same magnification, scale bar = 20 μm.

(1 and 2) *Lycopodiumsporites gristhorpensis* Couper 1958. Egerten SF 530-o-1, H42/1-3.

(3) *Perinopollenites elatoides* Couper 1958. Egerten SF 530-o-1, H40/3-4.

(4) *Quadraeculina anellaeformis* Maljavkina 1949. Egerten SF 538-o-1, X32/3-4.

(5) *Tasmanites* sp. Badenweiler SF 554-o-1, H28/1.

(6 and 7) *Botryococcus* sp. Descending focal series. Badenweiler SF 554-o-1, H26/1-2.

(Plate 5(5); Feist-Burkhardt, 1990). In ornamented morphotypes the ornamental elements may be more or less aligned, not along the sutures, but next to the sutures, thus faintly indicating the tabulation. In *Durotrigia* the sutures are indicated by sutural crests that are markedly elevated. This definition was introduced by Feist-Burkhardt and Monteil (2001). They defined *Dissiliodinium* as a genus without indication of tabulation on the cyst or with negative sutural features, whereas *Durotrigia* has positive (elevated) sutural features.

An easily recognizable and often abundant end member in *Dissiliodinium* is *D. giganteum*. In the Demissusbänke Member, *D. giganteum* is common at the base of the Badenweiler section (sample 548). Its top occurrence is within the same interval, with one specimen in Egerten (SF 530) and one questionable specimen in Badenweiler (SF 551). Its range top is therefore considered to occur within the late Sauzei Zone. *D. giganteum* is known to show a prominent acme further down in the Wedelsandstein Formation in southwest Germany and northern Switzerland, in the Laeviuscula Zone (Feist-Burkhardt, 1990, 1992; pers. obs.; Hostettler et al., 2016).

Durotrigia daveyi is very abundant in the lower part of the Demissusbänke Member (samples SF 548 to SF 551) in Badenweiler, reaching up to 64% of the dinoflagellate cyst assemblage. The top of its occurrence acme is in the later part of the Sauzei Zone (sample SF 551), just before the first occurrence of *C. hansgochtii* in sample SF 553. *D. daveyi* does occur later on in the succession, but it never becomes as prominent again.

Whereas *Dissiliodinium* and *Durotrigia* are described as acavate taxa, the present material contains species that are markedly cavate. The oldest of these taxa is *C. hansgochtii* described by Feist-Burkhardt and Monteil (2001) based on the same material as the present biostratigraphical study. This species is an end member of a morphological plexus and is identified quite easily and with confidence. It is characterized by a thin, unornamented periphragm and typically has verrucae on the endophragm that are clustered in the center of the plates (Plate 2(1–6)). Other morphotypes of *Cavatodissiliodinium* (sp. 1, sp. 2, sp. 3) have a more elaborate cavate wall structure and marked sutural crests which makes them morphologically closer to *Durotrigia. Cavatodissiliodinium* sp. 1, sp. 2, and sp. 3 all have younger first occurrences, in the Humphriesianum Zone.

C. hansgochtii has its base occurrence in Badenweiler in the upper part of the Demissusbänke Member (sample SF 553), above the top of the *D. daveyi* acme. Its top occurrence is also in the Demissusbänke Member (sample SF 552 in Badenweiler; sample SF 530 in Egerten), in the Sauzei Zone. It does not reach the Humphriesianum Zone.

The top occurrence of *C. hansgochtii* is considered a very good indicator for the top Sauzei Zone. The succession of events—acme of *D. giganteum*, acme of *D. daveyi*, occurrence of *C. hansgochtii*, and finally, top of *C. hansgochtii*—has been observed in several sections spanning the Laeviuscula and Sauzei zones from southwest Germany and northern Switzerland (Burkhalter et al., 1997; Feist-Burkhardt, 1990, 1992; pers. obs.; Hostettler et al., 2016).

Abundant in the Demissusbänke Member, especially in Badenweiler, are specimens of *Nannoceratopsis*. Most of them cannot be attributed to a described species. Their morphology does not agree completely, but is close to *Nannoceratopsis gracilis* s.l. (Plate 14(1, 3, 4, 6–8)). Rare specimens of *N. gracilis* s.l. occur right to the top of the sampled section.

Rare, but considered stratigraphically important, is the occurrence of *Evansia? spongogranulata*. It occurs with only a few specimens in the Demissusbänke Member, with a range top in the Sauzei Zone.

Other well-known dinoflagellate cysts that occur, albeit rarely, in the Demissusbänke Member (Sauzei Zone) are *Gongylodinium erymnoteichon* (all specimens identified only questionably because of unfavorable preservation and/or orientation), *Mancodinium semitabulatum*, *Pareodinia* spp., *Valvaeodinium cavum*, *Valvaeodinium euareatum*, *Valvaeodinium sphaerechinatum*, and *Valvaeodinium vermipellitum*.

6.2.2 Gosheim Formation, Humphriesioolith Member, and Blagdenischichten Member (Humphriesianum Zone, Pinguis to Blagdeni Subzone)

Development of the dinoflagellate cyst assemblages in the Gosheim Formation and the temporal succession of events are slightly different in Badenweiler and in Egerten. Therefore we describe the sections separately.

In **Badenweiler** there is only one sample from the very base the Humphriesieoolith Member. This corresponds to the very base of the Pinguis Subzone (SF 555). The sample is dominated by *Tasmanites* (52%) and *Botryococcus* (25%) with only relatively low proportions of dinoflagellate cysts (6%) and pollen and spores (15%). Composition of the dinoflagellate cyst assemblage is very similar to the underlying Demissusbänke Member of the Sauzei Zone. The quantitative composition of the palynomorph assemblage and especially the high proportion of green algae indicate a strong freshwater influx and a significantly different paleoenvironment to the underlying section. After an interval void of samples, the next assemblages are from the top Romani Subzone (SF 556) and the Humphriesianum Subzone (samples SF

557 to SF 559). Quantitative composition of these four palynomorph assemblages is quite similar, but contains the first occurrences of some new dinoflagellate cysts. In sample SF 557 (Humphriesianum Subzone) we record the first occurrence of several undescribed morphotypes of *Wanaea* (*Wanaea* sp. 1, *Wanaea* sp. 2, *Wanaea* spp.) and of *Cavatodissiliodinium* sp. 1. All of these taxa are rare, with only one or two specimens recorded in each sample. In sample SF 559 (also Humphriesianum Subzone) we record the first occurrence of a few more taxa: *Cavatodissiliodinium* sp. 3 and some questionable specimens of *Acanthaulax crispa* and *Valensiella/Ellipsoidictyum* spp. These newly incoming taxa are extremely rare. The dinoflagellate cyst assemblages are otherwise dominated by *Nannoceratopsis* spp. and *N. gracilis* s.l., and rich and diverse *Batiacasphaera* spp. Other rare dinoflagellate cysts that occur in this interval include *Dissiliodinium* spp., *D. daveyi*, *Durotrigia* spp., questionable *G. erymnoteichon*, *K. hypornatum*, *Kallosphaeridium* spp., questionable *M. semitabulatum*, *Pareodinia* spp., and *V. sphaerechinatum*.

At **Egerten**, the four basal Humphriesioolith Member samples (SF 531 to SF 534, Pinguis Subzone) are barren of palynomorphs. This may be due to strongly oxidative conditions and therefore destruction of the organic material in the depositional environment at the onset of the iron oolite deposition. All the samples from SF 535 up are productive again, irrespective of oolite sedimentation or not.

After the interval with barren samples, the next assemblages are from the top of Pinguis Subzone to the very base of the Blagdeni Subzone (samples SF 535 to SF 543). Quantitative composition of these nine palynomorph assemblages is quite similar, with up to 45% dinoflagellate cysts and between 45% and 59% pollen and spores. Prasinophytes are common with *Tasmanites* spp. peaking at 14%, while *Botryococcus* spp. are rare.

From the top of Pinguis Subzone to the very base of Blagdeni Subzone (samples SF 535 to SF 543) we record the first occurrences of several new dinoflagellate cysts. In sample SF 535 (top Pinguis Subzone) we record the first representatives of *Wanaea* (*Wanaea* sp. 1). This is followed in the next sample (SF 536, top Pinguis Zone) by the first occurrences of *B. laevigata*, the first representatives of *Meiourogonyaulax* spp., *Pareodinia* sp. 2 of Feist-Burkhardt and Monteil 1997, *Protobatioladinium* sp., *Valensiella/Ellipsoidictyum* spp., and *V. euareatum*. The overlying sample (SF 537), at the very base of the Romani Subzone, contains the first occurrences of additional *Wanaea* morphotypes (*Wanaea* sp. 2 and *Wanaea* spp.), while further first occurrences in the next sample (SF 538, Romani Subzone) include *Cavatodissiliodinium* sp. 1, *Meiourogonyaulax* sp. 2 (a precursor to the well-known

M. valensii) and *Phallocysta thomasii*. The presence of a single specimen of *Reutlingia gochtii* is not considered stratigraphically useful in this interval as it is known to potentially occur in older and younger strata. The next set of overlying samples (SF 539 to SF 543) range from the top of the Romani Subzone, through the Humphriesianum Subzone and to the very base of the Blagdeni Subzone. These samples contain yet further dinoflagellate cyst first occurrences. In sample SF 539 the following first occurrences were noted: *A. crispa*, *Cavatodissiliodinium* sp. 2 and sp. 3, *Durotrigia filapicata*, *Durotrigia* cf. *omentifera*, and questionable *Orobodinium* spp. Two single specimens of the generally rare *Nannoceratopsis triceras* occur in SF 540 and SF 542. *A. crispa* is a well-known marker species with a first appearance in the Humphriesianum Zone (Feist-Burkhardt and Wille, 1992; Poulsen and Riding, 2003). An important stratigraphical event is considered the last occurrence of *P. thomasii* in the Humphriesianum Subzone (sample SF 542).

In the higher part of the Blagdenischichten Member there are four samples (SF 544 to SF 547) that correspond to the later part of the Blagdeni Subzone. The palynomorph assemblages of this interval differ from the samples below. We record between 27% and 40% dinoflagellate cysts, between 19% and 49% pollen and spores, but an abundance of foraminiferal test linings reaching between 17% and 37% of the total assemblage. *Tasmanites* spp. are much rarer than before, with only rare to occasional occurrences. Acritarchs are slightly more abundant than before, but reaching not more than 4.5% *Micrhystridium/Baltisphaeridium* spp. and 3.5% *Kalyptea halosa*. The abundance of foraminiferal test linings indicates oxygenated conditions at the sediment surface.

Dinoflagellate cyst assemblages in this sample interval are largely similar to before. Minor differences include the lower abundance of *Nannoceratopsis* spp. and only a single specimen of *N. gracilis* s.l. was recorded in sample SF 545. *G. erymnoteichon* and *Valensiella/Ellipsoidictyum* spp. are common to abundant. There are only a few single specimens of *Wanaea* sp. 1 (in SF 546 and SF 547) and one specimen of *Cavatodissiliodinium* sp. 3 (in SF 547). *Cavatodissiliodinium* sp. 1 and 2 are not recorded.

M. valensii first occurs questionably in SF 545, with confidently identified specimens first recorded in the following sample (SF 546). In this interval, another set of first occurrences of dinoflagellate cyst species are recorded. *Rhynchodiniopsis*? sp. 1 (a precursor form of *Rhynchodiniopsis? regalis*) first occurs in sample SF 545, while the overlying sample (SF 546) contains the first occurrence of *Atopodinium* sp. 1 (a precursor of *A. polygonale*) and of *Nannoceratopsis spiculata*.

In the **Hohlebach** section, the boundary between the Gosheim and Hauptrogenstein formations is covered by one sample below (SF 560) and one above (SF 561) the boundary. The two samples correspond to the later part of the Blagdeni Subzone.

Despite the different lithologies, composition of the palynomorph assemblages of the two samples is quite similar. The proportion of dinoflagellate cysts is 28% in SF 560, and 38% in SF 561, pollen and spores reach 37% and 47%, respectively. The main difference between the samples is the high abundance of foraminiferal test linings. Sample SF 560 from the Gosheim Formation contains 29% of the total palynomorph assemblage, whereas sample SF 561 from the Hauptrogenstein Formation contains only 5% foraminiferal test linings.

Dinoflagellate cyst assemblages in the two samples are similar to the last samples discussed from the Gosheim Formation at Egerten. The samples contain common to abundant *A. crispa*, *Batiacasphaera* spp., *Durotrigia* spp., *G. erymnoteichon*, *K. hypornatum*, *Rhynchodiniopsis*? sp. 1, and *Valensiella/Ellipsoidictyum* spp. Rare taxa include i.a. *B. laevigata*, *Cavatodissiliodinium* sp. 1, *Dissiliodinium* spp., *Kallosphaeridium* spp., *Meiourogonyaulax* spp., *M. valensii*, *N. gracilis* s.l., *Nannoceratopsis* spp., and *V. sphaerechinatum*. *N. gracilis* s.s., *V. cavum*, and *Wanaea* sp. 1 are only recorded in the lower sample, while *Atopodinium* sp. 1 and *N. spiculata* were only noted in the upper sample.

6.3 Summary of Stratigraphical Events

The Sauzei and Humphriesianum zones of the Upper Rhine area are characterized by relatively diverse dinoflagellate cyst assemblages. Despite its short duration, the interval contains a wealth of first and last occurrences of dinoflagellate cysts. Some are new undescribed taxa in open nomenclature, some are well-known species. Several of these first and last occurrences are considered to have stratigraphical value. Based on the occurrence data from the present study, we suggest the following succession of events of selected dinoflagellate cyst taxa, in ascending order.

Sauzei Zone (pars):

- Top *D. giganteum* acme
- Base *C. hansgochtii*
- Top *Evansia? spongogranulata*, top *D. giganteum*
- Top *C. hansgochtii*

Pinguis Subzone, Humphriesianum Zone:

- Base *Wanaea* sp. 1
- Base *Valensiella/Ellipsoidictyum* spp., base *Meiourogonyaulax* spp.

Romani Subzone, Humphriesianum Zone:

- Base *Wanaea* sp. 2, base *Wanaea* spp.
- Base *Cavatodissiliodinium* sp. 1, base *Meiourogonyaulax* sp. 2, base *P. thomasii*
- Base *A. crispa*, base *Cavatodissiliodinium* sp. 2 and sp. 3, base *D. filapicata*

Humphriesianum Subzone, Humphriesianum Zone:

- Top *P. thomasii*

Blagdeni Subzone, Humphriesianum Zone:

- Base *M. valensii*, base *Rhynchodiniopsis*? sp. 1
- Base *Atopodinium* sp. 1, base *N. spiculata*

The aforementioned dinoflagellate cyst events may reflect the true range bases or range tops of taxa, but some may only reflect localized or regional ranges. The present records reflect at least the minimum or maximum age of the events, but some of the taxa are rare, they were recorded with only a few specimens per sample, so the true range bases or range tops may have been missed. The study of further uncondensed, independently age-calibrated sections without hiatuses is required to confirm the timing of the events noted in these successions. This would verify the high-resolution (subzonal level) palynostratigraphy described herein.

7. DISCUSSION

7.1 Evolutionary Precursors of Well-Known Dinoflagellate Cyst Taxa

It is striking that in the studied samples there are so many precursor forms of dinoflagellate cyst taxa present.

Atopodinium sp. 1 is a precursor of *A. polygonale*. In the present material we can observe the gradual change in morphology from *K. hypornatum* via *Atopodinium* sp. 1 leading to *A. polygonale*. *K. hypornatum* is a typical component of the Sauzei Zone and Humphriesianum Zone assemblages in the Upper Rhine area samples. Its range base has been observed somewhat earlier in neighboring Switzerland in the lower part of the Wedelsandstein Formation, in the Laeviuscula Zone at Benken and Bözberg (Feist-Burkhardt, pers. obs.). The species ranges up to at least the end of the Bajocian, Parkinsoni Zone or the Middle Bathonian (Feist-Burkhardt, pers. obs.; Wille, 1995; pers. comm.). *Atopodinium* sp. 1 occurs in the present material in the later part of the Humphriesianum Zone, the Blagdeni Subzone. *A. polygonale* has its first appearance in southwest Germany slightly later, in the Niortense Zone (Feist-Burkhardt and Wille, 1992; as *Bejuia polygonalis*).

Cavatodissiliodinium sp. 3 is a precursor of *A. aldorfensis*. *Cavatodissiliodinium* sp. 3 has the same type of cyst wall as in *Aldorfia*, but the archeopyle type is that of earlier evolved genera such as *Dissiliodinium* and *Durotrigia* that also lose more than one precingular plate. *Cavatodissiliodinium* sp. 3 has its first occurrence in the Upper Rhine area samples in the Humphriesianum Zone, Romani Subzone. The first appearance of *A. aldorfensis* in southwest Germany has been reported slightly later, in the latest part of the Humphriesianum Zone by Feist-Burkhardt and Wille (1992); however, those studies may not have distinguished the precursor forms.

Meiourogonyaulax sp. 2 is a precursor of *M. valensii*. The well-known species *M. valensii* has sutural crests and a "boxy," polygonal cyst shape. In the present material we can observe the stepwise inception of tabulated cysts with an apical archeopyle (*Meiourogonyaulax* spp.), then *Meiourogonyaulax* sp. 2 with crests similar to *M. valensii* but a more rounded cyst shape, and finally *M. valensii* s.s. with its typical crests and "boxy," polygonal cyst shape. In the present material *Meiourogonyaulax* sp. 2 first occurs in the Romani Subzone, and *M. valensii* s.s. in the latest Humphriesianum Zone, Blagdeni Subzone. Feist-Burkhardt and Wille (1992) give a first appearance datum for *M. valensii* sensu lato in southwest Germany in the middle of the Humphriesianum Zone. Their connotation *sensu lato* indicates the inclusion of slightly differing morphotypes that may be similar to *Meiourogonyaulax* sp. 2.

Some ornamented forms of the *Batiacasphaera* plexus may be considered precursor forms of the genus *Meiourogonyaulax*. These ornamented *Batiacasphaera* (treated as *Sentusidinium* spp. or *Escharisphaeridia* spp. by some authors) include some specimens that have structural elements arranged in a manner that hints at sutural crests. These morphotypes represent the precursor forms to early *Meiourogonyaulax* spp. As for *Batiacasphaera*, species of *Meiourogonyaulax* may have ortho- or meta-type apical arrangements (e.g., Feist-Burkhardt and Monteil, 1997, p. 40). *Batiacasphaera* is a long-ranging genus that first appears in the Aalenian (Feist-Burkhardt, 1990; Feist-Burkhardt and Pross, 2010) and is a typical component of dinoflagellate cyst assemblages in the Middle Jurassic and younger. In the Early Bajocian they are abundant and morphologically very diverse, as is also observed in the present material from the Sauzei and Humphriesianum zones of the Upper Rhine area. The first specimens that can be assigned to *Meiourogonyaulax* spp. are recorded in the basal Humphriesianum Zone, Romani Subzone. In published accounts, Feist-Burkhardt and Wille (1992) list *M. valensii* sensu lato as the first species of the genus with a first appearance in the middle of the Humphriesianum Zone. The genus *Meiourogonyaulax* is rich in species and is particularly abundant in younger Middle Jurassic strata worldwide (e.g., Feist-Burkhardt and Wille, 1992; Mantle and Riding, 2012; Prauss, 1989; Riding, 1987).

Rhynchodiniopsis? sp. 1 is a precursor of *Rhynchodiniopsis*? *regalis*. Ornamentation in the two species is conspicuous and unique in Bajocian dinoflagellate cysts. In the present material there is no obvious precursor to *Rhynchodiniopsis*? sp. 1 with similar ornamentation. Ornamentation in the well-known *Rhynchodiniopsis*? *regalis* clearly derived from the reduction of the number of ornamental elements from *Rhynchodiniopsis*? sp. 1. In the present material only *Rhynchodiniopsis*? sp. 1 is recorded, from the Blagdeni Subzone of the Humphriesianum Zone. The first appearance of *Rhynchodiniopsis*? *regalis* in southwest Germany is also reported to be from the latest part of the Humphriesianum Zone by Feist-Burkhardt and Wille (1992). However, in that compilation of studies ranges may include precursor forms.

The first representatives of *Wanaea* recorded in the present study (*Wanaea* sp. 1, sp. 2, and *Wanaea* spp.) may be considered to represent evolutionary precursor forms of *W. indotata* Drugg 1978. In addition to their variable ornamentation, some of the specimens recorded here have an elongated conical hypocyst, as in *W. indotata* Drugg 1978, while other specimens have a shorter, almost hemispherical hypocyst, but all always with a short antapical horn. Forms with a hemispherical hypocyst are close to the Australasian *Dissiliodinium caddaense* (Filatoff, 1975) Stover and Helby 1987. The specimens of *Wanaea* in the present study also have the tendency for splits to form between the precingular plates, thus again appearing similar to *Dissiliodinium* with its multiplate precingular archeopyle. In the present material, the first occurrence of a representative of *Wanaea* is *Wanaea* sp. 1 in the Pinguis Subzone, at the base of Humphriesianum Zone. The other species of *Wanaea* follow marginally later. The first appearance of *W. indotata* in southwest Germany is reported to be at the base of the Humphriesianum Zone by Feist-Burkhardt and Wille (1992, cited as *Energlynia acollaris*); however, this includes precursor forms treated outside of *W. indotata* s.s. herein. The holotype of *W. indotata* is from the Parkinsoni Zone of the United Kingdom (Drugg, 1978) and unequivocal specimens of the species are illustrated from the Parkinsoni Zone of the Bajocian stratotype in France

(Feist-Burkhardt and Monteil, 1997). *D. caddaense* occurs in Australasia in the Early Bajocian *D. caddaense* Zone (Laeviuscula Chronozone) and the next younger *Nannoceratopsis deflandrei* Zone (interpreted as late Early to Late Bajocian, Riding et al., 2010). It has its last appearance in the *N. deflandrei* Zone, before the inception of the first representative of *Wanaea*, which is *W. verrucosa*, in the *W. verrucosa* Zone (interpreted as possibly latest Early Bajocian or Late Bajocian to Early Bathonian, Mantle and Riding, 2012). *Wanaea* sp. 1 from the Upper Rhine area is actually very similar to the Australasian *W. verrucosa*, and may even be conspecific.

7.2 Middle Bajocian "Explosion" of Dinoflagellates and Experimentation in Archeopyle Styles

All of the cases of precursors of well-known taxa cited earlier belong to evolutionary lineages within the Gonyaulacaceae family. Members of the Gonyaulacaceae have their first appearance at the base of the Aalenian (Feist-Burkhardt and Wille, 1992; Feist-Burkhardt and Pross, 2010). Evolutionary progression is slow in the Aalenian, before increasing substantially in the Bajocian. The tremendous radiation of Gonyaulacaceae has a peak in the "middle" Bajocian and has been discussed as the middle Bajocian "explosion" of dinoflagellates (Feist-Burkhardt and Monteil, 1997; Riding, 1987). From the Bajocian onwards the Gonyaulacaceae are the most important family of cyst forming dinoflagellates in the fossil record, and they are still today. The dominance of gonyaulacacean dinoflagellate cysts in the Mesozoic, Cenozoic, and today is visible on the diversity plot of Fensome et al. (1996, Fig. 1).

The "explosion" of gonyaulacacean dinoflagellates in the Bajocian goes along with the experimentation in archeopyle styles (Feist-Burkhardt and Monteil, 2001; Feist-Burkhardt and Wille, 1992; Riding, 1987). Variable multiplate precingular archeopyles, as in the genera *Dissiliodinium*, *Durotrigia*, *Cavatodissiliodinium*, may on one hand evolve to single or double plate precingular archeopyles as in *Acanthaulax*, *Aldorfia*, *Gongylodinium* (2P), *Rhynchodiniopsis*, and later *Gonyaulacysta*, or alternatively may evolve to epicystal archeopyles as in *Wanaea*, and later *Ctenidodinium* and *Korystocysta*. The cases of *Cavatodissiliodinium* sp. 3 and *Wanaea* spp. described above are two examples from the present material.

The middle Bajocian "explosion" of dinoflagellates is not the first radiation event in dinoflagellate evolution. After the first appearance of unequivocal dinoflagellate cysts in the Late Triassic Carnian, diversity stays moderate during the Late Triassic. Dinoflagellates take a long time to diversify after the Triassic–Jurassic boundary mass extinction. During the Early Jurassic, diversification is slow and takes place in several pulses, the most obvious ones in the Pliensbachian, in the late Early Toarcian and in the Late Toarcian/Aalenian. The Lower Jurassic radiation pulses of dinoflagellates have been discussed by Van de Schootbrugge et al. (2005) in relation to oceanographic and climatic conditions, long-term eustatic rise and overall increase in the areal extent of continental shelves due to the breakup of Pangea and the opening of the Atlantic ocean. The authors suggested that the rise in dinoflagellate diversity later in the Jurassic appears to correspond to deep water ventilation as a result of the opening of the Atlantic seaway which seems to be reflected in the Bajocian radiation event as documented in the present study.

7.3 Time Estimates

Using absolute ages to measure the speed of evolutionary processes is a somewhat arbitrary process, but geological time scales have greatly improved over the last 35 years and can now provide very good estimates. As already stated, the evolution of Gonyaulacaceae started with the first few genera in the Aalenian and became very fast in the Bajocian, with a peak in the "middle" Bajocian. In the latest Geologic Time Scale GTS 2012 (Gradstein et al., 2012), the duration of the Aalenian is estimated at 3.85 Ma and the duration of the Bajocian at 2.02 Ma. The duration of Sauzei and Humphriesianum zones together is 420,000 years, that of the Humphriesianum Zone alone, the time interval when most of the described dinoflagellate events happen, is only 250,000 years. Figures have changed quite a lot in comparison to the Geologic Time Scale GTS 2004 (Gradstein et al., 2004) because the duration of the whole of the Middle Jurassic has been reduced in the updated version. Using the GTS 2004, duration of the Aalenian is estimated at 4 Ma, duration of the Bajocian at 3.9 Ma, that of the Sauzei and Humphriesianum zones together at 980,000 years, and Humphriesianum Zone alone at 590,000 years. The spread in time estimates between GTS 2004 and 2012 is large, but using either, it becomes clear that the evolutionary processes during the middle Bajocian "explosion" of dinoflagellates were extremely fast.

7.4 Coeval Palynofloras and Palynostratigraphy

Middle Jurassic dinoflagellate cysts and their stratigraphical distribution are best known from the classical study areas of the Jurassic in Northwest and Central Europe, especially the United Kingdom, Germany, and also France (e.g., Feist-Burkhardt, 1992; Feist-Burkhardt and Wille, 1992;

Feist-Burkhardt and Monteil, 1997; Prauss, 1989; Riding, 1987). From these areas palynostratigraphical studies are available with independent age control, mainly by ammonites, that provide a reliable calibration of palynostratigraphical zonation schemes (Poulsen and Riding, 2003; Prauss, 1989; Riding and Thomas, 1992; Woollam and Riding, 1983). The studied interval of the Early Bajocian Sauzei and Humphriesianum zones, though, is much less well studied than other parts of the Middle Jurassic column. There are only a few published studies available, and the present study provides the most comprehensive published account with independent age control of this time interval.

7.4.1 Southwest Germany, Switzerland, and France

Southwest Germany is an area with a long tradition in palynological studies of the Jurassic. Feist-Burkhardt and Wille (1992) summarized the data of all palynological studies on the Jurassic in southwest Germany until 1990. This includes published as well as unpublished accounts such as PhD theses. Stratigraphical ranges of dinoflagellate cysts are correlated to ammonite zones and are provided in the form of a range chart (Feist-Burkhardt and Wille, 1992, Fig. 2).

In recent years new results and large data sets have become available from detailed studies of the Middle Jurassic of northern Switzerland, with many outcrop and well sections covering also the Sauzei and Humphriesianum zones (e.g., Wille in Gonzalez and Wetzel, 1996; Feist-Burkhardt, unpubl.; Hostettler et al., 2016).

The results of the present study from the Upper Rhine area fit naturally with the distribution data of southwest Germany and northern Switzerland. Species of *Cavatodissiliodinium* have so far not been reported in other areas than in southwest Germany and Switzerland. A formal zonation for southwest Germany and northern Switzerland is in progress (Feist-Burkhardt, in progress).

Feist-Burkhardt and Monteil (1997) studied the dinoflagellate cyst assemblages of the historical Bajocian stratotype section in the vicinity of Bayeux, Normandy, France. The Sauzei and Humphriesianum zones are very condensed in this area and the Humphriesianum Zone is part of the condensed Oolithe Ferrugineuse de Bayeux Formation. Despite the condensed nature of the section there are four consecutive samples from the Sauzei and Humphriesianum zones with independent age control down to subzone level. A total of 22 dinoflagellate cyst taxa are reported from the Sauzei and Humphriesianum zones from the stratotype section.

Composition of the assemblages and stratigraphical distribution of taxa agree with the results of the present study.

7.4.2 Northwest European Zonation and United Kingdom

The latest northwest European dinoflagellate cyst zonation is from Poulsen and Riding (2003) which is a revised and updated version of Riding and Thomas (1992) and Woollam and Riding (1983). The zonation is mainly based on stratigraphical data from the British Isles. Middle Jurassic dinoflagellate cyst assemblages from the North Sea differ, in parts greatly, from the onshore records because of the deposition of nonmarine strata during the Aalenian to Bathonian with only brief brackish or marine intercalations. The dinoflagellate cyst assemblages of the present study can be correlated to the DSJ12 and DSJ13 zones of Poulsen and Riding (2003), with *M. semitabulatum* and *A. crispa* as marker species. Many of the other taxa recorded in the present study are either absent in the UK or unaccounted for in the zonation.

Comparable case studies in the UK with similar assemblages covering the Early Bajocian Sauzei and Humphriesianum zones are those by Riding et al. (1991) of the Inner Hebrides, and Gowland & Riding (1991) and Butler et al. (2005) on the Ravenscar Group of Yorkshire, UK.

Dinoflagellate cyst assemblages and their succession in the Sauzei and Humphriesianum zones of the Inner Hebrides (Riding et al., 1991) are similar to the present material from the Upper Rhine area, but recovery of dinoflagellate cysts is relatively poor and the assemblages are comparably poorer and lower in diversity. Reported taxa include *Batiacasphaera* spp., *Dissiliodinium* spp., *G. erymnoteichon* (as *Dissiliodinium erymnoteichum*), *K. halosa* (as *Caddasphaera halosa*), *M. semitabulatum*, *N. gracilis*, *Meiourogonyaulax* sp., and *Valensiella ovulum*. The authors report a few specimens of *N. spiculata* and a questionable specimen of *Valvaeodinium spinosum* from the Sauzei and Humphriesianum zones. We doubt the presence of these two species as early in the succession and think they are probably misidentifications. The illustrated specimens are from the Bathonian.

Butler et al. (2005) published a succinct but important paper on the Aalenian to Early Bathonian Ravenscar Group from the Cleveland Basin in Yorkshire, England. They developed a palynostratigraphical scheme that has allowed a chronostratigraphical subdivision of the Ravenscar Group and that enables regional correlation of the onshore formations with coeval sediments of similar facies of the North Viking Graben (Brent Group) and offshore Mid-Norway (Fangst Group).

The dinoflagellate cyst assemblages and the succession of events in the Aalenian and Bajocian that they describe are very similar to what is known from southwest Germany (Feist-Burkhardt, 1990; Feist-Burkhardt and Wille, 1992; Feist-Burkhardt and Monteil, 2001). Of particular interest is their analysis of the Scarborough Formation at Hundale Point, which was earlier the subject of a less detailed study by Gowland and Riding (1991). Butler et al. (2005) stressed the stratigraphical value of *P. thomasii*. They carefully documented the distribution of *P. thomasii* at Hundale Point and found its first occurrence in the Pinguis Subzone of the Hundale Sandstone Member, and its acme and last occurrence in the Humphriesianum Subzone of the Spindlethorne Limestone Member and Ravenscar Shale Member, respectively. Our own studies of the same section (Feist-Burkhardt, unpublished) confirm the published account of Butler et al. (2005). These are important events, as base, acme, and top of this species are used for correlation with the Brent Province and mid-Norway. *P. thomasii* is also recorded in the Pinguis and Humphriesianum subzones of the present material of the Upper Rhine area. It has also been recorded in several sections from northern Switzerland (Feist-Burkhardt, pers. obs.; Hostettler et al., 2016) where it is used as a reliable marker species.

7.4.3 Australasia

In recent years, more and more detailed data from the southern hemisphere have become available. Chronostratigraphical age control is difficult in Australasia and in many cases ages assigned to palynostratigraphical zones rely on correlation with the palynostratigraphy in Europe and have been revised since their erection (Riding et al., 2010; Mantle and Riding, 2012; and literature therein).

The interval studied in the Upper Rhine area corresponds approximately to the *N. deflandrei* Zone and *W. verrucosa* Zone in Australasia. The actual species content of the Australasian zones is slightly different (Riding et al., 2010; Mantle and Riding, 2012), but it is astounding how similar the succession of morphological types of dinoflagellate cysts is in the northern and in the southern hemisphere. *D. caddaense* (Filatoff 1975) Stover and Helby 1987 is the abundant nominative species of the Australasian *D. caddaense* Zone that is independently dated to the Laeviuscula ammonite zone. An excellent marker species for the Laeviuscula and Sauzei zones in southwest Germany and Switzerland is *D. giganteum* where it can be very abundant. After the top of consistent *D. caddaense*, there are assemblages of the *N. deflandrei* Zone containing *Batiacasphaera* spp., *Dissiliodinium*, *Durotrigia* spp.,

Nannoceratopsis spp., and *Pareodinia* spp., which is again similar to Europe. The first representatives of *Wanaea* appear with several species in the subsequent *W. verrucosa* Zone. Early forms of *Wanaea* in both hemispheres are strongly ornamented. The Australasian *W. verrucosa* is very similar to the here recorded *Wanaea* sp. 1 and they may even be conspecific. Along with several species of *Wanaea* this zone has the first appearances of *K. hypornatum*, *Meiourogonyaulax* spp., *V. spinosum* (Fenton et al. 1980) Below 1987, and later within the zone, *Aldorfia*, *Endoscrinium*, and *Valensiella*. A good marker in the middle of the zone is an acme of *Phallocysta granosa* (Riding and Helby 2001). This succession is strikingly similar to the Upper Rhine area. Even the acme of *P. granosa* reminds the short range of *P. thomasii* in the Humphriesianum Zone, an event that can be correlated from southwest Germany/northern Switzerland to England and the North Sea. The two species are very similar in morphology and are surely related, they may even be conspecific.

8. CONCLUSION

The quantitative palynological analysis of the outcrop sections in the Upper Rhine area yielded very detailed data on the dinoflagellate cyst assemblages of the Sauzei and Humphriesianum zones at an unprecedented high resolution down to ammonite subzone level at a time of rapid dinoflagellate cyst evolution. Many stratigraphical events, base occurrences, acmes, and top occurrences of dinoflagellate cysts are carefully documented. The present study is the most comprehensive published account with independent age control of this time interval.

Most of the first appearances of dinoflagellate cyst taxa belong to the mid-Bajocian dinoflagellate explosion of the Gonyaulacaceae that has been recognized previously. The present study helps to further resolve this explosion event.

Some of the taxa recorded are well known and their stratigraphical distribution in other parts of the world is similar. For some we can add information on the range bases or tops. Other taxa recorded herein have not been previously recognized and are therefore left in open nomenclature pending further taxonomical study (Feist-Burkhardt, in progress). Further work is also needed on the diverse morphological plexuses of forms that so far have not been split into discrete species. A better understanding of these is needed, i.a. to clear up some misunderstandings between authors who may use different names for possibly the same taxa. This will clarify palynostratigraphical schemes and make them more comparable.

Resolution of dinoflagellate cyst palynostratigraphy is well known for being very high. The present study demonstrates that there is potential for even higher resolution in the Middle Jurassic, provided that we get the right sections, which ought to be uncondensed, independently age dated and without hiatuses, for the calibration of events.

ACKNOWLEDGMENTS

We would like to thank W. Ohmert (formerly Geological Survey of Baden-Württemberg in Freiburg) for guiding us in the field and joint sampling of the sections. Many thanks also go to the other members of the German Subcommission on Jurassic Stratigraphy for fruitful discussions and advice. The constructive reviews by D. Mantle (MGPalaeo Pty Ltd., Perth, Australia), N. Butler (PetroStrat Ltd., Conwy, UK), and M. Franz (LGRB, Landesamt für Geologie, Rohstoffe und Bergbau, Freiburg i.Br., Germany) are greatly appreciated and helped to improve the manuscript. Sample collection and processing of samples was within the frame of a research project funded by Deutsche Forschungsgemeinschaft (DFG grant Fe 435/1-2).

APPENDIX: LIST OF PALYNOMORPHS

List of all dinoflagellate cysts, pollen, spores, acritarchs, green algae and miscellaneous palynomorphs recorded in this study with full author citations for the species.

Dinoflagellate Cysts

Acanthaulax crispa (Wetzel 1967) Woollam & Riding 1983 (Plate 1(1–8))
Atopodinium sp. 1 (*Atopodinium polygonale* precursor) (Plate 11(5))
Batiacasphaera laevigata (Smelror 1988) Feist-Burkhardt & Monteil 1997 (Plate 8(1–4))
Batiacasphaera spp. (Plates 8(5–8), 9(1–9), 10(1 and 2))
Batiacasphaera spp. (smooth-scabrate) (Plate 8(5))
Cavatodissiliodinium hansgochtii Feist-Burkhardt & Monteil 2001 (Plate 2(1–6))
Cavatodissiliodinium sp. 1 (continuous periphragm) (Plates 3(1–6), 4(1–9))
Cavatodissiliodinium sp. 2 (open net periphragm) (Plate 5(1–4))
Cavatodissiliodinium sp. 3 (*Aldorfia aldorfensis* precursor) (Plate 6(1–6))
Dissiliodinium aff. *giganteum*
Dissiliodinium giganteum Feist-Burkhardt 1990 (Plate 5(5))
Dissiliodinium spp.
Durotrigia cf. *omentifera* Feist-Burkhardt & Monteil 2001 (Plate 7(5 and 6))

Durotrigia daveyi Bailey 1987 (Plate 7(1 and 2))
Durotrigia filapicata (Gocht 1970) Riding and Bailey 1991 (Plate 7(3 and 4))
Durotrigia spp.
Evansia? *spongogranulata* Below 1990 (Plate 15(1–6))
Gongylodinium erymnoteichon Fenton et al. 1980 (Plate 5(6))
Kallosphaeridium hypornatum Prauss 1989 (Plate 11(1–4))
Kallosphaeridium spp. (Plate 11(6–9))
Mancodinium semitabulatum Morgenroth 1970 (Plate 16(1 and 2))
Meiourogonyaulax sp. 2 (*Meiourogonyaulax valensii* precursor) (Plate 10(5))
Meiourogonyaulax spp. (Plate 10(3 and 4))
Meiourogonyaulax valensii Sarjeant 1966 (Plate 10(6–11))
Nannoceratopsis gracilis Alberti 1961 s.l. (Plate 14(1))
Nannoceratopsis gracilis Alberti 1961 s.s.
Nannoceratopsis sp. B in Feist (1987)
Nannoceratopsis spiculata Stover 1966 (Plate 14(5))
Nannoceratopsis spp. (Plate 14(3, 4, 6–8))
Nannoceratopsis triceras Drugg 1978 (Plate 14(2))
Orobodinium spp.
Pareodinia sp. 2 in Feist-Burkhardt and Monteil (1997) (Plate 15(8 and 9))
Pareodinia spp. (Plate 15(10))
Phallocysta thomasii Smelror 1991 (Plate 15(7))
Protobatioladinium sp. (Plate 15(11–13))
Reutlingia gochtii Drugg 1978 (Plate 16(3))
Rhynchodiniopsis? sp. 1 (*Rhynchodiniopsis*? *regalis* precursor) (Plate 12(1–4))
Valensiella/*Ellipsoidictyum* spp. (Plate 12(5–8))
Valvaeodinium cavum (Davies 1983) Below 1987 (Plate 16(4))
Valvaeodinium euareatum Prauss 1989 (Plate 16(8–11))
Valvaeodinium sphaerechinatum Below 1987 (Plate 16(5))
Valvaeodinium spp.
Valvaeodinium vermipellitum Below 1987 (Plate 16(6 and 7))
Wanaea sp. 1 (granulate-lichen like) (Plate 13(1–4))
Wanaea sp. 2 (lichen like-trabeculate) (Plate 13(5 and 6))
Wanaea spp.

Pollen

Araucariacites australis Cookson 1947
Bisaccate pollen
Callialasporites dampieri (Balme 1957) Sukh Dev 1961

Callialasporites microvelatus Schulz 1966
Callialasporites minor (Tralau 1968) Guy 1971
Callialasporites segmentatus (Balme 1957) Srivastava 1963
Callialasporites turbatus (Balme 1957) Schulz 1967
Cerebropollenites mesozoicus (Couper 1958) Nilsson 1958
Chasmatosporites apertus (Rogalska 1954) Nilsson 1958
Chasmatosporites hians Nilsson 1958
Classopollis spp. (tetrads)
Classopollis torosus (Reissinger 1950) Couper 1958
Exesipollenites scabratus (Couper 1958) Pocock 1970
Exesipollenites tumulus Balme 1957
Ovalipollis spp. (reworked)
Perinopollenites elatoides Couper 1958 (Plate 17(3))
Quadraeculina anellaeformis Malyavkina 1949 (Plate 17(4))
Ricciisporites tuberculatus Lundblad 1954 (reworked)
Striate bisaccate pollen (reworked)
Vitreisporites pallidus (Reissinger 1959) Nilsson 1958

Spores

Conbaculatisporites spp.
Concavisporites spp.
Deltoidospora minor (Couper 1953) Pocock 1970
Densoisporites velatus Weiland and Krieger 1953
Densosporites spp. (reworked)
Dictyophyllidites spp.
Duplexisporites problematicus (Couper 1958) Playford and Dettmann 1965
Foveosporites spp.
Ischyosporites variegatus (Couper 1958) Schulz 1967
Kekryphalospora distincta Fenton and Riding 1988
Leptolepidites spp.
Lycopodiacidites rugulatus (Couper 1958) Schulz 1967
Lycopodiumsporites austroclavatidites (Cookson 1953) Potonie 1956
Lycopodiumsporites clavatoides (Couper 1958) Tralau 1968
Lycopodiumsporites gristhorpensis Couper 1958 (Plate 17(1 and 2))
Lycopodiumsporites reticulumsporites (Rouse 1959) Dettmann 1963
Lycopodiumsporites semimuris (Danzé-Corsin and Laveine 1963) Reiser and Williams 1969
Matonisporites spp.
Osmundacidites wellmanii Couper 1953

Staplinisporites caminus (Balme 1957) Pocock 1962
Todisporites minor Couper 1958
Trilete spores indet.
Trilites rariverrucatus (Danze-Corsin & Laveine 1963) Tralau 1968
Uvaesporites argenteaeformis (Bolchovitina 1953) Schulz 1967

Acritarchs

Fromea tornatilis (Drugg 1978) Lentin and Williams 1981 (Plate 16(12))
Kalyptea halosa Filatoff 1975
Leiosphaeridia spp.
Lunulidia sp. in Gocht (1964)
Micrhystridium/Baltisphaeridium spp.

Green Algae

Botryococcus spp. (Plate 17(6 and 7))
Cymatiosphaera spp.
Pterospermopsis spp.
Tasmanites spp. (Plate 17(5))
Zygospore sp. 1 (Plate 16(13))

Miscellaneous Palynomorphs

Foraminiferal test linings (Plate 16(14))
Fungal remains

REFERENCES

Bloos, G., Dietl, G., Schweigert, G., 2005. Der Jura Süddeutschlands in der Stratigraphischen Tabelle von Deutschland 2002. Newsl. Stratigr. 41 (1–3), 263–277.

Burkhalter, R.M., Bläsi, H.-R., Feist-Burkhardt, S., 1997. Der Dogger ß (oberes Aalénien) in den Bohrungen Herdern-1, Berlingen-1 und Kreuzlingen-1 (Nordschweiz) und seine Beziehungen zu den gleichaltrigen Schichten im Nordjura. Eclogae Geol. Helv. 90, 269–291.

Butler, N., Charnock, M.A., Hager, K.O., Watkins, C.A., 2005. The Ravenscar group: a stratigraphic analogue for the Middle Jurassic reservoirs of the north sea and offshore Mid Norway. In: Powell, A.J., Riding, J.B. (Eds.), Recent Developments in Applied Biostratigraphy. The Micropalaeontological Society, Special Publications. The Geological Society, London, pp. 43–53.

Deutsche Stratigraphische Kommission (Eds.), 2002. Stratigraphische Tabelle von Deutschland 2002.

Dietze, V., Bosch, K., Franz, M., Kutz, M., Schweigert, G., Wannenmacher, N., Studer, S., 2013. Die Humphriesianum-Zone (Unter-Bajocium, Mitteljura) am Kahlenberg bei Ringsheim (Oberrheingraben, SW Deutschland). Palaeodiversity 6, 29–61.

Dietze, V., Schweigert, G., Dietl, G., 2011. Chrono-/Biostratigraphie im Mitteljura von SW-Deutschland. 1. Teil: Bajocium. Jh. Ges. Naturkde. Württemberg 167 (1), 49–71.

Dietze, V., Wannenmacher, N., Franz, M., Schweigert, G., 2015. Lithological section and biostratigraphy of the Gosheim Formation in its type area (Lower Bajocian, Swabian Alb, SW Germany). Palaeodiversity 8, 31–69.
Drugg, W.S., 1978. Some Jurassic dinoflagellate cysts from England, France and Germany. Palaeontographica Abt. B 168 (1–3), 61–79.
Evitt, W.R., Lentin, J.K., Millioud, M.E., Stover, L.E., Williams, G.L., 1977. Dinoflagellate cyst terminology. Geol. Surv. Can. Pap. 76-24, 1–11.
Feist, S., 1987. Palynologische Untersuchungen im Braunjura beta und unteren Braunjura gamma (oberes Aalenium bis unteres Bajocium) der Bohrungen Hausen, nordöstliche Schwäbische Alb (Diploma thesis). University of Tübingen. unpublished.
Feist-Burkhardt, S., 1990. Dinoflagellate cyst assemblages of the Hausen coreholes (Aalenian to Early Bajocian), southwest Germany. Bull. Centres Rech. Explor.-Prod. Elf-Aquitaine 14 (2), 611–633.
Feist-Burkhardt, S., 1992. Palynological Investigations in the Lower and Middle Jurassic of Switzerland, France and Germany: Palynofacies, and Type of Organic Matter, Dinoflagellate Cyst Morphology and Stratigraphy (PhD Thesis). Département de Géologie et de Paléontologie, Université de Genève, Geneva, pp. 1–330.
Feist-Burkhardt, S., Monteil, E., 1997. Dinoflagellate cysts from the Bajocian stratotype (Calvados, Normandy, western France). Bull. Centre Rech. Elf Explor. Prod. 21 (1), 31–105.
Feist-Burkhardt, S., Monteil, E., 2001. Gonyaulacacean dinoflagellate cysts with multi-plate precingular archaeopyle. N. Jb. Geol. Paläont., Abh. 219 (1/2), 33–81.
Feist-Burkhardt, S., Pross, J., 1999. Morphological analysis and description of Middle Jurassic dinoflagellate cyst marker species using confocal laser scanning microscopy, digital optical microscopy, and conventional light microscopy [1998] Bull. Centre Rech. Elf Explor. Prod. 22 (1), 103–145.
Feist-Burkhardt, S., Pross, J., 2010. Dinoflagellate cyst biostratigraphy of the Opalinuston Formation (Middle Jurassic) in the Aalenian type area in southwest Germany and north Switzerland. Lethaia 43, 10–31.
Feist-Burkhardt, S., Wille, W., 1992. Jurassic palynology in southwest Germany – state of the art. Cahiers de Micropaléontologie 7 (1/2), 141–164.
Fensome, R.A., MacRae, R.A., Moldowan, J.M., Taylor, F.J.R., Williams, G.L., 1996. The early Mesozoic radiation of dinoflagellates. Paleobiology 22 (3), 329–338.
Fensome, R.A., Williams, G.L., MacRae, R.A., 2009. Late Cretaceous and Cenozoic fossil dinoflagellates and other palynomorphs from the Scotian Margin, offshore eastern Canada. J. Syst. Palaeontol. 7, 1–79.
Fensome, R.A., MacRae, R.A., Williams, G.L., 2008. DINOFLAJ2, Version 1. American Association of Stratigraphic Palynologists. Data Series no. 1. http://dinoflaj.smu.ca/wiki/Main_Page.
Gassmann, G., Ohmert, W., 1990. Der Humphriesi-Oolith von Egerten im Wollbachtal (Oberrheingebiet N Lörrach). Jh. Geol. Landesamt Baden-Württemberg. 32, 159–170.
Gocht, H., 1964. Planktonische Kleinformen aus dem Lias-/Dogger-Grenzbereich Nord- und Süddeutschlands. N. Jb. Geol. Paläont. Abh. 119, 113–133.
Gonzalez, R., 1996. Response of shallow-marine carbonate facies to third-order and high-frequency sea-level fluctuations: Hauptrogenstein Formation, northern Switzerland. Sediment. Geol. 102, 111–130.
Gonzalez, R., Wetzel, A., 1996. Stratigraphy and paleogeography of the Hauptrogenstein and Klingau Formations (middle Bajocian to late Bathonian), northern Switzerland. Eclogae Geol. Helv. 89 (1), 695–720.
Gowland, S., Riding, J.B., 1991. Stratigraphy, sedimentology and palaeontology of the Scarborough Formation (Middle Jurassic) at Hundale Point, North Yokshire. Proc. Yorkshire Geol. Soc. 48 (4), 375–392.

Gradstein, F.M., Ogg, J.G., Schmitz, M.D., Ogg, G.M. (Eds.), 2012. The Geologic Time Scale 2012. Elsevier BV, Boston, USA.

Gradstein, F.M., Ogg, J.G., Smith, A.G. (Eds.), 2004. A Geologic Time Scale 2004. Cambridge University Press.

Hostettler, B., Reisdorf, A.G., Jaeggi, D., Deplazes, G., Bläsi, H., Morard, A., Feist-Burkhardt, S., Waltschew, A., Dietze, V., Menkveld-Gfeller, U., 2016. Litho- and biostratigraphy of the Opalinus Clay and bounding formations in the Mont Terri rock laboratory (Switzerland) (in press). Swiss J. Geosci.

LGRB, 2016. Symbolschlüssel Geologie Baden-Württemberg - Verzeichnis Geologischer Einheiten (aktualisierte Ausgabe 2016). In: Landesamt für Geologie, Rohstoffe und Bergbau (Ed.), Regierungspräsidium Freiburg, 1 tab. http://www.lgrb-bw.de.

Mantle, D.J., Riding, J.B., 2012. Palynology of the Middle Jurassic (Bajocian-Bathonian) *Wanaea verrucosa* dinoflagellate cyst zone of the North West Shelf of Australia. Rev. Palaeobot. Palynology 180, 41–78.

Ohmert, W., 1988. Das Unter-Bajocium von Badenweiler (Oberrhein), verglichen mit Nachbargebieten. Jh. Geol. Landesamt Baden-Württemberg 30, 315–347.

Ohmert, W., 1994. The Frechi horizon (Humphriesianum Zone, Lower Bajocian) from the Oberrhein (South west Germany). Géobios 27, 359–367.

Poulsen, N.E., Riding, J.B., 2003. The Jurassic dinoflagellate cyst zonation of Subboreal northwest Europe. In: Inesen, J.R., Surlyk, F. (Eds.), The Jurassic of Denmark and Greenland, Geological Survey of Denmark and Greenland Bulletin, 1, pp. 115–144.

Prauss, M., 1989. Dinozysten-Stratigraphie und Palynofazies im Oberen Lias und Dogger von NW-Deutschland. Palaeontographica Abt. B 214 (1–4), 1–124.

Riding, J.B., 1987. Dinoflagellate cyst stratigraphy of the Nettleton Bottom Borehole (Jurassic: Hettangian to Kimmeridgian), Lincolnshire, England. Proc. Yorkshire Geol. Soc. 46 (3), 231–266.

Riding, J.B., Helby, R., 2001. A selective reappraisal of *Wanaea* Cookson & Eisenack 1958 (Dinophyceae). In: Laurie, J.R., Foster, C.B. (Eds.), Studies in Australian Mesozoic Palynology II Memoir of the Association of Australasian Palaeontologists, 24, pp. 33–58.

Riding, J.B., Mantle, D.J., Backhouse, J., 2010. A review of the chronostratigraphical ages of middle Triassic to late Jurassic dinofagellate cyst biozones of the north west Shelf of Australia. Rev. Palaeobot. Palynol. 162, 543–575.

Riding, J.B., Thomas, J.E., 1992. Dinoflagellate cysts of the Jurassic System. In: Powell, A.J. (Ed.), A Stratigraphic Index of Dinoflagellate Cysts. Chapman & Hall, London, pp. 7–57.

Riding, J.B., Walton, W., Shaw, D., 1991. Toarcian to Bathonian (Jurassic) palynology of the inner Hebrides, northwest Scotland. Palynology 15, 115–179.

Rioult, M., Contini, D., Elmi, S., Gabilly, J., 1997. Bajocien. In: Cariou, E., Hantzpergue, P. (Eds.), Biostratigraphie du Jurassique Ouest-Européen et Méditerranéen, Bulletin du Centre de Recherches Elf Exploration Production Mémoire, 17, pp. 41–53.

StrataBugs, version 2.1 (June 2016). StrataData Ltd., UK. http://www.stratadata.co.uk.

TimeScale Creator, version 6.4 (February 2015). Geologic TimeScale Foundation. https://engineering.purdue.edu/Stratigraphy/tscreator/.

Van de Schootbrugge, B., Bailey, T.R., Rosenthal, Y., Katz, M.E., Wright, J.D., Miller, K.G., Feist-Burkhardt, S., Falkowski, P.G., 2005. Early Jurassic climate change and the radiation of organic-walled phytoplankton in the Tethys Ocean. Paleobiology 31 (1), 73–97.

Wood, G.D., Gabriel, A.M., Lawson, J.C., 1996. Palynological techniques - processing and microscopy. In: Jansonius, J., McGregor, D.C. (Eds.), Palynology, Principles and Applications, vol. 1. American Association of Stratigraphic Palynologists Foundation, pp. 29–50.

Woollam, R., Riding, J.B., 1983. Dinoflagellate cyst zonation of the English Jurassic. Rep. Inst. Geol. Sci. London 83 (2), 1–41.

CHAPTER SIX

The Relevance of Iberian Sedimentary Successions for Paleogene Stratigraphy and Timescales

A. Payros*[,1], V. Pujalte*, X. Orue-Etxebarria*, E. Apellaniz*, G. Bernaola*, J.I. Baceta*, F. Caballero*, J. Dinarès-Turell[§], S. Monechi[¶], S. Ortiz[||], B. Schmitz[#] and J. Tosquella**

*University of the Basque Country (UPV/EHU), Bilbao, Spain
[§]Istituto Nazionale di Geofisica e Vulcanologia, Roma, Italy
[¶]University of Florence, Florence, Italy
[||]PetroStrat Ltd., Conwy, Wales, United Kingdom
[#]University of Lund, Lund, Sweden
**University of Huelva, Huelva, Spain
[1]Corresponding author: E-mail: a.payros@ehu.eus

Contents

Stratigraphy & Timescales, Volume 1
ISSN 2468-5178
http://dx.doi.org/10.1016/bs.sats.2016.08.001

Abstract

Throughout most of the 19th and 20th centuries, studies of Paleogene stratigraphy carried out in the Iberian Peninsula lagged behind those of northern Europe. More recently, however, Iberian contributions on this topic have increased in both number and importance. This is highlighted by the definition of several global stratotype sections and points (GSSPs) and auxiliary stratotype sections and points (ASSPs). Geographically, the Basque–Cantabrian region of the western Pyrenees is especially significant for late Cretaceous–early Paleogene stratigraphy, as it hosts five GSSPs and two ASSPs in less than 9000 km^2. In addition, the potential of other prospective Paleogene GSSPs in the region has also been extensively evaluated. Another reference area is the NE part of the Betic Cordillera, where one ASSP has been defined and several internationally significant sections occur.

This review shows that to date most of the Iberian contributions to Paleogene stratigraphy have been limited to the Paleocene-middle Eocene interval, meaning that there are still promising opportunities for future research. More specifically, expanded upper Eocene–Oligocene marine successions are well exposed in some western Pyrenean areas and, more extensively, in the Betic Cordillera. Many of these sections have not yet been studied in detail. Similarly, the Paleogene successions from the Iberian interior basins also have great potential for continental chronostratigraphic refinements. It is therefore hoped that this review will create new interest in the area and will be of practical help to potential researchers in selecting the most appropriate sections for their work.

1. INTRODUCTION

In order to improve the reliability of existing geological timescales, the International Commission on Stratigraphy (ICS) is defining global stratotype sections and points (GSSPs) for the bases of all standard chronostratigraphic stages. In addition, the ICS accepts that, with the aim of achieving a better understanding of chronostratigraphic boundaries, auxiliary stratotype sections and points (ASSPs) can also be defined, which should be intrinsically subordinate to their respective GSSPs (Remane et al., 1996). To achieve these goals, international subcommissions for each period were set up within the ICS. During the 28th International Geological Congress held in Washington in 1989, the International Subcomission on Paleogene Stratigraphy (ISPS) agreed on a set of nine Paleogene Stages. Subsequently, official Working Groups were set up to define their corresponding GSSPs.

The contribution of stratigraphic studies carried out in Iberian sedimentary successions to standard Paleogene chronostratigraphy has been significant in recent years, as highlighted by the definition of three out of nine Paleogene GSSPs in coastal outcrops of the Basque Country (Molina et al., 2011; Schmitz et al., 2011; Vandenbergue et al., 2012). In addition, the potential of other prospective Iberian sedimentary successions for the definition of other Paleogene GSSPs was also evaluated, albeit unsuccessfully. However, some of these successions were recognized as ASSPs.

Behind this recognition, there is of course a large body of lesser known studies carried out over the last five decades. Despite perhaps not being the headline material, many of these works paved the way for later breakthroughs in the general knowledge of Paleogene stratigraphy. Therefore, this review is partly intended as a tribute to the work of those researchers who have contributed to our current knowledge of the Iberian Paleogene stratigraphy. Additionally, a second aim is to review the most relevant results, in the hope that this will create new interest and will help researchers to select the most appropriate areas for their objectives.

2. GEOLOGICAL SETTING

The most extensive Iberian Paleogene outcrops are located in the South Pyrenean area and include both marine and terrestrial deposits (Fig. 1A). Marine and transitional deposits can also be found in the Betic Cordillera and in some coastal areas of Portugal. Scattered outcrops of Paleogene terrestrial deposits can also be found in many other areas throughout the Iberian Peninsula. The current distribution of these deposits shows the geometry and characteristics of the original sedimentary basins in

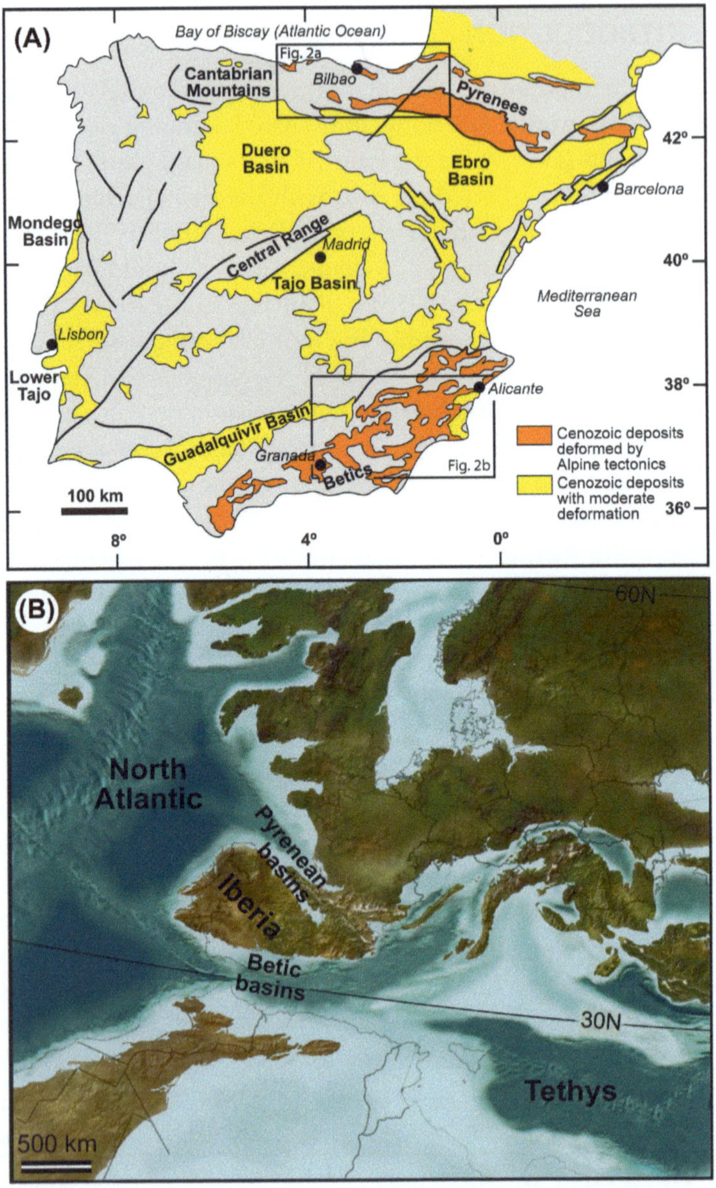

Figure 1 (A) Simplified geological map of the Iberian Peninsula, showing the main morphostructural units and Cenozoic basins. The basins deformed by Alpine tectonics are mainly Paleogene in age, whereas those with moderate deformation mainly contain Neogene successions, Paleogene deposits being restricted to their marginal areas. Based on Pujalte et al. (2002, Fig. 13.1) and Pais et al. (2012, Fig. 1B). (B) 50 Ma paleogeographic location of the Iberian Peninsula. *Available at http://www.cpgeosystems.com, modified with permission; © Ron Blakey, 2011, Colorado Plateau Geosystems, Inc.*

which they were accumulated (Banda, 1996; Saenz de Galdeano, 1996). In late Cretaceous and Paleogene times Iberia was a microplate located between the larger African and European plates. Paleogeographically, the occurrence of marine and transitional Paleogene deposits in the South Pyrenean area, in the Betic Cordillera, and in the Portuguese Mondego coastal plain shows the location of the northern, southern, and western Iberian continental margins, respectively, at that time (Fig. 1B).

The Iberian plate progressively rotated anticlockwise and approached the southwestern margin of the European plate (Roest and Srivastava, 1991; Smith, 1996), which eventually resulted in N–S directed collision between both plates, complex subduction of the Iberian plate beneath Europe in Eocene times and the uplift of the Pyrenean–Cantabrian orogen (Muñoz, 2002; Pedreira et al., 2007, 2015; Teixell, 2004). These processes started in the northeastern part of the Iberian plate, causing the creation of a narrow interplate marine gulf, which opened northwestwards into the Atlantic Ocean at approximately 30–35°N paleolatitude (Baceta et al., 2004; Barnolas et al., 2004; Plaziat, 1981; Pujalte et al., 2002). The gulf was flanked by shallow water–carbonate platforms and accumulated deep-sea hemipelagic sediments at the bottom of the basin. Tectonic compression progressively extended westwards, causing the growth of the Pyrenees and the subdivision of the interplate marine gulf into two foreland basins located to the north and south of the uplifting orogen (Fig. 1B). The combination of increasing thrusting, uplift, denudation, and sediment influx caused the infilling of the basins and the expansion of terrestrial conditions in late Eocene–Oligocene times, marine conditions only remaining in the westernmost part of the Pyrenean area (i.e., in the so-called Basque–Cantabrian region; Fig. 2A). This evolution further resulted in the tectonic inversion of the foreland basins and their transformation into piggy-back systems, which produced the concurrent removal of some of the original deposits and the exposure of others. Paleogene deposits scarcely crop out in the northern Pyrenees today, but are widely exposed on the southern side of the range (Fig. 1A). They constitute very thick successions (hundreds of meters thick in marginal areas but >6000-m thick at the basin axis) comprising a wide range of facies, from terrestrial to deep-marine.

The tectonic inversion of the Eocene South Pyrenean foreland basin brought about the creation of a terrestrial, endorheic foreland basin further southwards, the so-called Ebro basin (Fig. 1A; Muñoz et al., 2002; Pardo et al., 2004). This basin was characterized by the formation of alluvial fans on its northern and southern margins and the occurrence of extensive lakes in the center. Periods of increasing sediment supply and subsidence resulted in alluvial fan progradation, whereas the axial lakes expanded during periods

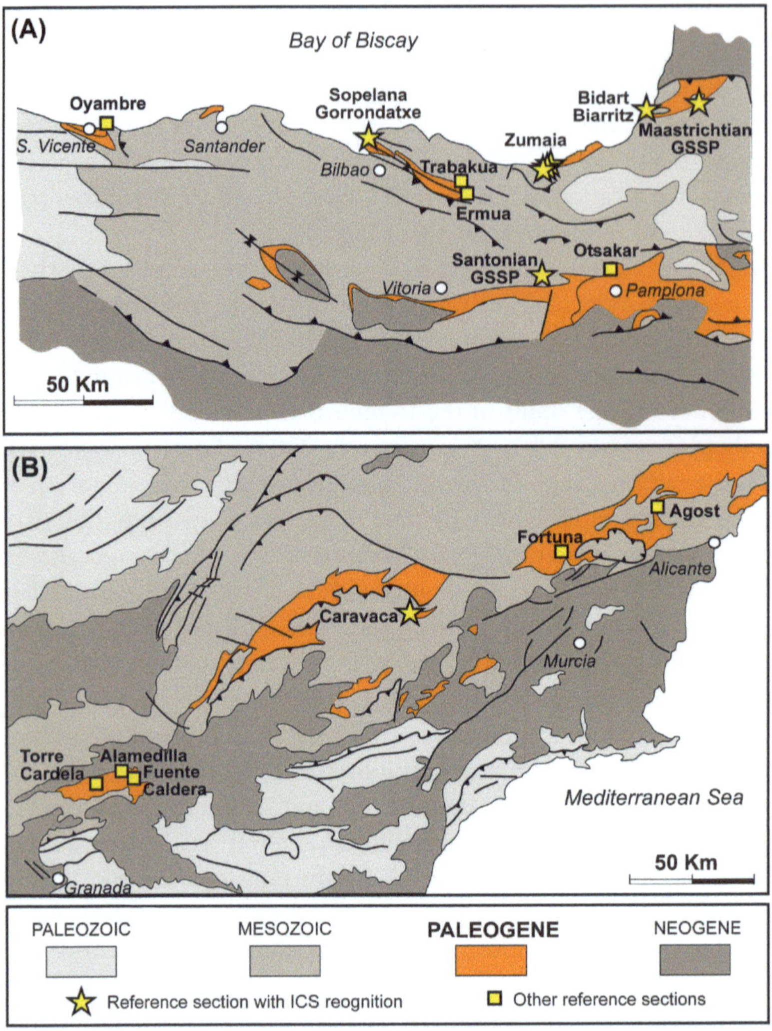

Figure 2 (A) Geological maps of the Basque–Cantabrian area in the western Pyrenees and (B) of the central and eastern parts of the Betic Cordillera (see Fig. 1A for location). Location of GSSPs and ASSPs (*stars*) and other reference sections discussed in the text (*squares*) is shown. *ASSPs*, auxiliary stratotype sections and points; *GSSPs*, global stratotype sections and points.

of relative stability. Although many of the Paleogene strata are covered by younger deposits, extensive outcrops occur throughout the Ebro basin, especially in its eastern part, sedimentary successions being thicker than 3000 m.

Marine Paleogene deposits from the external zones of the Betic Cordillera accumulated on the southern continental margin of the Iberian Peninsula (Figs. 1 and 2B; Braga et al., 2002; Vera, 2004). Overall, the northern part of the External Zones corresponded to terrestrial and shallow marine—carbonate platform environments throughout the Paleogene, whereas deeper marine sediments accumulated southwards. However, faulted blocks were inherited from previous rifting phases, which were reactivated in Paleogene times due to far-field responses to the Pyrenean compression, causing local variations in subsidence and sedimentation. A similar paleogeographic configuration was mirrored on the northern margin of the Mesomediterranean Alboran microplate, an independent wedge of continental crust which was originally located further eastwards, but today constitutes the Internal Zones of the Betic Cordillera. Progressive convergence of both margins caused subduction below the Mesomediterranean microplate and, eventually, the uplift of the area in Miocene times and the formation of the Guadalquivir foreland basin to the north (Fig. 1A). At the same time, deep-sea Paleogene deposits that had originally accumulated between the Mesomediterranean and African plates (in the so-called flysch trough of the Gibraltar complex) were squeezed and tectonically superposed on both the Internal and External Zones. Such complex evolution, combined with partial erosion and sedimentary cover in Neogene times, has resulted in an array of discontinuous Paleogene outcrops extending from the Gibraltar Strait in the west to the Alicante province in the east. They comprise sedimentary successions several hundreds to thousands of meters thick and display facies from varied depositional environments.

The position of the western Iberian Paleogene shore is shown by the occurrence of Campanian to lower Eocene coastal plain deposits, less than 200-m thick, in the Portuguese Mondego area (Fig. 1), where meandering fluvial systems drained towards the northwest (Pais et al., 2012). However, by middle Eocene times the intense Pyrenean compression induced lithospheric folding and uplift of Iberia. Consequently, middle Eocene to Oligocene deposits are continental in this area. Similarly, the middle Eocene to Oligocene successions exposed in the lower Tajo-Sado basin further southwards, which are up to 400 m thick, are also composed of alluvial deposits sourced from the northeast.

In addition to the Paleogene deposits accumulated on the margins of the Iberian plate, Eocene—Oligocene terrestrial sedimentation also occurred in several interior basins. The largest are the Duero and Tajo basins (Fig. 1A; Alonso-Gavilan et al., 2004; Alonso-Zarza and Calvo, 2002; Armenteros et al., 2002; Pais et al., 2012), but many other smaller basins also occur (Anadon et al., 2004; Armenteros et al., 2004; Pais et al., 2012). Their evolution was controlled by intracontinental transfer of tectonic compression from both the northern and southern margins of the Iberian plate in mid-late Eocene times. This compression reactivated older faults and caused the uplift of several intracontinental mountain belts, one of which (Central

Range) separated the Duero and Tajo basins (Fig. 1A). Deformation intensified in Oligocene times and further uplifted the margins of the basins, increasing marginal denudation and sediment supply. In the Duero basin, the northern part of which constitutes the foreland basin of the Cantabrian Mountains, alluvial sediment dispersal was mainly northeastwards, but northwesterly derived influx also occurred. Correspondingly, coarse-grained alluvial systems prograded northeastwards, where carbonate lakes existed. In the Tajo basin, marginal alluvial systems also prograded onto lacustrine systems located in the basin center. Subsequently, Paleogene deposits were covered by thick Neogene successions in both basins. Consequently, Eocene–Oligocene successions, which are hundreds to thousands of meters thick, are mainly exposed on marginal areas of the Duero and Tajo basins and contain significant unconformities in their upper levels.

3. HISTORICAL BACKGROUND

The contribution of Iberian sedimentary successions to international Paleogene stratigraphy has mainly come from work carried out over the last five decades. However, the development of both international and Iberian Paleogene stratigraphy started much earlier. It is therefore timely to review the historical development of modern stratigraphy at both international and Iberian levels.

Most of the stratigraphic terms used today to name the Paleogene Period, Epochs and Stages were coined at the end of the 19th century by pioneering European geologists who worked mostly in northern France, southern England, Belgium, Germany, and Denmark. The only exception is the term Priabonian, which despite being first used by French geologists, was derived from a north Italian region (Berggren, 1971, 1998; Vandenbergue et al., 2012). At that time, geological research in the Iberian Peninsula was mainly carried out by engineers engaged in mining activities (Julivert, 2014). Stratigraphic aspects were generally overlooked or, at best, limited to establishing the ages of the rocks. The only exceptions of note were the dissertations presented by some French researchers (e.g., Carez, 1881; Larrazet, 1896; Seunes, 1890; Roussel, 1893; Vezian, 1856) and the lesser known works of some local geologists (e.g., Pere Alsius, 1839–1915; Lucas Mallada, 1841–1921; Jaume Almera, 1845–1919; Ramon Adan de Yarza, 1848–1917). Consequently, the stratigraphic terms derived from studies carried out in the Iberian Peninsula did not make their way into the standard stratigraphic classification.

The coming of the 20th century brought new debates and improvements in knowledge to the European geological community. In the field of tectonics, Marcel Bertrand (1847–1907) presented his then controversial ideas on thrust sheets, while Alfred Wegener (1880–1930) focused on continental drift. Stratigraphic knowledge and understanding also improved significantly, as shown by the formulation of the "law of the correlation of facies" by German geologist Johannes Walther (1860–1937). In the Iberian Peninsula more progress was made in the biostratigraphic characterization of rock units. With regard to Paleogene terrains, French researchers presented several dissertations (e.g., Birot, 1934; Doncieux, 1933; Douville, 1922; Ciry, 1940; Lamare, 1936; Mengaud, 1920; Pfender, 1926). Additional work was carried out by Federico Gomez Llueca (1889–1960) on nummulitids. Under the leadership of Eduardo Hernández-Pacheco (1872–1965) significant effort was made on the stratigraphic characterization of terrestrial deposits and their fossil content in the interior basins, which include Paleogene deposits in their lowermost parts. However, the relatively modest role of stratigraphy over the first decades of the 20th century is illustrated by its presence in the 14th International Geological Congress held in Madrid in 1926: no scientific session was devoted to this field, stratigraphy only being represented in a specific workshop on "International Stratigraphic Lexicon" and, more indirectly, in some of the related field trips and guidebooks (Ayala-Carcedo et al., 2005; Julivert, 2014). Nevertheless, this event prompted the internationalization of the work of several local researchers and helped to further the interest of the international geological community (especially French and German) in the geology of the Iberian Peninsula, which obtained greater presence in international publications.

Unfortunately, events outside science interrupted the internationalization of Iberian geology. More specifically, the Spanish Civil War (1936–39) and the World War II (1939–45) caused the political and scientific isolation of Spain from the rest of Europe during the early Franco years (Julivert, 2014). Many geologists left the country, while those who stayed worked under the conditions inherent to military dictatorship. Consequently, little geological work was carried out and very few results found their way into the international arena. However, some geologists were still capable of carrying out research, maintained international relationships and made significant contributions to Paleogene stratigraphy. Examples include the work of Ramon Bataller (1890–1962), Maximo Ruiz de Gaona (1902–71), Valenti Masachs (1915–80), and Antonio Almela (1903–86), who described many Paleogene successions throughout the South Pyrenean area and made novel contributions to Eocene larger foraminiferal stratigraphy. Joaquin Gomez de Llarena (1891–1979) carried out pioneering research on Paleogene flysch deposits from the Basque coast, which would later be described as one of the first onshore-exposed submarine fan successions identified in the world

(Kruit et al., 1972). Jaume Marcet Riba (1894–1963), despite devoting most of his career to petrology, also made significant stratigraphic contributions from some Catalan Paleogene successions. Guillem Colom (1900–93) was a micropaleontologist who age-dated innumerable Iberian sections, in addition to defining numerous new Paleogene species; furthermore, being in contact with Joseph A. Cushman and other colleagues for foraminiferal studies, he became a member of the Society for Economic Paleontologists and Mineralogists. In addition, Lluis Sole Sabaris (1908–85), after beginning his career as a researcher with his thesis on Eocene corals from Catalonia, continued dedicating part of his research to general Cenozoic stratigraphy along with some of his students at the University of Barcelona (e.g., Miquel Crusafont, 1910–83; Oriol Riba, 1923–2011).

The onset of the Cold War in the 1950s gave Spain an unexpected strategic significance, leading indirectly to its reincorporation into some international institutions. This opened the door to international business partnerships and external financial support. Importantly for geology, this included the admission of multinational oil companies (Lanaja, 1987). Consequently, a new generation of more specialized European geologists arrived in the Iberian Peninsula, both for exploratory and academic purposes. With regard to Paleogene stratigraphy, the studies by Jean Phillippe Mangin, which culminated with the presentation of his thesis in 1958 (Mangin, 1960), must be highlighted. His work rediscovered the exceptional quality and value of many South Pyrenean sedimentary successions for the international geological community, thus paving the way for many other researchers. In this context, new universities were established in Spain in the early 1960s, some including degrees in geology, and fundamental research began to be valued and fomented. Consequently, the number of local geologists increased, as did the number of field specific studies about Paleogene stratigraphy. Due to these changes, the body of work carried out by both local and external researchers in the last five decades outweighs that of previous decades manyfold.

This prolific recent activity resulted in the successful internationalization of the research carried out in Paleogene successions of the Iberian Peninsula, leading to their current recognition as reference sections for global Paleogene stratigraphy. The following is a summary of the main stratigraphic research lines, organized in chronological order:

1. The first internationally significant stratigraphic contributions derived from Iberian Paleogene successions were published in the 1960s and 1970s and dealt with marine biostratigraphy and the calibration of timescales derived from different fossils. Subsequent development of this research line has been rather discontinuous but is still ongoing and has extended into terrestrial Paleogene biostratigraphy (e.g., mammals,

charophytes, palynology), making significant improvements to the international Paleogene timescale (see, for instance, Barbera et al., 2001; Beamud et al., 2003; Cascella and Dinarès-Turell, 2009; Costa et al., 2010, 2013; Feist et al., 1994; Lévêque, 1993; Mochales et al., 2012; Rodriguez-Pinto et al., 2012, 2013; Serra-Kiel et al., 1998).

2. During the 1960s, in close relationship with the previous research line, some attempts were also made to contribute to international Paleogene chronostratigraphy with the definition of two new Paleogene stages, the so-called Ilerdian and Biarritzian stages, which were largely based on studies carried out in the Pyrenean area. Neither of these stages was included in the set of nine official Paleogene stages approved by the ISPS in 1989, partly due to the application of the Priority Principle, as both Pyrenean stages overlapped others previously defined in the North Sea area. However, both informal stages are still in use at a regional (i.e., Tethyan and Caribbean) level and, at least in the case of the Ilerdian, still have some global stratigraphic significance.
3. An independent line of research was later initiated by Riba (1973, 1976a,b) in Paleogene terrestrial deposits from the Southern Pyrenees, which resulted in the definition of syntectonic unconformities and the establishment of a new conceptual model on stratigraphic architecture.
4. Since the 1980s and early 1990s magnetostratigraphic analysis has become a common technique in studies carried out in Paleogene successions of the Iberian Peninsula (e.g., Groot et al., 1989; Bentham and Burbank, 1996; Mary et al., 1991; Moreau et al., 1994; Pujalte et al., 1995). Along with advanced sequence stratigraphic and cyclostratigraphic studies carried out since the late 1990s, Iberian magnetostratigraphy has contributed to the development of a Paleogene astronomical timescale and has been of significant help in improving geochronological aspects of the general Paleogene timescale (e.g., Batenburg et al., 2012, 2014; Dinarès-Turell et al., 2002, 2003, 2007, 2010, 2012, 2014a; Ten Kate and Sprenger, 1993).
5. The culmination of the above research lines was the definition of GSSPs for the bases of three Paleogene stages (i.e., Selandian, Thanetian, and Lutetian) in the western Pyrenees (Molina et al., 2011; Schmitz et al., 2011). In addition, prospective studies of the same stratigraphic intervals were also carried out in other successions. Furthermore, the potential of Iberian sections for the GSSPs of other Paleogene stages was also considered, but they did not succeed. However, all of these sections can still be considered auxiliary stratotype sections of reference.

Given the enormous volume of internationally relevant information produced over the last five decades in the Iberian Peninsula, it is not possible to include here every single achievement on Paleogene stratigraphy. Consequently, this review will focus exclusively on the Iberian Paleogene sections which have contributed to international chronostratigraphy. The reason is that chronostratigraphy is the most multidisciplinary and transversal specialty of stratigraphy. Some of the sections included in this review encompass the whole Paleogene succession and, in addition to their relevance for the specific stratigraphic intervals for which they gained recognition, are also significant for the results they have yielded in other intervals; in such sections, all the stratigraphically significant intervals of the Paleogene succession will be reviewed herein. The sections hosting GSSPs are considered the most important and are therefore the first to be described following an ascending stratigraphic order, from oldest to the youngest. Subsequently, other sections will be presented according to decreasing significance and ascending stratigraphic order. If pertinent, the description of each section will be subdivided into subsections following the stratigraphic order from older to younger.

4. THE ZUMAIA–GETARIA SECTION: THE LARGEST CONCENTRATION OF GSSPs AND ASSPs WORLDWIDE

The Maastrichtian–Ypresian succession exposed along the 6-km-long coastal cliff between Zumaia and Getaria (Fig. 2A; Gipuzkoa province, Basque Country, western Pyrenees; from 43°17′58″N 02°16′08″W to 43°18′41″N 02°13′03″W) is one of the best known and most studied land-based Paleogene sections in the world. The sedimentary succession, >2500-m thick, is composed of deep-marine deposits accumulated at the bottom of the Pyrenean paleogulf at approximately 1000–1500 m water depth (Alegret and Ortiz, 2010; Pujalte et al., 1998).

Pioneering studies by Stuart-Menteath (1881), Adan de Yarza (1884), Kindelan (1919), and Mendizabal (1923) highlighted the expanded character of the Basque sections, notably that occurring between Zumaia and Getaria, and their relevance to determine the position of the Cretaceous/Paleogene boundary. Subsequent studies of the Zumaia–Getaria section focused on the paleontological content of the succession and highlighted its value for both biostratigraphic and ichnological analysis (Azpeitia, 1933; Gomez de Llarena, 1946, 1949, 1954–1956; Ruiz de Gaona, 1948; Ruiz de Gaona and Colom, 1950). These early studies allowed the initial internationalization of the Zumaia–Getaria section, as shown by its portrayal on the covers of reference textbooks (e.g., Pettijohn and Potter, 1964) and contributed to the arrival of other researchers from abroad (Hillebrant, 1965; Kapellos, 1974; Seilacher, 1962; Wiedmann, 1969). However, it was in the 1970s and early 1980s that the Zumaia–Getaria section obtained definitive international relevance. On the one hand, the Eocene succession was among the first onshore-exposed submarine fan successions ever described (Kruit et al., 1972; Van Vliet, 1978). In addition, the section was extensively studied in the context of the heated controversy around the mass extinction at the Cretaceous/Paleogene boundary (Alvarez et al., 1982; Percival and Fischer, 1977; Smit and Ten Kate, 1982).

Subsequently, the stratigraphic, paleontological, and sedimentological studies carried out in this section are too numerous to cite (for recent reviews, see Baceta et al., 2012; Bernaola et al., 2006c; Pujalte et al., 2016). For the purposes of this review, the most relevant works are those which dealt with chronostratigraphy and led to the evaluation of the Zumaia section for the definition of the four GSSPs related to the Paleocene Epoch. The golden spikes for the bases of the Selandian and Thanetian stages were finally placed at this section, whereas those for the bases of the Danian and Ypresian stages were defined elsewhere, as explained below. However, the Zumaia–Getaria section can be considered auxiliary stratotype for both. As a result, this section holds the largest concentration of GSSPs and ASSPs in the world, at least for the Paleogene Period and probably so for the entire geological timescale. In addition, the section has also become a reference for the Maastrichtian and Paleocene astrochronological timescale. It is also worth noting that the Zumaia section is relevant for the definition of Maastrichtian–Paleocene regional depositional sequences (e.g., Baceta et al., 2004; Pujalte et al., 1995, 1998), which can be correlated with coeval sequences worldwide suggesting that they were probably eustatic in origin.

4.1 The Maastrichtian–Danian Boundary: ASSP and Astrochronology

The Maastrichtian–Danian (i.e., Cretaceous/Paleogene, K/Pg) boundary coincides with one of the most and best-studied mass extinction events in the geological record. It caused the complete disappearance of many marine and continental biotas, such as nonavian dinosaurs and ammonites, as well as dramatic changes in many other groups, including marine plankton (Schulte et al., 2010). According to the most commonly accepted hypothesis, this biotic turnover was caused by the impact of a large asteroid (Alvarez et al., 1980), but the influence of intense volcanic activity has also been suggested

(Renne et al., 2015). Evidence for the former was found in the worldwide occurrence of a thin isochronous, rusty colored clay layer with high concentrations of extraterrestrial iridium, microtektites, microspherules with Ni-rich spinel crystals and shocked quartz (Alvarez et al., 1982), across which the abovementioned dramatic paleontological changes are recorded. In 1982, a specific working group was set up within the ICS with the aim of defining the GSSP for the K/Pg boundary (Molina et al., 2006a). By 1988 the choice of prospective sections was narrowed to four localities, and a postal ballot resulted in 10 votes in favor of El Kef (Tunisia), 6 for Zumaia, 2 for Brazos (United States), and 2 for Stevns Klint (Denmark). In 1989 the working group agreed to define the GSSP for the base of the Danian Stage at the base of the boundary clay of the El Kef section, a decision ratified by the ICS and the International Union of Geological Sciences (IUGS) in 1990 and 1991, respectively (Molina et al., 2006a). Subsequent deterioration of the El Kef section forced definition of auxiliary stratotypes, which would in addition allow the study of the K/Pg boundary in different facies and paleogeographic contexts (Molina et al., 2009). Taking into account the process and results of the GSSP definition, an obvious auxiliary section was Zumaia.

In Zumaia the K/Pg boundary is exposed at the Algorri cove (43°17′56″N 2°16′04″W) and practically separates the underlying 240-m-thick Zumaia–Algorri marl Formation from the overlying Aitzgorri limestone Formation (Fig. 3). Unfortunately, interbed sliding during tectonic tilting concentrated at the boundary clay, as shown by the occurrence of calcite slickensides. Consequently, the boundary clay is only 1-cm thick and is not laterally continuous, but occurs only in small patches (Apellaniz et al., 1997; MacLeod and Keller, 1991). However, comparison with other nearby sections showed that the K/Pg boundary clay is best characterized as a tripartite layer, up to 8-cm thick, which includes a lower gray silt that contains partly dissolved foraminifera, an intermediate brownish siltstone where the asteroid-impact indicators occur, and an upper thinly laminated dark gray siltstone. Biostratigraphic studies carried out in Zumaia dealt, among others, with ammonoids (e.g., Ward and Kennedy, 1993; Ward et al., 1991; Wiedmann, 1988), calcareous nannofossils (e.g., Bernaola, 2002; Bernaola et al., 2006c; Burnett et al., 1992; Lamolda and Gorostidi, 1992; Percival and Fischer, 1977), and planktonic foraminifera (e.g., Apellaniz et al., 1997, 2002; Arenillas et al., 1998; Arz et al., 1999; Caballero, 2007; Lamolda, 1990; Pérez-Rodríguez et al., 2012), all of which underwent sudden extinction at the boundary clay.

The Maastrichtian and Danian successions underlying and overlying the K/Pg boundary consist of 30–50-cm-thick hemipelagic limestone-marl couplets (Fig. 3), the origin of which was soon attributed to astronomical cycles in the Milankovitch band (Mount and Ward, 1986; Pujalte et al., 1995, 1998; Ten Kate and Sprenger, 1993). This interpretation was confirmed in subsequent, more in-depth cyclostratigraphic studies which, focusing on physical and geochemical parameters in combination with detailed biomagnetostratigraphic analyses, identified precession cycles of 21 kyr, short eccentricity cycles of 100 kyr, long eccentricity cycles of 405 kyr, obliquity modulation cycles of 1.2 Myr, and very long eccentricity modulation cycles of 2.8 Myr (Batenburg et al., 2012, 2014; Dinarès-Turell et al., 2003, 2010, 2012, 2013, 2014a; Hilgen et al., 2010; Kuiper et al., 2008; Westerhold et al., 2008). These studies tackled the tuning of the cyclostratigraphic record with available astronomical solutions, concluding that the K/Pg boundary occurred at a distinctive 405 kyr eccentricity minimum (Fig. 3). This approach allowed astronomically calibrated ages to be calculated for successive biomagnetostratigraphic events, which proved crucial to determining accurately the current official age of 66 Ma for the K/Pg boundary (Fig. 4; Renne et al., 2013; Vandenberghe et al., 2012).

4.2 A Prospective Global Unit Stratotype for the Danian Interval

The Danian Aitzgorri Formation (Bernaola et al., 2009) is a 49-m-thick alternation of reddish limestones and marls with some intercalations of thin-bedded turbidites (Figs 3 and 4). Many early studies overlooked the effect of a normal fault that duplicates a significant portion of the succession, thus overestimating the true thickness of the Danian succession (Fig. 5). This problem was solved through the correlation of distinctive beds in the hangingwall and footwall, which allowed a detailed stratigraphy to be established (Fig. 4; Baceta, 1996). Integrated biomagnetostratigraphic studies were carried out by, among others, Pujalte et al. (1995, 1998), Dinarès-Turell et al. (2003, 2010, 2012, 2014a), Bernaola et al. (2006c, 2009), Arenillas et al. (2008), and Monechi et al. (2013), who demonstrated that the Danian succession is complete and more expanded than in other sections worldwide. An interesting result from the upper part of the Aitzgorri Formation was the discovery of a hyperthermal event, which is commonly referred to as the Latest Danian Event (Dinarès-Turell et al., 2012; Monechi et al., 2013). This event was a relatively short-term (200 kyr long) global warming

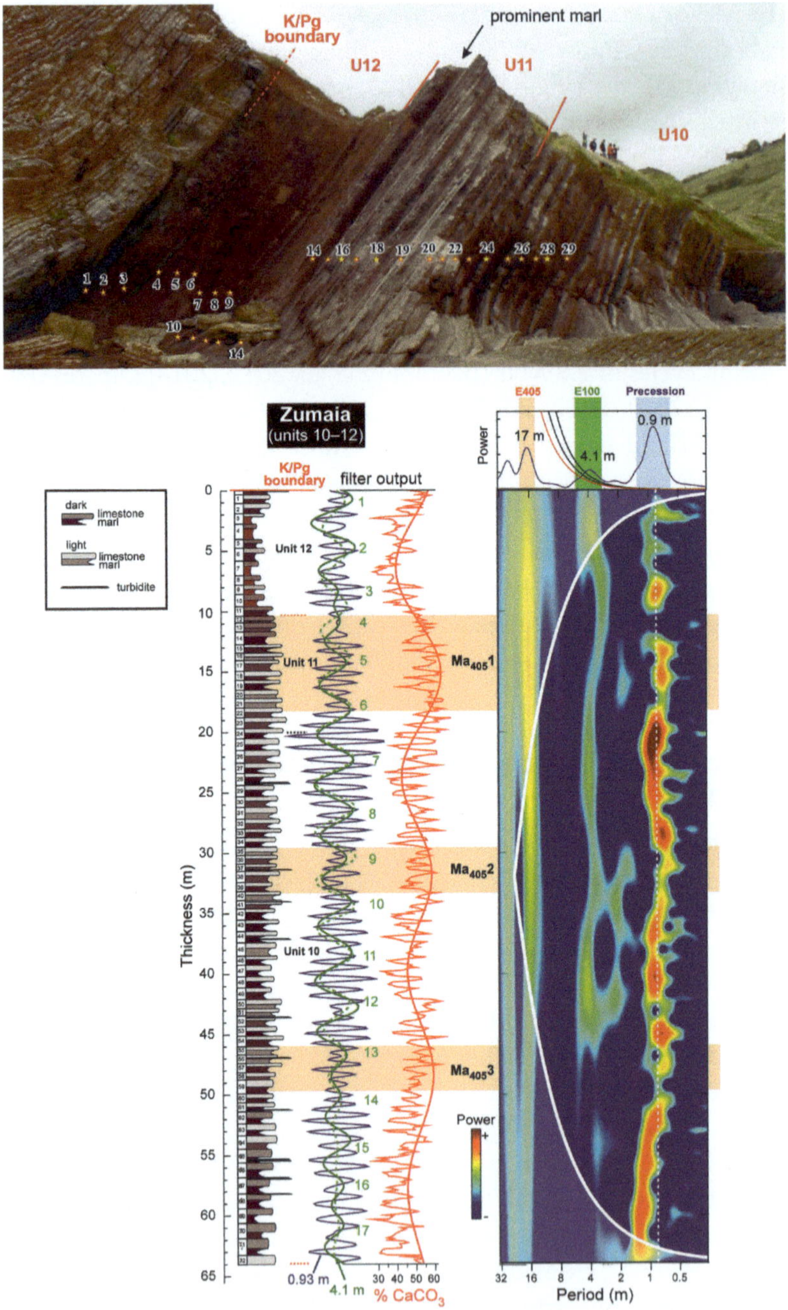

Figure 3 Field view of the Cretaceous/Paleogene (K/Pg) transition at Zumaia. The precession-related lithological couplets (*stars*) are numbered from the K/Pg boundary

episode caused by the intense greenhouse effect derived from a global perturbation in the carbon cycle.

The most distinctive characteristic of the Aitzgorri Formation is the clearly rhythmic pattern of the lithological alternation. Pujalte et al. (1995, 1998) pointed out that the origin of the limestone—marl alternation may be related to 21-kyr-long precession cycles, whereas bundles composed of approximately five couplets could represent 100-kyr-long eccentricity cycles. In addition to using biostratigraphic age estimates, their interpretation was based on the fact that the same lithological cycles were found in sections more than 100 km apart. Furthermore, they developed a sequence stratigraphic model which allows the delimitation of third-order (i.e., Myr-long) depositional sequences and systems tracts in hemipelagic successions based on the variable lithological expression of Milankovitch cycles. The significance of the lithological cycles was further investigated by Dinarès-Turell et al. (2003, 2010, 2012, 2014a) using lithology-related proxy records, which were processed through spectral analysis (Fig. 4). This approach allowed the identification of precession-related cycles of 21 kyr, short eccentricity-related cycles of 100 kyr, long eccentricity-related cycles of 405 kyr, and 2.8 Myr-long eccentricity modulation cycles. Dinarès-Turell et al. (2003) fine-tuned the cyclostratigraphic record using available astronomical solutions, thus obtaining astronomically calibrated ages for many Danian biomagnetostratigraphic events for the first time and improving the accuracy of the ages previously assumed in existing standard timescales. However, alternative tunings were later suggested by other authors in the light of more recent astronomical solutions (Hilgen et al., 2010; Kuiper et al., 2008; Westerhold et al., 2008). The discrepancies in the different

downwards. The detailed lithological log includes the corresponding carbonate ($CaCO_3$) record. The filter output of Gaussian filters centered at 0.93 m (*blue line*, 1.074 ± 0.5 bandpass), 4.1 m (*solid green line*, 0.244 ± 0.02 bandpass), and 17 m (*red line*, 0.058 ± 0.015 bandpass) are also shown. In addition to numbered precession-related lithological couplets, the green numbers indicate lithological bundles corresponding to the short-eccentricity (100 kyr) minima and the *black labels* indicate the long eccentricity (405 kyr) minima (Ma_{405}) relative to the K/Pg boundary. Wavelet and global power spectrum with a lag-1 *red-noise* estimate and respective 80 and 95% significance levels (right-hand graphs). *Reprinted with permission from Dinarès-Turell, J., Pujalte, V., Stoykova, K., Elorza, J., 2013. Detailed correlation and astronomical forcing within the upper Maastrichtian succession in the Basque Basin. Bol. Geol. Min. 124, 253–282, Figs. 5A and 7 (© Instituto Geologico y Minero de España, 2013).*

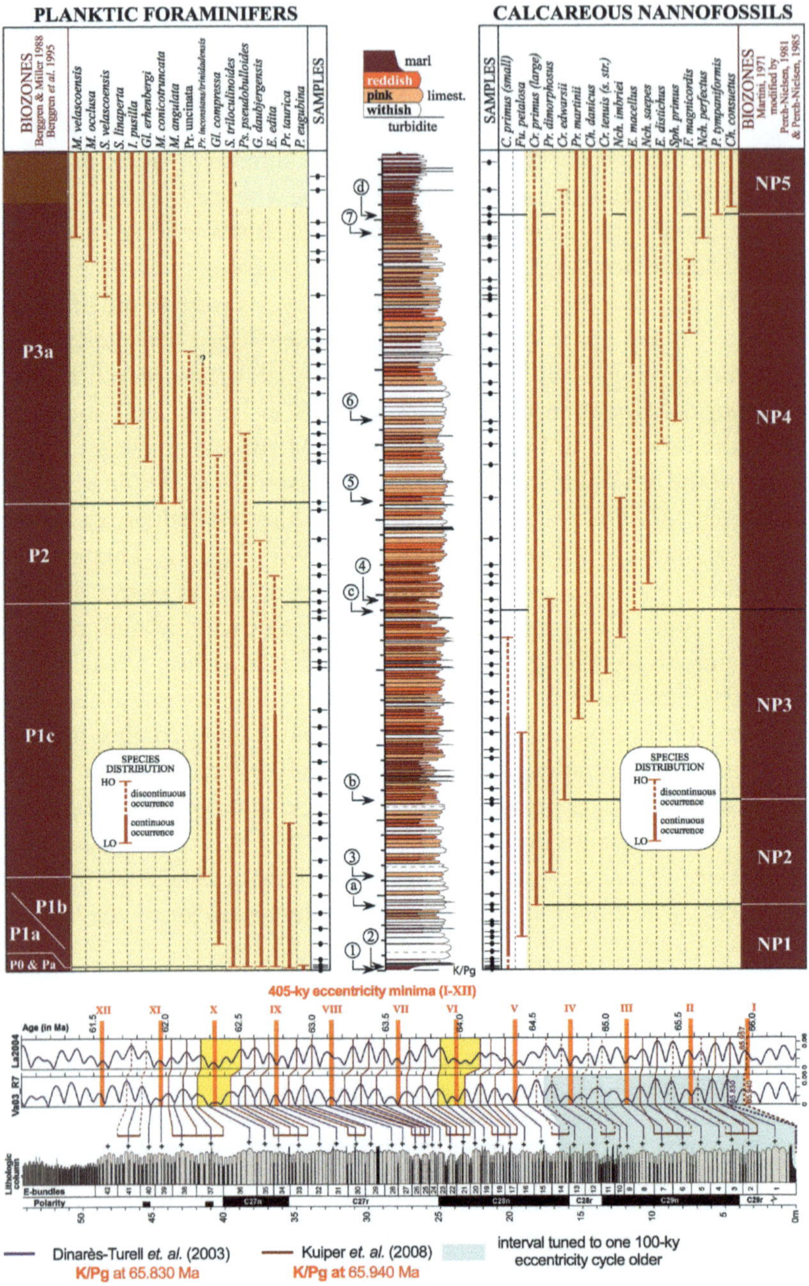

Figure 4 Selected planktonic foraminifera and calcareous nannofossil distribution across the Danian succession of Zumaia *(Modified with permission from Dinarès-Turell, J., Baceta, J.I., Pujalte, V., Orue-Etxebarria, X., Bernaola, G., Lorito, S., 2003. Untangling*

astronomical tunings were finally solved by Dinarès-Turell et al. (2014a). They reliably determined that eleven 405-kyr-long eccentricity cycles occurred in the Danian succession, which lasted 4.415 Myr ± 80 kyr (Fig. 4). In addition, they correlated the Zumaia data with other Basque sections and oceanic deep-sea cores to obtain an integrated stratigraphic framework for the Danian Stage, with precision at the 100-kyr-short eccentricity level. This achievement helped to extend the astronomical polarity timescale into the early Paleogene and the Maastrichtian, thus leading to unprecedented resolution in geological timescales and increasing the potential for accurate global correlations in Danian successions. Taking everything into account, it was suggested that the Zumaia section should be designated Global Unit Stratotype for the Danian Stage (Dinarès-Turell et al., 2014a).

4.3 The Selandian GSSP

In 1993 the ISPS commissioned the Paleocene Working Group to define GSSPs for the bases of the Selandian and Thanetian stages. There was a general consensus that Zumaia would be one of the prime candidates for the GSSPs (Schmitz et al., 1998), but detailed or pilot studies were undertaken on many other sections, most of which were located in peri-Mediterranean countries (Schmitz et al., 2011). After 14 years of research, the discussion on the Selandian GSSP was narrowed down to the comparison of two sections, Zumaia and Qreiya (Egypt). In June 2007, 23 members of the Working Group attended a meeting held in Zumaia and unanimously agreed that

the Palaeocene climatic rhythm: an astronomically calibrated Early Palaeocene magnetostratigraphy and biostratigraphy at Zumaia (Basque basin, northern Spain). Earth Planet. Sci. Lett. 216, 483–500; Fig. S1; © Elsevier, 2003.). Main events: (1) LO *Parvulorugoglobigerina longiapertura*; (2) LO *Parasubbotina pseudobulloides* and *Subbotina triloculinoides*; (3) LO *Praemurica trinidadensis/Praemurica inconstans*; (4) LO *Morozovella uncinata*; (5) LO *Morozovella angulata*; (6) LO *Igorina pusilla*; (7) LO *Morozovella velascoensis*; (a) LO *Cruciplacolithus primus* (large); (b) LO *Cruciplacolithus edwardsii*; (c) FO *Ellipsolithus macellus*; (d) FO *Fasciculithus tympaniformis*. The lower part shows the tuning of the Danian succession to the Va03_R7 and La04 orbital solutions *(Reprinted with permission from Dinarès-Turell et al., 2014; Fig. S1; © Elsevier, 2014.)*. Crosses above the lithological column mark midpoints of limestone-dominated parts of the ~100-kyr eccentricity bundle or E-cycles numbered as in Dinarès-Turell et al. (2003) that correlate with 100-kyr eccentricity minima. *Brown lines* represent the tuning proposed by Kuiper et al. (2008) that takes the expression of the ~405-kyr eccentricity cycle into account, while *blue lines* represent the tuning to the Va03_R7 astronomical solution anticipated in Dinarès-Turell et al. (2003). Note that both tuning schemes coincide for the upper eccentricity bundles (15–42) while cycles 1–14 are correlated to one eccentricity cycle older in the Kuiper et al. (2008) scheme leading to a slightly older age estimate for the K/Pg boundary.

this location excelled through better accessibility to the outcrop and a wider gamut of scientific information (availability of magnetostratigraphic and cyclostratigraphic data), in addition to being more expanded than Qreiya. Furthermore, due its intermediate location between the North Sea (boreal) region, where the original unit stratotypes for both the Danian and Selandian stages had been defined, and the southerly Tethys region, fossil elements from both regions cooccur in Zumaia, a circumstance that facilitates global correlation. A precise correlation between Zumaia, Qreiya, and the classic unit stratotypes in Denmark was established (Bernaola et al., 2009), which determined that the former and the latter shared the same sea level history and, therefore, display similar lithological successions. Based on the evaluation of extensive and detailed data sets, the Zumaia section was unanimously considered the most suitable section to host the GSSP for the base of the Selandian Stage. The decision was ratified by the IUGS on September 23, 2008 (Schmitz et al., 2011).

The GSSP for the base of the Selandian Stage is located at the sharp lithological boundary between the underlying limestone-dominated Aitzgorri Formation and the overlying marl-dominated Itzurun Formation exposed at the Itzurun beach, Zumaia (43°17′57.1″N 2°15′39.6″W), approximately 49 m above the K/Pg boundary (Fig. 5). The formation of this boundary was related to a long lasting (>2 Myr) prominent (c. 80 m) sea level drop. Therefore, it corresponds to a sequence boundary (Baceta et al., 2004; Pujalte et al., 1995, 1998), but no hiatus or unconformity affects the succession at Zumaia due to its basinal setting. The primary marker for the GSSP is the onset of the second radiation (or diversification) of the calcareous nannofossil group fasciculithus (Fig. 5). Thus, species such as *Fasciculithus ulii*, *Fasciculithus billii*, *Fasciculithus janii*, *Fasciculithus involutus*, *Fasciculithus pileatus* and *Fasciculithus tympaniforms* first occurred in the interval from a few decimeters below the GSSP level up to 1.1 m above it; in particular, the marker species for nannofossil Zone NP5, *F. tympaniforms*, first occurs 1.1 m above the GSSP level. The end of an acme of the nannofossil family *Braarudosphaeracea*, along with the cessation of long-term carbonate deposition, can also be used for correlation with North European regions. More specific information, along with many other secondary markers, can be found in Schmitz et al. (1998) for carbon isotope stratigraphy, Bernaola et al. (2009) and Monechi et al. (2013) for calcareous nannofossils, Caballero (2007) and Arenillas et al. (2008) for planktonic foraminifera, and Dinarès-Turell et al. (2003, 2007, 2010, 2012, 2014a) and Ellwood et al. (2008) for magnetostratigraphy and cyclostratigraphy (most of this information was compiled, reviewed, and summarized by Bernaola et al., 2006c, Schmitz et al., 2011, and Pujalte et al., 2016). According to these works, the GSSP for the base of the Selandian Stage is included in the uppermost part of calcareous nannofossil Zone NP4 of Martini (1971) and Zone NTp8b of Varol (1989), in the upper part of planktonic foraminiferal Zone P3a of Wade et al. (2011; approximately midway between the lowest occurrences of *Morozovella occlusa* and *Morozovella velascoensis*, and that of *Igorina albeari*), and in the lower part of magnetochron C26r (Fig. 5). According to state of the art cyclostratigraphy and astronomical tuning carried out by Dinarès-Turell et al. (2014a), the age of the GSSP was calculated to be 61.6075 Ma $\pm$ 40 kyr.

The overlying Itzurun Formation is affected by a normal fault, but beds from both the hangingwall and footwall can be readily correlated (Fig. 5). The Itzurun Formation spans the entire Selandian and continues up through the Thanetian, showing vertical variations in the relative proportion of hemipelagic marls and limestones. Lower Member A is largely dominated by marls, whereas upper Member B includes greater proportion of limestones. The boundary between both members was arbitrarily defined at the point in which limestone beds reach and maintain $CaCO_3$ values higher than 60%. This boundary is located 22.5 m above the base of the Itzurun Formation and 6.5 m below the top of the Selandian succession. It can be visually approximated using a distinctive, 1-m-thick clay-rich layer 2.5 m below the top of the Selandian succession. This layer, on the basis of its paleontological and geochemical characteristics, was referred to as the Mid-Paleocene Biotic Event, another Paleocene hyperthermal event (Bernaola et al., 2007).

4.4 The Thanetian GSSP

The Paleocene Working Group meeting held in Zumaia in June 2007 was also the forum for deliberation about the GSSP for the base of the Thanetian Stage. The Working Group carefully evaluated the standing of the Zumaia section in relation to the ICS requirements for GSSPs and found that it is close to ideal. Accordingly, the 23 members who attended the meeting unanimously agreed that Zumaia was the most suitable section to host the GSSP for the base of the Thanetian Stage. The decision was ratified by the IUGS on September 23, 2008 (Schmitz et al., 2011).

The GSSP for the base of the Thanetian Stage was defined 29 m above the GSSP for the base of the Selandian Stage (Figs. 5 and 6), at a marlstone bed included in the lower part of Member B of the Itzurun Formation (cliff of the Itzurun beach in Zumaia; 43°17′58.4″N 2°15′39.1″W). The GSSP

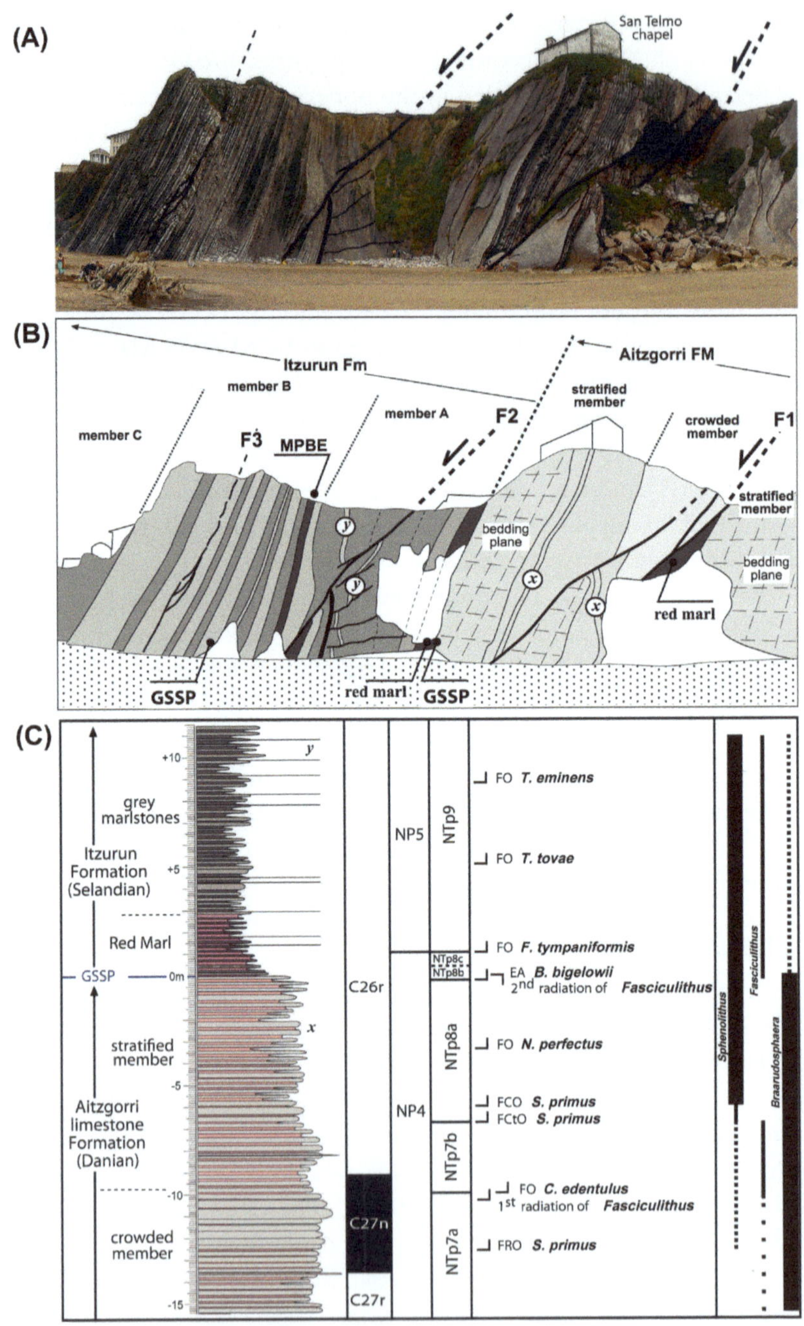

Figure 5 Panoramic view (A) and synthetic sketch (B) of the Paleocene succession exposed at the Itzurun beach (Zumaia), showing lithological units and the main fault

level contains the C26r/C26n magnetochron boundary (Dinarès-Turell et al., 2007) and correlates with the middle part of calcareous nannofossil Zone NP6 and with planktonic foraminiferal Zone P4 (Schmitz et al., 2011). In terms of cyclostratigraphy, the GSSP occurs eight precession-driven limestone-marl couplets (2.8 m) above the Mid-Paleocene Biotic Event, which correlates with a distinctive eccentricity maximum. The base of the Thanetian GSSP is located 105 precession cycles above the base of the Itzurun Formation, meaning that the Selandian Stage lasted 2.103 Myr. According to cyclostratigraphic calculations by Dinarès-Turell et al. (2007), the duration of the overlying Thanetian succession was estimated to be 3.129 Myr. Regarding sequence stratigraphy, the GSSP level correlates with the transgressive systems tract of a depositional sequence encompassing the Selandian and part of the Thanetian. The historical unit stratotype for the Thanetian, which is located in southern England, also reflects a major transgression (Schmitz et al., 2011).

4.5 The Paleocene/Eocene Boundary: Prospective ASSP and PETM

The establishment of the Working Group on the Paleocene/Eocene boundary (i.e., Thanetian/Ypresian) at the 28th International Geological Congress coincided with the discovery of significant, stratigraphically linked events in the biotic and isotopic records from deep-sea sediments (Aubry et al., 2007). Therefore, the search for a GSSP for the base of the Ypresian Stage became intertwined with the effort to document and explain the significance of those events. According to current hypotheses, the release of

systems (F1, F2, and F3). The location of the Selandian and Thanetian GSSPs is also shown. *GSSPs*, global stratotype sections and points. (C) Calcareous nannofossil biostratigraphy of the Danian/Selandian transition (*FO*, first occurrence; *FCO*, first common occurrence; *FRO*, first rare occurrence; *FCtO*, first continuous occurrence; *EA*, end of acme). NP scale: calcareous nannofossil zonation by Martini (1971); NTp scale: calcareous nannofossil zonation by Varol (1989). Note that additional calcareous nannofossil details were later provided by Monechi et al. (2013, Fig. 4). *(B) Modified with permission from Dinarès-Turell, J., Baceta, J.I., Bernaola, G., Orue-Etxebarria, X., Pujalte, V., 2007. Closing the mid-Paleocene gap: toward a complete astronomically tuned Paleocene Epoch and Selandian and Thanetian GSSPs at Zumaia (Basque basin, W Pyrenees). Earth Planet. Sci. Lett. 262, 450–467, Fig. 1 (© Elsevier, 2007). (C) Modified with permission from Bernaola, G., Martín-Rubio, M., Baceta, J.I., 2009. New high resolution calcareous nannofossil analysis across the Danian/Selandian transition at the Zumaia section: comparison with South Tethys and Danish sections. Geol. Acta 7, 79–92, Fig. 3 (© Geologica Acta, 2009).*

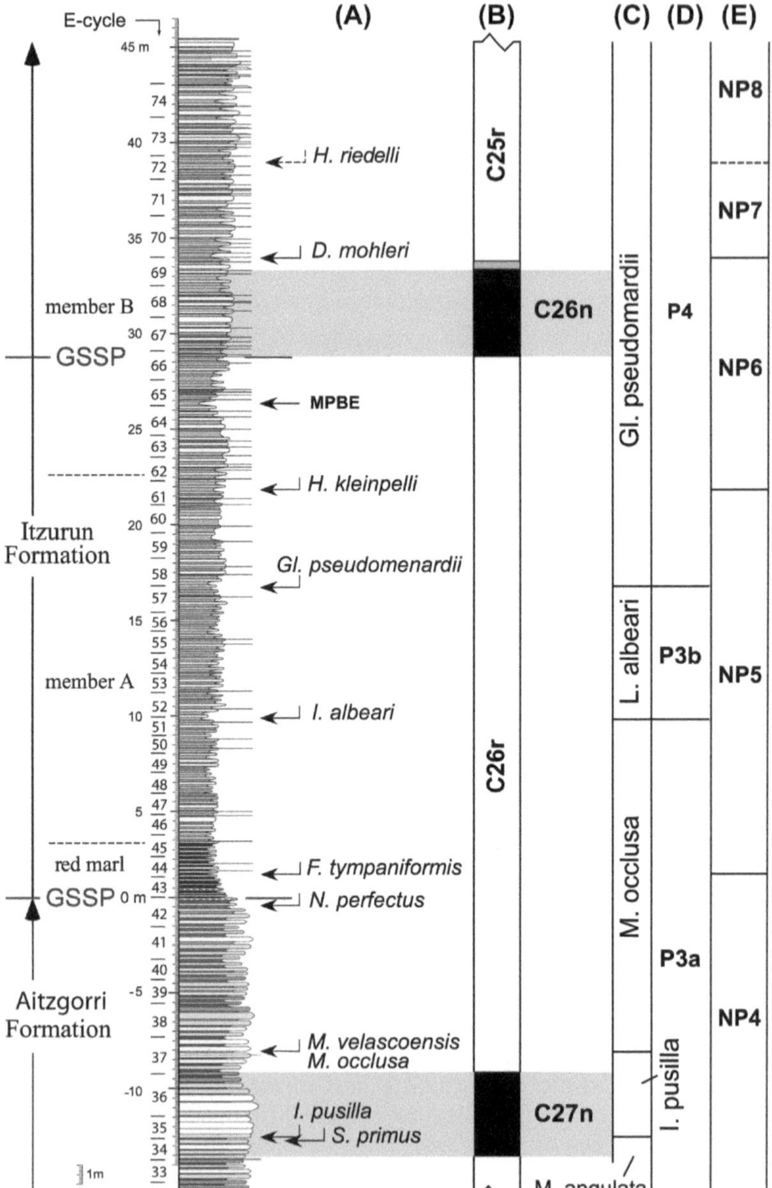

Figure 6 Integrated lithostratigraphy, biostratigraphy, and magnetostratigraphy of the mid-Paleocene Zumaia section, showing (A) main calcareous plankton events, (B) magnetic polarity stratigraphy, (C) local planktonic foraminiferal biozonation, (D) standard planktonic foraminiferal biozonation of Berggren et al. (1995), and (E) calcareous nannofossil biozonation of Martini (1971). E-cycles refer to short-eccentricity bundles of 100 kyr and are numbered from the K/Pg boundary in ascending stratigraphic order. *Reprinted with permission from Dinarès-Turell, J., Baceta, J.I., Bernaola, G., Orue-Etxebarria, X., Pujalte, V., 2007. Closing the mid-Paleocene gap: toward a complete astronomically tuned Paleocene Epoch and Selandian and Thanetian GSSPs at Zumaia (Basque basin, W Pyrenees). Earth Planet. Sci. Lett. 262, 450–467, Fig. 5 (© Elsevier, 2007).*

>2000 gigatons of ^{13}C-depleted carbon into the ocean-atmosphere system 56 Ma caused a transient global warming episode, commonly referred to as the Paleocene–Eocene Thermal Maximum (PETM), when temperatures rose 5 to 8°C (see, for example, Stott et al., 1990; Zachos et al., 2001; for a more recent review, see McInerney and Wing, 2011). The carbon cycle perturbation is recorded by a large (3–5‰) negative carbon isotope excursion (CIE) found in both marine and terrestrial successions. In addition to physical and geochemical environmental perturbations, the PETM had a significant impact on terrestrial and marine biotas, affecting land mammals and floras, shallow marine larger foraminifera, open ocean planktonic organisms (dinoflagellates, coccolithoporids, foraminifera), and deep-sea benthic foraminifera, among others.

Twenty-three sections were investigated by the Paleocene/Eocene Working Group, nine of which were considered possible candidates for the GSSP of the Ypresian Stage (Aubry et al., 2007). After closer examination, the selection was narrowed down to two sections, an abandoned quarry in Dababiya (Egypt) and the coastal cliff in Zumaia. A ballot was organized in a meeting held in Luxor (Egypt) in February 2002, and the members of the Working Group voted unanimously in favor of placing the GSSP for the base of the Ypresian Stage in Dababiya. The GSSP level is the base of the Dababiya Quarry Beds of the Esna Formation and records the basal inflection of the prominent negative CIE that marks the onset of the PETM. It may also be correlated on the basis of (1) the mass extinction of abyssal and bathyal benthic foraminifera, (2) the transient occurrence of characteristic excursion taxa among planktonic foraminifera, such as *Acarinina africana*, *Acarinina sibaiyaensis*, and *Morozovella allisonensis*, (3) the transient occurrence of the *Rhomboaster* spp.—*Discoaster araneus* calcareous nannofossil assemblage, and (4) an acme of the dinoflagellate *Apectodinium* complex. The proposal was accepted by the ISPS in May 2003, by the ICS in August 2003, and was ratified by the IUGS in August 2004. However, concerns have subsequently arisen about the suitability of the GSSP selection, as it was later found that the GSSP level was formed at the base of a laterally restricted submarine channel that eroded away part of the underlying Paleocene deposits (e.g., Ellwood et al., 2010; Khozyem et al., 2014). Furthermore, the current sociopolitical situation in Egypt has complicated accessibility to the Dababiya area, which is a requisite for any GSSP. Consequently, Zumaia, the runner-up in the selection process for the Ypresian GSSP, becomes the obvious choice to be designated auxiliary stratotype section.

In Zumaia the Paleocene/Eocene boundary occurs on the cliff at the entrance to the Itzurun beach (Fig. 7; 43°18′00.55″N 2°15′35.50″W), 142 m above the K/Pg boundary. The first studies of the Paleocene/Eocene boundary were carried out by Canudo et al. (1995), Ortiz (1995), and Schmitz et al. (1997). They determined that the Paleocene/Eocene CIE occurs at the base of a 40-cm-thick marl bed, the top of which records the benthic foraminiferal extinction. This marl is located just below a conspicuous 4-m-thick reddish lutite (the lowermost 1.3 m of which are practically devoid of carbonate), which spans the whole PETM. The Paleocene/Eocene section is affected by several faults in Zumaia (Fig. 7), but a complete, 70 m-thick bed-by-bed succession was reconstructed by Baceta et al. (2000) after restoring the tectonic complexity. Identical coeval successions have been reported from many other sections in the Basque–Cantabrian area, which showed that a distinctive 80-cm-thick greenish limestone bed systematically underlies the Paleocene/Eocene boundary (Baceta et al., 2000; Orue-Etxebarria et al., 1996; Pujalte et al., 1995, 1998; Schmitz et al., 2001).

Subsequently, the Paleocene/Eocene boundary interval of Zumaia has been the objective of many specific studies, which have dealt, among others, with geochemistry (e.g., Manners et al., 2013; Schmitz et al., 2001, 2004; Storme et al., 2012), calcareous plankton (Angori et al., 2007; Arenillas and Molina, 2000; Orue-Etxebarria et al., 2004; Pardo et al., 1999), benthic foraminifera (Alegret and Ortiz, 2010; Alegret et al., 2009a; Arreguin-Rodriguez and Alegret, 2015; Kuhnt and Kaminski, 1997; Zili et al., 2009), magneto-cyclostratigraphy (Dinarès-Turell et al., 2002, 2007; Ellwood et al., 2010), and ichnology (Giannetti, 2010; Giannetti and McCann, 2010; Rodriguez-Tovar et al., 2011a), to cite just the most recent. These studies contributed to elucidating the environmental impact of the PETM and helped to constrain the chronostratigraphy of the Paleocene/Eocene boundary within magnetochron C24r (Fig. 7). Regarding calcareous plankton biostratigraphy, in addition to reassessing the stratigraphic range of several key species, Orue-Etxebarria et al. (2004) specified that this boundary practically coincides with the base of the regional *Acarinina wilcoxensis–M. occlusa* planktonic foraminiferal biozone (thus correlating with Zone E1 of Wade et al., 2011) and with that of calcareous nannofossil Zone NP10 (as amended by Bybell and Self-Trail, 1995). Cyclostratigraphic studies showed that the Paleocene/Eocene boundary occurs 47 precession couplets above the top of magnetochron C25n (Dinarès-Turell et al., 2002, 2007), and that the PETM lasted eight

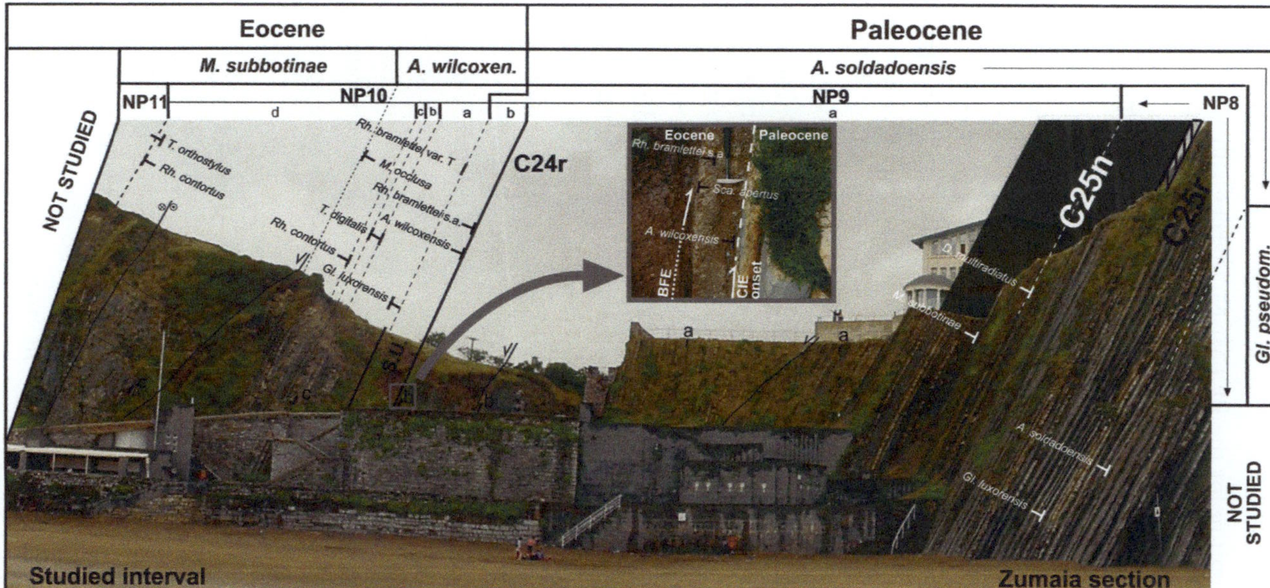

Figure 7 The Paleocene/Eocene transition in Zumaia, showing the main tectonic structures and the location of magnetostratigraphic, calcareous plankton, and isotopic events (a, b, c on the main photo: distinctive beds; CIE, carbon isotope excursion, i.e., base of the Eocene; BFE, benthic foraminiferal extinction). *Reprinted with permission from Orue-Etxebarria, X., Bernaola, G., Baceta, J.I., Angori, E., Caballero, F., Monechi, S., Pujalte, V., Dinarès-Turell, J., Apellaniz E., Payros A., 2004. New constraints on the evolution of planktic foraminifers and calcareous nannofossils across the Paleocene-Eocene boundary interval: the Zumaia section revisited. Neues Jahr. Geol. Paläontol. Abh. 234, 223–259, Fig. 8 (© Echweiserbart, 2004).*

precessional cycles (Storme et al., 2012). These studies, combined with others carried out elsewhere, helped to calculate the official 56 Ma age for the Paleocene/Eocene boundary (Vandenberghe et al., 2012) and the 170–220 kyr time duration for the PETM (McInerney and Wing, 2011).

5. THE SOPELANA–BARINATXE–GORRONDATXE–GALEA SECTION: A GSSP AND MORE

This area has long been renowned internationally for the occurrence of a well-preserved K/Pg boundary section in Sopelana. More recently, the definition of the GSSP for the base of the Lutetian Stage in Gorrondatxe has increased its stratigraphic value. However, few nonlocal stratigraphers have probably realized that both sections are part of one single succession which includes intervening Ypresian deposits in Barinatxe (Fig. 8A).

The Sopelana–Barinatxe–Gorrondatxe–Galea section is exposed along 4 km of coastal cliffs, approximately 15 km northwest of Bilbao (Biscay Province, Basque Country, western Pyrenees). The sedimentary succession, composed of deep-marine deposits which accumulated at the bottom of the Paleogene Pyrenean gulf at approximately 1500 m water depth, extends from the Maastrichtian (43°23′21″N 2°59′38″W) to the middle-late Lutetian (43°22′27″N 03°02′07″W). The Maastrichtian–Danian interval, approximately 200-m thick, is affected by several faults, which caused its detachment from the overlying 2300-m-thick Eocene succession, leaving only some discontinuous Selandian–Thanetian slices in between. The first studies which identified and analyzed Cretaceous and Paleogene deposits in the area were those by Kindelan (1919), Jorge (1926), Ruiz de Gaona (1946), and Ruiz de Gaona and Colom (1950). Subsequently, the Paleogene succession was the objective of planktonic foraminiferal biostratigraphic studies by Orue-Etxebarria (1983). More recent studies have focused on specific intervals of the succession, as detailed below.

5.1 Biostratigraphy and Cyclostratigraphy of the Maastrichtian–Danian Interval

The Maastrichtian–Danian succession is best exposed on the cliff of the Arrietara beach, in Sopelana (Fig. 8B). It displays a regular alternation of hemipelagic limestones and marls, which can be correlated bed-to-bed with those found in Zumaia and in many other coeval sections throughout the Basque Country (Baceta, 1996; Batenburg et al., 2012, 2014; Dinarès-Turell et al., 2013; Pujalte et al., 1998). As in the Zumaia section, the origin of this alternation was attributed to astronomically driven cycles of global climate change (Milankovitch cycles). Despite the practically identical lithostratigraphic successions of the Sopelana and Zumaia sections, the former offers two significant advantages: on the one hand, it does not contain turbidites, meaning that the cyclostratigraphic record can be more reliably obtained (Pujalte et al., 1998); on the other hand, problems for Maastrichtian magnetostratigraphic analysis have long been recognized in Zumaia, but the paleomagnetic signal can be more accurately retrieved in Sopelana (Batenburg et al., 2012, 2014; Dinarès-Turell et al., 2013; Mary et al., 1991; Moreau et al., 1994). Consequently, the Sopelana section proved invaluable for the astronomical tuning of the Maastrichtian—Danian succession and for the calculation of an astronomically calibrated age for the K/Pg boundary (Dinarès-Turell et al., 2013; Batenburg et al., 2014).

Regarding the K/Pg boundary, tectonic problems were found at Zumaia. Consequently, the records from more than 20 sections were examined in the Basque Country (Apellaniz et al., 1997). One of them was the Sopelana section, which had previously been studied by Lamolda et al. (1983) and had in fact been the basis for the definition of new planktonic foraminiferal species relevant to the Maastrichtian and Danian stages, such as *Hedbergella hillebrandti*, *Eoglobigerina ferreri*, and *Civisina euskalherriensis* (Apellaniz et al., 2002; Orue-Etxebarria, 1985a; Orue-Etxebarria and Apellaniz, 1991, 2000). Interestingly, in addition to the typical extraterrestrial and impact material found in K/Pg boundary clays elsewhere (i.e., iridium, Ni-rich spinels, etc.), the Sopelana section also provided a high concentration of soot (Rocchia et al., 1988). However, the K/Pg boundary section exposed on the cliff of Sopelana was also found to be incomplete (MacLeod and Keller, 1991). Subsequent studies on top of the cliff (Sopelana III section) found one of the thickest and most complete K/Pg sections of the Basque area (Fig. 8C), which in addition provided the richest planktonic foraminiferal assemblages with up to 110,000 specimens per gram of dry processed sample (Apellaniz et al., 1997; Baceta et al., 2005). Furthermore, ichnological studies showed that disruption due to bioturbation is less pronounced in Sopelana than in more proximal sections (Rodriguez-Tovar et al., 2011b). Taking everything into account, it was concluded that, despite not being classified as an auxiliary section for the GSSP of the Danian Stage at El Kef (Molina et al., 2009), the Sopelana III section, along with Bidart, displays the best K/Pg section in the Basque Country (Apellaniz et al., 1997). According to these studies, some amendments

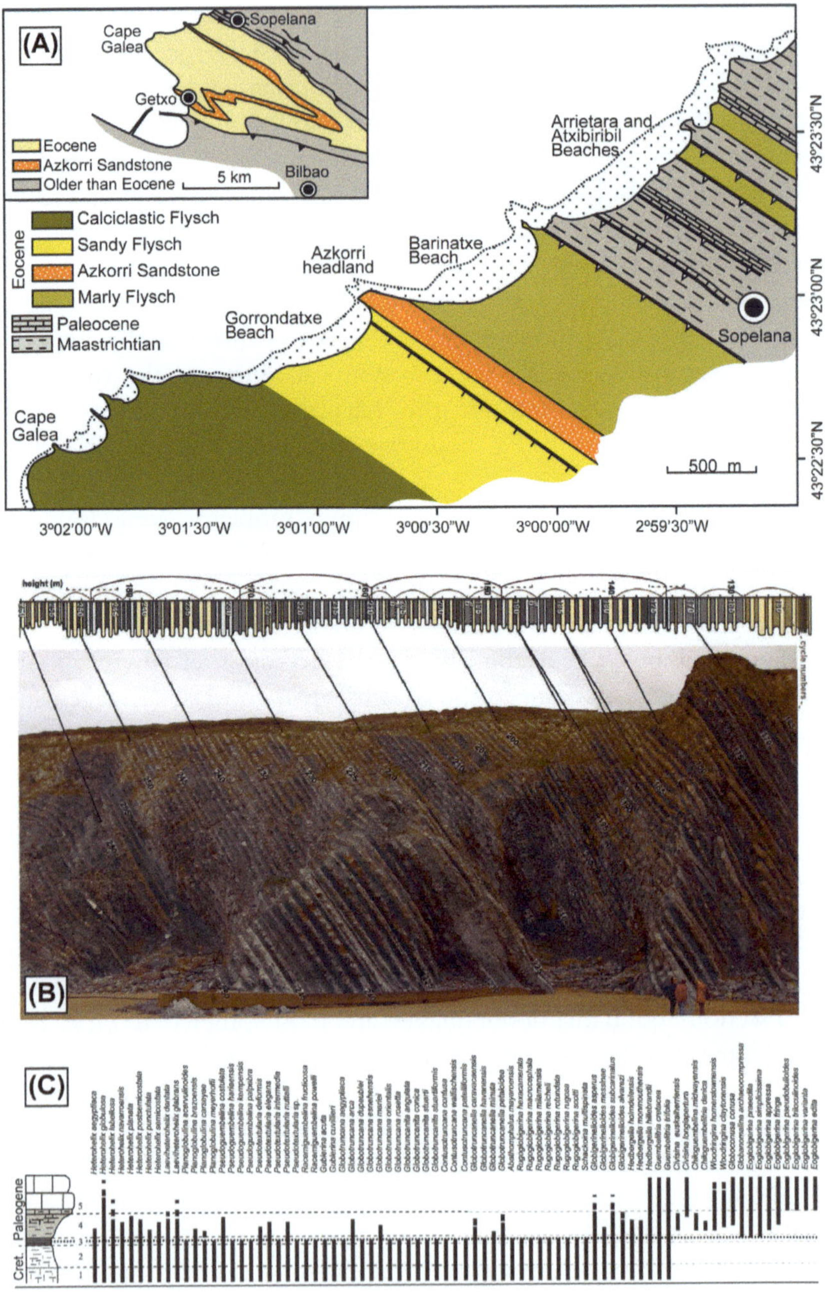

Figure 8 (A) Geological map of the Sopelana–Barinatxe–Gorrondatxe–Galea area. (B) Field view and stratigraphic log of the Maastrichtian succession exposed at the Arrietara beach (Sopelana), showing the hierarchic arrangement of cyclostratigraphic units

were suggested for the standard sequence of bioevents across the K/Pg boundary.

In addition to the abovementioned studies, the Sopelana section also proved significant for the cyclostratigraphic reconstruction of the Danian Stage (Dinarès-Turell et al., 2014a; Pujalte et al., 1998). Although the Danian succession is not complete, the limestone—marl couplets were correlated bed-to-bed with those found in sections more than 100 km apart. Furthermore, the Danian/Selandian transition is also exposed in Sopelana, being identical to, and thus complementing the observations made at the GSSP for the base of the Selandian Stage in Zumaia (Rodriguez-Tovar et al., 2013).

5.2 The Ypresian Interval in Barinatxe: Long-Term Global Warming and Cooling

The Ypresian succession is approximately 1000-m thick. It is exposed on the cliffs of the Barinatxe beach, the Azkorri headland, and the northeastern part of the Gorrondatxe beach (Fig. 8A; Payros et al., 2006, 2009a, 2015a). The succession is mostly made up of hemipelagic limestones and marls with intercalations of turbidites. Three lithostratigraphic units were defined based on lithological characteristics, namely the Marly Flysch, Azkorri Sandstone, and Sandy Flysch (Figs. 8A and 9). Planktonic foraminiferal biostratigraphic studies were carried out by Orue-Etxebarria (1983) and Orue-Etxebarria et al. (1984), with later additions by Bernaola et al. (2006a) and Payros

of different orders. (C) Biostratigraphic distribution of planktonic foraminifers from the Cretaceous/Paleogene boundary interval of the Sopelana section. Layers 1 and 2 correspond to the Maastrichtian *A. mayaroensis* biozone and are characterized by a gradual decrease in carbonate content. Layer 3, 7-cm-thick, corresponds to the K/Pg boundary clay and contains a lower gray silt with partly dissolved foraminifera, an intermediate brownish siltstone where asteroid-impact indicators occur, and an upper thinly laminated dark gray siltstone. Layer 4 is composed of gray marls which progressively increase in carbonate content (Danian *Gb. Cretacea* and *Civisina euskalherriensis* biozones), whereas layer 5 (*P. pseudobulloides* biozone) is mainly composed of limestones. *(B) Reprinted with permission from Batenburg, S.J., Gale, A.S., Sprovieri, M., Hilgen, F.J., Thibault, N., Boussaha, M., Orue-Etxebarria, X., 2014. An astronomical time scale for the Maastrichtian based on the Zumaia and Sopelana sections (Basque country, northern Spain). J. Geol. Soc. Lond. 171, 165—180; Fig. 1; © Geological Society of London, 2014. (C) Reprinted with permission from Baceta, J.I., Bernaola, G., Orue-Etxebarria, X., Apellaniz, E., Pujalte, V., Caballero, F., 2005. Contribuciones del Pirineo Vasco al conocimiento de la crisis biológica del límite Cretácico-Terciario. Munibe 57, 55—66; Fig. 3B; © Aranzadi Zientzia Elkartea, 2005.*

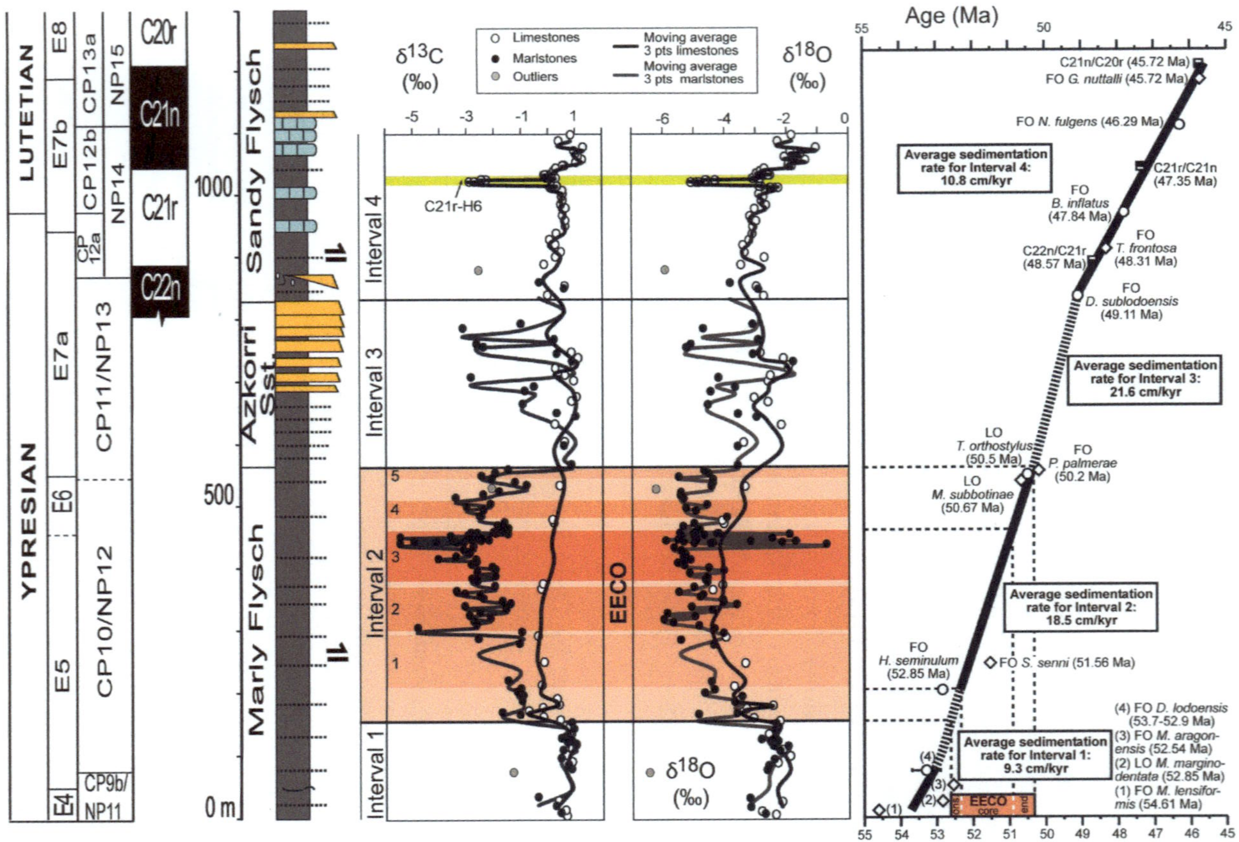

Figure 9 Lithochronostratigraphy of the Ypresian succession exposed at the Barinatxe and Gorrondatxe beaches. The upper part of the Marly Flysch (interval 2) is characterized by the occurrence of siderite-rich layers, their abundance being represented herein by the intensity of the reddish background. Stable isotope and biostratigraphic data show that this interval was formed during the early Eocene climatic optimum (EECO). The age model shows that the EECO extended from 52.6 to 50.3 Ma (*FO*, first occurrence; *LO*, last occurrence). The overlying Azkorri Sandstone was accumulated as a consequence of a glacioeustatic sea level fall. *Reprinted with permission from Payros, A., Ortiz, S., Millan, I., Arostegi, J., Orue-Etxebarria, X., Apellaniz, E., 2015a. Early Eocene climatic optimum: environmental impact on the North Iberian continental margin. Geol. Soc. Amer. Bull. 127, 1632–1644, Figs. 2 and 5. (© Geological Society of America, 2015).*

et al. (2009b) from the interval exposed at Gorrondatxe. They defined seven Ypresian biozones, which correlate with Zones E2 to E7b of Wade et al. (2011). Calcareous nannofossil studies (Bernaola et al., 2006a,b) allowed identification of Zones NP11 to NP14a of Martini (1971). In addition, Rodriguez-Lazaro and Garcia-Zarraga (1996) identified the ostracod *Rimacytheropteron* cf. *rotundapunctata* in Barinatxe, which is a biostratigraphic index for the Ypresian. According to the available sedimentological and biostratigraphic data, the Ypresian succession is practically continuous, with the exception of two faulted and distorted intervals in the lower and upper levels.

An interesting result derived from the correlation of planktonic foraminiferal and calcareous nannofossil data is that their biostratigraphic calibration in Barinatxe shows some discrepancies with that in standard timescales (Fig. 9; Vandenberghe et al., 2012). This mismatch has not yet been examined in detail, but it is worth mentioning that this is not the only case from the Basque–Cantabrian area in which Eocene planktonic foraminiferal biostratigraphic results have been at odds with standard geologic timescales (Payros et al., 2007a, 2015b). Interestingly, the previous mismatches, despite seeming questionable when first presented, were later confirmed by independent studies and eventually led to the revision of the standard Eocene biomagnetochronostratigraphic framework (Vandenberghe et al., 2012; Wade et al., 2011). Therefore, it is plausible that the Ypresian calcareous plankton biostratigraphic calibration derived from Barinatxe constitutes the correct framework.

Payros et al. (2006) quantified vertical variations in turbidite content throughout the Eocene succession, thus defining three turbidite-poor intervals and three turbidite-rich intervals. Furthermore, they observed that the turbidite-poor intervals are typified by warm water planktonic foraminiferal assemblages, whereas the turbidite-rich intervals are generally characterized by cool water associations. The Marly Flysch exposed in Barinatxe corresponds to the first turbidite-poor, warm water interval. Detailed multidisciplinary studies carried out in this unit (Payros et al., 2015a) allowed identification of the Early Eocene Climatic Optimum (EECO) interval, the period with the warmest sustained temperatures of the past 80 My (Zachos et al., 2001). In Barinatxe the EECO is represented by a 410-m-thick interval characterized by the scarcity of hemipelagic limestones, the abundance of dark marls, which record a reduction in calcium carbonate content and an increase in kaolinite, and the occurrence of conspicuous red layers with a high siderite and pyrite content (Fig. 9). The base of the

EECO occurs in the lower part of the calcareous nannofossil Zone NP12 and slightly postdates the first occurrences of *Chiasmolithus grandis* and *Lophodolithus reniformis* and the last occurrence of *Discoaster diastypus*. Additionally, it slightly predates the first occurrence of *Helicosphaera seminulum* and *Scyphosphaera columella* and the last occurrence of *Discoaster pacificus* (Bernaola et al., 2006a). In the planktonic foraminiferal sequence of events the base of the EECO occurs in the lower part of Zone E5, midway between the first occurrences of *Morozovella aragonensis* and *Subbotina senni* (Orue-Etxebarria et al., 1984). According to the age model implemented for the Barinatxe section (Fig. 9), the EECO started at 52.6 Ma and lasted ~2.3 Myr. Its onset was characterized by the addition of ^{13}C-depleted carbon into the ocean-atmosphere system for 80 kyr, which caused a hotter climate and a perennial rainfall regime, thus increasing the supply of terrestrial clays, organic matter, and iron oxides into the sea. Eventually, these changes affected the deep-sea bottom 270 kyr after the onset of the EECO, creating conditions in which opportunistic benthic foraminifera thrived and leading to increased methanogenesis in the subsurface, which caused the widespread formation of siderite. A subsequent gradual recovery brought about the end of the EECO, which is marked by the first occurrence of the cool water calcareous nannofossil *Reticulofenestra dictyoda* (Bernaola et al., 2006a) and the rapid transition into the overlying Azkorri Sandstone (Fig. 9), which constitutes the first turbidite-rich interval characterized by cool water planktonic foraminiferal assemblages (Payros et al., 2006, 2015a). These changes practically coincided with the NP12/13 and E6/7a boundaries at 50.3 Ma.

Payros et al. (2009a) demonstrated that the Azkorri Sandstone represents a lowstand deposit. Furthermore, they added that a similar sea level situation can be deduced from another six coeval sections throughout the Pyrenean area, despite their being up to 350 km apart from each other, containing deposits of different origins (intrabasinal carbonate sediments, siliciclastic sediments sourced from the Iberian plate, and terrigenous sediments sourced from the uplifting Pyrenees) and having been accumulated in different sedimentary environments (from continental to transitional, shallow marine, and deep marine) and in different geodynamic settings (piggy-back basin, foreland basin, and cratonic margin). This common evolution was interpreted in terms of a sea level–driven depositional sequence that encompassed the late Ypresian to early Lutetian interval. Its correlation with other sequences from North European areas, the Antarctic Ocean and New Jersey suggested that the depositional sequence could extend worldwide. Furthermore, the association of the lowstand Azkorri Sandstone with cool water conditions and the correlation of the Ypresian/Lutetian depositional sequence with oceanic temperature variations recorded globally suggested that the depositional sequence might be the result of climatically driven glacioeustatic sea level changes.

5.3 The Lutetian Interval: Biostratigraphic Amendments for a GSSP in Gorrondatxe

The quality of the Eocene Sandy Flysch succession exposed at Gorrondatxe (Fig. 8A) for biostratigraphic studies and its potential to become a reference section for the Ypresian/Lutetian transition was recognized long ago (Orue-Etxebarria and Apellaniz, 1985). However, specific studies were not carried out until the GSSP approach was adopted by the ICS, and a working group for the Ypresian/Lutetian boundary stratotype was created within the ISPS. Interestingly, the results obtained at Gorrondatxe exceeded the definition of the GSSP for the base of the Lutetian Stage. In the following review, these results will be described separately.

5.3.1 Reassessment of the Ypresian/Lutetian Biomagnetochronology

Before the integrated stratigraphic studies carried out in Gorrondatxe around 2005, the calibration of geomagnetic polarity and calcareous plankton biostratigraphic scales at the Ypresian/Lutetian transition had been recognized as a difficult task, as shown by the variable correlation schemes produced over more than 50 years (e.g., Aubry et al., 1988; Berggren, 1969, 1972; Berggren et al., 1985, 1995; Cavelier and Pomerol, 1986; Hardenbol and Berggren, 1978; Luterbacher et al., 2004; Martini, 1970; Odin and Luterbacher, 1992; Vandenberghe et al., 2012). This was probably so because the study of the Ypresian/Lutetian transition is hampered in many deep-marine successions by the occurrence of stratigraphic hiatuses or the lack of reliable biomagnetostratigraphic results (e.g., Aubry, 1995). Despite these problems, the criterion most commonly used to identify the base of the Lutetian was the first appearance of specimens belonging to the planktonic foraminiferal genus *Hantkenina*, which was later suggested to be approximately simultaneous with the first appearance of the species *Guembelitrioides nuttalli* (E7/E8 zonal boundary; Berggren and Pearson, 2005). According to the available biomagnetostratigraphic calibrations (Berggren et al., 1995; Luterbacher et al., 2004), which were mostly derived from stratigraphic sections near Gubbio (Italy), it was considered that the first appearance of *Hantkenina* correlated with calcareous nannofossil Zone

CP12a and coincided with the boundary between magnetozones C22n and C21r. However, it had long been recognized that some levels of the Italian reference sections presented problems for biostratigraphic studies due to poor fossil preservation.

In this uncertain scenario, the 700-m-thick Gorrondatxe section (Fig. 10), which was revealed as the most expanded Ypresian/Lutetian succession worldwide (Bernaola et al., 2006a), was found to be a notable exception and allowed the precise calibration of biostratigraphic and magnetostratigraphic scales. Surprisingly, the biomagnetostratigraphic calibration of the Ypresian/Lutetian planktonic foraminiferal events from the Gorrondatxe section was at odds with that shown in standard geological timescales (Fig. 10). Thus, the planktonic foraminiferal events which in the standard biomagnetostratigraphic scheme were correlated with magnetic polarity Chron C22r could be assigned to Chron C21r in the Gorrondatxe section; the events formerly correlated with Chron C22n were included within Chron C21n in the Gorrondatxe section; and the events considered to have occurred during the time span of Chron C21r were found within Chron C20r in the Gorrondatxe section. Regarding more specifically the planktonic foraminiferal events that had been used to pinpoint the Ypresian/Lutetian boundary, the lower occurrence of *G. nuttalli* (E7/E8 boundary) was found close to the boundary between magnetic polarity Chrons C21n/C20r and within calcareous nannofossil Zone CP13a, and not at the C22n/C21r boundary or within Zone CP12a as depicted in the standard timescales. The lowest occurrence of hantkeninids was found even higher in the succession, correlating with Chron C20r and Zone CP13b; however, it was acknowledged that, taking into account the scarcity of hantkeninids in Gorrondatxe, their lowest occurrence might not correspond to their true first appearance in the stratigraphic record.

Payros et al. (2007a) analyzed the abovementioned discrepancies between the Gorrondatxe section and the standard timescales. To this end, they evaluated the robustness of the original data obtained from the Gorrondatxe section and the Italian reference sections. In addition, they compiled biomagnetostratigraphic information from another 12 Ypresian/Lutetian successions which agreed with the Gorrondatxe calibration. Taking everything into account, they concluded that the correlation scheme derived from Gorrondatxe might well constitute the correct biomagnetostratigraphic framework. On that assumption, Payros et al. (2007a) elaborated on the Gorrondatxe data with the aim of amending the Ypresian/Lutetian biomagnetochronology. They acknowledged that, given the common occurrence of turbidites in the Gorrondatxe section, the distance between successive biomagnetostratigraphic events does not directly reflect the time lapse involved, but mainly the frequency and magnitude of sediment gravity flows. With the aim of analyzing the Gorrondatxe events from a true temporal perspective, they deducted the thickness of all the turbiditic deposits and obtained a hypothetical 195-m-thick pelagic-only succession, in which the distance between successive events should quite reliably reflect the time span between them (Fig. 10). Using the absolute ages provided for calcareous nannofossil and magnetostratigraphic events in standard timescales, and interpolating the planktonic foraminiferal events of the pelagic-only Gorrondatxe section, Payros et al. (2007a) determined that the lowest occurrence of the planktonic foraminifer *G. nuttalli* (E7/E8 boundary) was 3.1 Myr younger than formerly considered. The Gorrondatxe results were later confirmed by independent studies carried out in other areas (Coccioni and Bancala, 2012; Jovane et al., 2010; Larrasoaña et al., 2008; Ortiz et al., 2008; Payros et al., 2011; Pearson and Coxall, 2014; Rögl and Egger, 2010; Weinbaum-Hefetz and Benjamini, 2011; Zakrevskaya et al., 2011) and eventually led to the revision of the standard Eocene biomagnetochronostratigraphic framework (Vandenberghe et al., 2012; Wade et al., 2011).

The Gorrondatxe section was also the type area for the definition of the planktonic foraminiferal species *Morozovella gorrondatxensis* (Orue-Etxebarria, 1985b). This species was later recategorized as being merely a junior synonym of *Morozovella crater* (Pearson et al., 2006), but a thorough review by Orue-Etxebarria et al. (2014) showed that *M. gorrondatxensis* is in fact a separate species. Furthermore, they demonstrated that *M. gorrondatxensis* was distributed worldwide and that it has great potential for early Lutetian biostratigraphic refinements, as its lowest occurrence can be used to approximate the Ypresian/Lutetian boundary and further subdivide Zone E7b of Wade et al. (2011). Another slightly younger species defined in the Gorrondatxe area is *Globigerina (Eoglobigerina) bizkaiensis* (Orue-Etxebarria, 1985b), which was later deemed as a synonym of *Subbotina jacksonensis* (Pearson et al., 2006).

5.3.2 The GSSP for the Base of the Lutetian Stage

After nearly 20 years of stratigraphic research since the foundation of the ISPS Working Group for the Ypresian/Lutetian Boundary Stratotype, a workshop was held in Getxo (Biscay) in September 2009, which was attended by 17 researchers from 6 countries (Molina et al., 2011; Orue-Etxebarria et al., 2009). On the basis of the biomagnetostratigraphic

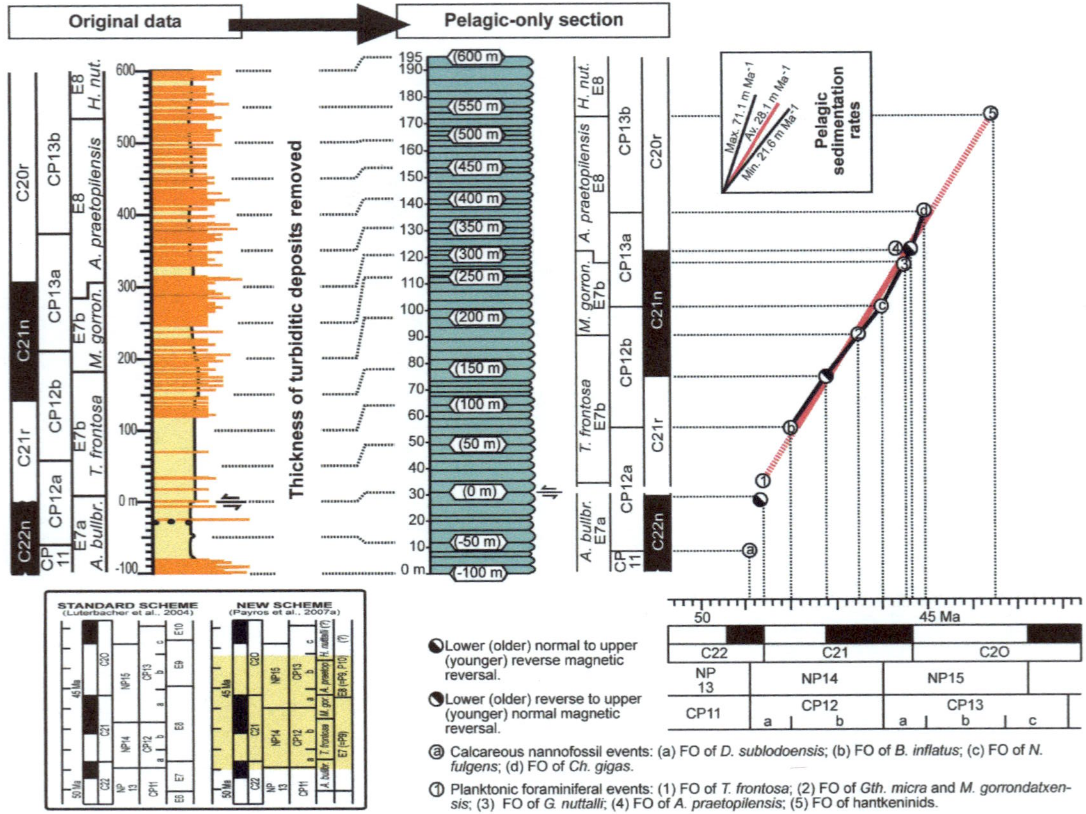

Figure 10 Actual stratigraphic log of the Gorrondatxe section (0 m is equivalent to 900 m in the general Eocene log in Fig. 9) and transformation into a hypothetical pelagic-only section by removal of the thickness of turbiditic deposits. The age model defined using magnetostratigraphic and calcareous nannofossil biostratigraphic data allowed interpolation of planktonic foraminiferal bioevents and reassessment of the Ypresian/Lutetian chronostratigraphy. *Compiled and modified with permission from Payros, A., Bernaola, G., Orue-Etxebarria, X., Dinarès-Turell, J., Tosquella, J. Apellaniz, E., 2007a. Reassessment of the early-middle Eocene biomagnetochronology based on evidence from the Gorrondatxe section (Basque country, W Pyrenees). Lethaia 40, 183–195, Figs. 5–7 (© Wiley, 2007).*

information available from the two historical unit stratotypes under consideration (Ypresian from the Ieper area, Belgium, and Lutetian from the Paris Basin, France), it was agreed that the GSSP for the base of the Lutetian Stage should be defined in the stratigraphic interval between the lower occurrence of the planktonic foraminifer *Turborotalia frontosa* and that of the calcareous nannofossil *Blackites inflatus*. In the light of the amended Ypresian/Lutetian biomagnetochronology derived from the Gorrondatxe section (Fig. 10; Payros et al., 2007a, 2009b), this meant that the GSSP would be included within Chron C21r and, consequently, that some of the criteria traditionally used to approach the Ypresian/Lutetian boundary (e.g., the C22n/C21r Chron boundary and the first occurrences of hantkeninids and *G. nuttalli*) had to be definitively abandoned. It was also agreed that, to fulfill the chronostratigraphic principle that the base of the upper stage defines the boundary, the Lutetian GSSP should preferably coincide with an event that, being included within the Ypresian/Lutetian boundary interval, was as close as possible to the base of the historical Lutetian unit stratotype from Paris. All the events near the base of the Lutetian unit stratotype were discussed, and it was finally concluded that the lowest occurrence of *B. inflatus*, marker of the base of calcareous nannofossil Subzone CP12b (Okada and Bukry, 1980), would best define the Ypresian/Lutetian boundary. Finally, the stratigraphic sections suitable for the Lutetian GSSP were appointed. In this regard, a major problem was that there were not many suitable sections, as many successions included unconformities at the Ypresian/Lutetian transition, were too condensed, lacked important fossil groups or did not provide reliable paleomagnetic data. Consequently, the discussion was restricted to the Agost section in the Betic Cordillera and the Gorrondatxe section in the Basque coast. After evaluation of the characteristics of both sections, a ballot was organized and the result was unanimous in favor of Gorrondatxe (Fig. 11). Thus, the GSSP for the base of the Lutetian Stage was defined at 67.85 m of the Gorrondatxe section (equivalent to 967.85 m of the Eocene Barinatxe–Gorrondatxe section), in a dark marly level in which the calcareous nannofossil *B. inflatus* has its lowest occurrence (43°22′47″N, 3°00′51.61″W; Molina et al., 2011). The proposal was approved by the ISPS in February 2010, by the ICS in January 2011, and was ratified by the IUGS in April 2011.

In addition to the primary correlation marker for the GSSP (i.e., the lowest occurrence of *B. inflatus*), many other biomagnetostratigraphic events, including all that had traditionally been used to approach the Ypresian/Lutetian boundary, were pinpointed in

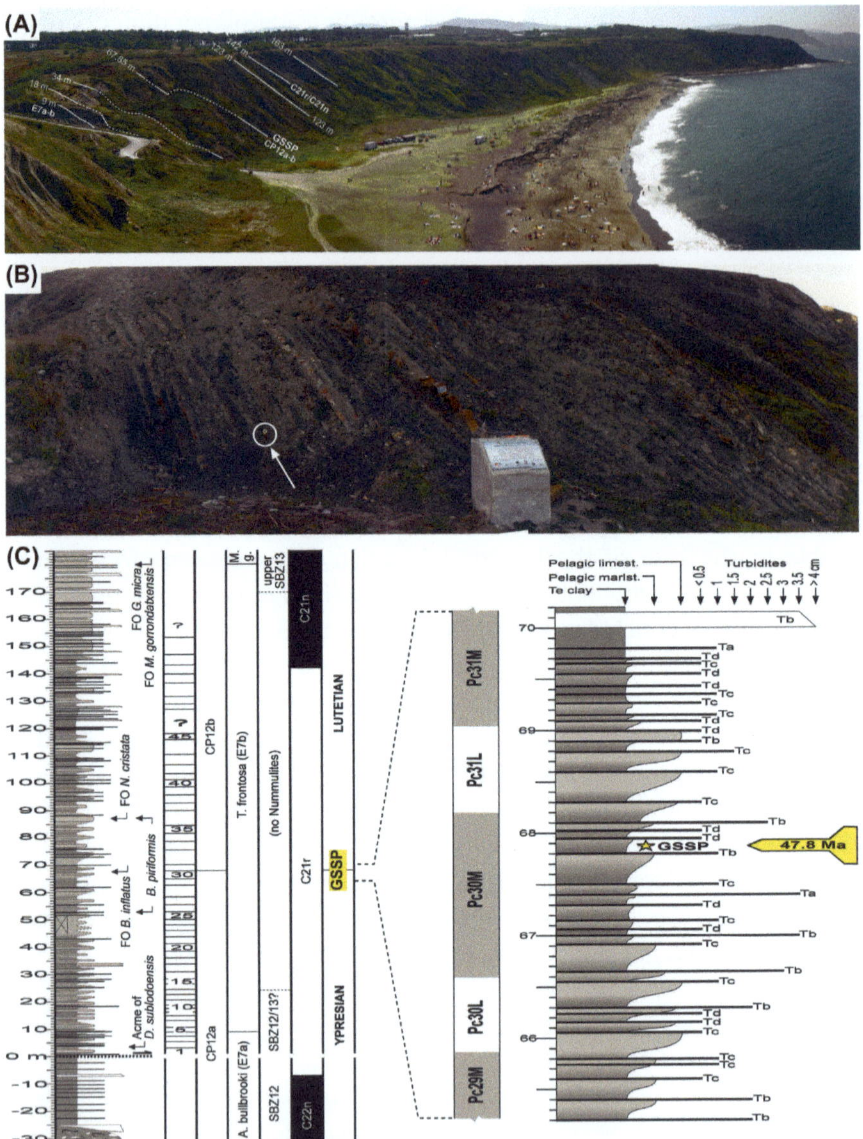

Figure 11 (A) Panoramic view of the Gorrondatxe beach, showing the position and lateral extent of some key levels (including the Lutetian GSSP) numbered according to their position in the stratigraphic log. (B) Detailed view of the GSSP for the base of the Lutetian Stage (*golden spike* is circled). (C) General and detailed stratigraphic logs of the Ypresian/Lutetian transition (0 m is equivalent to 900 m in the general Eocene log in Fig. 9). I, 21-kry-long precession couplets; II, calcareous nannofossil biostratigraphy; III, planktonic foraminifera biostratigraphy; IV, nummulitid biostratigraphy; V, magnetostratigraphy; VI, chronostratigraphy; *Pc*, precession couplet; M, marly

Gorrondatxe (Bernaola et al., 2006a; Molina et al., 2011; Orue-Etxebarria and Apellaniz, 1985; Payros et al., 2007a, 2009b, 2011), facilitating thus the identification of the GSSP-based Ypresian/Lutetian boundary elsewhere. Furthermore, an astronomically controlled cyclostratigraphic framework was provided for the Ypresian/Lutetian boundary interval (Payros and Martinez-Braceras, 2014; Payros et al., 2009b, 2011), which allowed the high-precision age-dating of most chronostratigraphic events with respect to the GSSP (Fig. 11). Thus, the C22n/C21r magnetic reversal lies 39 precession-related mudstone—marl couplets below the Lutetian GSSP, which represent 819 kyr (assuming a modern mean precession period of 21 kyr); by the same means the C21r/C21n magnetic reversal is 507 kyr younger than the GSSP. These calculations allowed the GSSP to be age dated at 47.8 Ma. The acme of the calcareous nannofossil *Discoaster sublodoensis* and the lowest occurrence of the planktonic foraminifer *T. frontosa* were found 26 precession couplets (546 kyr) below the GSSP. The lowest occurrence of the calcareous nannofossil *Blackites piriformis* was found to be 105 kyr (five precession couplets) older than the GSSP, whereas its highest occurrence, as well as the lowest occurrence of *Nannotetrina cristata*, is 115 kyr younger than the GSSP. The lowest occurrence of the planktonic foraminifers *M. gorrondatxensis* and *Globigerinatheka micra* are 1.25 Myr younger than the GSSP. The shallow marine larger foraminiferal SBZ12/13 boundary is located close to the GSSP (Payros et al., 2011). The onset of the global warming related to the C21r-H6 event (Fig. 9; Sexton et al., 2011; Payros et al., 2012) occurred 15 precession couplets (315 kyr) after the formation of the GSSP layer (Fig. 11). Furthermore, the GSSP corresponds to the maximum flooding surface of a global Ypresian/Lutetian depositional sequence (Payros et al., 2009a), the lower and upper boundaries of which correlate with oxygen isotope global events Ei5 and Ei6 (Pekar et al., 2005). The correlation of the maximum flooding surface between marine and coastal/terrestrial successions showed that the

hemicouplet; L, limy hemicouplet; T, turbidite, with its basal Bouma division. *Reprinted with permission from Payros, A., Dinarès-Turell, J., Bernaola, G., Orue-Etxebarria, X., Apellaniz, E., Tosquella, J., 2011. On the age of the Early/Middle Eocene boundary and other related events: cyclostratigraphic refinements from the Pyrenean Otsakar section and the Lutetian GSSP. Geol. Mag. 148, 442—460, Fig. 1 (© Cambridge University Press, 2011) and Payros, A., Martinez-Braceras, N., 2014. Orbital forcing in turbidite accumulation during the Eocene greenhouse interval. Sedimentology 61, 1411—1432, Fig. 3 (© Wiley, 2014).*

Lutetian GSSP is younger than the base of mammal Zone MP10 and older than MP14 (Payros et al., 2009a).

6. THE BIDART–BIARRITZ SECTION: ASSP FOR THE BASE OF THE DANIAN STAGE AND UNIT STRATOTYPE FOR THE BIARRITZIAN STAGE

Despite not being located within the Iberian Peninsula, the Upper Cretaceous/Oligocene succession exposed at the Bidart–Biarritz section (Labourd province, northern Basque Country) is included in this review for both geographic and paleogeographic reasons: geographically, it lies just at the western end of the Pyrenean isthmus and can therefore be considered to be located on the edge of the Iberian Peninsula (Fig. 2A); and, paleogeographically, it was formed within the same sedimentary basin as the Zumaia–Getaria and Sopelana–Barinatxe–Gorrondatxe–Galea sections (i.e., in the Basque–Cantabrian basin), being therefore genetically connected to them.

The Bidart–Biarritz section has been well known since the 19th century as the home to important Paleogene fossil sites (e.g., Archiac, 1846, 1850; Bouille, 1876; Boussac, 1906, 1908, 1911; Douville, 1903, 1905; Jacquot, 1864; Harpe, 1879; Rouault, 1850; Seunes, 1890). It is in fact the type locality of several stratigraphically relevant species, such as the middle-late Eocene calcareous nannofossil *Pemma basquense* (Martini, 1959), the Bartonian larger foraminifera *Nummulites biarritzensis* (Archiac, 1837), or the Rupelian *Nummulites vascus* (Joly and Leymerie, 1848), among others (see Mathelin and Sztrakos (1993) for a comprehensive review of species defined in the Bidart-Biarritz section). However, the section first gained global relevance when it was designated unit stratotype for a new chronostratigraphic stage (the so-called Biarritzian; Hottinger and Schaub, 1960), which later fell into disuse. More recently, the discovery of an excellent K/Pg boundary section revived the stratigraphic value of the succession (Fig. 12).

The Bidart–Biarritz succession extends from the Maastrichtian to the Rupelian along approximately 4 km of coastal cliffs. However, the succession is not continuously exposed because the cliffs are locally interrupted by beaches and coves. Furthermore, the stratigraphic continuity of the Paleogene succession has been a matter of debate due to the occurrence of several intervals with Triassic clays and evaporites. These intervals were first related with diapirs and thrust sheets, which would imply that the succession is not continuous. However, the Triassic deposits were later interpreted as being resedimented by submarine debris flows in Eocene times (Fondecave-Wallez et al., 1995), meaning that the Paleogene succession could be more or less continuous.

6.1 ASSP for the Base of the Danian Stage in Bidart

The K/Pg succession is exposed north of Bidart, on the cliffs between the Erretegia and Caseville–Pavillon Royal beaches (43°26′59.50″N 1°35′11″W). The section is identical to those found in Zumaia, Sopelana, and other Basque sections (Apellaniz et al., 1997; Pujalte et al., 1998), being composed of Maastrichtian gray marls and marlstones (Bidart marls, equivalent to the Algorri Formation in Zumaia) and Danian white to pinkish limestones (equivalent to the Aitzgorri Formation in Zumaia). However, Apellaniz et al. (1997) showed that the K/Pg boundary clay in Bidart is the thickest and most complete in the Basque–Cantabrian area (Fig. 12B). In fact, the boundary clay interval is considered to be one of the most complete sections in Europe and has consequently been designated ASSP for the base of the Danian Stage (Molina et al., 2009).

The K/Pg interval has been thoroughly studied for geochemistry, including extraterrestrial iridium and Ni-rich spinel content (Bonte et al., 1984; Font et al., 2016; Nelson et al., 1991; Renard et al., 1982; Rocchia et al., 2002; Romein and Smit, 1981; Smit and Ten Kate, 1982; Vonhof and Smit, 1997), ammonites (Ward and Kennedy, 1993), calcareous nannofossils (Burnett et al., 1992; Gorostidi and Lamolda, 1995; Martini, 1971; Minoleotti et al., 2004; Perch-Nielsen, 1979; Seyve, 1990; Thibault et al., 2004), planktonic foraminifera (Apellaniz et al., 1997; Gallala, 2014; Gallala and Zaghbib-Turki, 2010; Gallala et al., 2009; Haslett, 1994; Punekar et al., 2016), benthic foraminifera (Alegret, 2007; Alegret et al., 2004), bioturbation (Alegret et al., 2015; Rodriguez-Tovar et al., 2010, 2011b), and magnetostratigraphy and magnetic susceptibility (Abrajevitch et al., 2015; Delacotte et al., 1985; Font et al., 2011; Galbrun and Gardin, 2004). These studies have sometimes led to heated discussions among researchers, but have eventually helped to increase our current knowledge of the global processes behind the K/Pg boundary event, their physical and chemical effects on pelagic and deep-sea ecosystems, and the pattern of biotic extinction, rate of survivorship and subsequent polyphyletic radiation, and evolutionary lineages, thus providing the K/Pg transition with a global stratigraphic framework of unprecedented accuracy (Fig. 12).

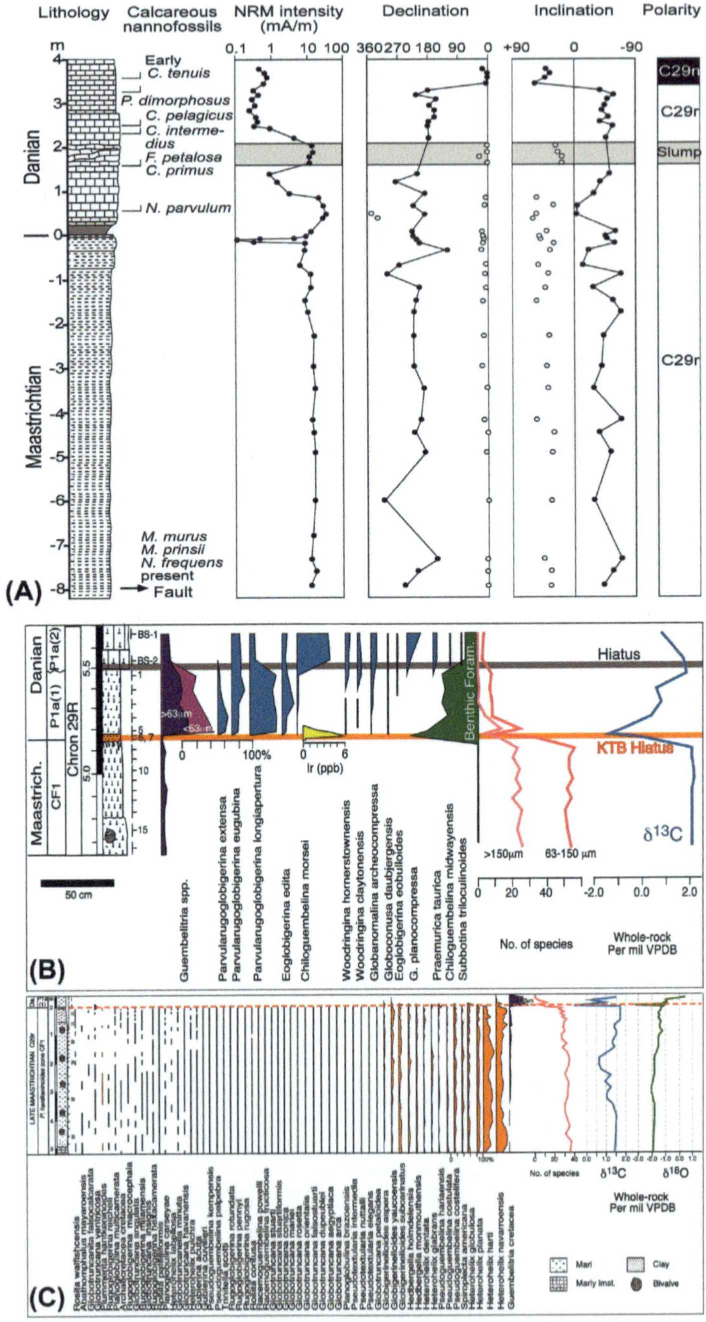

Figure 12 (A) Lithological log of the K/Pg boundary section at Bidart, showing calcareous nannofossil events, paleomagnetic results, and Chron assignment (ChRM

6.2 The Biarritzian Stage

Hottinger and Schaub (1960) observed that a significant stratigraphic interval between the Lutetian and the Ledian (one of the terms used at that time to refer to the late Eocene) was missing in northern European sections of reference. They added that significant steps in larger foraminifera evolution, such as the extinction of *Alveolina*, *Assilina*, and large *Nummulites*, occurred during that stratigraphic interval. The interval had commonly been referred to as Auversian in the Alps and Italy, but Hottinger and Schaub (1960) realized that the stratigraphic range of the Auversian did not match in age with the Auvers sands in the Paris basin and did not coincide with what had been interpreted as Auversian in the Biarritz section. Therefore, Hottinger and Schaub (1960) coined a new term, the Biarritzian, which was regarded as an extended version of the Auversian, and designated Biarritz unit stratotype (Fig. 13). The base of the unit stratotype was defined at the Peyreblanque limestones, which are exposed discontinually at the Ilbarritz and Milady beaches (between 43°27′33″N 1°34′47″W and 43°28′01″N 1°34′31″W), whereas the top is included in the Villa Marbella marls (43°28′24″N 1°34′08″W).

Opinions against the Biarritzian Stage were published in the following years, as it was considered that it correlated with part of the Lutetian (see Berggren, 1971; Cavelier and Pomerol, 1986; Pomerol, 1981). Consequently, the ISPS did not include the Biarritzian in the set of nine official Paleogene stages approved in the 28th International Geological Congress held in Washington in 1989, and its validity was thus restricted to regional usage. However, the term is still widely used by some stratigraphers,

directions, i.e., middle temperature component in the Maastrichtian marly limestones and lower Danian limestones and high temperature component in the Danian limestone, are plotted as *black circles*; the high temperature normal overprint directions in the Maastrichtian marly limestones, the lower Danian limestones, and the Danian slump are plotted as *open circles*). (B) Key foraminifera and geochemical characteristics of the K/Pg boundary. (C) Abundance of late Maastrichtian planktonic foraminifera of the 63–150 μm fraction. The $\delta^{13}C$ record shows a characteristic negative shift at the K/Pg. *(A) Modified with permission from Galbrun, B., Gardin, S., 2004. New chronostratigraphy of the Cretaceous-Paleogene boundary interval at Bidart (France). Earth Planet. Sci. Lett. 224, 19–32, Fig. 7 (© Elsevier, 2004). (B and C) Reprinted with permission from Punekar, J., Keller, G., Khozyem, H.M., Adatte, T., Font, E., Spangenberg, J., 2016. A multi-proxy approach to decode the end-Cretaceous mass extinction. Palaeogeogr. Palaeoclimatol. Palaeoecol. 441, 116–136; Fig. 5; © Elsevier, 2016.*

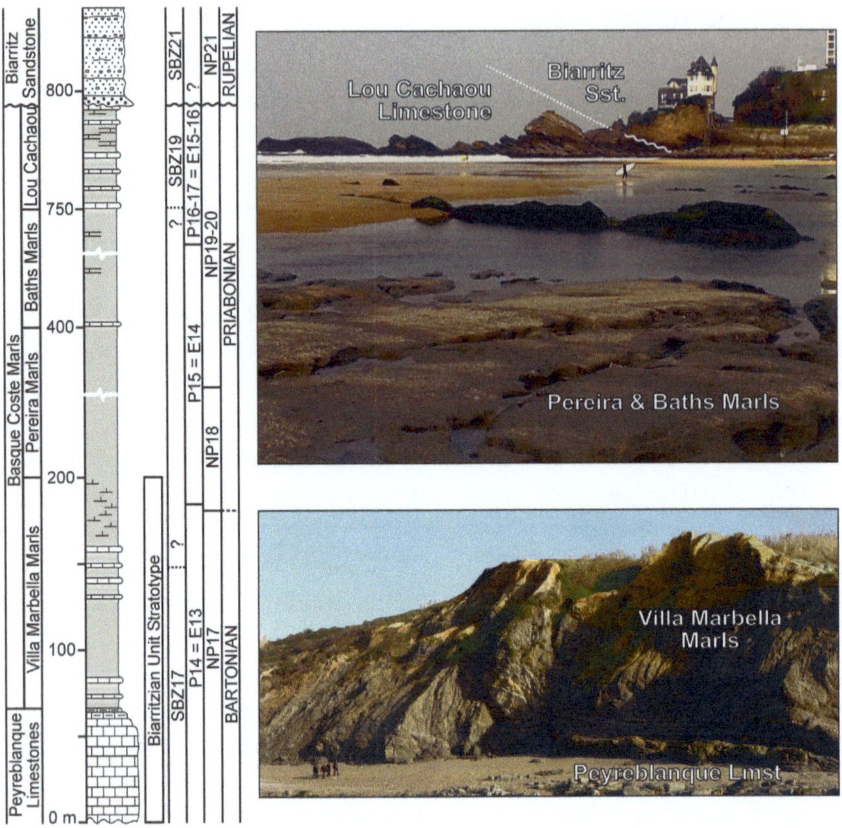

Figure 13 Synthetic lithochronostratigraphy of the middle-upper Eocene and lower Oligocene succession of Biarritz, showing the extent of the Biarritzian Unit Stratotype. The lower photo shows the middle part of the Biarritzian succession at the Handia cliff (Ilbarritz beach). The upper photo shows the location of the Eocene/Oligocene boundary unconformity at the northern side of the Basque Coast beach. *Data compiled from Mathelin, J.C., Sztrakos, K., 1993. L'Eocène de Biarritz (Pyrénées Atlantiques, SW France): stratigraphie et paléoenvironment, monographie des foraminifères. Cah. Micropaleontol. 8, 1–85.*

especially by larger foraminiferal paleontologists (see Luterbacher et al., 2004), for whom the Biarritzian represents a distinct stratigraphic interval.

Mathelin and Sztrakos (1993) showed that the historical Biarritzian unit stratotype contains upper SBZ17 larger foraminifera, NP17-18 calcareous nannofossils, and E13-14 planktonic foraminifera (Fig. 13), thus including the Bartonian/Priabonian transition. Serra-Kiel et al. (1998) later suggested that the Biarritzian Stage should encompass the whole SBZ17 (equivalent to the upper parts of calcareous nannofossil Zone NP16 and planktonic foraminiferal Zone E11), therefore also including the lower part of the Bartonian stage.

7. THE CARAVACA SECTION: ASSP FOR THE BASE OF THE DANIAN STAGE

Like Zumaia and Bidart, the Maastrichtian/Danian succession exposed in Caravaca (Murcia province, NE part of the Betic Cordillera) is an auxiliary stratotype section for the base of the Danian Stage (Molina et al., 2009), being therefore subordinate to the GSSP defined at El Kef, Tunisia (Molina et al., 2006a). Definition of Danian ASSPs was justified on the basis of subsequent deterioration of the El Kef section and the need of K/Pg boundary sections in different facies and paleogeographic contexts (Molina et al., 2009). The Caravaca section was not included in the final ballot for the selection of the GSSP for the base of the Danian Stage (Molina et al., 2006a), but it does deserve special historical recognition. In fact, it was using data from the Caravaca section that Smit and Hertogen (1980) produced the first international publication in which an extraterrestrial iridium anomaly was described at the K/Pg boundary clay (Fig. 14). This work was published one month before the seminal paper by Alvarez et al. (1980), but priority was not claimed due to the fact that in 1979 the Alvarez team had presented the extraterrestrial theory in a congress.

The K/Pg boundary is exposed at the Gredero ravine, on the NW side of the C-336 road, 4 km south of Caravaca village (38°054′36.39″N 1°52′41.45″W). The section is included in the 225-m-thick Maastrichtian-lower Eocene Quipar–Jorquera Formation, which is mostly composed of marls and marly limestones, interspersed with turbidites, accumulated in a middle bathyal environment (Rodriguez-Tovar and Uchman, 2006; Smit, 2004; and references therein). At the boundary interval, Maastrichtian light gray marls grade upwards into a green transitional layer, 3-mm thick, which in turn is sharply capped by a rust-red layer, 2- to 3-mm thick, containing extraterrestrial and impact-derived material. The K/Pg boundary is therefore located between the green and rusty layers. The latter is overlain by a 7- to 10-cm-thick blackish layer of laminated clays, followed by gray argillaceous marls (Fig. 14).

After the presentation of the asteroid-impact hypothesis, most studies carried out in the Caravaca K/Pg section have dealt either with the characterization of the extraterrestrial and impact-derived material (Bohor et al.,

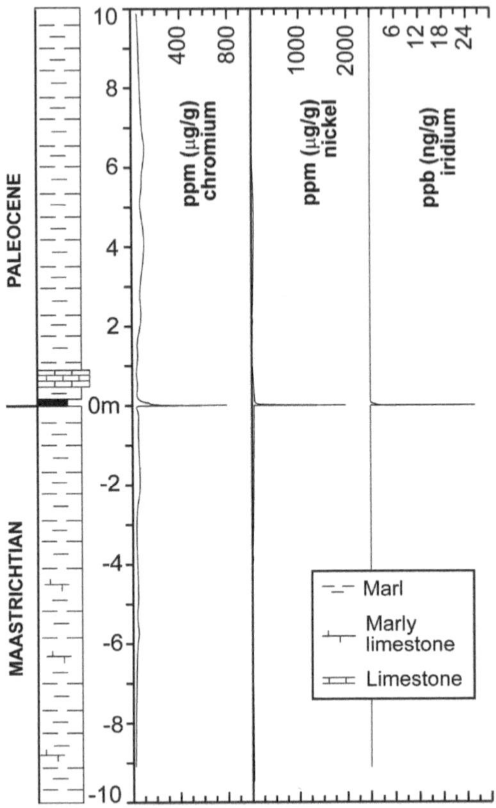

Figure 14 Concentration profiles of the siderophile elements Cr, Ni, and Ir in the Caravaca section. Anomalous values of all three elements in the same (K/Pg) level indicate a common source. *Redrawn with permission from Smit, J., 2004. The section of the Barranco de Gredero (Caravaca, Spain): a crucial section for the Cretaceous/Tertiary boundary impact extinction hypothesis. J. Iber. Geol. 31, 179–191, Fig. 8 (© Universidad Complutense de Madrid, 2004).*

1986; De Paolo et al., 1983; Izett, 1987; Kamo et al., 2011; Kyte et al., 1985; Martinez-Ruiz et al., 1997, 1999, 2006; Robin et al., 1991; Shukolyukov and Lugmair, 1998; Smit and Klaver, 1981; Smit and Ten Kate, 1982; Trinquier et al., 2006; Villasante-Marcos et al., 2007) or with biostratigraphy. In this regard, initial planktonic foraminiferal biostratigraphic studies carried out in Caravaca produced significant improvements in the biozonation across the K/Pg boundary (Hillebrandt, 1974; Smit, 1977, 1979). Subsequent research focused on the pattern of biological extinctions, rate of survivorship, and succeeding diversification trends of planktonic foraminifera (Arenillas et al., 2006; Arz et al., 2000; Canudo et al., 1991;

Coccioni and Luciani, 2005; Gallala, 2014; Kaiho and Lamolda, 1999; MacLeod and Keller, 1991; Olsson and Liu, 1993) and calcareous nannofossils (Gardin and Monechi, 1998; Lamolda et al., 2005). The influence of postdepositional bioturbation on the stratigraphic record was evaluated by Rodriguez-Tovar et al. (2006), Kedzierski et al. (2011) and Sosa-Montes et al. (2013). In addition, the collateral environmental disturbance caused by the asteroid impact (such as climate change, acid rain, wildfires, temporal darkness, collapse of ecological communities, and redox anomalies) has also been analyzed in Caravaca (Alegret et al., 2003; Arinobu et al., 1999, 2005; Brinkhuis et al., 1998; Coccioni and Galeotti, 1994; Kaiho et al., 1999; Mizukami et al., 2013; Sosa-Montes et al., 2013, 2016).

The overlying Paleogene part of the Quipar–Jorquera Formation has attracted less attention, but planktonic foraminiferal assemblages from the Danian/Selandian transition were analyzed by Arenillas (2012), who compared the results with those from the GSSP at Zumaia. Additionally, the Paleocene/Eocene boundary and the PETM were studied by Molina et al. (1994), Canudo et al. (1995), Ortiz (1995), Angori and Monechi (1996), Guernet and Molina (1997), and Angori et al. (2007).

8. THE AGOST SECTION: PROSPECTIVE ASSP FOR THE BASE OF THE LUTETIAN STAGE

The Loma de la Beata hill north of Agost (Alicante province, northeastern Betic Cordillera) has long been known for its Paleogene successions rich in microfossils (Hillebrandt, 1974, 1976; Isuman, 1983; Leclerc, 1971) and became globally relevant when a well-preserved K/Pg boundary section was found (Groot et al., 1989). The area gained official recognition when it was designated auxiliary stratotype section for the base of the Lutetian Stage in a joint work by most members of the Ypresian/Lutetian Boundary Working Group (Molina et al., 2012).

8.1 The Cretaceous/Paleogene Boundary

The K/Pg boundary is exposed on the eastern side of the Loma de la Beata hill, 1 km from Agost on the road CV-827 (38°27′07″N 0°38′12″W). Soon after its discovery (Groot et al., 1989; Smit, 1990), it was regarded as one of the most complete K/Pg boundary sections in the world (MacLeod and Keller, 1991; Olsson and Liu, 1993). Being included in the Quipar–Jorquera Formation, the lithological features of the Agost section are very similar to those in Caravaca, which is located approximately 100 km westwards. Thus, the succession is composed of an alternation of silty marls and calcisiltites with intercalated calciturbidites. Interestingly, Ten Kate and Sprenger (1993) showed that the Agost Maastrichtian/Danian lithological alternation can be correlated cyclostratigraphically to that in Zumaia, more than 550 km apart. More specifically for the K/Pg boundary, the Maastrichtian gray marls of Agost are overlain by a 2-mm-thick reddish layer which includes evidence of the asteroid impact (Ni-rich spinel; spherules; iridium and osmium concentrations; platinum group elements; and anomalies in Ni, Cr, Co, and Fe), and the following Danian deposits consist of a 10-cm-thick dark clay that gradually increases in carbonate content upwards (Fig. 15).

After the initial description of the K/Pg boundary clay, its geochemistry and mineralogy were thoroughly analyzed to determine its origin (Diaz-Martinez et al., 2002; Martinez-Ruiz et al., 1992, 1997, 1999, 2006; Villasante-Marcos et al., 2007). Paleontological studies mainly focused on the behavior of foraminiferal assemblages across the K/Pg boundary (Fig. 15). After considerable debate about the pattern of extinction and rate of survivorship, it can now be considered that a catastrophic mass extinction affected 70–90% of the planktonic foraminiferal species (Arenillas et al., 2004; Canudo et al., 1991; Gallala, 2014; Molina et al., 1996, 1998, 2005), whereas benthic foraminifera were less drastically affected (Alegret and Thomas, 2005; Alegret et al., 2003; Pardo et al., 1996). The effect on marine macrobenthos and the timing of subsequent seabed recolonization was analyzed by means of bioturbation analysis (Rodriguez-Tovar, 2005; Rodriguez-Tovar and Uchman, 2004a,b; Rodriguez-Tovar et al., 2002, 2004, 2006). Calcareous nannofossils were studied by Lamolda et al. (2016), who, in addition to reporting that 86% of the species vanished at the K/Pg boundary, also assessed the influence of climate and sea level changes (Fig. 15).

8.2 The Ypresian/Lutetian Interval

The Loma de la Beata hill area in Agost provides an auxiliary stratotype section for the base of the Lutetian Stage (Molina et al., 2012), which is subordinate to the GSSP defined at Gorrondatxe at the marl layer that contains the lowest occurrence of the calcareous nannofossil *B. inflatus* (i.e., CP12a/b subzone boundary; Molina et al., 2011). In fact, Gorrondatxe and Agost were the two contenders in the final ballot held for the definition of the Lutetian GSSP in the 2009 meeting of the Working Group for the Ypresian/Lutetian boundary stratotype.

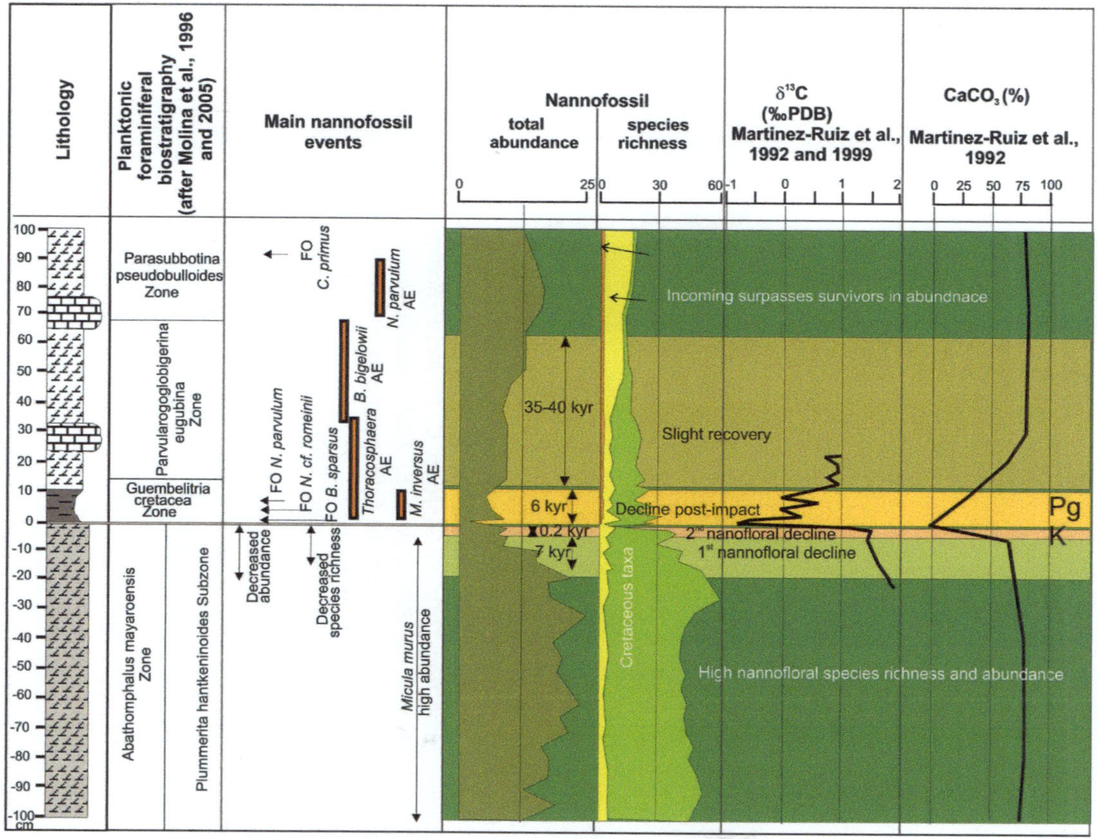

Figure 15 Stratigraphic log, planktonic foraminifera biostratigraphy, and nannofossil events, abundance and species richness across the K/Pg boundary interval at Agost. Ages are approximated from sedimentation rate. The level 0 m marks the base of the Paleocene. *FO*, first occurrence; *AE*, acme event. *Modified with permission from Lamolda, M.A., Melinte-Dobrinescu, M.C., Kaiho, K., 2016. Calcareous nannoplankton assemblage changes linked to paleoenvironmental deterioration and recovery across the Cretaceous-Paleogene boundary in the Betic Cordillera (Agost, Spain). Palaeogeog. Palaeoclimatol. Palaeoecol. 441, 438–452, Fig. 7 (© Elsevier, 2016).*

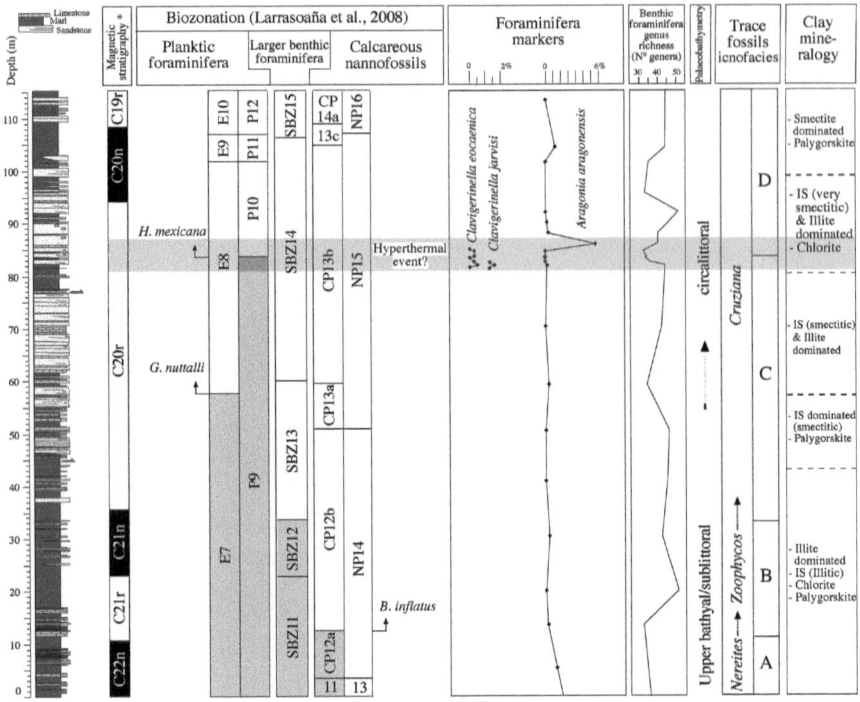

Figure 16 Integrated stratigraphy of the Ypresian/Lutetian transition at Agost. The gray bars show the extent of the Ypresian deposits according to different fossil groups before definition of the GSSP for the base of the Lutetian stage at Gorrondatxe. *Reprinted with permission from Ortiz, S., Gonzalvo, C., Molina, E., Rodríguez-Tovar, F.J., Uchman, A., Vandemberghe, N., Zeelmaekers, E., 2008. Palaeoenvironmental turnover across the Ypresian-Lutetian transition at the Agost section, southeastern Spain: in search of a marker event to define the Stratotype for the base of the Lutetian Stage. Mar. Micropaleontol. 69, 297–313, Fig. 7 (© Elsevier, 2008).*

The auxiliary stratotype section for the base of the Lutetian Stage is located in the NW part of the Loma de la Beata hill, 1.5 km from Agost (38°27′30″N 0°38′48″W). Another Ypresian/Lutetian section had previously been studied approximately 500 m further south (Gonzalvo and Molina, 1998; Molina et al., 2000), but it offered poorer exposure and fossil preservation. The auxiliary stratotype section is composed of 115 m of marls with intercalated limestone and sandstone beds accumulated on the south-facing slope of a carbonate platform (Fig. 16; Larrasoaña et al., 2008; Ortiz et al., 2008). The integrated biomagnetostratigraphic study showed that the amended Ypresian/Lutetian chronostratigraphy developed at the Gorrondatxe section (Fig. 10; Payros et al., 2007a) was valid for the Agost section, despite the fact that both sections had different paleogeographic affinities in Eocene times (Fig. 1B; Atlantic for Gorrondatxe, Tethyan for Agost), thus offering a sound basis for subsequent modifications of the standard Eocene timescale (Vandenberghe et al., 2012; Wade et al., 2011). Furthermore, the Agost section, despite (or because of) being less expanded than Gorrondatxe, encompasses a larger time span than Gorrondatxe, allowing the chronostratigraphic amendments to be extended almost to the entire Lutetian interval. In this regard, Larrasoaña et al. (2008) were the first to suggest that the boundary between larger foraminiferal zones SBZ12 and SBZ13, which had commonly been used to approximate the Ypresian/Lutetian boundary, correlates with calcareous nannofossil Zone CP12b and not with the CP12a/b boundary as shown in previous standard calibrations (Fig. 16). Subsequently, Tori and Monechi (2013) elaborated on the upper Ypresian and Lutetian calcareous nannofossils of the Agost section, defining numerous bioevents that may improve stratigraphic resolution.

9. THE OTSAKAR SECTION: PROSPECTIVE ASSP FOR THE BASE OF THE LUTETIAN STAGE

In a joint work by most members of the Ypresian/Lutetian Boundary Working Group, the Otsakar section was presented as an auxiliary stratotype section for the base of the Lutetian Stage (Molina et al., 2012), being therefore subordinate to the GSSP defined at Gorrondatxe (Molina et al., 2011). The need for an ASSP for the base of the Lutetian Stage was justified because the chronostratigraphic position of the GSSP level, which is located within magnetic polarity Chron C21r, could not be cyclostratigraphically calibrated with well-dated magnetostratigraphic events in Gorrondatxe. First, the underlying C22n/C21r chron boundary is separated from the GSSP-bearing succession by a fault, which causes the loss of an unknown thickness of the succession (Fig. 11; Payros et al., 2011). Second, the age difference with the overlying C21r/C21n chron boundary could not be accurately calculated due to the partial absence of astronomical cycles in a 24 m-thick interval attributed to the C21r-H6 global warming event (Payros et al., 2011, 2012). This meant that, although the exact age difference between other events could be established in Gorrondatxe by means of cyclostratigraphy, their absolute ages could not be determined. The only approach available was to assume that the 48 Ma age estimated previously for the CP12a/b subzone boundary (Luterbacher et al., 2004) was correct. However, it had long been acknowledged that this was one of the most poorly dated

Eocene events (Berggren et al., 1995). These problems were overcome with data from the Otsakar section (Payros et al., 2011).

Otsakar is a village located 15 km northwest of Pamplona (Navarre province) in the western Pyrenees (42°54′53″N, 1° 44′04″W), approximately 120 km ESE from the GSSP at Gorrondatxe. The 120-m-thick Ypresian/Lutetian succession is exposed along a paved track that leads from Otsakar to a telecommunication mast near the Arzelaieta hill, the main outcrop being a soil-stripped, gullied hillside near the village (Fig. 17). The succession is mostly composed of hemipelagic mudstone–marl alternations and calciclastic sediment gravity flow deposits accumulated at approximately 500 m water depth on the slope of a distally steepened carbonate ramp (Payros, 1997; Payros et al., 2007b, 2009a, 2011). A detailed study of the main outcrop revealed that it is affected by several normal faults, which made the Otsakar section unsuitable for the Lutetian GSSP (Molina et al., 2011). Notwithstanding the tectonic disturbance, a bed-by-bed succession was reconstructed by correlating characteristic beds throughout the outcrop.

The integrated bio-, magneto-, and cyclostratigraphic study showed that successive biostratigraphic events are separated by a practically identical number of mudstone–marl couplets both in Otsakar and Gorrondatxe, thus demonstrating that all couplets had the same astronomical precession-related origin (Payros et al., 2011). Furthermore, by counting 39 precession-related mudstone–marl couplets of 21 kyr between the C22n/C21r chron boundary and the calcareous nannofossil CP12a/b zonal boundary in Otsakar, the time lapse between both events was calculated to be 819 kyr (Fig. 17). Similar studies from Gorrondatxe suggested that the CP12a/b boundary was 507 kyr older than the C21r/C21n chron boundary. The combination of the Otsakar and Gorrondatxe results indicated that the whole duration of Chron C21r was of 1.326 Myr, which is very similar to other estimates (Luterbacher et al., 2004; Vandenberghe et al., 2012; Westerhold and Rohl, 2009). More importantly, according to these findings the age of the CP12a/b boundary, and hence that of the GSSP for the base of the Lutetian Stage, was established at 47.76 Ma of the Luterbacher et al. (2004) timescale available then, being therefore practically 250 kyr younger than previously thought. Subsequent publications rounded this age at 47.8 Ma (Molina et al., 2011; Vandenberghe et al., 2012).

The Otsakar results provided further refinements to the Eocene bio-magnetochronology (Payros et al., 2011). Thus, these results were among the first to confirm the amendments made by Payros et al. (2007a, 2009b)

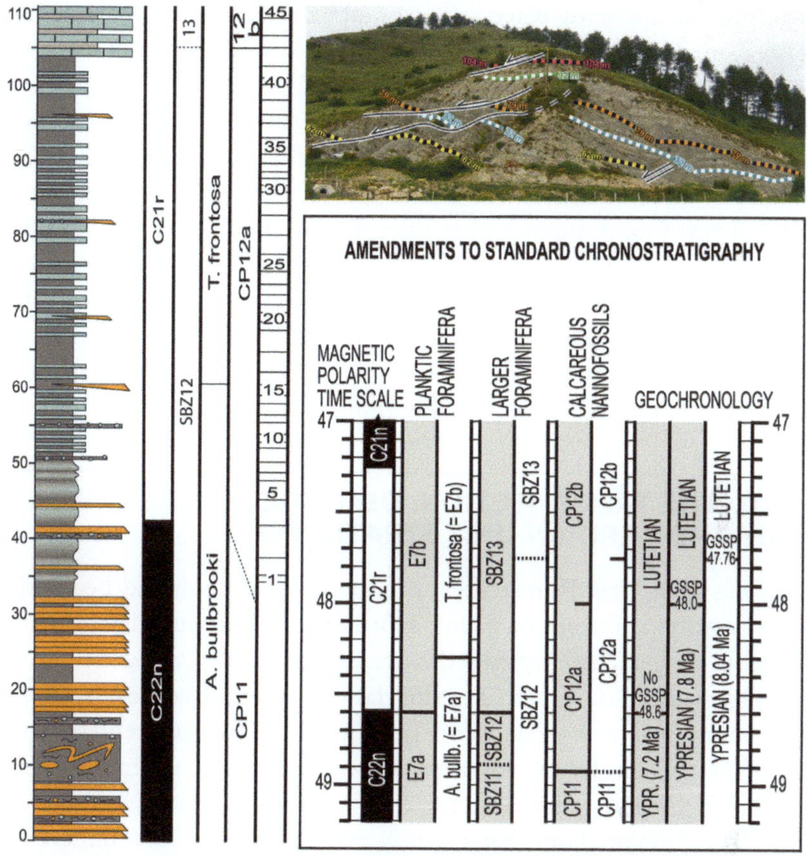

Figure 17 Field view of the main outcrop at Otsakar, which is affected by several normal faults. Despite the tectonic disturbance, a bed-by-bed succession was reconstructed. Columns to the right of the log show, from left to right, magnetostratigraphy, nummulitid biostratigraphy, planktonic foraminiferal biostratigraphy, calcareous nannofossil zonation, and 21-kyr-long precession couplets. Inset shows the resulting amendments to the standard chronostratigraphy of Luterbacher et al. (2004, gray columns). *Reprinted with permission from Payros, A., Dinarès-Turell, J., Bernaola, G., Orue-Etxebarria, X., Apellaniz, E., Tosquella, J., 2011. On the age of the Early/Middle Eocene boundary and other related events: cyclostratigraphic refinements from the Pyrenean Otsakar section and the Lutetian GSSP. Geol. Mag. 148, 442–460; © Cambridge University Press, 2011.*

to the Ypresian/Lutetian biomagnetostratigraphic calibration on the basis of data from Gorrondatxe (Fig. 10). More novel, the results showed that the chronostratigraphic position of the nummulitid Zone SBZ12 in Otsakar was at odds with that assumed in the standard scheme of Serra-Kiel et al.

(1998). First, this scheme placed the base of SBZ12 within calcareous nannofossil Zone CP12a. Nevertheless, in Otsakar the latter part of calcareous nannofossil Zone CP11 correlated with SBZ12 (Fig. 17). Second, and more importantly, the results did not support the position of the top of SBZ12 within calcareous nannofossil Zone CP12a and at the C22n/C21r chron boundary, as shown by Serra-Kiel et al. (1998). In Otsakar the top of SBZ12 was found much higher than the C22n/C21r chron boundary, close to the base of Zone CP12b. Payros et al. (2011) compiled similar correlations from other sections. Taking everything into account, they concluded that the new position of the top of SBZ12, as approximately correlated in Otsakar with the calcareous nannofossil CP12a/b zonal boundary and with Chron C21r, is probably more accurate.

10. THE TREMP SECTION: DEFINITION AND RENEWAL OF THE ILERDIAN UNIT STRATOTYPE

Long after most Paleogene chronostatigraphic units had been established in northern European basins, Hottinger and Schaub (1960) defined the Ilerdian Stage. They justified the need for this new Paleogene chronostatigraphic unit on the fact that in the classic northern European sections, a significant interval of the Paleocene/Eocene transition is not represented by marine deposits, but by the so-called Sparnacian terrestrial deposits (Aubry et al., 2005; Cavelier and Pomerol, 1986; Dollfus, 1880). Hottinger and Schaub (1960) pointed out that a significant step in larger foraminifera evolution occurred in many Tethyan areas during that stratigraphic interval. Later, Luterbacher (1973) added that the interval represented by the Ilerdian Stage was also significant in terms of zonal successions based on pelagic microfossils. Taking everything into account, the International Working Group on Paleogene Stratigraphy concluded, in a meeting held in Paris in 1974, that the Ilerdian is not only a stage that allows a clarification of the Mesogean Paleogene stratigraphy, but may also serve as the base to build up a stratigraphic succession better defined than that from northern European basins (Pomerol, 1975). In their original definition, Hottinger and Schaub (1960) designated the stratigraphic succession exposed in Tremp (Lleida province, Catalonia, southeastern Pyrenees) type section for the Ilerdian Stage, which was actually named after the Latin name (Ilerda) for Lleida. In addition, Schaub (1969) defined the unit stratotype in the 800-m-thick littoral and shallow marine—marly deposits exposed along the C1311 road between Tremp and the Montllobat mountain pass (Fig. 18;

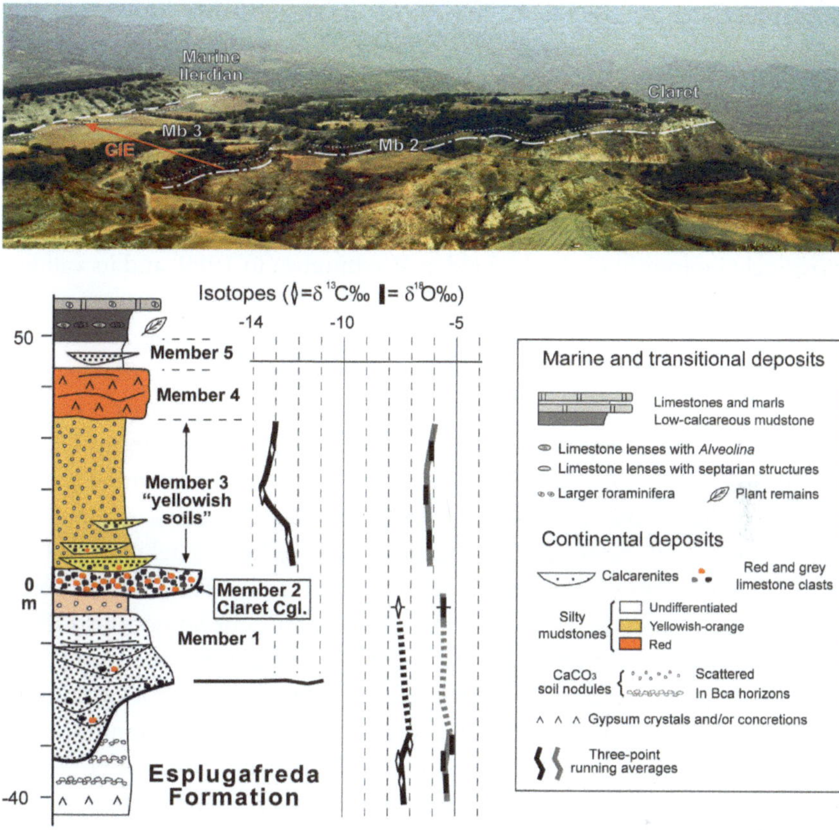

Figure 18 Panoramic view and lithostratigraphic log of the lower part of the Ilerdian (early Ypresian) unit stratotype near Claret (road C1311 from Tremp to the Montllobat mountain pass). A prominent carbon isotope excursion shows that the base of the Ypresian, and thus that of the Ilerdian, lies at the Claret conglomerate, 50 m below the classic Ilerdian marine deposits. *Modified with permission from Pujalte, V., Baceta, J.I., Schmitz, B., Orue-Etxebarria, X., Payros, A., Bernaola, G., Apellaniz, E., Caballero, F., Robador, A., Serra-Kiel, J., Tosquella, J., 2009a. Redefinition of the Ilerdian stage (early Eocene). Geol. Acta 7, 77–94, Figs. 4B and 5C (© Geologica Acta, 2009).*

approximately between 42°9′12″N 0°52′4″E and 42°9′19″N 0°47′57″E). Initial biostratigraphic studies of the Ilerdian Unit Stratotype at Tremp dealt, among others, with larger foraminifera (*Nummulites* and *Alveolina* by Hottinger and Schaub, 1960), planktonic foraminifera (Ferrer et al., 1973; Gartner and Hay, 1962; Hillebrandt, 1965, 1975; Luterbacher, 1969), and calcareous nannofossils (Kapellos and Schaub, 1973; Wilcoxon, 1973).

The Ilerdian Stage became widely used by stratigraphers working in the Alpine and Mediterranean areas but found opposition in northern European stratigraphers, who continued using the classic stages from the Paris Basin. Some of the criticism on the Ilerdian Stage came from the fact that the marine deposits of the stratotypic Tremp section are encased between terrestrial deposits, making it difficult to define the stage boundaries; in addition, open marine fossils were found to be scarce, which hampered its comparison with other chronostratigraphic units. Consequently, the ISPS did not include the Ilerdian in the set of nine official Paleogene stages approved in the 28th International Geological Congress held in Washington in 1989, and its validity became restricted to regional usage.

Interest in the Ilerdian was revived in connection with the activities of the IGCP projects 286 (Early Paleogene Benthos) and 308 (Paleocene/Eocene boundary events). Several integrated biomagnetostratigraphic studies carried out in the South Pyrenean area revisited the Tremp section (e.g., Molina et al., 1992; Serra-Kiel et al., 1994), which confirmed the results from earlier studies and provided further details. These studies identified larger foraminiferal zones SBZ5-9 (but no SBZ6) of Serra-Kiel et al. (1998), the transition between planktonic foraminiferal zones E2 and E3 (sensu Wade et al., 2011) in the middle part of the Ilerdian Unit Stratotype, calcareous nannofossil zones NP9-11 of Martini (1971), and parts of chrons C24r and C24n. It is interesting to note that, additionally, the terrestrial succession underlying the Ilerdian deposits of the Tremp area has proven significant for the characterization of Maastrichtian and Paleocene continental biotas (Canudo et al., 2016; Lopez-Martinez and Pelaez-Campomanes, 1999; Lopez-Martinez et al., 2001).

The Ilerdian Stage gained new relevance after the definition of the GSSP for the base of the Ypresian Stage (Paleocene/Eocene boundary) in the Dababiya Quarry (Egypt) at a level characterized by a negative 3–5‰ excursion in $\delta^{13}C$ (Aubry et al., 2007), which can be correlated globally from deep-marine to terrestrial records. The relationship between this CIE and the larger foraminiferal turnover that characterizes the onset of the Ilerdian Stage at the SBZ4/5 zonal boundary had been long discussed, but it was finally demonstrated that both events were simultaneous (Orue-Etxebarria et al., 2001; Pujalte et al., 2003b). This relationship reinforced the usefulness of the Ilerdian Stage as a chronostratigraphic unit, as its base became directly correlatable with that of the Ypresian Stage. This situation led Pujalte et al. (2009a) to redefine the Ilerdian Stage following a procedure similar to that used by the ICS to establish global chronostratigraphic standards, namely by defining its lower boundary. Surprisingly, stable isotope analyses carried out in the Tremp section identified the onset of the Paleocene/Eocene boundary $\delta^{13}C$ excursion 50 m below the base of the Ilerdian marine deposits, in a terrestrial conglomeratic unit named Claret Conglomerate after the village built on it (Fig. 18; Pujalte and Schmitz, 2005; Pujalte et al., 2009a; Schmitz and Pujalte, 2003, 2007). Therefore, Pujalte et al. (2009a) suggested redefining the base of the Ilerdian Stage at the base of the Claret Conglomerate in the Tremp section, so that the historical Ilerdian Unit Stratotype maintains its relevance.

11. THE CAMPO SECTION: UNIT "PARASTRATOTYPE" FOR THE ILERDIAN AND CUISIAN STAGES

Some stratigraphers leveled criticism at the Ilerdian Stage defined in the Tremp section (Hottinger and Schaub, 1960; Schaub, 1969) due to the fact that the stratotypic marine deposits are encased between terrestrial deposits, making it difficult to define the stage boundaries; in addition, open marine fossils were found to be scarce, which hampered chronostratigraphic correlations. To overcome these problems, Schaub (1969) suggested that the coeval succession exposed in Campo (Esera valley, Huesca province, central Pyrenees), approximately 45 km to the northwest of Tremp, be designated "parastratotype", as the Ilerdian interval of this section is encased between marine deposits and contains abundant open marine microfossils (Figs. 19 and 20). Schaub (1973) further argued that the Campo section is rather unique because various groups of microfossils used to zone the Paleogene of western Europe and the Mediterranean area occur together. The proposal was accepted by the International Working Group on Paleogene Stratigraphy in a meeting held in Paris in 1974 (Bignot, 1975). Subsequently, Schaub (1992) considered that the Campo section should also be designated Tethyan "parastratotype" of the northern European unit stratotype of the Cusian Stage. However, the ICS stated that the term "parastratotype" is not acceptable, but rather auxiliary stratotypes may be defined (Cowie et al., 1986; Harland, 1992; Remane et al., 1996). Consequently, the Campo section can now be regarded as auxiliary to the regional Ilerdian and Cuisian unit stratotypes. In addition, it is worth mentioning that the Campo section has also provided significant information regarding the K/Pg boundary in marine-terrestrial transitional environments (e.g., Canudo et al., 2016; Fernandez-Marron et al., 2004), Paleocene biomagnetostratigraphic calibrations (e.g., Pujalte et al., 2003a; Serra-Kiel et al., 1994, 1998), and Lutetian continental biomagnetochronostratigraphy (Beamud et al., 2003).

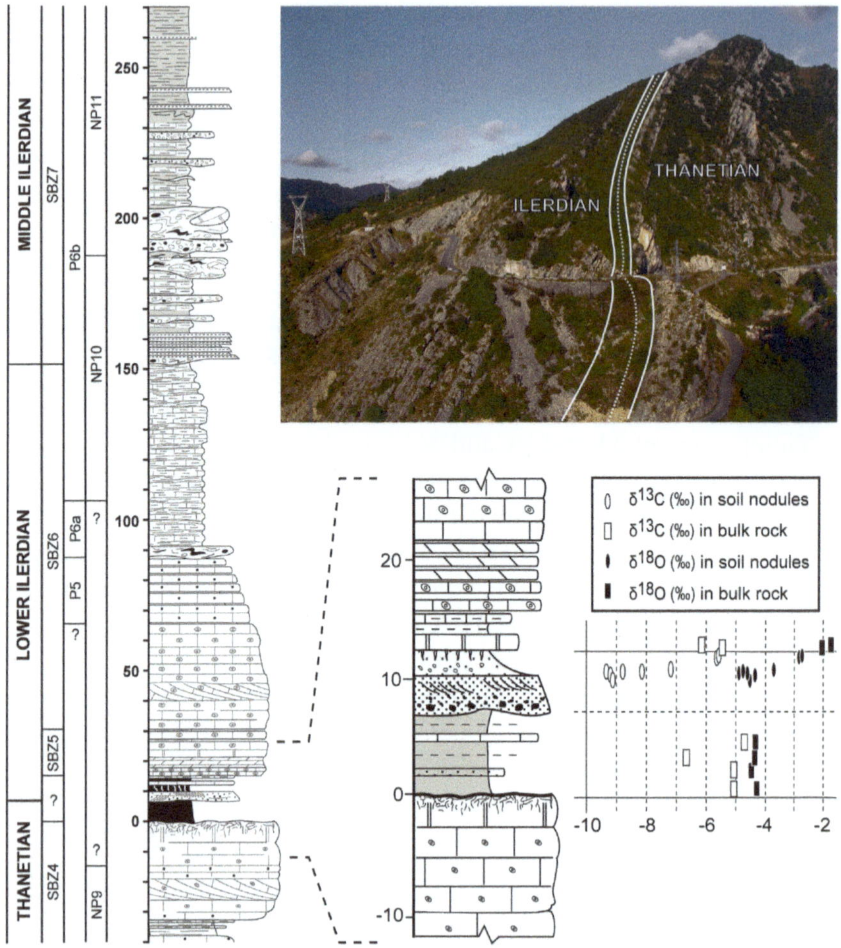

Figure 19 Field view and stratigraphic log of the Thanetian/Ilerdian succession of the Campo section *(Modified with permission from Orue-Etxebarria, X., Pujalte, V., Bernaola, G., Apellaniz, E., Baceta, J.I., Payros, A., Núñez-Betelu, K., Serra-Kiel, J., Tosquella, J., 2001. Did the Late Paleocene thermal máximum affect the evolution of larger foraminifers? Evidence from calcareous plakton of the Campo Section (Pyrenees, Spain). Mar. Micropaleontol. 41, 45–71; © Elsevier, 2001.)*. Stable isotope analysis *(Data compiled from Schmitz, B., Pujalte, V., 2003. Sea-level, humidity, and land-erosion records across the initial Eocene thermal máximum from a continental-marine transect in northern Spain. Geology 31, 689–692; Pujalte, V., Baceta, J.I., Schmitz, B., Orue-Etxebarria, X., Payros, A., Bernaola, G., Apellaniz, E., Caballero, F., Robador, A., Serra-Kiel, J., Tosquella, J., 2009a. Redefinition of the Ilerdian stage (early Eocene). Geol. Acta 7, 77–94; Pujalte, V., Schmitz, B., Baceta, J.I., Orue-Etxebarria, X., Bernaola, G., Dinarès-Turell, J., Payros, A., Apellaniz, E., Caballero, F., 2009b. Correlation of the Thanetian-Ilerdian turnover of larger foraminifera and the Paleocene-Eocene thermal maximum: confirming evidence from the Campo section (Pyrenees, Spain). Geol. Acta 7, 161–175.)* showed that the carbon isotope excursion that marks the base of the Eocene is related to a terrestrial pebbly sandstone, which correlates with the Claret conglomerate in Tremp (see Fig. 18). The electric transmission line in the photo lies on the lower/middle Ilerdian boundary (approximately 150 m in the stratigraphic log).

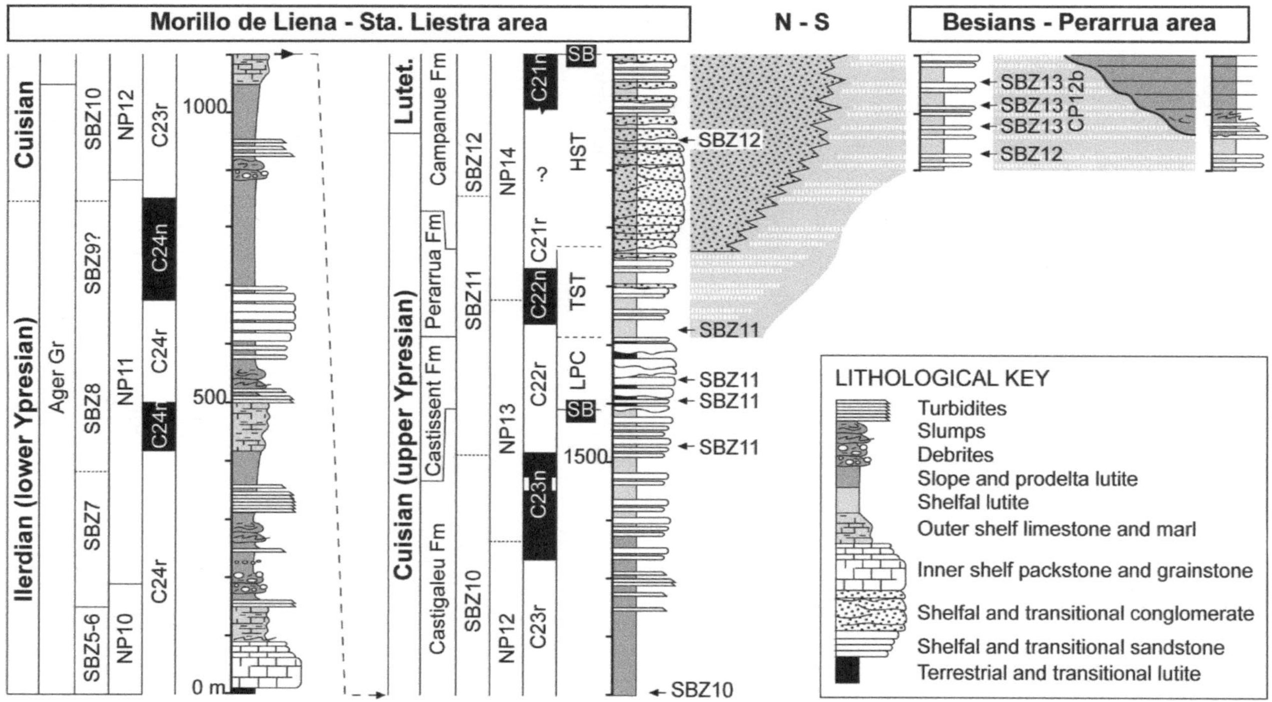

Figure 20 General lithological log and chronostratigraphy of the Campo section, showing the extension of the Ilerdian and Cuisian "parastratotypes". *Modified with permission from Payros, A., Tosquella, J., Bernaola, G., Dinarés-Turell, J., Orue-Etxebarria, X., Pujalte, V., 2009a. Filling the north European early/middle Eocene (Ypresian/Lutetian) boundary gap: insights from the Pyrenean continental to deep-marine record. Palaeogeogr. Palaeoclimatol. Palaeoecol. 280, 313–332, Fig. 9 (© Elsevier, 2009).*

11.1 The Ilerdian Interval: Significance for the Paleocene/Eocene Boundary

The lower and middle parts of the Ilerdian succession are best exposed along the now abandoned road from Campo to Ainsa (Fig. 19; beginning at 42°23′20″N 0°23′47″E), whereas the middle and upper parts extend along the A139 road to Graus and the margins of the Esera river (outcrops south of Navarri and Morillo de Liena villages, at approximately 42°22′39″N 0°23′25″E). The approximately 800-m-thick Ilerdian succession begins with a thin terrestrial interval (Pujalte et al., 2009a, 2009b; Schmitz and Pujalte, 2003, 2007) but mostly displays varied marine facies including the characteristic shallow ramp *Alveolina* grainstone in the lowermost part, which is followed, in ascending stratigraphic order, by middle ramp nummulitid packstones, outer ramp mudstone-marl alternations, slope marls with interbedded breccias and turbidites, and finally prodelta lutites and sandstones (Figs. 19 and 20; Payros et al., 2000; Schaub, 1973; Serra-Kiel et al., 1994). Initial biostratigraphic studies of these deposits dealt, among others, with larger foraminifera (*Nummulites* and *Alveolina* by Hottinger and Schaub, 1960), planktonic foraminifera (Hillebrandt, 1965, 1975; Luterbacher, 1969), and calcareous nannofossils (Kapellos and Schaub, 1973; Wilcoxon, 1973). Interestingly, these studies were among the first in the world to tackle calibrations between different biostratigraphic scales. More recent studies improved and complemented these calibrations with magnetostratigraphy (e.g., Molina et al., 1992; Serra-Kiel et al., 1994), identifying larger foraminiferal zones SBZ5-9 of Serra-Kiel et al. (1998), planktonic foraminiferal zones E2 to E5 (sensu Wade et al., 2011), calcareous nannofossil zones NP9-11 of Martini (1971), and parts of chrons C24r and C24n (Fig. 20).

However, the calibrations from the lowermost part of the Ilerdian succession were later reassessed in the context of the definition of the Paleocene/Eocene boundary. The base of the Eocene is now defined by the prominent negative excursion in $\delta^{13}C$ that marks the onset of the PETM event (Aubry et al., 2007). This warming event had a major impact in many ecosystems from deep-marine to terrestrial (McInerney and Wing, 2011). Thus, deep-sea benthic foraminifera suffered the greatest extinction in the last 90 Myr (Thomas, 1990), calcareous plankton organisms experienced transient diversification (Aubry, 1998; Kelly et al., 1996), archaic mammals were replaced by modern groups (Clyde and Gingerich, 1998), and the flora was subject to significant transient changes (Wing et al.,

2005). Long before the discovery of the PETM and its marker events, Hottinger and Schaub (1960) had partly justified the definition of the Ilerdian Stage on the fact that larger foraminifera (mainly nummulitids and alveolinids) underwent a rapid turnover at its base, which corresponds to the boundary between zones SBZ4 and SBZ5 (Serra-Kiel et al., 1998). Based on data mainly derived from the Campo section, where both larger foraminifera and calcareous planktonic microfossils occur together, it was long held by many authors (Cavelier and Pomerol, 1986; Kapellos and Schaub, 1973; Luterbacher, 1973; Molina et al., 1992; Serra-Kiel et al., 1994, 1998) that the larger foraminiferal turnover at the base of the Ilerdian Stage correlated with the lower part of the Paleocene calcareous nannofossil Zone NP9 of Martini (1971) and with the planktonic foraminiferal Zone P5 of Wade et al. (2011). Accordingly, the Paleocene/Eocene boundary was thought to occur around 150–200 m (or even higher) of the Ilerdian Campo section, therefore correlating with the larger foraminiferal Zone SBZ7 (Fig. 19); in some cases it was even suggested that the entire Ilerdian Stage should be conceptually included in the Paleocene Epoch (e.g., Hillebrandt, 1965, 1975; Luterbacher et al., 2004; their Fig. 20.2). In conclusion, the larger foraminiferal turnover at the base of the Ilerdian stage was considered to be older than, and unrelated to, the PETM and the Paleocene/Eocene boundary.

This long-held calibration was challenged when Orue-Etxebarria et al. (2001) revisited the calcareous plankton of the Campo section, as their results rather suggested that the lowermost Ilerdian SBZ5 deposits approximated the calcareous nannofossil NP9/10 zonal boundary and the planktonic foraminiferal Zone E1, therefore postdating the Paleocene/Eocene boundary (Fig. 19). In support of their alternative calibration, they correlated the Campo section with the deep-marine Trabakua and Ermua sections in the Basque Country, where, in addition to biostratigraphic data, stable isotope and magnetostratigraphic information was also available. This alternative calibration was later confirmed with more accurate bio-magnetostratigraphic data from the uppermost Paleocene deposits of the Campo section (Pujalte et al., 2003a) and with additional biostratigraphic and stable isotope information from the Urrobi and Mintxate sections, which are located midway between Campo and the Basque sections (Pujalte et al., 2003b); the latter authors also specified that the base of the Ilerdian is best typified by the occurrence of the characteristic larger foraminifera *Nummulites* aff. *minervensis*. However, the definitive confirmation of the alternative calibration came with the identification of the Paleocene/Eocene boundary $\delta^{13}C$ excursion in the terrestrial deposits underlying the Ilerdian marine successions of the Tremp and Campo areas (Fig. 19; Pujalte et al., 2009a,b; Schmitz and Pujalte, 2003, 2007). Furthermore, similar findings were reported from several sections in Egypt (Scheibner and Speijer, 2009; Scheibner et al., 2005) and Slovenia (Drobne et al., 2014; Zamagni et al., 2012). In conclusion, the alternative stratigraphic calibration originally suggested by Orue-Etxebarria et al. (2001) using data from the Campo section, which correlates the base of the Ilerdian Stage (marked by the larger foraminiferal turnover at the SBZ4/5 boundary) with the Paleocene/Eocene boundary (marked by the onset of the negative $\delta^{13}C$ excursion and several coeval biostratigraphic events), was finally accepted by the stratigraphic community and included in standard timescales (Vandenberghe et al., 2012). Consequently, the Ilerdian Stage still remains a chronostratigraphic unit of reference at a Tethyan scale and the Campo section maintains its original stratigraphic relevance at a global scale.

11.2 The Cuisian Interval: Significance for the Ypresian/Lutetian Boundary

Dollfus (1880) erected the Cuisian Stage (unit stratotype at the Cuise Sands in the Paris Basin) as an equivalent to the Belgian Ypresian (Dumont, 1850). Due to this originally acknowledged duplication and the uneven acceptance of the term by the stratigraphic community, Bignot (1981) recommended abandoning the Cuisian stage. Consequently, the ISPS did not include the Cuisian in the set of nine official Paleogene stages approved in 1989, and its validity was restricted to regional usage. However, the term is still used by some stratigraphers, especially by larger foraminiferal paleontologists (see Vandenberghe et al., 2012), for whom the Cuisian represents a distinct stratigraphic interval between the well-defined Ilerdian and Lutetian stages.

Comprehensive reviews of the Cuisian unit stratotype (Aubry, 1986; Bignot, 1981; Cavelier and Pomerol, 1986) showed that it is characterized by the occurrence of the larger foraminifera *Nummulites planulatus* (i.e., SBZ10 of Serra-Kiel et al., 1998), NP11-12 calcareous nannofossils, and E3-5 planktonic foraminifera. However, Schaub (1992) argued that the Cuisian unit stratotype is incomplete and does not provide a valid reference record of the homonym stage. Consequently, he suggested defining a "parastrotype" (or auxiliary unit stratotype, according to the formal terminology of the ICS; Remane et al., 1996) at the Campo section (Fig. 20). The selection of the Campo section was based on its completeness, its accessibility, and the cooccurrence of larger foraminifera and calcareous open marine planktonic microfossils in many levels, which allowed calibrations between different biostratigraphic scales. Initial biostratigraphic studies of these deposits dealt, among others, with larger foraminifera (*Nummulites* and *Alveolina* by Hottinger and Schaub, 1960) and calcareous nannofossils (Kapellos and Schaub, 1973; Wilcoxon, 1973). Subsequently, nummulitids were further studied by Tosquella (1995), and magnetostratigraphic analyses were carried out by Bentham and Burbank (1996). The information was integrated by Serra-Kiel et al. (1998) and revised by Payros et al. (2009a).

The base of the Cuisian succession is exposed south of Navarri and Morillo de Liena villages (approximately 42°22′39″N 0°23′25″E), whereas the transition to the Lutetian occurs at several localities further south, such as Besians (42°17′16″N 0°21′01″E; Payros et al., 2009a; Tosquella, 1995) and the Barranco de Cambra, south of Perarrua, at approximately 42°15′31″N 0°20′23″E (Kapellos and Schaub, 1973; Schaub, 1992). The approximately 1300-m thick Cuisian succession is mainly composed of prodelta and delta-front deposits (the so-called Castigaleu, Castissent, and Perarrua formations), but also includes a thick package of fan-delta conglomerates (Campanue Formation). According to the available bio-magnestostratigraphic information, the Cuisian succession of the Campo section encompasses the larger foraminiferal SBZ10-12 zones of Serra-Kiel et al. (1998) extends from the calcareous nannofossil NP11/12 zonal transition to the CP12a/b zonal boundary and from Chron C23r to C21r (Fig. 20). These findings demonstrate that the initial calibration of the SBZ12/13 zonal boundary (i.e., the Ypresian/Lutetian boundary as originally conceived by larger foraminifera paleontologists) with the C22n/C21r chron boundary, as suggested by Serra-Kiel et al. (1998), was incorrect, as confirmed by similar findings in other Iberian sections (Bernaola et al., 2006a; Larrasoaña et al., 2008; Payros et al., 2011). Therefore, the Cuisian "parastrotype" at the Campo section, despite referring to an obsolete chronostratigraphic concept, is still a valuable reference in terms of Ypresian biomagnetostratigraphic calibrations.

The overlying Lutetian deposits of the Campo section yielded SBZ13 nummulitids and CP12b calcareous nannofossils, and were correlated with the C21r/C21n chron transition (see review by Payros et al., 2009a). These deposits are overlain by a terrestrial succession (Capella Formation) which proved useful for improving the standard Lutetian continental biomagnetochronostratigraphy (Beamud et al., 2003).

12. OTHER SECTIONS

The abovementioned Iberian sections have contributed to increase the current knowledge of global Paleogene stratigraphy and have consequently received formal international recognition, being designated GSSPs, ASSPs, unit stratotypes, or auxiliary sections for standard and regional stages. In addition to these, there are many other Iberian sections which have also made significant contributions to global Paleogene stratigraphy but have not achieved the same level of official recognition. Most are successions which were presented as prospective stratotypes at some point but, after evaluation by the working groups in charge, did not make their way into the final selection of sections to be voted, meaning that they were subsequently consigned to oblivion. This does not imply, however, that their contributions to international Paleogene stratigraphy are no longer valid or they are not suitable for further studies. Consequently, they will be briefly reviewed below.

12.1 The Paleocene/Eocene Transition at Trabakua

The Paleocene/Eocene transition at the Trabakua mountain pass (province of Biscay, Basque Country, western Pyrenees; approximately 43°18′00.55″N 2°15′35.50″W) was very well exposed near km 41 of road Bi633 in the early 1990s, when the road was widened and a new drain channel was excavated. More than 20 years later, however, the outcrop has deteriorated considerably due to pervasive growth of vegetation. The basin floor succession, being identical to that of the Paleocene/Eocene transition in Zumaia (Fig. 7), which is 25 km apart (Fig. 2A), is composed of hemipelagic marls and limestones but includes a conspicuous 4-m-thick red clay interval. Biomagnetostratigraphic and geochemical results showed that, as in Zumaia, the red clay interval corresponds to PETM (Bolle et al., 1998; Orue-Etxebarria et al., 1992, 1996; Pujalte et al., 1995, 1998, 2015; Schmitz et al., 2001). On the basis of the positive initial results obtained by Orue-Etxebarria et al. (1992, 1996), the Trabakua section was considered a prospective GSSP for the base of the Ypresian Stage, but it was disregarded later on account of its deep burial diagenesis and poor fossil preservation (Bolle et al., 1998).

Interestingly, however, the planktonic foraminiferal studies carried out in Trabakua (Orue-Etxebarria et al., 1992) proved valuable for the biostratigraphic characterization of the PETM and the accurate definition of the

Paleocene/Eocene boundary. According to Berggren and Miller (1988) and all previous stratigraphic frameworks, the Paleocene/Eocene boundary was equated with the planktonic foraminiferal P6a/b zonal boundary. The older P5 planktonic foraminiferal zone was defined as the partial range of *M. velascoensis* between the last occurrence of *Globanomalina pseudomenardii* and the first occurrence of *Morozovella subbotinae*, an interval that was estimated to last approximately 500 kyr. This stratigraphic framework was first questioned by Orue-Etxebarria et al. (1992; see also 1996), who reported the simultaneous occurrence of the latter two species at the Trabakua section. This overlapping implied that Zone P6 directly succeeded Zone P4. Subsequent confirmation of the alternative calibration in other areas induced the redefinition of Zone P5 as the interval between the last occurrence of *G. pseudomenardii* and that of *M. velascoensis* (Berggren et al., 1995), thus encompassing former P5 and P6a zones. Consequently, the Paleocene/Eocene boundary was redefined as occurring at the revised P5/P6 zonal boundary (Berggren et al., 1995), which in current zonations (Wade et al., 2011) is equivalent to the P5/E1 zonal boundary.

12.2 The Paleocene/Eocene Transition at Ermua

The Paleocene/Eocene succession of Ermua (province of Biscay, Basque Country, western Pyrenees; approximately 43°10′40″N 2°29′55″W) was initially described as auxiliary to the basin floor Trabakua section, which is located less than 10 km northwestwards from Ermua (Fig. 2A) and had been presented as a prospective GSSP for the base of the Ypresian Stage (Orue-Etxebarria et al., 1992, 1996). Despite being so close, both sections display very different facies, as the Ermua deposits consist of marls, calciturbidites, breccias, and slumps accumulated on a base-of-slope apron (Baceta, 1996; Orue-Etxebarria et al., 1992, 1996; Pujalte et al., 1998) and therefore provide Paleocene/Eocene boundary information from a different environment. According to stable isotope and biostratigraphic results from Ermua, the PETM interval is represented by 20.5 m of clays and thin-bedded turbidites bearing larger foraminifera (Bolle et al., 1998; Orue-Etxebarria et al., 1992, 1996; Schmitz et al., 2001). Interestingly, the juxtaposition of stable isotope and larger foraminiferal data showed for the first time that the marked drop in carbon isotope values that defines the onset of the PETM and the Paleocene/Eocene boundary practically coincides with the larger foraminiferal SBZ4/5 zonal boundary. The significance of this coincidence was further analyzed by Orue-Etxebarria et al. (2001) at the Campo section and concluded that, contrary to what had been thought before (Serra-Kiel et al., 1998), the basal Ilerdian larger foraminiferal turnover was genetically related to the environmental perturbations driven by the PETM (Fig. 19).

12.3 The Paleocene/Eocene Transition at Alamedilla

Most studies of the Paleocene/Eocene transition in the Betic Cordillera have focused on the succession exposed at the Valencianos ravine, 1 km south of Alamedilla (Fig. 2B; Granada province; 37°34′5″N 3°14′29″W). Pujalte et al. (2012, 2014) showed that the Paleogene succession in the Alamedilla area is incomplete, as it is affected by several unconformities of laterally variable significance, but reported more complete and expanded Paleogene successions from the Gor river area, approximately 15 km to the southeast. In Alamedilla the Maastrichtian and upper Ypresian strata correspond to reddish marls, whereas the intervening 40-m-thick upper Thanetian-lower Ypresian succession is mostly composed of gray marls accumulated at lower bathyal-upper abyssal depths. A conspicuous 2-m-thick reddish interval located approximately 10 m from the base of the gray marls corresponds to the PETM, as shown by biostratigraphic and stable isotope studies (Angori et al., 2007; Arenillas and Molina, 1996; Lu et al., 1996, 1998; Monechi et al., 2000; Pardo et al., 1999; Pujalte et al., 2012, 2014). These studies showed that the PETM record is more complete in Alamedilla than in other areas and thus contributed to improve standard planktonic foraminiferal and calcareous nannofossil biozonations of this critical interval. In addition, insightful clues into the environmental perturbance during the PETM were derived from the study of benthic foraminifera, as 37% of the species (especially k-selection estrategists with calcareous tests) underwent extinction but others, such as the agglutinated *Glomospira* and r-selection estrategists, peaked in abundance (Alegret et al., 2009b, 2010; Arreguin-Rodriguez et al., 2013, 2014). The results showed that the oceanic carbonate dissolution associated to the PETM was less severe in Alamedilla, as carbonate-poor clays only occur in a less than 30-cm-thick layer at the lower part of the PETM deposits. Taking everything into account, the Alamedilla section has been regarded as one of the best sections for the study of the PETM.

12.4 The Ypresian/Lutetian Transition at Fortuna

For some years, the Fortuna section was the leading candidate for the base-Lutetian GSSP (Luterbacher et al., 2004; Molina et al., 2006b, 2011). The Ypresian/Lutetian succession is exposed 14 km north of Fortuna on the Mu17A road cut, midway between Fuente Blanca and Peña Zafra de Abajo villages, approximately 300 m NE from the La Rauda ravine (Fig. 2B; province of Murcia, NE part of the Betic Cordillera; 38°16′12″N 1°08′49″W). The section is composed of marls interspersed with limestone and sandstone beds, which accumulated between lower upper bathyal and sublittoral water depths (Molina et al., 2006b; Ortiz and Thomas, 2006). Integrated stratigraphic studies showed that it spans the calcareous nannofossil zones CP12b–CP14a and the planktonic foraminiferal zones E7b to E9 (Molina et al., 2006b). When these studies were carried out, the criterion most commonly used to identify the base of the Lutetian was the first appearance of specimens of the planktonic foraminiferal genus *Hantkenina*, which was later suggested to be approximately simultaneous with the first appearance of the species *G. nuttalli* (E7b/E8 zonal boundary of Wade et al., 2011). However, subsequent studies initiated in Gorrondatxe demonstrated that the standard Ypresian/Lutetian biomagnetostratigraphic calibration needed amendments (Fig. 10; Payros et al., 2007a) and added that the GSSP for the base of the Lutetian Stage should not be younger than the calcareous nannofossil CP12a/b zonal boundary (Payros et al., 2009a). According to these conclusions, the stratigraphic interval studied in the Fortuna section was too young to host the GSSP for the base of the Lutetian Stage. It is currently unknown whether older strata from the Fortuna section may offer a good record of the Ypresian/Lutetian transition as defined in the Gorrondatxe GSSP (Molina et al., 2011).

12.5 The Lutetian/Bartonian Transition at Oyambre

The GSSP for the base of the Bartonian Stage has not yet been defined (Vandenberghe et al., 2012). The potential of the western Pyrenees was analyzed by Payros et al. (2014). They ruled out the outcrops near Biarritz and Pamplona (Fig. 2A) on account of incomplete successions and/or poor exposure conditions. However, they obtained positive preliminary results from the succession exposed on the eastern side of the Cape of Oyambre in San Vicente de la Barquera (Fig. 2A; Cantabria province, Basque–Cantabrian region, western Pyrenees; 43°24′13″N 4°20′24″W), which is composed of limestone–marl alternations with interbedded turbidites (Dinarès-Turell et al., 2014b). Unfortunately, more in-depth studies by Payros et al. (2015b) concluded that Chron C19n, which is the most commonly used marker for the Lutetian/Bartonian boundary (Vandenberghe et al., 2012), could not be indisputably identified in Oyambre, meaning that this section is not suitable for the GSSP for the base of the Bartonian Stage.

An unexpected finding was that the chronostratigraphic calibration of the Oyambre planktonic foraminiferal events revealed major discrepancies with the standard timescale of Wade et al. (2011) and Vandenberghe et al. (2012). The lowest occurrence of *Globigerinatheka index* was found in the upper part of Chron C20r, and not in the upper part of C20n as depicted in the standard chart, being therefore dated at 43.68 Ma and not at 42.64 Ma. The lowest occurrences of *Turborotalia pomeroli* and *Globigerinatheka euganea* were found in the upper part of calcareous nannofossil Zone CP14a (40.92 Ma), and not in the upper part of Zone CP13c (42.21 Ma). Finally, the highest occurrences of *G. nuttalli* and *Morozovelloides bandyi* were also found in the upper part of Zone CP14a (40.68 Ma), and not in the upper part of Zone CP13c (42.07 Ma). The accuracy of these anomalous results was substantiated with data from other areas, strongly suggesting that the standard Lutetian/Bartonian chronostratigraphy was in need of amendment (Payros et al., 2015b).

Other significant mismatches between the biomagnetostratigraphic results from Basque—Cantabrian sections and the standard timescale had previously been identified at the Ypresian/Lutetian transition (Fig. 10; Payros et al., 2007a, 2009b), which were later confirmed by independent studies elsewhere (Vandenberghe et al., 2012; Wade et al., 2011; and references therein). Therefore, Payros et al. (2015b) considered that the standard chronostratigraphic framework could have simply transferred the mistake previously made at the Ypresian/Lutetian calibration to the Lutetian/Bartonian transition. The Oyambre findings, if proven significant at a global scale, would have two important chronostratigraphic outcomes. First, the standard Eocene planktonic foraminiferal biostratigraphy would need revision, as there would be practically no space to accommodate zones E9 and E11 of Wade et al. (2011). Second, an array of sound biostratigraphic markers would now be defined around Chron C19n, thus facilitating correlation of a potential Bartonian GSSP.

12.6 The Bartonian/Priabonian and Eocene/Oligocene Transitions at Torre Cardela

The GSSP for the base of the Priabonian Stage has not yet been officially defined, the leading candidate being the Alano di Piave section in Italy (Vandenberghe et al., 2012). The prospective GSSP level is the base of the Tiziano volcanic crystal tuff layer, which lies immediately above the lowest occurrence of the calcareous nannofossil *Chiasmolithus oamaruensis* (NP17/18 zonal boundary) and within Chron C17n (Agnini et al., 2011).

Among the sections studied in the Iberian Peninsula, Torre Cardela (Granada province, central part of the Betic Cordillera) was found to be the most suitable on the basis of planktonic foraminiferal data (Fig. 2B; Gonzalvo and Molina, 1996; Molina, 2015). The 145-m-thick Bartonian/Priabonian succession, composed of hemipelagic marls with interbedded turbidites (Cardela Formation), was originally described as being exposed on the western trench of the N-324 road at 185 km. However, this road was later rebuilt and the original geographic reference points are therefore no longer valid. The upper Eocene—Oligocene is now exposed at approximately 72 km of road A401 (37°28′55″N 3°20′41″W). The planktonic foraminiferal studies carried out in Torre Cardela allowed identification of significant bioevents, which helped to refine biostratigraphic zonations (Gonzalvo and Molina, 1996; Molina, 2015). Furthermore, it was recognized that the extinction of middle Eocene muricate planktonic foraminifera, which had previously been regarded as being abrupt, was in fact gradual in the Torre Cardela section and lasted 2 Myr.

The planktonic foraminifera of the Torre Cardela section were also analyzed along the 110-m-thick Eocene/Oligocene transition (Gonzalvo and Molina, 1992; Molina et al., 1993). The most significant bioevents were defined, which allowed a high-resolution biozonation to be obtained.

12.7 The Eocene/Oligocene and Rupelian/Chattian Transitions at Fuente Caldera

The stratigraphic level that records the extinction of planktonic foraminifera *Hantkenina* and *Cribrohantkenina* in the Massignano section (Italy) constitutes the GSSP for the base of the Rupelian Stage (Eocene/Oligocene boundary; Premoli Silva and Jenkins, 1993). The cause of these extinctions was subject to considerable discussion, climatic and extraterrestrial factors being most commonly considered. In fact, the onset of the present icehouse climate occurred in Priabonian times and accelerated at the Eocene/Oligocene boundary (Zachos et al., 2001). On the other hand, multiple approximately coeval meteorite-impact structures were found in Popigai, Siberia, and Chesapeak, United States (Schmitz et al., 2015); the Azuara structure in Spain was initially regarded as another impact crater (Ernston and Fiebag, 1992), but this interpretation was questioned later (Aurell et al., 1993; Cortes et al., 2002; Diaz-Martinez et al., 2002).

The potential of the central sector of the Betic Cordillera for the study of the Eocene/Oligocene transition was analyzed in the middle 1980s (Molina, 1986; Molina et al., 1986, 1988; Monechi, 1986), concluding that the best succession is that exposed at the Gavilan ravine near the Fuente Caldera farmhouse (Pedro Martinez village, Granada province; 37°33′07″N 3°11′19″W), approximately 6 km southeast from Alamedilla (Fig. 2B). Additional reference sections in the area are Molino Cobo (Gobernador village, 12 km to the southwest) and Torre Cardela (see above), all of which were found to be more expanded than the stratotypic section at Massignano. The 200-m-thick Fuente Caldera section is composed of hemipelagic marls and interbedded turbidites (Cañada Formation), which accumulated on the southern slope of the Iberian continental margin at approximately 1000 m water depth (Molina et al., 2006c). Evidence for meteorite impacts was found by Robin and Molina (2006), who described Ni-rich spinel deposits most likely derived from the Popigai impact crater. A precise chronostratigraphy of the section was established using planktonic and benthic foraminifera, which allowed correlation with the Rupelian GSSP at Massignano (Alegret et al., 2008; Molina et al., 2006c). The results showed that the Ni-rich spinel deposits accumulated in mid-Priabonian times, approximately 1 Myr before the Eocene/Oligocene boundary, thus demonstrating that the meteorite impact was not related to the biotic events at the Eocene/Oligocene boundary. This suggested that the foraminiferal extinction was most likely caused by the cooling at the onset of the Oligocene, which eliminated most of the warm water and surface-dwelling foraminifera.

The Oligocene part of the Fuente Caldera section is 370-m thick. The environmental effects of glaciations (Oi1 and Oi2 events) and of the so-called Late Oligocene Warming Event were analyzed by Alegret et al. (2008) and Fenero et al. (2012). In addition, they provided planktonic foraminiferal biostratigraphic data which allowed identification of the last common occurrence of *Chiloguembelina cubensis* (O4/O5 zonal boundary of Wade et al., 2011). This event has traditionally been used as the main criterion for the definition of the Rupelian/Chattian boundary. Currently, the leading candidate to host the GSSP for the Chattian Stage is the Monte Cagnero section in Italy (Vandenberghe et al., 2012).

13. CONCLUSIONS

During the 19th century in northern Europe modern stratigraphy was first being developed. However, in the Iberian Peninsula stratigraphic work was based around more traditional geology related to mining activities. In fact, it was not until the 1960s that modern research into Iberian stratigraphy became more significant in the international arena. But this relatively late start does not detract from the stratigraphic importance of the area. For example, three Paleogene GSSPs have now been defined in Iberian sections, alongside several auxiliary stratotype sections and regional unit stratotypes.

Geographically, the Basque–Cantabrian region of the western Pyrenees is especially significant for international Paleogene stratigraphy (Fig. 2A). This area hosts the GSSPs for the bases of the Selandian and Thanetian Stages in Zumaia and the GSSP for the base of the Lutetian Stage in Gorrondatxe (Molina et al., 2011; Schmitz et al., 2011). It is worth noting that the western Pyrenees also include the GSSPs for the bases of the Santonian and Maastrichtian stages, which were defined in Tercis-les-bains and Olazagutia, respectively (Lamolda et al., 2014; Odin and Lamaurelle, 2001). Thus, five late Cretaceous–Paleogene GSSPs are located in a semicircular area with a radius of less than 75 km, yielding one of the largest concentrations of GSSPs in the world. Furthermore, Zumaia and Bidart are official auxiliary stratotype sections for the base of the Danian Stage (Cretaceous/Paleogene boundary; Molina et al., 2009). In fact, Zumaia was the runner-up in the selection of the GSSPs for the Danian and Ypresian stages. In addition, Otsakar has been appointed auxiliary section for the GSSP for the base of the Lutetian Stage (Molina et al., 2012). Finally, the Biarritz section was the unit stratotype for the now disused Biarritzian Stage (Hottinger and Schaub, 1960). Another reference area is the NE part of the Betic Cordillera (Fig. 2B), which hosts auxiliary stratotype sections for the base of the Danian Stage in Caravaca (Molina et al., 2009) and for the base of the Lutetian Stage in Agost (Molina et al., 2012). In addition, the Alamedilla section is a global reference for the Paleocene/Eocene boundary.

It follows from the above that to date most of the internationally relevant contributions to Paleogene stratigraphy have been limited to the Paleocene–middle Eocene interval of Iberian sections. The reason for this stratigraphic constraint might partly be that upper Eocene and Oligocene marine successions are scarcer in the Pyrenean area, the region which has been most thoroughly studied. However, this means that there are still promising opportunities for future research on Paleogene stratigraphy in the Iberian successions. More specifically, expanded upper Eocene–Oligocene marine successions are well exposed but relatively poorly known in the Biarritz and San Vicente de la Barquera areas, in the western Pyrenees (Fig. 2A). More importantly, deposits of that age are widespread in the Betic Cordillera (Fig. 2B) and include facies from a wide range of depositional environments (Braga et al., 2002; Vera, 2004). The integrated stratigraphic studies carried out in the Torre Cardela and Fuente Caldera sections (central part of the Betic Cordillera) have already contributed to improve the late Eocene–Oligocene chronostratigraphy, but many other sections have not yet been examined, meaning that the area has potential to produce further improvements.

The Iberian Paleogene successions also have great potential for continental chronostratigraphic refinements. Significant work has been carried out (e.g., Barbera et al., 2001; Beamud et al., 2003; Feist et al., 1994; Lévêque, 1993), but there is room for future improvement. In this regard, the extensive Paleogene terrestrial successions of the Iberian interior basins may still have stratigraphic treasures to discover (Fig. 1A). It is therefore hoped that the last five decades of fruitful research in Iberian Paleogene stratigraphy can be built on.

ACKNOWLEDGMENTS

This work is a contribution to research projects CGL2015-65404-R (MINECO/FEDER, Spanish Government, with support by the ERDF/FEDER fund of the European Union) and IT-930-16 (Basque Government). This review was intended as a tribute to the researchers who have contributed to our current knowledge of the Iberian Paleogene stratigraphy. Thanks are due to all of them, while apologies are owed to those who, unintentionally, were not mentioned. We are grateful to the following publishers for permission to use copyrighted material: Aranzadi Zientzia Elkartea, Cambridge University Press, Echweiserbart, Elsevier, Geologica Acta, Geological Society of America, Geological Society of London, Instituto Geologico y Minero de España, Universidad Complutense de Madrid, Wiley. Thanks are due to Mary A. Zimmerman and Helene Kabes (Elsevier) for their invitation to contribute to this volume. Carl Sheaver kindly helped with the English language.

REFERENCES

Abrajevitch, A., Font, E., Florindo, F., Roberts, A.P., 2015. Asteroid impact vs. Deccan eruptions: the origin of low magnetic susceptibility beds below the Cretaceous-Paleogene boundary revisited. Earth Planet. Sci. Lett. 430, 209–223.

Adan de Yarza, R., 1884. Descripción física y geológica de la provincia de Guipúzcoa. Mem. Com. Mapa Geol. España 12, 1–175.

Agnini, C., Fornaciari, E., Giusberti, L., Grandesso, P., Lanci, L., Luciani, V., Muttoni, G., Palike, H., Rio, D., Spofforth, D.J.A., Stefani, C., 2011. Integrated bio-magnetostratigraphy of the Alano section (NE Italy): a proposal for defining the middle-late Eocene boundary. Geol. Soc. Amer. Bull. 123, 841–872.

Alegret, L., 2007. Recovery of the deep-sea floor after the Cretaceous-Paleogene boundary event: the benthic foraminiferal record in the Basque-Cantabrian basin and in South-eastern Spain. Palaeogeogr. Palaeoclimatol. Palaeoecol. 255, 181–194.

Alegret, L., Cruz, L.E., Fenero, R., Molina, E., Ortiz, S., Thomas, E., 2008. Effects of the Oligocene climatic events on the foraminiferal record from Fuente Caldera section (Spain, western Tethys). Palaeogeogr. Palaeoclimatol. Palaeoecol. 269, 94–102.

Alegret, L., Kaminski, M.A., Molina, E., 2004. Paleoenvironmental recovery after the Cretaceous/Paleogene boundary crisis: evidence from the marine Bidart section (SW France). Palaios 19, 574–586.
Alegret, L., Molina, E., Thomas, E., 2003. Benthic foraminiferal turnover across the Cretaceous/Paleogene boundary at Agost (southeastern Spain): paleoenvironmental inferences. Mar. Micropaleontol. 48, 251–279.
Alegret, L., Ortiz, S., 2010. El corte de Zumaya (España): registro de los foraminíferos bentónicos del Paleógeno inferior. Rev. Mex. Cienc. Geol. 27, 477–489.
Alegret, L., Ortiz, S., Molina, E., 2009b. Extinction and recovery of benthic foraminifera across the Paleocene–Eocene thermal maximum at the Alamedilla section (southern Spain). Palaeogeogr. Palaeoclimatol. Palaeoecol. 279, 186–200.
Alegret, L., Ortiz, S., Orue-Etxebarria, X., Bernaola, G., Baceta, J.I., Monechi, S., Apellaniz, E., Pujalte, V., 2009a. The Paleocene–Eocene thermal maximum: new data on microfossil turnover at the Zumaia section, Spain. Palaios 24, 318–328.
Alegret, L., Ortiz, S., Arenillas, I., Molina, E., 2010. What happens when the ocean is overheated? The foraminiferal response across the Paleocene–Eocene Thermal Maximum at the Alamedilla section (Spain). Geol. Soc. Amer. Bull. 122, 1616–1624.
Alegret, L., Rodriguez-Tovar, F., Uchman, A., 2015. How bioturbation obscured the Cretaceous-Palaeogene boundary record. Terra Nova 27, 225–230.
Alegret, L., Thomas, E., 2005. Cretaceous/Paleogene boundary bathyal paleo-environments in the central north pacific (DSDP site 465), the northwestern Atlantic (ODP site 1049), the gulf of Mexico and the Tethys: the benthic foraminiferal record. Palaeogeog. Palaeoclimatol. Palaeoecol. 224, 53–82.
Alonso-Gavilan, G., Armenteros, I., Carballeira, J., Corrochano, A., Huerta, P., Rodriguez, J.M., 2004. Cuenca del Duero. In: Vera, J.A. (Ed.), Geología de España. Soc. Geol. España & Inst. Geol. Min. España, Madrid, pp. 550–556.
Alonso-Zarza, A.M., Calvo, J.P., 2002. Tertiary: Tajo basin. In: Gibbons, W., Moreno, T. (Eds.), The Geology of Spain. The Geological Society, London, pp. 315–320.
Alvarez, L.W., Alvarez, W., Asaro, F., Michel, H.V., 1980. Extraterrestrial cause for the Cretaceous-Tertiary extinction. Science 208, 1095–1108.
Alvarez, W., Alvarez, L.W., Asaro, F., Michel, H.V., 1982. Current status of the impact theory for the terminal Cretaceous extinction. Geol. Soc. Amer. Spec. Pap. 190, 305–315.
Anadon, P., Alcala, L., Alonso-Zarza, A.M., Calvo, J.P., Orti, F., Rosell, L., Sanz-Rubio, E., 2004. Cuencas Cenozoicas de las Cordilleras Ibérica, Costero Catalana y Pirenaica. In: Vera, J.A. (Ed.), Geología de España. Soc. Geol. España & Inst. Geol. Min. España, Madrid, pp. 562–576.
Angori, E., Bernaola, G., Monechi, S., 2007. Calcareous nannofossil assemblages and their response to the Paleocene-Eocene Thermal Maximum event at different latitudes: ODP Site 690 and Tethyan section. Geol. Soc. Amer. Spec. Pap. 424, 69–85.
Angori, E., Monechi, S., 1996. High-resolution calcareous nannofossil biostratigraphy across the Paleocene/Eocene boundary at Caravaca (southern Spain). Isr. J. Earth Sci. 44, 197–206.
Apellaniz, E., Baceta, J.I., Bernaola, G., Núñez-Betelu, K., Orue-Etxebarria, X., Payros, A., Pujalte, V., Robin, E., Rocchia, R., 1997. Analysis of uppermost Cretaceous-lowermost Tertiary hemipelagic successions in the Basque country (Western Pyrenees): evidence of a sudden extinction of more than half planktic foraminifer species at the K/T boundary. Bull. Soc. Geol. Fr. 168, 783–793.
Apellaniz, E., Orue-Etxebarria, X., Luterbacher, H., 2002. Evolution of the Early Paleocene planktonic foraminifera: a Basque point of view. Neues Jahrb. Geol. Paläontol. Abh. 225, 157–194.
Archiac, A.d', 1837. Mémoire sur la formation crétacée du Sud-Ouest de la France. Mem. Soc. Geol. Fr. 2, 157–192.

Archiac, A.d', 1846. Description des fossils du groupe nummulitique recueileeis par M. Thorent dans les couches à nummulines des environs de Bayonne. Mem. Soc. Geol. Fr. 2, 189–217.

Archiac, A.d', 1850. Description des fossils du groupe nummulitique recueileeis par M. S.P. Pratt et M.J. Delbos aux environs de Bayonne et Dax. Mem. Soc. Geol. Fr. 3, 397–456.

Arenillas, I., 2012. Patterns of spatio-temporal distribution as criteria for the separation of planktic foraminiferal species across the Danian-Selandian transition in Spain. Acta Paleontol. Pol. 57, 401–422.

Arenillas, I., Arz, J.A., Grajales-Nishimura, J.M., Murillo-Muñeton, G., Alvarez, W., Camargo-Zanoguera, A., Molina, E., Rosales-Dominguez, C., 2006. Chicxulub impact event is Cretaceous/Paleogene boundary in age: new micropaleontological evidence. Earth Planet. Sci. Lett. 249, 241–257.

Arenillas, I., Arz, J.A., Molina, E., 1998. El límite Cretácico-Terciario en Zumaia, Osinaga y Musquiz (Pirineos): control bioestratigráfico y cuantitativo de hiatos con foraminíferos planctónicos. Rev. Soc. Geol. España 11, 127–138.

Arenillas, I., Arz, J.A., Molina, E., 2004. A new high-resolution planktic foraminiferal zonation and subzonation for the lower Danian. Lethaia 37, 79–95.

Arenillas, I., Molina, E., 1996. Bioestratigrafía y evolución de las asociaciones de foraminíferos planctónicos del tránsito Paleoceno–Eoceno en Alamedilla (Cordilleras Béticas). Rev. Esp. Micropaleontol. 28, 75–96.

Arenillas, I., Molina, E., 2000. Reconstrucción paleoambiental con foraminíferos planctónicos y cronoestratigrafía del tránsito Paleoceno-Eoceno de Zumaya (Guipúzcoa). Rev. Esp. Micropaleontol. 32, 283–300.

Arenillas, I., Molina, E., Ortiz, S., Schmitz, B., 2008. Foraminiferal and $\delta^{13}C$ isotopic event stratigraphy across the Danian–Selandian transition at Zumaya (northern Spain): chronostratigraphic implications. Terra Nova 20, 38–44.

Arinobu, T., Ishiwatari, R., Kaiho, K., Lamolda, M.A., Seno, H., 2005. Abrupt and massive influx of terrestrial biomarkers into the marine environment at the Cretaceous-Tertiary boundary, Caravaca, Spain. Palaeogeogr. Palaeoclimatol. Palaeoecol. 224, 108–116.

Arinobu, T., Ishiwatari, R., Kaiho, K., Lamolda, M.A., 1999. Spike of pyrosynthetic polycyclic aromatic hydrocarbons associated with an abrupt decrease in delta C-13 of a terrestrial biomarker at the Cretaceous-Tertiary boundary at Caravaca, Spain. Geology 27, 723–726.

Armenteros, I., Alonso-Gavilan, G., Carballeira, J., Corrochano, A., Huerta, P., Rodriguez, J.M., 2004. Cuencas Cenoizicas del Macizo Iberico. In: Vera, J.A. (Ed.), Geologia de España. Soc. Geol. España & Inst. Geol. Min. España, Madrid, pp. 581–584.

Armenteros, I., Corrochano, A., Alonso-Gavilan, G., Carballeira, J., Rodriguez, J.M., 2002. Tertiary: Duero basin (northern Spain). In: Gibbons, W., Moreno, T. (Eds.), The Geology of Spain. The Geological Society, London, pp. 309–315.

Arreguín-Rodríguez, G.J., Alegret, L., 2015. Experimentos de disolución de $CaCO_3$ en foraminíferos bentónicos aglutinados del Paleoceno-Eoceno. Est. Geol. 71, e023.

Arreguin-Rodriguez, G., Alegret, L., Ortiz, S., 2013. *Glomospira* acme during the Paleocene–Eocene thermal maximum: response to $CaCO_3$ dissolution or to ecological forces? J. Foram. Res. 43, 37–49.

Arreguin-Rodriguez, G., Alegret, L., Sepulveda, J., Newman, S., Summons, R.E., 2014. Enhanced terrestrial input supporting the *Glomospira* acme across the Paleocene-Eocene boundary in Southern Spain. Micropaleontology 60, 43–51.

Arz, J.A., Arenillas, I., Molina, E., Sepulveda, R., 2000. Planktonic foraminiferal stability in the Upper Maastrichtian and the catastrophic mass extinction at the Cretaceous-Tertiary (K/T) boundary at Caravaca (Spain). Rev. Geol. Chile 27, 27–47.

Arz, J.A., Arenillas, I., Molina, E., 1999. Extinción de foraminíferos planctónicos en el tránsito Cretácico-Terciario de Zumaya (Guipúzcoa): ¿supervivencia o reelaboración? Rev. Española Micropaleontol. 31, 297–304.
Aubry, M.P., 1986. Paleogene calcareous nannoplankton biostratigraphy of Northwestern Europe. Palaeogeogr. Palaeoclimatol. Palaeoecol. 55, 267–334.
Aubry, M.P., 1995. From chronology to stratigraphy: interpreting the Lower and Middle Eocene stratigraphic record in the Atlantic ocean. SEPM Spec. Publ. 54, 213–274.
Aubry, M.P., 1998. Early Paleogene calcareous nannoplankton evolution: a tale of climatic deterioration. In: Aubry, M.P., Lucas, S., Berggren, W.A. (Eds.), Late Paleocene-early Eocene Climatic and Biotic Events in the Marine and Terrestrial Record. Columbia Univ. Press, New York, pp. 158–203.
Aubry, M.P., Berggren, W.A., Kent, D.V., Flynn, J.J., Klitgord, K., Obradovich, J.D., Prothero, D.R., 1988. Paleogene geochronology: an integrated approach. Paleoceanography 3, 707–742.
Aubry, M.P., Ouda, K., Dupuis, C., Berggren, W.A., Van Couvering, J.A., the members of the Working Group on the Paleocene/Eocene Boundary, 2007. The global standard stratotype-section and point (GSSP) for the base of the Eocene series in the Dababiya section (Egypt). Episodes 30, 271–286.
Aubry, M.P., Thiry, M., Dupuis, C., Berggren, W.A., 2005. The Sparnacian deposits of the Paris Basin: a lithostratigraphic classification. Stratigraphy 2, 65–100.
Aurell, M., Gonzalez, A., Perez, A., Guimera, J., Casas, A., Salas, R., 1993. Discussion of 'The Azuara impact structure (Spain): new insights from geophysical and geological investigations' by K. Ernston and J. Fiebag. Geol. Rundsch. 82, 750–755.
Ayala-Carcedo, F.J., Perejón, A., Jordá, L., Puche, O., 2005. The XIV international geological congress of 1926 in Spain. Episodes 28, 42–47.
Azpeitia, F., 1933. Datos sobre el studio paleontológico del Fysch de la costa cantábrica y de algunos otros puntos de España. Bol. Inst. Geol. Min. España 53, 1–65.
Baceta, J.I., 1996. El Maastrichtiense superior, Paleoceno e Ilerdiense inferior de la Región Vasco-Cantábrica: secuencias deposicionales, facies y evolución paleogeográfica (Ph.D. thesis). Univ. Basque Country, Leioa.
Baceta, J.I., Bernaola, G., Orue-Etxebarria, X., Apellaniz, E., Pujalte, V., Caballero, F., 2005. Contribuciones del Pirineo Vasco al conocimiento de la crisis biológica del límite Cretácico-Terciario. Munibe 57, 55–66.
Baceta, J.I., Orue-Etxebarria, X., Apellaniz, E., Martin-Rubio, M., Bernaola, G., 2012. El flysch del litoral Deba-Zumaia: una "ventana" a los secretos de nuestro pasado geologic. Univ. Basque Country, Leioa.
Baceta, J.I., Pujalte, V., Dinarès-Turell, J., Payros, A., Orue-Etxebarria, X., Bernaola, G., 2000. The Paleocene-Eocene boundary interval in the Zumaia section (Gipuzkoa, Basque basin): magnetostratigraphy and high-resolution lithostratigraphy. Rev. Soc. Geol. España 13, 375–391.
Baceta, J.I., Pujalte, V., Serra-Kiel, J., Robador, A., Orue-Etxebarria, X., 2004. El Maastrichtiense final, Paleoceno e Ilerdiense inferior de la Cordillera Pirenaica. In: Vera, J.A. (Ed.), Geología de España. Soc. Geol. España & Inst. Geol. Min. España, Madrid, pp. 308–313.
Banda, E., 1996. Deep crustal expression of Tertiary basins in Spain. In: Friend, P.F., Dabrio, C.J. (Eds.), Tertiary Basins of Spain: The Stratigraphic Record of Crustal Kinematics. Cambridge Univ. Press, Cambridge, pp. 15–19.
Barbera, X., Cabrera, L., Marzo, M., Pares, J.M., Agusti, J., 2001. A complete terrestrial Oligocene magnetobiostratigraphy from the Ebro basin, Spain. Earth Planet. Sci. Lett. 187, 1–16.

Barnolas, A., Payros, A., Sanso, J.M., serra-Kiel, J., Tosquella, J., 2004. La Cuenca Surpireniaca desde el Ilerdiense medio al Priaboniense. In: Vera, J.A. (Ed.), Geología de España. Soc. Geol. España & Inst. Geol. Min. España, Madrid, pp. 313–320.
Batenburg, S.J., Gale, A.S., Sprovieri, M., Hilgen, F.J., Thibault, N., Boussaha, M., Orue-Etxebarria, X., 2014. An astronomical time scale for the Maastrichtian based on the Zumaia and Sopelana sections (Basque country, northern Spain). J. Geol. Soc. Lond. 171, 165–180.
Baternburg, S.J., Sprovieri, M., Gale, A.S., Hilgen, F.J., Hüsing, S., Laskar, J., Liebrand, D., Lirer, F., Orue-Etxebarria, X., Pelosi, N., Smit, J., 2012. Cyclostratigraphy and astronomical tuning of the late Maastrichtian at Zumaia (Basque country, northern Spain). Earth Planet. Sci. Lett. 359–360, 264–278.
Beamud, E., Carces, M., Cabrera, L., Muñoz, J.A., Almar, Y., 2003. A new middle to late Eocene continental chronostratigraphy from NE Spain. Earth Planet. Sci. Lett. 216, 501–514.
Bentham, P., Burbank, D.W., 1996. Chronology of Eocene foreland basin evolution along the western oblique margin of the South-Central Pyrenees. In: Friend, P.F., Dabrio, C. (Eds.), Tertiary Basins of Spain: The Stratigraphic Record of Crustal Kinematics. Cambridge Univ. Press, Cambridge, pp. 144–152.
Berggren, W.A., 1969. Cenozoic chronostratigraphy, planktonic foraminiferal zonation and the radiometric time scale. Nature 224, 1072–1075.
Berggren, W.A., 1971. Tertiary boundaries and correlations. In: Funnell, B.M., Riedel, W.R. (Eds.), The Micropaleontology of Oceans. Cambridge Univ. Press, Cambridge, pp. 693–809.
Berggren, W.A., 1972. A Cenozoic time-scale: some implications for regional geology and paleobiogeography. Lethaia 5, 195–215.
Berggren, W.A., 1998. The Cenozoic Era: Lyellian (chrono)stratigraphy and nomenclatural reform at the millennium. Geol. Soc. Lond. Spec. Publ. 143, 111–132.
Berggren, W.A., Kent, D.V., Flynn, J.J., 1985. Jurassic to Paleogene: Part 2-Paleogene geochronology and chronostratigraphy. Geol. Soc. Lond. Mem. 10, 141–195.
Berggren, W.A., Kent, D.V., Swisher III, C.C., Aubry, M.P., 1995. A revised Cenozoic geochronology and chronostratigraphy. SEPM Spec. Publ. 54, 129–212.
Berggren, W.A., Miller, K.G., 1988. Paleogene tropical planktonic foraminiferal biostratigraphy and magnetobiochronology. Micropaleontology 34, 362–380.
Berggren, W.A., Pearson, P.N., 2005. A revised tropical to subtropical Paleogene planktonic foraminiferal zonation. J. Foram. Res. 35, 279–298.
Bernaola, G., 2002. Los nannofósiles calcáreos del Paleoceno en el dominio Pirenaico: bioestratigrafía, cronoestratigrafía y paleoecología (Ph.D. thesis). Univ. Basque Country, Leioa.
Bernaola, G., Baceta, J.I., Orue-Etxebarria, X., Alegret, L., Martín-Rubio, M., Arostegui, J., Dinarès-Turell, J., 2007. Evidence of an abrupt environmental disruption during the mid-Paleocene biotic event (Zumaia section, western Pyrenees). Geol. Soc. Amer. Bull. 119, 785–795.
Bernaola, G., Baceta, J.I., Payros, A., Orue-Etxebarria, X., Apellaniz, E., 2006c. The Paleocene and Lower Eocene of the Zumaia Section (Basque Basin). Climate and Biota of the Early Paleogene, Post-conference Fieldtrip Guidebook, Bilbao.
Bernaola, G., Martín-Rubio, M., Baceta, J.I., 2009. New high resolution calcareous nannofossil analysis across the Danian/Selandian transition at the Zumaia section: comparison with South Tethys and Danish sections. Geol. Acta 7, 79–92.
Bernaola, G., Nuño-Arana, Y., Payros, A., 2006b. Análisis bioestratigráfico del Eoceno Inferior de la sección de Barinatxe (Pirineos occidentales) mediante nanofósiles calcáreos. Geogaceta 40, 175–178.
Bernaola, G., Orue-Etxebarria, X., Payros, A., Dinarès-Turell, J., Tosquella, J., Apellaniz, E., Caballero, F., 2006a. Biomagnetostratigraphic analysis of the Gorrondatxe section

(Basque country, western Pyrenees): its significance for the definition of the Ypresian/Lutetian boundary stratotype. Neues Jahr. Geol. Paläontol. Abh. 241, 67–109.
Bignot, G., 1975. Extraits de la discussion qui suivit la presentation des communications sur le contenu de l'Ilerdien et sa place dans le Paléogène. Bull. Soc. Geol. Fr. 7, 218–221.
Bignot, G., 1981. Cuisian. Bull. Inf. Geol. Bass. Paris Mem. 2, 63–75.
Birot, P., 1934. Sur les poudingues Nummulitiques dans leurs rapports avec les mascifs calcaires Sud-Pyrénéens, entre la Segre et la Llobregat. C. R. Somm. Soc. Geol. Fr. 4, 11–13.
Bohor, B.F., Foord, E.E., Ganapathy, R., 1986. Magnesioferrite from the Cretaceous-Tertiary boundary, Caravaca, Spain. Earth Planet. Sci. Lett. 81, 57–66.
Bolle, M.P., Adatte, T., Keller, G., von Salis, K., Hunziker, J., 1998. Biostratigraphy, mineralogy and geochemistry of the Trabakua Pass and Ermua sections in Spain. Eclog. Geol. Helv. 91, 1–25.
Bonte, P., Delacotte, O., Renard, M., Laj, C., Boclet, D., Jehanno, C., Rocchia, R., 1984. An iridium rich layer at the Cretaceous/Tertiary boundary in the Bidart section (southern France). Geophys. Res. Lett. 11, 473–476.
Bouille, R., 1876. Paléontologie de Biarritz et de quelques autres localités des Basses-Pyrénées. Imprimerie Lithogr. Veronese. Pau.
Boussac, J., 1906. Sur le terrain nummulitique á Biarritz et dans le Vicentin. Bull. Soc. Geol. Fr. 2, 555–560.
Boussac, J., 1908. Not esur la succession des faunes nummulitiques á Biarritz. Bull. Soc. Geol. Fr. 4, 237–255.
Boussac, J., 1911. Etudes stratigraphiques et paléontologiques sur le Nummulitique de Biarritz. Ann. Hebert Paris 5, 1–96.
Braga, J.C., Martin, J.M., Aguirre, J., 2002. Tertiary: southern Spain. In: Gibbons, W., Moreno, T. (Eds.), The Geology of Spain. The Geological Society, London, pp. 320–327.
Brinkhuis, H., Bujak, J.P., Smit, J., Versteegh, G.J.M., Visscher, H., 1998. Dinoflagellate-based sea surface temperature reconstructions across the Cretaceous-Tertiary boundary. Palaeogeogr. Palaeoclimatol. Palaeoecol. 141, 67–83.
Burnett, J.A., Kennedy, W.J., Ward, P.D., 1992. Maastrichtian nannofossil biostratigraphy in the Biscay region (south-western France, northern Spain). Newslett. Strat. 26, 145–155.
Bybell, L.M., Self-Trail, J.M., 1995. Evolutionary, biostratigraphic, and taxonomic study of calcareous nannofossils from a continuous Paleocene/Eocene boundary section in New Jersey. U.S. Geol. Surv. Prof. Pap. 1554, 1–36.
Caballero, F., 2007. Análisis micropaleontológico de los límites Cretácico/Terciario, Daniense/Selandiense y Paleoceno/Eoceno en la Cuenca Vasco-Cantábrica, a través de foraminíferos planctónicos (Ph.D. thesis). Univ. Basque Country, Leioa.
Canudo, J.I., Keller, G., Molina, E., 1991. Cretaceous-Tertiary boundary extinction pattern and faunal turnover at Agost and Caravaca, SE Spain. Mar. Micropaleontol. 17, 319–341.
Canudo, J.I., Keller, G., Molina, E., Ortiz, N., 1995. Planktic foraminiferal turnover and $\delta^{13}C$ isotopes across the Paleocene-Eocene transition at Caravaca and Zumaya, Spain. Palaeogeogr. Palaeoclimatol. Palaeocol. 114, 75–100.
Canudo, J.I., Oms, O., Vila, B., Galobart, A., Fondevilla, V., Puertolas-Pascual, E., Selles, A.G., Cruzado-Caballero, P., Dinarès-Turell, J., Vicens, E., Castanera, D., Company, J., Burrel, L., Estrada, R., marmi, J., Blanco, A., 2016. The upper Maastrichtian dinosaur fossil record from the southern Pyrenees and its contribution to the topic of the Cretaceous-Palaeogene mass extinction event. Cretac. Res. 57, 540–551.
Carez, L., 1881. Études des terrains crétacés et tertiaires du nord de l'Espagne (Ph.D. thesis). Fac. Sci. Paris, Paris.
Cascella, A., Dinarès-Turell, J., 2009. Integrated calcareous nannofossil biostratigraphy and magnetostratigraphy from the uppermost marine Eocene deposits of the southeastern

Pyrenean foreland basin: evidences for marine Priabonian deposition. Geol. Acta 7, 281–296.
Cavelier, C., Pomerol, C., 1986. Stratigraphy of the Paleogene. Bull. Soc. Geol. Fr. 8, 255–265.
Ciry, R., 1940. Etude géologique d'une partie des provinces de Burgos, Palencia, León et Santander. Bull. Soc. Hist. Natur. 74, 1–519. Toulouse.
Clyde, W.C., Gingerich, P.D., 1998. Mammalian community response to the latest Paleocene thermal maximum: an isotaphonomic study from the northern Bighorn basin, Wyoming. Geology 26, 1011–1014.
Coccioni, R., Bancala, C., 2012. New insights into the pattern, timing, and duration of the evolutionary origin of the foraminiferal genus *Hantkenina*. Rev. Micropaleontol. 55, 71–81.
Coccioni, R., Galeotti, S., 1994. K-T boundary extinction: geologically instantaneous or gradual event? Evidence from deep-sea benthic foraminífera. Geology 22, 779–782.
Coccioni, R., Lucian, V., 2005. Guembelitria irregularis bloom at the K-T boundary: morphological abnormalities induced by impact-related extreme environmental stress? In: Cockell, C., Koeberl, C., Gilmour, I. (Eds.), Biological Processes Associated With Impact Events. Springer, Berlin, pp. 179–196.
Cortes, A.L., Diaz-Martinez, E., Sanz-Rubio, E., Martinez-Frias, J., Fernandez, C., 2002. Cosmic impact versus terrestrial origin of the Azuara structure (Spain): a review. Meteorit. Planet. Sci. 37, 875–894.
Costa, E., Garces, M., Lopez-Blanco, M., Beamud, E., Gomez-Paccard, M., Larrasoaña, J.C., 2010. Closing and continentalization of the South Pyrenean foreland basin (NE Spain): magnetochronological constraints. Basin Res. 22, 904–917.
Costa, E., Garces, M., Lopez-Blanco, M., Serra-Kiel, J., Bernaola, G., Cabrera, L., Beamud, E., 2013. The Bartonian-Priabonian marine record of the eastern South Pyrenean foreland basin (NE Spain): a new calibration of the larger foraminifers and calcareous nannofossil zonation. Geol. Acta 11, 177–193.
Cowie, J.W., 1986. Guidelines for boundary stratotypes. Episodes 9, 78–82.
De Paolo, et al., 1983. Rb-Sr, Sm-Nd, K-Ca, O, and H isotopic study of Cretaceous-Tertiary boundary sediments, Caravaca, Spain: evidence for an oceanic impact site. Earth Planet. Sci. Lett. 64, 356–373.
Delacotte, O., Renard, M., Laj, C., Perch-Nielsen, K., Premoli-Silva, I., Clauser, S., 1985. Magnèstostratigraphie et biostratigraphie du passage Crétacé-Tertiaire de la coupe de Bidart (Pyrénées Atlantiques). Geol. Fr. 3, 243–254.
Diaz-Martinez, E., Sanz-Rubio, E., Martinez-Frias, J., 2002. Sedimentary record of impact events in Spain. Geol. Soc. Amer. Spec. Pap. 356, 551–562.
Dinarès-Turell, J., Baceta, J.I., Bernaola, G., Orue-Etxebarria, X., Pujalte, V., 2007. Closing the mid-Paleocene gap: toward a complete astronomically tuned Paleocene Epoch and Selandian and Thanetian GSSPs at Zumaia (Basque basin, W Pyrenees). Earth Planet. Sci. Lett. 262, 450–467.
Dinarès-Turell, J., Baceta, J.I., Pujalte, V., Orue-Etxebarria, X., Bernaola, G., 2002. Magnetostratigraphic and cyclostratigraphic calibration of a prospective Paleocene-Eocene stratotype at Zumaia (Basque basin, Northern Spain). Terra Nova 34, 279–290.
Dinarès-Turell, J., Baceta, J.I., Pujalte, V., Orue-Etxebarria, X., Bernaola, G., Lorito, S., 2003. Untangling the Palaeocene climatic rhythm: an astronomically calibrated Early Palaeocene magnetostratigraphy and biostratigraphy at Zumaia (Basque basin, northern Spain). Earth Planet. Sci. Lett. 216, 483–500.
Dinarès-Turell, J., Payros, A., Monechi, S., Orue-Etxebarria, X., Ortiz, S., Apellaniz, E., Bernaola, G., 2014b. In search of the Bartonian (middle Eocene) GSSP (II): preliminary

results from the Oyambre section (N Spain). In: Rocha, R., Pais, J., Kullberg, J.C., Finney, S. (Eds.), Strati 2013. Springer, Heidelberg, pp. 79–83.

Dinarès-Turell, J., Pujalte, V., Stoykova, K., Baceta, J.I., Ivanov, M., 2012. The Paleocene "top chron C27n" transient greenhouse episode: evidences from marine pelagic Atlantic and peri-Tethyan sections. Terra Nova 24, 477–486.

Dinarès-Turell, J., Pujalte, V., Stoykova, K., Elorza, J., 2013. Detailed correlation and astronomical forcing within the upper Maastrichtian succession in the Basque Basin. Bol. Geol. Min. 124, 253–282.

Dinarès-Turell, J., Stoykova, K., Baceta, J.I., Ivanov, M., Pujalte, V., 2010. High-resolution intra- and interbasinal correlation of the Danian-Selandian transition (early Paleocene): the Bjala section (Bulgaria) and the Selandian GSSP at Zumaia (Spain). Palaeogeogr. Palaeoclimatol. Palaeoecol. 297, 511–533.

Dinarès-Turell, J., Westerhold, T., Pujalte, V., Röhl, U., Kroon, D., 2014a. Astronomical calibration of the Danian stage (early Paleocene) revisited: settling chronologies of sedimentary records across the Atlantic and Pacific Oceans. Earth Planet. Sci. Lett. 405, 119–131.

Dollfus, G.F., 1880. Essai sur l'extension des terrains tertiaires dans le basin anglo-parisien. Bull. Soc. Geol. Normandie 6, 584–605.

Doncieux, L., 1933. Notes paléontologiques sur quelques Nummulites de l'Espagne. Bull. Inst. Cat. Hist. Nat. 33, 175–182.

Douville, H., 1903. Sur le terrain nummulitique à Biarritz et dans les Alpes. Bull. Soc. Geol. Fr. 2, 149–154.

Douville, H., 1905. Sur le terrain nummulitique du Bassin de l'Adour. Bull. Soc. Geol. Fr. 4, 9–54.

Douville, H., 1922. Le Nummulitique au sud des Pyrénées. C. R. Ac. S. C. 174, 507–510.

Drobne, K., Jez, J., Cosovic, V., Ogorelec, B., Stenni, B., Zakrevskaya, E., Hottinger, L., 2014. Identification of the Palaeocene-Eocene boundary based on larger foraminifers in deposits of the Palaeogene Adriatic carbonate platform, southwestern Slovenia. In: Rocha, R., Pais, J., Kullberg, J.C., Finney, S. (Eds.), Strati 2013. Springer, Heidelberg, pp. 83–93.

Dumont, A., 1850. Rapport sur la carte geologique du Royaume. Bull. Acad. Roy. Sci. Lett. B. Arts. Belg. 16, 351–371.

Ellwood, B.B., Kafafy, A.Z., Kassab, A., Tomkin, J.H., Abdeldayem, A., Obaidalla, N., Randall, K.W., Thompson, D.E., 2010. Magnetostratigraphy susceptibility used for high-resolution correlation among Paleocene-Eocene boundary sequences in Egypt, Spain, and the U.S.A. SEPM Spec. Publ. 94, 167–179.

Ellwood, B.B., Tomkin, J.H., Febo, L.A., Stuart Jr., C.N., 2008. Time series analysis of magnetic susceptibility variations in deep marine sedimentary rocks: a test using the upper Danian–Lower Selandian proposed GSSP, Spain. Palaeogeogr. Palaeoclim. Palaeocol. 261, 270–279.

Ernston, K., Fiebag, J., 1992. The Azuara impact structure (Spain): new insights from geophysical and geological investigations. Geol. Rundsch. 81, 403–427.

Feist, M., Anadon, P., Cabrera, L., Choi, S.J., Colombo, F., Saez, A., 1994. Upper Eocene-lowermost Miocene charophyte succession in the Ebro basin (Spain). Contribution to the charophyte biozonation in western Europe. Newslett. Strat. 30, 1–32.

Fenero, R., Thomas, E., Alegret, L., Molina, E., 2012. Oligocene benthic foraminifera from the Fuente Caldera section (Spain, western Tethys): taxonomy and paleoenvironmental inferences. J. Foram. Res. 42, 286–304.

Fernandez-Marron, M.T., Lopez-Martinez, N., Fonolla-Ocete, J.F., Valle-Hernandez, M.F., 2004. The palynological record across the Cretaceouse-Tertiary

boundary in two different palaeogeographic settings from the Southern Pyrenees (Spain). Geol. Soc. Lond. Spec. Pub 230, 243–255.
Ferrer, J., Le Calvez, Y., Luterbacher, H., Premoli Silva, I., 1973. Contribution à l'étude des foraminifères iledіens de la region de Tremp (Catalogne). Mem. Mus. Nation. Hist. Natur. Sci. Terre 29, 1–80.
Fondecave-Wallez, M.J., Peybernès, B., Eichène, P., 1995. Mise en éidence micropaléontologique d'un flysch ypréso-lutétien, ex-crétacé, à materiel triasique resédimenté, dans la coupe de Bidart-Nord (Pyrénées Atalntiques, sud-ouest de la France). Geol. Fr. 3, 47–56.
Font, E., Adatte, T., Sial, A.N., Lacerda, L.D., de, Keller, G., Punekar, J., 2016. Mercury anomaly, Deccan volcanism, and the end-Cretaceous mass extinction. Geology 44, 171–174.
Font, E., Nedelec, A., Ellwood, B., Mirao, J., Silva, P.F., 2011. A new sedimentary benchmark for the Deccan Traps volcanism? Geophys. Res. Lett. 38, L24309.
Galbrun, B., Gardin, S., 2004. New chronostratigraphy of the Cretaceous-Paleogene boundary interval at Bidart (France). Earth Planet. Sci. Lett. 224, 19–32.
Gallala, N., 2014. Biostratigraphy, paleoecology and acme-zones of planktic foraminifera of the Cretaceous-Paleogene transition in the Tethys (Tunisia and Spain) and the Atlantic realm (France). Ann. Paleontol. 100, 193–215.
Gallala, N., Zaghbib-Turki, D., 2010. High resolution biostratigraphy based on planktic foraminifera across the Cretaceous-Paleogene transition at the Bidart section (SW France). Acta Geol. Pol. 60, 243–255.
Gallala, N., Zaghbib-Turki, D., Arenillas, I., Arz, J.A., Molina, E., 2009. Catastrophic mass extinction and assemblage evolution in planktic foraminifera across the Cretaceous/Paleogene (K/Pg) boundary at Bidart (SW France). Mar. Micropaleontol. 72, 196–209.
Gardin, S., Monechi, S., 1998. Palaeoecological change in middle to low latitude calcareous nannoplankton at the Cretaceous/Tertiary boundary. Bull. Soc. Geol. Fr. 169, 709–723.
Gartner, S., Hay, W.W., 1962. Planktonic foraminifera from the type Ilerdian. Eclog. Geol. Helv. 55, 553–572.
Giannetti, A., 2010. Influence of climate, sea-level changes and tectonics on ichnoassemblages distribution in a carbonate-dominated, deep-marine environment (Upper Paleocene, Zumaya section). Palaeogeogr. Palaeoclimatol. Palaeocol. 285, 104–118.
Giannetti, A., McCann, T., 2010. The upper Paleocene of the Zumaya section (northern Spain): review of the ichnological content and preliminary palaeoecological interpretation. Ichnos 17, 137–161.
Gomez de Llarena, J., 1946. Revisión de algunos datos paleontológicos del Flysch Cretáceo y Nummulítico de Guipúzcoa. Notas Com. Inst. Geol. Min. España 15, 109–165.
Gomez de Llarena, J., 1949. Datos paleoicnológicos. Notas Com. Inst. Geol. Min. España 19, 115–127.
Gomez de Llarena, J., 1954–1956. Observaciones geológicas en el flysch Cretácico-Nummulítico de Guipúzcoa. Monogr. Inst. Lucas Mallada, Madrid.
Gonzalvo, C., Molina, E., 1992. Bioestratigrafía y cronoestratigrafía del tránsito Eoceno–Oligoceno en Torre Cardela y Massignano (Italia). Rev. Esp. Paleontol. 7, 109–126.
Gonzalvo, C., Molina, E., 1996. Bioestratigrafía y cronoestratigrafía del tránsito Eoceno medio-Eoceno superior en la Cordillera Bética. Rev. Esp. Micropaleontol. 27, 25–44.
Gonzalvo, C., Molina, E., 1998. Planktic foraminiferal biostratigraphy across the lower-middle Eocene transition in the betic Codillera (Spain). Neues Jahr. Geol. Palaontol. Monats 11, 671–693.
Gorostidi, A., Lamolda, M., 1995. La nannoflora calcárea y el tránsito KT de la sección de Bidart (SW de Francia). Rev. Esp. Paleontol. no Homen. Guillermo Colom. 153–168.

Groot, J.J., de Jonge, R.B.G., Langereis, C.G., ten Kate, W.G., Smit, J., 1989. Magnetostratigraphy of the Cretaceous-Tertiary boundary at Agost (Spain). Earth Plant Sci. Lett. 94, 385–397.

Guernet, C., Molina, E., 1997. Ostracodes across the Paleocene-Eocene boundary in the betic Cordillera (Caravaca section, Spain). Geobios 30, 31–43.

Hardenbol, J., Berggren, W.A., 1978. A new Paleogene numerical time scale. AAPG Stud. Geol. 6, 213–234.

Harland, W.B., 1992. Stratigraphic guidance and regulation: a critique of current tendencies in stratigraphic codes and guides. Geol. Soc. Amer. Bull. 104, 231–1235.

de la Harpe, P., 1879. Description des nummulites apparteant á la zone supérieure des falaises de Biarritz. Bull. Soc. Borda 2, 137–156.

Haslett, S.K., 1994. Planktonic foraminiferal biostratigraphy and paleoceanography of the Cretaceous-Tertiary boundary section at Bidart, south-west France. Cretac. Res. 15, 179–192.

Hilgen, F.J., Kuiper, K.F., Lourens, L.J., 2010. Evaluation of the astronomical time scale for the Paleocene and earliest Eocene. Earth Planet. Sci. Lett. 300, 139–151.

Hillebrandt, A.v., 1965. Foraminiferen – stratigraphie in alttertiár von Zumaya (provinz Guipúzcoa, NW Spanien) und ein Vergleich mit anderen Tethys-Gebieten. Bay. Akad. D. Wissen. Math-Naturw. Kl. Abh. N. F. 123, 1–62.

Hillebrandt, A.v., 1974. Bioestratigrafía del Paleógeno en el sureste de España (prov. De Murcia y Alicante). Cuad. Geol. 5, 135–153.

Hillebrandt, A.v., 1975. Corrélation entre les biozones de grans foraminifères et de foraminifères planctoniques de l'Ilerdien. Bull. Soc. Geol. Fr. 17, 162–167.

Hillebrandt, A.v., 1976. Los foraminíferos planctónicos, nummulítidos y coccolitofóridos de la zona de *Globorotalia palmerae* del Cuisiense (Eoceno inferior) en el SE de España (provincias de Murcia y Alicante). Rev. Esp. Micropaleontol. 8, 323–394.

Hottinger, L., Schaub, H., 1960. Zur Stufeneinteilung des Paleocaens und des Eocaens. Einführung der Stufen Ilerdien und Biarritzien. Eclog. Geol. Helv. 53, 453–479.

Isuman, N., 1983. Mikropaläontologische Untersuchungen von Großforami-niferen (Nummuliten und Assilinen) im Alttertiär von Südostspanien (Aspe und Agost in der Provinz Alicante). Berl. Geowiss. Abh 49, 61–170.

Izett, G.A., 1987. Authigenic spherules in K-T boundary sediments at Caravaca, Spain, and Raton Basin, Colorado and New Mexico, may not be impact derived. Geol. Soc. Amer. Bull. 99, 78–86.

Jacquot, M., 1864. Description géologique des falaises de Biarritz, Bidard-Guetáry et St-Jean-de-Luz. Actes Soc. Linnénne Bordx. 25, 1–58.

Joly, N., Leymerie, A., 1848. Mémoire sur les Nummulites considérées zoologiquement et géologiquement. Mem. Acad. Sci. Toulouse 3, 1–70.

Jorge, E.de, 1926. El Eoceno de Vizcaya. Not. Com. Inst. Geol. Min. España 6, 91–101.

Jovane, L., Sprovieri, M., Coccioni, R., Florindo, F., Marsili, A., Laskar, J., 2010. Astronomical calibration of the middle Eocene contessa Highway section (Gubbio, Italy). Earth Planet. Sci. Lett. 298, 77–88.

Julivert, M., 2014. Una historia de la geología en España en su contexto socioeconómico, cultural y politico, y en el marco de la geología internacional. Univ. Barcelona, Barcelona.

Kaiho, K., Kajiwara, Y., Tazaki, K., Ueshima, M., Takeda, N., Kawahata, H., Arinobu, T., Ishiwatari, R., Hirai, A., Lamolda, M.A., 1999. Oceanic primary productivity and dissolved oxygen levels at the Cretaceous/Tertiary boundary: their decrease, subsequent warming, and recovery. Paleoceanography 14, 511–524.

Kaiho, K., Lamolda, M.A., 1999. Catastrophic extinction of planktonic foraminifera at the Cretaceous-Tertiary boundary evidenced by stable isotopes and foraminiferal abundance at Caravaca, Spain. Geology 27, 355–358.
Kamo, S., Cristiano, L., Morgan, J.V., 2011. U-Pb ages of shocked zircon grains link distal K-Pg boundary sites in Spain and Italy with the Chicxulub impact. Earth Planet. Sci. Lett. 310, 401–408.
Kapellos, C.v., Schaub, H., 1973. Zur Korrelation von Biozonierungen mit Grossforaminiferen und Nannoplankton im Paläogen der Pyrenäen. Eclog. Geol. Helv. 66, 687–737.
Kapellos, C.H., 1974. Über das Nannoplankton im Alttertiär des Profils von Zumaya-Guetaria (Provinz Guipuzcoa, Nordspanien). Eclog. Geol. Helv. 67, 435–444.
Kedzierski, M., Rodriguez-Tovar, F., Uchman, A., 2011. Vertical displacement and taphonomic filtering of nannofossils by bioturbation in the Cretaceous-Palaeogene boundary section at Caravaca, SE Spain. Lethaia 44, 321–328.
Kelly, D.C., Bralower, T.J., Zachos, J.C., Premoli Silva, I., Thomas, T., 1996. Rapid diversification of planktonic foraminifera in the tropical Pacific (ODP Site 865) during the late Paleocene thermal maximum. Geology 24, 423–426.
Khozyem, H., Adatte, T., Keller, G., Tantawy, A.A., Spangenberg, J.E., 2014. The Paleocene-Eocene GSSP at Dababiya, Egypt – revisited. Episodes 37, 78–86.
Kindelan, V., 1919. Notas sobre el Cretáceo y el Eoceno de Guipúzcoa. Bol. Inst. Geol. Min. España 40, 165–203.
Kruit, C., Brouwer, J., Ealey, P., 1972. A deep-water sand fan in the Eocene Bay of Biscay. Nature 240, 59–61.
Kuhnt, W., Kaminski, M.A., 1997. Cenomanian to lower Eocene deep-water agglutinated foraminífera from the Zumaya section, northern Spain. Ann. Soc. Geol. Pol. 67, 257–270.
Kuiper, K.F., Deino, A., Hilgen, F.J., Krijgsman, W., Renne, P.R., Wijbrans, J.R., 2008. Synchronizing rock clocks of Earth history. Science 320, 500–504.
Kyte, F.T., Smit, J., Wasson, J.T., 1985. Siderophile interelement variations in the Cretaceous-Tertiary boundary sediments from Caravaca, Spain. Earth Planet. Sci. Lett. 73, 183–195.
Lamare, P., 1936. Recherches géologiques dans les Pyrénées basques d'Espagne. Mem. Soc. Geol. Fr. XII-27 1–465.
Lamolda, M.A., 1990. The Cretaceous-Tertiary boundary crisis at Zumaya (northern Spain): micropaleontological data. Lect. Not. Earth Sci. 30, 393–399.
Lamolda, M.A., Gorostidi, A., 1992. Nannofossil stratigraphic record in upper Maastrichtian-lowermost Danian at Zumaya. Mem. Ist. Geol. Min. Univ. Padova 43, 149–161.
Lamolda, M.A., Melinte, M.C., Kaiho, K., 2005. Nannofloral extinction and survivorship across the K/T boundary at Caravaca, southeastern Spain. Palaeogeogr. Palaeoclimatol. Palaeoecol. 224, 27–52.
Lamolda, M.A., Melinte-Dobrinescu, M.C., Kaiho, K., 2016. Calcareous nannoplankton assemblage changes linked to paleoenvironmental deterioration and recovery across the Cretaceous-Paleogene boundary in the Betic Cordillera (Agost, Spain). Palaeogeog. Palaeoclimatol. Palaeoecol. 441, 438–452.
Lamolda, M.A., Orue-Etxebarria, X., Proto-Decima, F., 1983. The Cretaceous-Tertiary boundary in Sopelana (Biscay, Basque country). Zitteliana 10, 663–670.
Lamolda, M.A., Paul, C.R.C., Peryt, D., Pons, J.M., 2014. The Global Boundary Stratotype and Section Point (GSSP) for the base of the Santonian Stage, "Cantera de Margas", Olazagutia, Northern Spain. Episodes 37, 2–13.
Lanaja, J.M., 1987. Contribución de la exploración petrolífera al conocimiento de la geología de España. Inst. Geol. Min. España, Madrid.
Larrasoaña, J.C., Gonzalvo, C., Molina, E., Monechi, S., Ortiz, S., Tori, F., Tosquella, J., 2008. Integrated magnetobiochronology of the Early/Middle Eocene transition at Agost (Spain): implications for defining the Ypresian/Lutetian boundary stratotype. Lethaia 41, 395–415.

Larrazet, M., 1896. Recherches géologiques sur la région orientale de la province de Burgos et sur quelques points des provinces d'Alava et de Logroño (Ph.D. thesis), Fac. Sciences de Paris, Lille.

Leclerc, J., 1971. Etude géologique du massif du Maigmó et de ses abords (Ph.D. thesis). Univ. Paris, Paris.

Lévêque, F., 1993. Corrrelating the Eocene-Oligocene mammalian biochronological scale from SW Europe with the marine magnetic anomaly sequence. J. Geol. Soc. Lond. 150, 661−664.

Lopez-Martinez, N., Canudo, J.I., Ardevol, L., Pereda-Suberbiola, X., Orue-Etxebarria, X., Cuenca-Bescos, G., Ruiz-Omeñaca, J.I., Murelaga, X., Feist, M., 2001. New dinosaur sites correlated with upper Maastrichtian pelagic deposits in the Spanish Pyrenees: implications for the dinosaur extinction pattern in Europe. Cretac. Res. 22, 41−61.

Lopez-Martinez, N., Pelaez-Campomanes, P., 1999. New mammals from south-central Pyrenees (Tremp Formation, Spain) and their bearing on late Paleocene marine-contenental correlations. Bull. Soc. Geol. Fr. 170, 681−696.

Lu, G., Adatte, T., Keller, G., Ortiz, N., 1998. Abrupt climatic, oceanographic and ecologic changes near the Paleocene-Eocene transition in the deep Tethys basin: the Alamedilla section, southern Spain. Ecog. Geol. Helv. 91, 293−306.

Lu, G., Keller, G., Adatte, T., Ortiz, N., Molina, E., 1996. Long-term (10^5) or short-term (10^3) 13C excursion near the Paleocene-Eocene transition: evidence from the Tethys. Terra Nova 8, 347−355.

Luterbacher, H.P., 1969. Remarques sur la position stratigraphique de la formation d'Ager (Pyrénées méridionales). BRGM Mem. 69, 225−232.

Luterbacher, H.P., 1973. La sección tipo del piso Ilerdiense. In: Enadimsa, C.N.G. (Ed.), Libro guía del XIII Coloquio Europeo de Micropaleontología. Imp. Pablo Lopez, Madrid, pp. 101−128.

Luterbacher, H.P., Ali, J.R., Brinkhuis, H., Gradstein, F.M., Hooker, J.J., Monechi, S., Ogg, J.G., Powell, J., Röhl, U., Sanfilippo, A., Schmitz, B., 2004. The Paleogene period. In: Gradstein, F.M., Ogg, J.G., Smith, A.G. (Eds.), A Geologic Time Scale 2004. Cambridge Univ. Press, Cambridge, pp. 384−408.

MacLeod, N., Keller, G., 1991. How complete are Cretaceous/Tertiary boundary sections? A chronostratigraphic estimate based on graphic correlation. Geol. Soc. Amer. Bull. 103, 1439−1457.

Mangin, J.P., 1960. Le Nummulitique sud-pyrénéen à l'Ouest de I'Aragon. Pirineos 51-58, 1−656.

Manners, H.R., Grimes, S.T., Sutton, P.A., Domingo, L., Leng, M.J., Twitchett, R.J., Harta, M.B., Jones, T.D., Pancost, R.D., Duller, R., López-Martínez, N., 2013. Magnitude and profile of organic carbón isotope records from the Paleocene−Eocene Thermal Maximum: evidence from northern Spain. Earth Planet. Sci. Lett. 376, 220−230.

Martinez-Ruiz, F., Ortega-Huertas, M., Palomo, I., 1999. Positive Eu anomaly development during diagenesis of the K/T boundary ejecta layer in the Agost section (SE Spain): implications for trace-element remobilization. Terra Nova 11, 290−296.

Martinez-Ruiz, F., Ortega-Huertas, M., Palomo, I., Barbieri, M., 1992. The geochemistry and mineralogy of the Cretaceous-Tertiary boundary at Agost (southeast Spain). Chem. Geol. 95, 265−281.

Martinez-Ruiz, Ortega-Huertas, M., Palomo, I., Acquafredda, P., 1997. Quench textures in altered spherules from the Cretaceous-Tertiary boundary layer at Agost and Caravaca, SE Spain. Sedim. Geol. 113, 137−147.

Martinez-Ruiz, F., Ortega-Huertas, M., Rivas, P., 2006. Rare earth element composition as evidence of the precursor material of Cretaceous-Tertiary boundary sediments at distal sections. Chem. Geol. 232, 1−11.

Martini, E., 1959. *Pemma angulatum* und *Micrantholithus basquensis*, zwei neue Coccolithoporiden-Arten aus dem Eozän. Senck. Leth. Frankf. 40, 415–421.
Martini, E., 1970. Standard Palaeogene calcareous nannoplankton zonation. Nature 226, 560–561.
Martini, E., 1971. Standard Tertiary and Quaternary calcareous nannoplankton zonation. Tecnosci. Roma 2, 739–785.
Mary, C., Moreau, M.G., Orue-Etxebarria, X., Apellaniz, E., Courtillot, V., 1991. Biostratigraphy and magnetostratigraphy of the Cretaceous/Tertiary Sopelana section (Basque country). Earth Planet. Sci. Lett. 106, 133–150.
Mathelin, J.C., Sztrakos, K., 1993. L'Eocène de Biarritz (Pyrénées Atlantiques, SW France): stratigraphie et paléoenvironment, monographie des foraminifères. Cah. Micropaleontol. 8, 1–85.
McInerney, F.A., Wing, S.L., 2011. The Paleocene Eocene Thermal Maximum: a perturbation of carbon cycle, climate, and biosphere with implications for the future. Ann. Rev. Earth Planet. Sci. Lett. 39, 489–516.
Mendizabal, J., 1923. Deslinde del Eoceno en la provincial de Guipúzcoa. Bol. Inst. Geol. España 64, 449–453.
Mengaud, L., 1920. Recherches géologiques dans la région cantabrique. Bull. Soc. Hist. Natur. Toulouse 48, 73–272.
Minoleotti, F., Rafaeklis, M., renard, M., Gardin, S., 2004. Reworking of Maastrichtian-like alcareous nannofossils in the lowermost Danian sediments of Bidart section (France): isotopic evidence (carbon and oxygen). Rev. Micropaleontol. 47, 145–152.
Mizukami, T., Kunio, K., Masahiro, O., 2013. Significant changes in land vegetation and oceanic redox across the Cretaceous/Paleogene boundary. Palaeogeogr. Palaeoclimatol. Palaeoecol. 369, 41–47.
Mochales, T., Barnolas, A., Pueyo, E.L., Serra-Kiel, J., Casas, A.M., Samso, J.M., Ramajo, J., Sanjuan, J., 2012. Chronostratigraphy of the Boltaña anticline and the Ainsa basin (southern Pyrenees). Soc. Geol. Amer. Bull. 124, 1229–1250.
Molina, E., 1986. Description and biostratigraphy of the main reference section of the Eocene/Oligocene boundary in Spain: Fuente Caldera section. Dev. Paleontol. Strat. 9, 49–52.
Molina, E., 2015. Evidence and causes of the main extinction events in the Paleogene record based on extinction and survival patterns of foraminifera. Earth Sci. Rev. 140, 166–181.
Molina, E., Alegret, L., Apellaniz, E., Bernaola, G., Caballero, F., Monechi, S., Ortiz, S., Orue-Etxebarria, X., Payros, A., Tosquella, J., 2012. The Global Stratotype Section and Point for the Base of the Lutetian Stage: Auxiliary Sections and Correlation in Spain. Abstracts of the 34th International Geological Congress, Brisbane, Australia, p. 1508.
Molina, E., Alegret, L., Apellaniz, E., Bernaola, G., Caballero, F., Dinarès-Turell, J., Hardenbol, J., Heilmann-Clausen, C., Larrasoaña, J.C., Luterbacher, H., Monechi, S., Ortiz, S., Orue-Etxebarria, X., Payros, A., Pujalte, V., Rodríguez-Tovar, F.J., Tori, F., Tosquella, J., Uchman, A., 2011. The global stratotype section and point (GSSP) for the base of the Lutetian stage at the Gorrondatxe section, Spain. Episodes 34, 86–108.
Molina, E., Alegret, L., Arenillas, I., Arz, J.A., Gallala, N., Grajales-Nishimura, M., Murillo-Muñeton, G., Zaghbib-Turni, D., 2009. The global boundary stratotype section and point for the base of the Danian stage (Paleocene, Paleogene, "Tertiary", Cenozoic): auxiliary sections and correlation. Episodes 32, 84–95.
Molina, E., Alegret, L., Arenillas, I., Arz, J.A., Gallala, N., Hardenbol, J., von Salis, K., Steurbaut, E., Vandenberghe, N., Zaghbib-Turni, D., 2006a. The global boundary stratotype section and point for the base of the Danian stage (Paleocene, Paleogene, "Tertiary", Cenozoic) at el Kef, Tunia: original definition and revision. Episodes 29, 263–273.

Molina, E., Arenillas, I., Arz, J.A., 1996. The Cretaceous/Tertiary boundary mass extinction in planktic foraminifera at Agost, Spain. Rev. Micropaleontol. 39, 225–243.
Molina, E., Arenillas, I., Arz, J.A., 1998. Mass extinction in planktic foraminifera at the Cretaceous/Tertiary boundary in subtropical and temperate latitudes. Bull. Soc. Geol. Fr. 169, 351–363.
Molina, E., Arenillas, I., Arz, J.A., 2005. The Cretaceous/Paleogene boundary at the Agost section revisited: paleoenvironmental reconstruction and mass extinction pattern. J. Iber. Geol. 31, 135–148.
Molina, E., Canudo, J.I., Guernet, C., McDoudall, K., Ortiz, N., Pacual, J.O., Pares, J.O., Samso, J.M., Serra-Kiel, J., Tosquella, J., 1992. The stratotypic Ilerdian revisited: integrated stratigraphy across the Paleocene/Eocene boundary. Rev. Micropaleontol. 35, 143–156.
Molina, E., Canudo, J.I., Martinez-Ruiz, F., Ortiz, N., 1994. Integrated stratigraphy across the Paleocene/Eocene boundary at Caravaca, southern Spain. Eclog. Geol. Helv. 87, 47–61.
Molina, E., Cosovic, V., Gonzalvo, C., Von Salis, K., 2000. Integrated biostratigraphy across the Ypresian/Lutetian boundary at Agost, Spain. Rev. Micropaleontol. 43, 381–391.
Molina, E., Gonzalvo, C., Keller, G., 1993. The Eocene-Oligocene planktic foraminiferal transition: extinctions, impacts and hiatuses. Geol. Mag. 130, 483–499.
Molina, E., Gonzalvo, C., Mancheño, M.A., Ortiz, S., Schmitz, B., Thomas, E., Von Salis, K., 2006b. Integrated stratigraphy and chronostratigraphy across the Ypresian-Lutetian transition in the Fortuna section (betic Cordillera, Spain). Newslett. Strat. 42, 1–19.
Molina, E., Gonzalvo, C., Ortiz, S., Cruz, L.E., 2006c. Foraminiferal turnover across the Eocene–Oligocene transition at Fuente Caldera, southern Spain: no cause–effect relationship between meteorite impacts and extinctions. Mar. Micropaleontol. 58, 270–286.
Molina, E., Keller, G., Madile, M., 1988. Late Eocene to Oligocene events: Molino de Cobo, Betic Cordillera, Spain. Rev. Esp. Micropaleontol. 20, 491–514.
Molina, E., Monaco, P., Nocchi, M., Parisi, G., 1986. Biostratigraphic correlation between the central Subbetic (Spain) and Umbro-Marchean (Italy) pelagic sequences at the E/O boundary using foraminifera. Dev. Paleontol. Strat. 9, 75–86.
Monechi, S., 1986. Calcareous nannofossils of the Fuente Caldera section. Dev. Paleontol. Strat. 9, 65–70.
Monechi, S., Angori, E., Von Salis, K., 2000. Calcareous nannofossil turnover around the Paleocene/Eocene transition at Alamedilla (southern Spain). Bull. Soc. Geol. Fr. 171, 477–489.
Monechi, S., Reale, V., Bernaola, G., Balestra, B., 2013. The Danian/Selandian boundary at Site 1262 (South Atlantic) and in the Tethyan region: biomagnetostratigraphy, evolutionary trends in fasciculiths and environmental effects of the Latest Danian Event. Mar. Micropaleontol. 98, 28–40.
Moreau, M.G., Cojan, I., Ory, J., 1994. Mechanisms of remanent magnetization acquisition in marl and limestone alternations. Case study: upper Cretaceous (Chron 31–30), Sopelana, Basque Country. Earth Planet. Sci. Lett. 123, 15–37.
Mount, J.F., Ward, P., 1986. Origin of limestone/marl alternations in the upper Maastrichtian of Zumaya. Spain. J. Sedim. Res. 56, 228–236.
Muñoz, A., Arenas, C., Gonzalez, A., Luzon, A., Pardo, G., Perez, A., Villena, J., 2002. Tertiary: Ebro basin (northeastern Spain). In: Gibbons, W., Moreno, T. (Eds.), The Geology of Spain. The Geological Society, London, pp. 301–309.
Muñoz, J.A., 2002. Alpine tectonics I: the Alpine system north of the betic Cordillera; the Pyrenees. In: Gibbons, W., Moreno, T. (Eds.), The Geology of Spain. The Geological Society, London, pp. 370–385.

Nelson, B.K., MacLeod, G.K., Ward, P.D., 1991. Rapid change in strontium isotopic composition of sea-water before the Cretaceous-Tertiary boundary. Nature 351, 644–647.
Odin, G.S., Lamaurelle, M.A., 2001. The global Campanian-Maastrichtian stage boundary. Episodes 24, 229–238.
Odin, G.S., Luterbacher, H., 1992. The age of Paleogene Stage boundaries. Neues Jahr. Geol. Palaontol. Abh. 186, 21–48.
Okada, H., Bukry, D., 1980. Supplementary modificatiun and introduction of code numbers to the low-latitude coccolith biostratigraphic zonation (Bukry, 1973; 1975). Mar. Micropaleontol. 5, 321–325.
Olsson, R.K., Liu, C.J., 1993. Controversies on the placement of Cretaceous-Paleogene boundary and the K/P mass extinction of planktonic-foraminifera. Palaios 8, 127–139.
Ortiz, N., 1995. Differential patterns of benthic foraminiferal extinctions near the Paleocene/Eocene boundary in the North Atlantic and the western Tethys. Mar. Micropaleontol. 136, 17–30.
Ortiz, S., Gonzalvo, C., Molina, E., Rodríguez-Tovar, F.J., Uchman, A., Vandemberghe, N., Zeelmaekers, E., 2008. Palaeoenvironmental turnover across the Ypresian-Lutetian transition at the Agost section, southeastern Spain: in search of a marker event to define the Stratotype for the base of the Lutetian Stage. Mar. Micropaleontol. 69, 297–313.
Ortiz, S., Thomas, E., 2006. Lower-middle Eocene benthic foraminífera from the Fortuna section (betic Cordillera, southestern Spain). Micropaleontology 52, 97–150.
Orue-Etxebarria, X., 1983. Los foraminíferos plantónicos del Paleógeno del Sinclinorio de Bizkaia (corte de Soeplana-Punta de la Galea). Kobie 13–14, 175–429.
Orue-Etxebarria, X., 1985a. Descripción de *Globigerina hillebrandt* n. sp. En el límite Cretácico/Terciario de la sección de Sopelana (País Vasco): evolución de los primeros foraminíferos planctónicos al comienzo del Terciario. Newslett. Strat. 15, 71–80.
Orue-Etxebarria, X., 1985b. Descripción de dos nuevas especies de foraminíferos planctónicos en el Eoceno costero de la provincia de Bizkaia. Rev. Esp. Micropaleontol. 17, 467–477.
Orue-Etxebarria, X., Apellaniz, E., 1985. Estudio del límite Cusiense-Luteciense en la costa vizcaína por medio de los foraminíferos planctónicos. Newslett. Stratigr. 15, 1–12.
Orue-Etxebarria, X., Apellaniz, E., 1991. Descripción de *Eoglobigerina ferreri*, un nuevo foraminífero planctónico del Paleoceno. Estud. Geol. 47, 97–101.
Orue-Etxebarria, X., Apellaniz, E., 2000. Modification of the original stratigraphic distribution of *Globigerina hillebrandti* Orue-Etxebarria, 1985 and its inclusion in another genus: one more planktic foraminifer species surviving the mass extinction of the K/T boundary. Bull. Soc. Geol. Fr. 171, 291–293.
Orue-Etxebarria, X., Apellaniz, E., Baceta, J.I., Coccioni, R., Di Leo, R., Dinarès-Turell, J., Galeotti, S., Monechi, S., Núñez-Betelu, K., Pares, J.M., Payros, A., Pujalte, V., Samso, J.M., Serra-Kiel, J., Schmitz, B., Tosquella, J., 1996. Physical and biostratigraphic analysis of two prospective Paleocene-Eocene boundary stratotypes in the intermediate-deep water Basque Basin, western Pyrenees: the Trabakua Pass and Ermua sections. Neues Jahr. Geol. Paläontol. Abh. 201, 179–242.
Orue-Etxebarria, X., Baceta, J.I., Pujalte, V., Nunez-Betelu, K., Serra-Kiel, J., Apellaniz, E., Payros, A., 1992. Evaluating prospective Paleocene-Eocene boundary stratotypes in the deep water Basque basin, western Pyrenees: Ermua and Trabakua Pass sections, Biscay. In: Molina, E., Canudo, J.I., Cuenca, G., Pardo, A. (Eds.), IGCP Project 308: Paleocene/Eocene Boundary Events, Field-trip Guidebook. Univ. Zaragoza, Zaragoza, pp. 1–32.
Orue-Etxebarria, X., Bernaola, G., Baceta, J.I., Angori, E., Caballero, F., Monechi, S., Pujalte, V., Dinarès-Turell, J., Apellaniz, E., Payros, A., 2004. New constraints on the

evolution of planktic foraminifers and calcareous nannofossils across the Paleocene-Eocene boundary interval: the Zumaia section revisited. Neues Jahr. Geol. Paläontol. Abh. 234, 223–259.

Orue-Etxebarria, X., Lamolda, M., Apellaniz, E., 1984. Bioestratigrafía del Eocene Vizcaino por medio de los foraminíferos planctónicos. Rev. Esp. Micropaleontol. 16, 241–263.

Orue-Etxebarria, X., Payros, A., Caballero, F., Apellaniz, E., Pujalte, V., Ortiz, S., 2014. *Morozovella gorrondatxensis* (Orue-Etxebarria 1985) vs *M. crater* (Hornibrook 1958): taxonomy and significance for Early/Middle Eocene boundary biostratigraphy. Stratigraphy 11, 173–183.

Orue-Etxebarria, X., Payros, A., Caballero, F., Molina, E., Apellaniz, E., Bernaola, G., 2009. The Ypresian/Lutetian Transition in the Gorrondatxe Beach (Getxo, Western Pyrenees): Review, Recent Advances and Future Prospects. Univ. Basque Country, Leioa.

Orue-Etxebarria, X., Pujalte, V., Bernaola, G., Apellaniz, E., Baceta, J.I., Payros, A., Núñez-Betelu, K., Serra-Kiel, J., Tosquella, J., 2001. Did the Late Paleocene thermal máximum affect the evolution of larger foraminifers? Evidence from calcareous plakton of the Campo Section (Pyrenees, Spain). Mar. Micropaleontol. 41, 45–71.

Pais, J., Cunha, P.P., Pereira, D., Legoinha, P., Dias, R., Moura, D., Brum da Silveira, A., Kullberg, J.C., Gonzalez-Delgado, J.A., 2012. The Paleogene and Neogene of Western Iberia (Portugal): A Cenoizic Record in the European Atlantic Domain. Springer, Heidelberg.

Pardo, A., Keller, G., Oberhansli, H., 1999. Paleoecologic and paleoceanographic evolution of the Tethyan realm during the Paleocene-Eocene transition. J. Form. Res. 29, 37–57.

Pardo, A., Ortiz, N., Keller, G., 1996. Latest Maastrichtian and Cretaceous–Tertiary boundary foraminiferal turnover and environmental changes at Agost, Spain. Geol. Soc. Spec. Publ. 228, 29–62.

Pardo, G., Arenas, C., Gonzalez, A., Luzon, A., Muñoz, A., Perez, A., Perez-Rivares, F.J., Vazquez-Urbez, M., Villena, J., 2004. La Cuenca del Ebro. In: Vera, J.A. (Ed.), Geología de España. Soc. Geol. España & Inst. Geol. Min. España, Madrid, pp. 533–543.

Payros, A., 1997. El Eoceno de la Cuenca de Pamplona: estratigrafía, facies y evolución paleogeográfica (Ph.D. thesis). Univ. Basque Country, Leioa.

Payros, A., Bernaola, G., Orue-Etxebarria, X., Dinarès-Turell, J., Tosquella, J., Apellaniz, E., 2007a. Reassessment of the early-middle Eocene biomagnetochronology based on evidence from the Gorrondatxe section (Basque country, W Pyrenees). Lethaia 40, 183–195.

Payros, A., Dinarès-Turell, J., Bernaola, G., Orue-Etxebarria, X., Apellaniz, E., Tosquella, J., 2011. On the age of the Early/Middle Eocene boundary and other related events: cyclostratigraphic refinements from the Pyrenean Otsakar section and the Lutetian GSSP. Geol. Mag. 148, 442–460.

Payros, A., Dinarès-Turell, J., Monechi, S., Orue-Etxebarria, X., Ortiz, S., Apellaniz, E., Martinez-Braceras, N., 2015b. The Lutetian/Bartonian transition (middle Eocene) at the Oyambre section (northern Spain): implications for standard chronostratigraphy. Palaeogeogr. Palaeoclimatol. Palaeoecol. 440, 234–248.

Payros, A., Dinarès-Turell, J., Orue-Etxebarria, X., Monechi, S., Ortiz, S., Apellaniz, E., Bernaola, G., 2014. In search of the Bartonian (middle Eocene) GSSP (I): potential in the Basque-Cantabrian and Aquitanian basins (W Pyrenees). In: Rocha, R., Pais, J., Kullberg, J.C., Finney, S. (Eds.), Strati 2013. Springer, Heidelberg, pp. 79–83.

Payros, A., Martinez-Braceras, N., 2014. Orbital forcing in turbidite accumulation during the Eocene greenhouse interval. Sedimentology 61, 1411–1432.

Payros, A., Ortiz, S., Alegret, L., Orue-Etxebarria, X., Apellaniz, E., Molina, E., 2012. An early Lutetian carbon-cycle perturbation: insights from the Gorrondatxe section (western Pyrenees, Bay of Biscay). Paleoceanography 27, PA2213.

Payros, A., Ortiz, S., Millan, I., Arostegi, J., Orue-Etxebarria, X., Apellaniz, E., 2015a. Early Eocene climatic optimum: environmental impact on the North Iberian continental margin. Geol. Soc. Amer. Bull. 127, 1632–1644.
Payros, A., Orue-Etxebarria, X., Pujalte, V., 2006. Covarying sedimentary and biotic fluctuations in Lower-Middle Eocene Pyrenean deep-sea deposits: Palaeoenvironmental implications. Palaeogeogr. Palaeoclimatol. Palaeoecol. 234, 258–276.
Payros, A., Orue-Etxebarria, X., Bernaola, G., Apellaniz, E., Dinarès-Turell, J., Tosquella, J., Caballero, F., 2009b. Characterization and astronomically calibrated age of the first occurrence of *Turborotalia frontosa* in the Gorrondatxe section, a prospective Lutetian GSSP: implications for the Eocene time scale. Lethaia 42, 255–264.
Payros, A., Pujalte, V., Baceta, J.I., Bernaola, G., Orue-Etxebarria, X., Apellaniz, E., Caballero, F., Ferràndez, C., 2000. Lithostratigraphy and sequence stratigraphy of the upper Thanetian to middle Ilerdian strata of the Campo section (southern Pyrenees, Spain): revision and new data. Rev. Soc. Geol. España 13, 213–226.
Payros, A., Pujalte, V., Orue-Etxebarria, X., 2007b. A point-sourced calciclastic submarine fan complex (Eocene Anotz Formation, western Pyrenees): facies architecture, evolution and controlling factors. Sedimentology 54, 137–168.
Payros, A., Tosquella, J., Bernaola, G., Dinarés-Turell, J., Orue-Etxebarria, X., Pujalte, V., 2009a. Filling the north European early/middle Eocene (Ypresian/Lutetian) boundary gap: insights from the Pyrenean continental to deep-marine record. Palaeogeogr. Palaeoclimatol. Palaeoecol. 280, 313–332.
Pearson, P.N., Coxall, H.K., 2014. Origin of the Eocene planktonic foraminifer *Hantkenina* by gradual evolution. Palaeontology 57, 243–267.
Pearson, P.N., Olsson, R.K., Hemblen, C., Huber, B.T., Berggren, W.A., 2006. Atlas of Eocene Planktonic Foraminifera. Cushman Foundation Special Publication 41, p. 513.
Pedreira, D., Pulgar, J.A., Gallart, J., Torne, M., 2007. Three-dimensional gravity and magnetic modeling of crustal indentation and wedging in the western Pyrenees-Cantabrian Mountains. J. Geophys. Res. 112, B12405.
Pedreira, D., Afonso, J.C., Pulgar, J.A., Gallastegui, J., Carballo, A., Fernandez, M., Garcia-Castellanos, D., Jimenez-Munt, I., Semprich, J., Garcia-Moreno, O., 2015. Geophysical-petrological modeling of the lithosphere beneath the Cantabrian Mopuntains and the North-Iberian margin: geodynamic implications. Lithos 230, 46–68.
Pekar, S.F., Hucks, A., Fuller, M., Li, S., 2005. Glacioeustatic changes in the early and middle Eocene (51-42 Ma): shallow-water stratigraphy from ODP Leg 189 Site 1171 (South Tasman Rise) and deep-sea $\delta^{18}O$ records. Geol. Soc. Amer. Bull. 117, 1081–1093.
Perch-Nielsen, K., 1979. Calcareous nannofossils at the Cretaceous/Tertiary boundary near Biarritz, France. In: Christensen, W.K., Birkelund, T. (Eds.), Cretaceous/Tertiary Boundary Events. Univ. Copenhagen, Copenhagen, pp. 151–155.
Percival Jr., S.F., Fischer, A.G., 1977. Changes in calcareous nan-noplankton in the Cretaceous-Tertiary biotic crisis at Zumaya, Spain. Evol. Theory 2, 1–35.
Pérez-Rodríguez, I., Lees, J.A., Larrasoaña, J.C., Arz, J.A., Arenillas, I., 2012. Planktonic foraminiferal and calcareous nannofossil biostratigraphy and magnetostratigraphy of the uppermost Campanian and Maastrichtian at Zumaia, northern Spain. Cretac. Res. 37, 100–126.
Pettijohn, F.J., Potter, P.E., 1964. Atlas and Glossary of Primary Sedimentary Structures. Springer-Verlag, New York.
Pfender, J., 1926. Sur les organismos du Nummulitique de la colline de San Salvador près de Camarasa (province de Lerida, Catalogne). Bol. R. Soc. Esp. Hist. Nat. 26, 321–330.
Plaziat, J.C., 1981. Late Cretaceous to late Eocene palaeogeographic evolution of southwest Europe. Palaeogeogr. Palaeoclimatol. Palaeoecol. 36, 263–293.
Pomerol, C., 1975. La signification de l'Ilerdien et l'intérêt de cet étage dans la stratigraphie du Paléogène mésogéen. B. S. G. F. 17, 213–217.

Pomerol, C., 1981. Stratotypes of Paleogene stages. Bull. Inf. Geol. Bass. Paris, Mem. 2, 1–302.
Premoli Silva, I., Jenkins, D.G., 1993. Decision on the Eocene–Oligocene boundary stratotype. Episodes 16, 379–382.
Pujalte, V., Baceta, J.I., Dinarès-Turell, J., Orue-Etxebarria, X., Parès, J.M., Payros, A., 1995. Biostratigraphic and magnetostratigraphic intercalibration of late Maastrichtian and Paleocene depositional sequences from the deep-water Basque basin, W Pyrenees, Spain. Earth Planet. Sci. Lett. 136, 17–30.
Pujalte, V., Baceta, J.I., Orue-Etxebarria, X., Payros, A., 1998. Paleocene strata of the Basque Country, western Pyrenees, northern Spain: facies and sequence development in a deep-water starved basin. SEPM Spec. Publ. 60, 311–325.
Pujalte, V., Baceta, J.I., Payros, A., 2002. Tertiary: western Pyrenees and Basque-Cantabrian region. In: Gibbons, W., Moreno, T. (Eds.), The Geology of Spain. The Geological Society, London, pp. 293–301.
Pujalte, V., Baceta, J.I., Schmitz, B., 2015. A massive input of coarse-grained siliciclastics in the Pyrenean Basin during the PETM: the missing ingredient in a coeval abrupt change in hydrological regime. Clim. Past. 11, 1653–1672.
Pujalte, V., Baceta, J.I., Schmitz, B., Orue-Etxebarria, X., Payros, A., Bernaola, G., Apellaniz, E., Caballero, F., Robador, A., Serra-Kiel, J., Tosquella, J., 2009a. Redefinition of the Ilerdian stage (early Eocene). Geol. Acta 7, 77–94.
Pujalte, V., Dinarès-Turell, J., Bernaola, G., Baceta, J.I., Payros, A., 2003a. A reappraisal of the position of chron C25n in the Campo section (Huesca province, south-central Pyrenees). Geogaceta 34, 155–158.
Pujalte, V., Orue-Etxebarria, X., Apellaniz, E., Caballero, C., 2014. El Paleógeno inferior de los Montes Orientales, proincia de Granada, Zona Subbética: revision y nuevos datos. Rev. Soc. Geol. España 27, 137–149.
Pujalte, V., Orue-Etxebarria, X., Caballero, F., Apellaniz, E., Hilario, A., 2016. The upper Maastrchtian, Paleocene and lowermost Ypresian of the Zumaia section (Basque basin, western Pyrenees). Diput. Foral Gipuzkoa – Gipuzkoako Foru Aldundia, Donostia – San Sebastián.
Pujalte, V., Orue-Etxebarria, X., Schmitz, B., Tosquella, J., Baceta, J.I., Payros, A., Bernaola, G., Caballero, F., Apellaniz, E., 2003b. Basal Ilerdian (earliest Eocene) turnover of larger foraminifera: age constraints based on calcareous plankton and $\delta^{13}C$ isotopic profiles from new southern Pyrenean sec tions (Spain). Geol. Soc. Amer. Spec. Pap. 369, 205–221.
Pujalte, V., Orue-Etxebarria, X., Apellaniz, E., Caballero, F., Robles, S., 2012. Río Gor, comarca de Guadix, provincia de Granada: una nueva sección de referencia del Paleógeno inferior Subbético. Geogaceta 52, 53–56.
Pujalte, V., Schmitz, B., 2005. Revisión de la estratigrafía del Grupo Tremp («Garumniense», Cuenca de Tremp-Graus, Pirineos meridionales). Geogaceta 38, 79–82.
Pujalte, V., Schmitz, B., Baceta, J.I., Orue-Etxebarria, X., Bernaola, G., Dinarès-Turell, J., Payros, A., Apellaniz, E., Caballero, F., 2009b. Correlation of the Thanetian-Ilerdian turnover of larger foraminifera and the Paleocene-Eocene thermal maximum: confirming evidence from the Campo section (Pyrenees, Spain). Geol. Acta 7, 161–175.
Punekar, J., Keller, G., Khozyem, H.M., Adatte, T., Font, E., Spangenberg, J., 2016. A multi-proxy approach to decode the end-Cretaceous mass extinction. Palaeogeogr. Palaeoclimatol. Palaeoecol. 441, 116–136.
Remane, J., Bassett, M.G., Cowie, J.W., Gohrbandt, K.H., Lane, H.R., Michelsen, O., Naiwen, W., 1996. Revised guidelines for the establishment of global chronostratigraphic standards by the International Commission on Stratigraphy (ICS). Episodes 19, 77–81.

Renard, M., Delacotte, O., Létolle, R., 1982. Le strontium et les isotopes stables dans les carbonates totaux de quelques sites de l'Aquitaine et de la Tethys. Bull. Soc. Geol. Fr. 14, 519–534.

Renne, P.R., Deino, A.L., Hilgen, F.J., Kuiper, K.F., Mark, D.F., Mitchell 3rd, W.S., Morgan, L.E., Mundil, R., Smit, J., 2013. Time scales of critical events around the Cretaceous-Paleogene boundary. Science 339, 684–687.

Renne, P.R., Sprain, C.J., Richards, M.A., Self, S., Vanderkluysen, L., Pande, K., 2015. State shift in Deccan volcanism at the Cretaceous-Paleogene boundary, possibly induced by impact. Science 350, 76–78.

Riba, O., 1973. Las dicordancias progresivas del Alto Cardener (Prepirineo catalán). Acta Geol. Hisp. 8, 90–99.

Riba, O., 1976a. Syntectonic unconformities of the Alto Cardener, spanish Pyrenees: a genetic interpretation. Sedim. Geol. 15, 213–233.

Riba, O., 1976b. Tectogènese et sedimentation: deux modèles de discordances syntectoniques pyrénénnes. Bull. Bur. Rech. Geol. Min. 4, 383–401.

Robin, E., Bocklet, D., Bonté, P., Froget, L., Jéhanno, C., Rocchia, R., 1991. The stratigraphic distribution of Ni-rich spinels in Cretaceous-Tertiary boundary rocks at el Kef (Tunisia), Caravaca (Spain) and Hole 761C (leg 122). Earth Planet. Sci. Lett. 107, 715–721.

Robin, E., Molina, E., 2006. Chronostratigraphy, composition and source of Ni-rich spinel in the late Eocene Fuente Caldera section (Spain): one impact or more? Meteorit. Planet. Sci. 41, 1231–1248.

Rocchia, R., Boclet, D., Bonté, P., Buffetaut, E., Orue-Etxebarria, X., Jaeger, J.J., Jéhanno, C., 1988. Structure de l'anomalie del'iridium à la limite Crétacé-Tertiaire du site de Sopelana (Pays Basque espagnol). C. R. Acad. Sci. Paris 307, 1217–1223.

Rocchia, R., Robin, E., Smit, J., Pierrard, O., Lefever, I., 2002. K/T impact remains in an ammonite from the uppermost Maastrichtian of Bidart section (French Basque Country). In: Buffetaut, E., Koeberl, C. (Eds.), Geological and Biological Effects of Impact Events. Springer-Verlag, Berlin, pp. 159–166.

Rodríguez-Lázaro, J., García-Zarraga, E., 1996. Paleogene deep-marine ostracodes from the Basque basin. In: Proc. 2nd Europ. Ostracod. Meet., Glasgow 1993. Brit. Micropaleontol. Soc, pp. 79–85.

Rodriguez-Pinto, A., Pueyo, E.L., Serra-Kiel, J., Barnolas, A., Samso, J.M., Pocovi, A., 2013. The upper Ypresian and Lutetian in San Pelegrin section (southwestern Pyrenean basin): magnetostratigraphy and larger foraminifera correlation. Palaeogeogr. Palaeoclimatol. Palaeoecol. 370, 13–29.

Rodriguez-Pinto, A., Pueyo, E.L., Serra-Kiel, J., Samso, J.M., Barnolas, A., Pocovi, A., 2012. Lutetian magnetostratigraphic calibration of larger foraminifera zonation (SBZ) in the Southern Pyrenees: the Isuela section. Palaeogeogr. Palaeoclimatol. Palaeoecol. 333, 107–120.

Rodriguez-Tovar, F., Uchman, A., 2006. Ichnological analysis of the Cretaceous-Palaeogene boundary interval at the Caravaca section, SE Spain. Palaeogeogr. Palaeoclimatol. Palaeoecol. 242, 313–325.

Rodriguez-Tovar, F., Uchman, A., Molina, E., Monechi, S., 2010. Bioturbational redistribution of Danian calcareous nannofossils in the uppermost Maastrichtian across the K-Pg boundary at Bidart, SW France. Geobios 43, 569–579.

Rodriguez-Tovar, F.J., 2005. Fe-oxide spherules infilling *Thalassinoides* burrows at the Cretaceous-Paleogene (K-P) boundary: evidence of a near-contemporaneous macrobenthic colonization during the K-P event. Geology 33, 585–588.

Rodriguez-Tovar, F.J., Martinez-Ruiz, F., Bernasconi, S.M., 2002. Carbon isotope composition of bioturbation infills as indication of the macrobenthic-colonization timing across the Cretaceous-Tertiary boundary (Agost section, SE Spain). Geochem. Cosmochi. Acta 66, A644.

Rodriguez-Tovar, F.J., Martinez-Ruiz, F., Bernasconi, S.M., 2004. Carbon isotope evidence for the timing of the Cretaceous-Palaeogene macrobenthic colonisation at the Agost section (southeast Spain). Palaeogeog. Palaeoclimatol. Palaeoecol. 203, 65–72.
Rodriguez-Tovar, F.J., Martinez-Ruiz, F., Bernasconi, S.M., 2006. Use of high-resolution ichnological and stable isotope data for assessing completeness of a K-P boundary section, Agost, Spain. Palaeogeog. Palaeoclimatol. Palaeoecol. 237, 137–146.
Rodriguez-Tovar, F.J., Uchman, A., 2004a. Trace fossils after the K-T boundary event from the Agost section, SE Spain. Geol. Mag. 141, 429–440.
Rodriguez-Tovar, F.J., Uchman, A., 2004b. Ichnotaxonomic analysis of the Cretaceous/Palaeogene boundary interval in the Agost section, south-east Spain. Cretac. Res. 25, 635–647.
Rodríguez-Tovar, F.J., Uchman, A., Alegret, L., Molina, E., 2011a. Impact of the Paleocene–Eocene Thermal Maximum on the macrobenthic community: ichnological record from the Zumaia section, northern Spain. Mar. Geol. 282, 178–187.
Rodriguez-Tovar, F.J., Uchman, A., Orue-Etxebarria, X., Apellaniz, E., Baceta, J.I., 2011b. Ichnological analysis of the Bidart and Sopelana Cretaceous/Paleogene (K/Pg) boundary sections (Basque basin, W Pyrenees): refining eco-sedimentary environment. Sedim. Geol. 234, 42–55.
Rodriguez-Tovar, F.J., Uchman, A., Orue-Etxebarria, X., Apellaniz, E., 2013. Palaeoenviromental changes during the Danian/Selandian boundary interval: the ichnological record at the Sopelana section (Basque basin, W Pyrenees). Sedim. Geol. 284–285, 106–116.
Roest, W.R., Srivastava, S.P., 1991. Kinematics of plate boundaries between Eurasia, Iberia, and Africa in the north Atlantic from the late Cretaceous to the present. Geology 19, 613–616.
Rögl, F., Egger, H., 2010. The missing link in the evolutionary origin of the foraminiferal genus *Hantkenina* and the problem of the lower-middle Eocene boundary. Geology 38, 23–26.
Romein, A.J.T., Smit, J., 1981. Carbon-oxygen stable isotope stratigraphy of the Cretaceous-Tertiary boundary interval – data from the Biarritz section (sw France). Geol. Mijnb. 60, 541–544.
Rouault, A., 1850. Description des fossils du terrain éocène des environs de Pau. Mem. Soc. Geol. Fr. 2, 457–507.
Roussel, J., 1893. Étude stratigraphique des Pyrénées (Ph.D. thesis). Univ. Lille, Paris. Baudry et Cie.
Ruiz de Gaona, M., 1946. El Terciario nummulítico fértil de Guecho (Vizcaya). Not. Com. Inst. Geol. Min. España 16, 183–214.
Ruiz de Gaona, M., 1948. La fauna principalmente nummulítica de la serie terciaria guipuzcoana. Estud. Geol. 9, 133–158.
Ruiz de Gaona, M., Colom, G., 1950. Estudios sobre las sinecias de los foraminíferos eocénicos de la vertiente meridional del Pirineo (Cataluña-Vizcaya). Estud. Geol. 12, 293–434.
Saenz de Galdeano, C.M., 1996. Tertiary tectonic framework of the Iberian Peninsula. In: Friend, P.F., Dabrio, C.J. (Eds.), Tertiary Basins of Spain: The Stratigraphic Record of Crustal Kinematics. Cambridge Univ. Press, Cambridge, pp. 9–14.
Schaub, H., 1969. L'Ilerdien: etat actuel du problem. BRGM Mem. 69, 259–266.
Schaub, H., 1973. La sección de Campo (Prov. de Huesca), Ilerdiense. In: Enadimsa, C.N.G. (Ed.), Libro guía del XIII Coloquio Europeo de Micropaleontología. Imp. Pablo Lopez, Madrid, pp. 139–158.
Schaub, H., 1992. The Campo section (NE Spain), a Tehtyan parastratotype of the Cuisian. Neues, Jahr. Geol. Palaontol. Abh. 186, 63–70.
Scheibner, C., Speijer, R.P., 2009. Recalibration of the Tethyan shallow-benthic zonation across the Paleocene-Eocene boundary: the Egyptian record. Geol. Acta 7, 195–214.

Scheibner, C., Speijer, R.P., Marzouk, A.M., 2005. Larger foraminiferal turnover during the Paleocene/Eocene thermal maximum and paleoclimatic control on the evolution of platform ecosystems. Geology 33, 493–496.
Schmitz, B., Boschi, S., Cronholm, A., Heck, P.R., Monechi, S., Montanari, A., Terfelt, F., 2015. Fragments of late Eocene Earth-impacting asteroids linked to disturbance of asterois belt. Earth Planet. Sci. Lett. 425, 77–83.
Schmitz, B., Asaro, F., Molina, E., Monechi, S., Von Salis, K., Speijer, R., 1997. High-resolution iridium, $\delta^{13}C$, $\delta^{18}O$, foraminifera and nannofossil profiles across the lates Paleocene benthic extinction event at Zumaia, Spain. Palaeogeogr. Palaeoclimatol. Palaeocol. 133, 49–69.
Schmitz, B., Molina, E., Von Salis, K., 1998. The Zumaya section in Spain: a possible global stratotype section for the Selandian and Thanetian stages. Newslett. Strat. 36, 35–42.
Schmitz, B., Peucker-Ehrenbrik, B., Heilmann-Clausen, C., Aberg, G., Asaro, F., Lee, C.T.A., 2004. Explosive volcanism, but no comet impact, at the Paleocene-Eocene boundary: high-resolution chemical and isotopic records from Egypt, Spain and Denmark. Earth Planet. Sci. Lett. 225, 1–17.
Schmitz, B., Pujalte, V., Núnez-Betelu, K., 2001. Climate and sea-level perturbations during the initial Eocene thermal maximum: evidence from siliciclastic units in the Basque basin (Ermua, Zumaya and Trabakua Pass), northern Spain. Palaeogeogr. Palaeoclimatol. Palaeocol. 165, 299–320.
Schmitz, B., Pujalte, V., 2003. Sea-level, humidity, and land-erosion records across the initial Eocene thermal máximum from a continental-marine transect in northern Spain. Geology 31, 689–692.
Schmitz, B., Pujalte, V., 2007. Abrupt increase in seasonal extreme precipitation at the Paleocene-Eocene boundary. Geology 35, 215–218.
Schmitz, B., Pujalte, V., Molina, E., Monechi, S., Orue-Etxebarria, X., Speijer, R.P., Alegret, L., Apellaniz, E., Arenillas, I., Aubry, M.P., Baceta, J.I., Berggren, W.A., Bernaola, G., Caballero, F., Clemmensen, A., Dinarès-Turell, J., Dupuis, C., Heilman-Clausen, C., Hilario Orus, A., Knox, R., Martin-Rubio, M., Ortiz, S., Payros, A., Petrizzo, M.R., Von Salis, K., Sprong, J., Steurbaut, E., Thomsen, E., 2011. The global stratotype sections and points for the bases of the Selandian (middle Paleocene) and Thanetian (upper Paleocene) stages at Zumaia, Spain. Episodes 34, 220–243.
Schulte, P., Alegret, L., Arenillas, I., Arz, J.A., Barton, P.J., Bown, P.R., Bralower, T.J., Christeson, G.L., Claeys, P., Cockell, C.S., Collins, G.S., Deutsch, A., Goldin, T.J., Goto, K., Grajales-Nishimura, J.M., Grieve, R.A.F., Gulick, S.P.S., Johnson, K.R., Kiessling, W., Koeberl, C., Kring, D.A., MacLeod, K.G., Matsui, T., Melosh, J., Montanari, A., Morgan, J.V., Neal, C.R., Nichols, D.J., Norris, R.D., Pierazzo, E., Ravizza, G., Rebolledo-Vieyra, M., Reimold, W.U., Robin, E., Salge, T., Speijer, R.P., Sweet, A.R., Urrutia-Fucugauchi, J., Vajda, V., Whalen, M.T., Willumsen, P.S., 2010. The Chicxulub asteroid impact and mass extinction at the Cretaceous–Paleogene boundary. Science 327, 1214–1218.
Seilacher, A., 1962. Paleontological studies on turbidite sedimentation and erosion. J. Geol. 70, 227–234.
Serra-Kiel, J., Canudo, J.I., Dinarés, J., Molina, E., Ortiz, N., Pascual, J.O., Samsó, J.M., Tosquella, J., 1994. Cronoestratigrafía de los sedimentos marinos del Terciario inferior de la Cuenca de Graus-Tremp (Zona Central Surpirenaica). Rev. Soc. Geol. España 7, 273–297.
Serra-Kiel, J., Hottinger, L., Caus, E., Drobne, K., Ferrandez, C., Jauhri, A.K., Less, G., Pavlovec, R., Pignatti, J., Samso, J.M., Schaub, H., Sirel, E., Strougo, A., Tambareau, Y., Tosquella, J., Zakrevskaya, E., 1998. Larger foraminiferal biostratigraphy of the Tethyan Paleocene and Eocene. Bull. Soc. Geol. Fr. 169, 281–299.

Seunes, J., 1890. Recherches géologiques sur les terrains secondaires et l'Eocène inférieur de la région sous-pyrénéene du Sud-Ouest de la France (Basses-Pyrenées et Landes). Dunod, Paris.
Sexton, P.F., Norris, R.D., Wilson, P.A., Palike, H., Westerhold, T., Rohl, U., Bolton, C.T., Gibbs, S., 2011. Eocene global warming events driven by ventilation of oceanic dissolved organic carbon. Nature 471, 349–352.
Seyve, C., 1990. Nannofossil biostratigraphy of the Cretaceous-Tertiary boundary in the French Basque country. Bull. Cent. Res. Expl. Prod. Elf Aquit. 14, 553–572.
Shukolyukov, A., Lugmair, G.W., 1998. Isotopic evidence for the Cretaceous-Tertiary impactor and its type. Science 282, 927–929.
Smit, J., 1977. Discovery of a planktonic foraminiferal association between the *Abanthomphalus mayaroensis* and the "*Globigerina*" *eugubina* zone at the Cretaceous/Tertiary boundary in the Barranco del Gredero (Caravaca, Spain): a preliminary report. Proc. Konnkl. Nederl. Akad. Wetensch. 80, 280–301.
Smit, J., 1979. The Cretaceous/Tertiary transition in the Barranco del Gredero, Spain. In: Christensen, W.K., Birkelund, T. (Eds.), Cretaceous-tertiary Boundary Events. Univ. Copenhagen, Copenhagen, pp. 156–163.
Smit, J., 1990. Meteorite impact, extinctions and the Cretaceous-Tertiary boundary. Geol. Mijnb. 69, 187–204.
Smit, J., 2004. The section of the Barranco de Gredero (Caravaca, Spain): a crucial section for the Cretaceous/Tertiary boundary impact extinction hypothesis. J. Iber. Geol. 31, 179–191.
Smit, J., Hertogen, J., 1980. An extraterrestrial event at the Cretaceous-Tertiary boundary. Nature 285, 198–200.
Smit, J., Klaver, G., 1981. Sanidine spherules at the Cretaceous-Tertiary boundary indicate a large impact event. Nature 292, 47–49.
Smit, J., Ten Kate, W.G.H.Z., 1982. Trace-element patterns at the Cretaceous/Tertiary boundary: consequences of a large impact. Cretac. Res. 3, 307–332.
Smith, A.G., 1996. Cenozoic latitudes, positions and topography of the Iberian Peninsula. In: Friend, P.F., Dabrio, C.J. (Eds.), Tertiary Basins of Spain: The Stratigraphic Record of Crustal Kinematics. Cambridge Univ. Press, Cambridge, pp. 6–8.
Sosa-Montes, C., Martinez-Ruiz, F., Rodriguez-Tovar, F., 2013. Bottom-water conditions in a marine basin after the Cretaceous-Paleogene impact event: timing the recovery of oxygen levels and productivity. PLoS One 8, e82242.
Sosa-Montes, C., Rodriguez-Tovar, F., Martinez-Ruiz, F., 2016. Geochemical and isotopic characterization of trace fossil infillings: new insights on tracemaker activity after the K/Pg impact event. Cretac. Res. 57, 391–401.
Storme, J.Y., Devleeschouwer, X., Schnyder, J., Cambier, G., Baceta, J.I., Pujalte, V., Di Matteo, A., Iacumin, P., Yans, J., 2012. The Palaeocene/Eocene boundary section at Zumaia (Basque-Cantabric Basin) revisited: new insights from high-resolution magnetic susceptibility and carbon isotope chemostratigraphy on organic matter (d13Corg). Terra Nova 24, 310–317.
Stott, L.D., Kennett, J.P., Shackleton, N.J., Corfield, R.C., 1990. The evolution of Antarctic surface waters during the Paleogene: inferences from stable isotopic composition of planktonic foraminifers, ODP leg 113. Proc. ODP Sci. Res. 113, 849–863.
Stuart-Menteath, P.W., 1881. Note preliminaire sur la géologie des Pyrénées, de la Navarre, du Guipúzcoa et du Labourd. Bull. Soc. Geol. Fr. 3, 304–333.
Teixell, A., 2004. Estructural cortical de la Cordillera Pirenaica. In: Vera, J.A. (Ed.), Geología de España. Soc. Geol. España & Inst. Geol. Min. España, Madrid, pp. 320–321.
Ten Kate, W.G., Sprenger, A., 1993. Orbital cyclicities above and below the Cretaceous/Paleogene boundary at Zumaya (N Spain), Agost and Relleu (SE Spain). Sedim. Geol. 87, 69–101.

Thibault, N., Minoletti, F., Gardin, S., Renard, M., 2004. Morphometry of selected calcareous nannofossils across the Cretaceous-Paleocene boundary at the Bidart (France) and Elles (Tunisia) sections. Comparison with carbon and oxygen stable isotope ratios. Bull. Soc. Geol. Fr. 175, 399–412.

Thomas, E., 1990. Late Cretaceous-early Eocene mass extinction in deep sea. Geol. Soc. Amer. Spec. Pub 247, 481–495.

Tosquella, J., 1995. Els Nummulitinae del Paleocè-Eocè Inferior de la Conca Sudpiranica (Ph.D. thesis). Univ. Barcelona, Barcelona.

Trinquier, A., Birck, J.L., Allegre, C.J., 2006. The nature of the KT impactor. A ^{54}Cr reappraisal. Earth Planet. Sci. Lett. 241, 780–788.

Van Vliet, A., 1978. Early Tertiary deep-water fans of Guipuzcoa, northern Spain. In: Stanley, D.J., Kelling, G. (Eds.), Sedimentation in Submarine Canyons, Fans and Trenches. Dowden, Hutchinson and Ross, Stroudsburg, Pennsylvannia.

Vandenberghe, N., Hilgen, F.J., Speijer, R.P., Ogg, J.G., Gradstein, F.M., Hammer, O., Hollis, C.J., Jooker, J.J., 2012. The Paleogene period. In: Gradstein, F.M., Ogg, J.G., Schmitz, M.D., Ogg, G.M. (Eds.), The Geologic Time Scale 2012. Elsevier, Amsterdam, pp. 855–922.

Varol, O., 1989. Palaeocene calcareous nannofossil biostratigraphy. In: Crux, J.A., van Heck, S.E. (Eds.), Nannofossils and Their Applications, 12. British Micropaleontol. Soc., pp. 267–310

Vera, J.A., 2004. Cordillera Bética y Baleares. In: Vera, J.A. (Ed.), Geología de España. Soc. Geol. España & Inst. Geol. Min, España, pp. 347–464.

Vezian, A., 1856. Mollusques et zoophytes des terrains Nummulitique et Tertiaire marin de la province de Barcelone (Ph.D. thesis). Fac. Sci. Montpellier, Montpellier.

Villasante-Marcos, V., Martinez-Ruiz, F., Osete, M.L., Urrutia-Fucugauchi, J., 2007. Magnetic characterization of Cretaceous-Tertiary boundary sediments. Meteor. Planet. Sci. 42, 1505–1527.

Vonhof, H.B., Smit, J., 1997. High-resolution late Maastrichtian early Danian oceanic Sr-87/Sr-86 record: implications for Cretaceous-Tertiary boundary events. Geology 25, 347–350.

Wade, B.S., Pearson, P.N., Berggren, W.A., Palike, H., 2011. Review and revision of Cenozoic tropical planktonic foraminiferal biostratigraphy and calibration to the geomagnetic polarity and astronomical time scale. Earth Sci. Rev. 104, 111–142.

Ward, P.D., Kennedy, W.J., 1993. Maastrichtian ammonites from the Biscay region (France, Spain). Paleontol. Soc. Mem. 34, 1–58.

Ward, P.D., Kennedy, W.J., MacLeod, K.G., Mount, J.L., 1991. Ammonite and inoceramid bivalve extinction patterns in Cretaceous/Tertiary boundary sections of the Biscay region (southwestern France, northern Spain). Geology 19, 1181–1184.

Weinbaum-Hefetz, M., Benjamini, C., 2011. Calcareous nannofossil assemblage changes from early to middle Eocene in the Levant margin of the Tethys, central Israel. J. Micropaleontol. 30, 129–139.

Westerhold, T., Röhl, U., 2009. High resolution cyclostratigraphy of the early Eocene: new insights into the origin of the Cenozoic cooling trend. Clim. Past. 5, 309–327.

Westerhold, T., Röhl, U., Raffi, I., Fornaciari, E., Monechi, S., Reale, V., Bowles, J., Evans, H.F., 2008. Astronomical calibration of the Paleocene time. Palaeogeogr. Palaeoclimatol. Palaeoecol. 257, 377–403.

Wiedmann, J., 1969. The heteromorphs and ammonoid extinction. Biol. Rev. 44, 563–602.

Wiedmann, J., 1988. The Basque coastal sections of the K/T boundary: a key to understanding "mass extinctions" in the fossil record. In: Lamolda, M.A., Kauffman, E.G., Walliser, O.H. (Eds.), Palaeontology and Evolution: Extinction Events. Rev. Esp. Paleontol, pp. 127–140 (Spec. Iss.).

Wilcoxon, J.A., 1973. Paleogene calcareous nannoplankton from the Campo and Tremp sections of the Ilerdian stage in NE Spain. Rev. Esp. Micropaleontol. 3, 773–780.

Wing, S.L., Harrington, G.J., Smith, F.A., Bloch, J.I., Boyer, D.M., Freeman, K.H., 2005. Transient floral change and rapid global warming at the Paleocene-Eocene boundary. Science 310, 993–996.

Zachos, J., Pagani, M., Sloan, L., Thomas, E., Billups, K., 2001. Trends, rhythms, and aberrations in the global climate 65 Ma to Present. Science 292, 686–693.

Zakrevskaya, E., Beniamovsky, V., Less, G., Báldi-Beke, M., 2011. Integrated biostratigraphy of Eocene deposits in the Gubs section (Northern Caucasus) with special attention to the Ypresian/Lutetian boundary and the peritethyan-Tethyan correlation. Turk. J. Earth Sci. 20, 753–792.

Zamagni, J., Mutti, M., Ballato, P., Kosir, A., 2012. The Paleocene-Eocene Thermal Maximum (PETM) in shallow-marine successions of the Adriatic carbonate platform (SW Slovenia). Geol. Soc. Amer. Bull. 124, 1071–1086.

Zili, L., Zaghbib-Turki, D., Alegret, L., Arenillas, I., Molina, E., 2009. Foraminiferal turnover across the Paleocene/Eocene boundary at the Zumaya section, Spain: recod of a bathyal gradual mass extinction. Rev. Mex. Cienc. Geol. 26, 729–744.

INDEX

'*Note:* Page numbers followed by "f" indicate figures and "t" indicate tables.'

C

D

H

I

N

O

Q

R

S

T

FSC
www.fsc.org
MIX
Papier | Fördert
gute Waldnutzung
FSC® C083411

Kolibri 360 GmbH
Ferdinand-Jühlke-Straße 7
99095 Erfurt, Deutschland
produktsicherheit@kolibri360.de